The Future Rice
Strategy for India

The Future Rice Strategy for India

Edited by

Samarendu Mohanty

P.G. Chengappa

Mruthyunjaya

J.K. Ladha

Sampriti Baruah

Elumalai Kannan

A.V. Manjunatha

ACADEMIC PRESS

An imprint of Elsevier

Academic Press is an imprint of Elsevier
125 London Wall, London EC2Y 5AS, United Kingdom
525 B Street, Suite 1800, San Diego, CA 92101-4495, United States
50 Hampshire Street, 5th Floor, Cambridge, MA 02139, United States
The Boulevard, Langford Lane, Kidlington, Oxford OX5 1GB, United Kingdom

Notices
Knowledge and best practice in this field are constantly changing. As new research and experience broaden our understanding, changes in research methods, professional practices, or medical treatment may become necessary.

Practitioners and researchers must always rely on their own experience and knowledge in evaluating and using any information, methods, compounds, or experiments described herein. In using such information or methods they should be mindful of their own safety and the safety of others, including parties for whom they have a professional responsibility.

To the fullest extent of the law, neither the Publisher nor the authors, contributors, or editors, assume any liability for any injury and/or damage to persons or property as a matter of products liability, negligence or otherwise, or from any use or operation of any methods, products, instructions, or ideas contained in the material herein.

Library of Congress Cataloging-in-Publication Data
A catalog record for this book is available from the Library of Congress

British Library Cataloguing-in-Publication Data
A catalogue record for this book is available from the British Library

ISBN: 978-0-12-805374-4 (paperback)
ISBN: 978-0-12-813936-3 (hardback)

For information on all Academic Press publications visit our website at
https://www.elsevier.com/books-and-journals

Working together
to grow libraries in
developing countries

www.elsevier.com • www.bookaid.org

Publisher: Andre G. Wolff
Acquisition Editor: Nancy Maragioglio
Editorial Project Manager: Billie Jean Fernandez
Production Project Manager: Nicky Carter
Designer: Matthew Limbert

Typeset by Thomson Digital

Contents

3. **Spatial and Temporal Patterns of Rice Production and Productivity**

Elumalai Kannan, Ambika Paliwal, Adam Sparks

4. **Ecological Footprints of and Climate Change Impact on Rice Production in India**

Rudrasamy Balasubramanian, Venkatachalam Saravanakumar, Kovilpillai Boomiraj

5.   Structural Transformation of the Indian Rice Sector

Humnath Bhandari, Praduman Kumar, Parshuram Samal

6.   Frontiers in Rice Breeding

Darshan S. Brar, Kuldeep Singh, Gurdev S. Khush

7.   Rice Varietal Development to Meet Future Challenges

Arvind Kumar, Nitika Sandhu, Shailesh Yadav,
Sharat Kumar Pradhan, Annamalai Anandan,
Elssa Pandit, Anumalla Mahender, Tilathoo Ram

8.   Growing Rice in Eastern India: New Paradigms of Risk Reduction and Improving Productivity

Sudhir-Yadav, Virender Kumar, Sudhanshu Singh, Rapolu M. Kumar,
Sheetal Sharma, Rahul Tripathi, Amaresh K. Nayak, J.K. Ladha

9. Technological Innovations, Investments, and Impact of Rice R&D in India

Alka Singh, Suresh Pal, Anbukkani Perumal

10. Intellectual Property Rights, Innovation, and Rice Strategy for India

Sachin Chaturvedi, Krishna Ravi Srinivas

11. Postharvest Management and Value Addition of Rice and Its By-Products

Arindam Samaddar, Mohammed Mohibbe Azam,
Kunjithapatham Singaravadivel, Natarajan Venkatachalapathy,
Braja Bandhu Swain, Purnananda Mishra

12. Institutional Innovations in Rice Production and Marketing in India: Experience and Strategies

Sukhpal Singh

13. The Rice Seed System in India: Structure, Performance, and Challenges

Aldas Janaiah, Behura Debdutt

14. Export Competitiveness of Indian Rice

A.V. Manjunatha, Lalith Achoth, Mamatha N.C.

15. Extension Policy Reforms in India: Implications for Rice Production Systems

Suresh Chandra Babu

16. Accelerating Impact Through Rice Innovation Systems: Integrating Knowledge, Technology, and Markets

Shaik N. Meera, Rikin Gandhi, Rabindra Padaria

Contributors

Lalith Achoth, Dairy Science College, Bengaluru, Karnataka, India

Annamalai Anandan, National Rice Research Institute, ICAR, Cuttack, Odisha, India

Suresh Chandra Babu, International Food Policy Research Institute, Washington, DC, United States

Vemuri Ravindra Babu, Indian Institute of Rice Research, ICAR, Hyderabad, Telangana, India

Rudrasamy Balasubramanian, Tamil Nadu Agricultural University, Coimbatore, Tamil Nadu, India

Humnath Bhandari, International Rice Research Institute, Dhaka, Bangladesh

Kovilpillai Boomiraj, Tamil Nadu Agricultural University, Coimbatore, Tamil Nadu, India

Darshan S. Brar, Punjab Agricultural University, Ludhiana, Punjab, India

Sachin Chaturvedi, Research and Information System for Developing Countries (RIS), New Delhi, India

Behura Debdutt, Odisha University of Agriculture and Technology, Bhubaneswar, Odisha, India

Rikin Gandhi, Digital Green, New Delhi, India

Aldas Janaiah, International Rice Research Institute, ICAR, New Delhi, India

Elumalai Kannan, Jawaharlal Nehru University (JNU), New Delhi, India

Gurdev S. Khush, University of California, Davis, Davis, CA, United States

Arvind Kumar, International Rice Research Institute, Los Baños, Philippines

Praduman Kumar, Indian Agricultural Research Institute, ICAR, New Delhi, India

Rapolu M. Kumar, Indian Institute of Rice Research, ICAR, Hyderabad, Telangana, India

Virender Kumar, International Rice Research Institute, Los Baños, Philippines

J.K. Ladha, International Rice Research Institute, Los Baños, Philippines

Anumalla Mahender, National Rice Research Institute, ICAR, Cuttack, Odisha, India

A.V. Manjunatha, Agricultural Development and Rural Transformation Centre, Institute for Social and Economic Change, Bengaluru, Karnataka, India

Shaik N. Meera, Indian Institute of Rice Research, ICAR, Hyderabad, Telangana, India

Purnananda Mishra, Indian Institute of Rice Research, ICAR, Cuttack, Odisha, India

Samarendu Mohanty, International Rice Research Institute, Los Baños, Philippines

Mohammed Mohibbe Azam, Indian Institute of Rice Research, Hyderabad, Telangana, India

Mamatha N.C., Agricultural Development and Rural Transformation Centre, Institute for Social and Economic Change, Bengaluru, Karnataka, India

Sudha Narayanan, Indira Gandhi Institute for Developmental Research, Mumbai, Maharashtra, India

Amaresh K. Nayak, National Rice Research Institute, ICAR, Cuttack, Odisha, India

Rabindra Padaria, Indian Agricultural Research Institute, ICAR, New Delhi, India

Suresh Pal, ICAR—National Institute of Agricultural Economics and Policy Research, New Delhi, India

Ambika Paliwal, International Rice Research Institute (IRRI), Los Baños, Philippines

Elssa Pandit, National Rice Research Institute, ICAR, Cuttack, Odisha, India

Anbukkani Perumal, ICAR—Indian Agricultural Research Institute, New Delhi, India

Sharat Kumar Pradhan, National Rice Research Institute, ICAR, Cuttack, Odisha, India

Tilathoo Ram, Indian Institute of Rice Research, ICAR, Hyderabad, Telangana, India

Arindam Samaddar, International Rice Research Institute, New Delhi, India

Parshuram Samal, National Rice Research Institute, Cuttack, Odisha, India

Nitika Sandhu, International Rice Research Institute, Los Baños, Philippines

Venkatachalam Saravanakumar, Tamil Nadu Agricultural University, Coimbatore, Tamil Nadu, India

Sheetal Sharma, International Rice Research Institute, Los Baños, Philippines

Kunjithapatham Singaravadivel, Indian Institute of Crop Processing Technology, Thanjavur, Tamil Nadu, India

Alka Singh, ICAR—Indian Agricultural Research Institute, New Delhi, India

Kuldeep Singh, Punjab Agricultural University, Ludhiana, Punjab, India

Sudhanshu Singh, International Rice Research Institute, Los Baños, Philippines

Sukhpal Singh, Centre for Management in Agriculture (CMA), Indian Institute of Management (IIM), Ahmedabad, Gujarat, India

Adam Sparks, Centre for Crop Health, University of Southern Queensland, Queensland, Australia

Krishna Ravi Srinivas, Research and Information System for Developing Countries (RIS), New Delhi, India

Braja Bandhu Swain, International Livestock Research Institute, New Delhi, India

Rahul Tripathi, National Rice Research Institute, ICAR, Cuttack, Odisha, India

Natarajan Venkatachalapathy, Indian Institute of Crop Processing Technology, Thanjavur, Tamil Nadu, India

Shailesh Yadav, International Rice Research Institute, Los Baños, Philippines

Sudhir- Yadav, International Rice Research Institute, Los Baños, Philippines

Takashi Yamano, International Rice Research Institute, Los Baños, Philippines

Editor Biographies

Dr. Samarendu Mohanty

Samarendu Mohanty is the Principal Scientist, Head of the Social Sciences Division, and Flagship Leader of "Accelerating Impact and Equity" at the International Rice Research Institute (IRRI), Los Baños, Laguna, Philippines. Prior to joining IRRI in 2008, he was an Associate Professor and the Associate Director of the Cotton Economics Research Institute in the Department of Agricultural and Applied Economics, Texas Tech University, Lubbock, TX, United States, from 2000 to 2008. He worked as a scientist at the Food and Agricultural Policy Research Institute, Iowa State University, Ames, IA, from 1994 to 2000.

He received his MS and PhD degrees in Agricultural Economics at the University of Nebraska-Lincoln, Lincoln, NE in 1995, and BS degree in Agricultural Marketing and Cooperation from the University of Agricultural Sciences (UAS), Bengaluru, Karnataka, India, in 1989. His research focus includes all aspects of economics of rice, including policy, market, and ex ante and ex post impact assessment. He has received several awards and recognitions for his teaching and research, including recognition from the Western Agricultural Economics Association and University of Nebraska-Lincoln for his contribution to agricultural, resource, and environmental economics in the Western States.

Dr. P.G. Chengappa

Dr. P.G. Chengappa is one of the leading agricultural economists of the country. He obtained his PhD degree in Agricultural Economics from IARI. He underwent a course on International Agricultural Marketing at German Foundation for International Development (DSE), Germany. Dr. Chengappa has over 37 years of experience in teaching, research, extension, and academic administration, including research carried out in international organizations. Dr. Chengappa started his career as a Research Assistant in UAS and served as an Associate Professor, Professor, and Head of the Department of Agricultural Marketing, Cooperation, and Agribusiness for over 2 decades. He rose to become the Director of Instruction (Agril) and Vice Chancellor of the University in 2007. During his tenure as the Vice Chancellor, the University was rated as the third best agricultural university in the country by the Times group. He was awarded the prestigious National Professor by ICAR, which he completed recently at ISEC. Prof. Chengappa has offered a wide range of courses in the areas of Agricultural Economics and Agribusiness for both graduate and postgraduate students. He has acted as the Chairman/Member, Board of Studies, Academic Council, and Board of Management in many universities. Prof. Chengappa is/was associated with several academic bodies and policy-making committees constituted by the Government: Member, Committee constituted by the Ministry of External Affairs, GOI, for extending agricultural assistance to Sri Lanka; Member, Scientific Committee, Food Safety and Standards Authority of India; Member, Knowledge Commission, State Farmers' Commission, and Biotechnology Task Force constituted by the Government of Karnataka, India; Member, Working Group of Planning Commission of India on Agricultural Marketing Infrastructure, Secondary Agriculture, and Policy, and subgroup on Agricultural Education and Research; Member, Governing Body of Indian Institute of Forest Management, Bhopal, Madhya Pradesh, India; Member, National Steering Committee of National Agricultural Innovation Project, ICAR; and Chairman, Review Committee on the functioning KVKs of ICAR located in the northeast region of India. He was a member of the Indian delegation convention on Biological Diversity held in Hyderabad during October 2012. He was nominated to the Roster of Experts as per the provisions of the Cartagena Protocol on Biosafety, GOI.

His international work experience includes working as a consultant at IFPRI; Socio Economist the International Plant Genetics Research Institute and International Crops Research Institute the Semi-Arid Tropics; Visiting Scientist, IRRI; Facilitator, International Course on Agricultural Marketing, DSE Germany; consultant, FAO Regional Office, Bangkok, Thailand; and Visiting Scientist at universities of Reading, Wales, and Purdue, and Iowa State University.

Dr. Chengappa's research efforts have resulted in 86 research articles in nationally and internationally peer-reviewed journals, 9 popular articles, 2 books, and 12 chapters in edited books.

He was President (Elect), ISAE, 2012, and President, Agricultural Economics Research Association, New Delhi, India, during 2013–16. He was conferred the National Fellow Award by the Soil Conservation Society of India in 2009.

Dr. Mruthyunjaya

Dr. Mruthyunjaya completed his school education in 1964, and graduation and postgraduation from UAS in 1969 and 1971, respectively. He received his doctoral degree from the Indian Agricultural Research Institute (IARI), New Delhi, India, in 1976 with high merit, distinction, recognition awards, several gold medals, and the Jawaharlal Nehru Outstanding PhD Thesis Award in Agricultural Economics from the Indian Council of Agricultural Research (ICAR) in 1976. After serving as a lecturer at UAS during 1971–73, he joined the Agricultural Research Service (ARS) as a scientist in 1976; served at the Indian Institute of Horticultural Research (IIHR), Bengaluru, Karnataka, up to 1981;

the Central Arid Zone Research Institute (CAZRI), Bhuj, Rajasthan, up to 1983; and IARI as a Professor and Principal Scientist till 1994. At these ICAR institutions, using advanced quantitative tools and techniques, he has made significant research contributions in measurement, modeling, economic, technological, institutional, and policy analysis of growth and instability in production, consumption, income, savings, investments, employment, supply, demand, storage, credit, marketing, transportation, storage of field, horticultural, and plantation crops, livestock, fisheries, and natural resource management (NRM) issues. He has also taught courses at graduate and postgraduate levels at UAS and IARI and guided MSc and PhD students at IARI. Dr. Mruthyunjaya has published more than 100 refereed research articles. He is a fellow of the National Academy of Agricultural Sciences (NAAS) and Indian Society of Agricultural Economics (ISAE), Mumbai, Maharashtra. During 2017, he was bestowed with the prestigious Dr. M.S. Randhwa Memorial Award by NAAS, India, for his lifelong contributions in agricultural administration, social sciences, and transfer of technology.

He served as the Assistant Director General (ADG) in Economics, Statistics, and Marketing (ESM); Policy and Planning (PP); and Planning and Monitoring and Implementation (PIM) at ICAR (1994–2000), and facilitated enhanced funding, formulation, monitoring, and evaluation of research programs of ESM section and ICAR as a whole as ADG (PIM). As ADG (PP), he helped ICAR to prepare the ICAR Vision 2020, South Asian Association for Regional Cooperation (SAARC) Perspective Plan, Asia-Pacific Association of Agricultural Research Institutions (APAARI) Vision 2025, and several O&M reforms at ICAR. He has also officiated as Deputy Director General (Education) [DDG (Edn)] in ICAR for about a year. He was selected as Director, National Centre for Agricultural Economics and Policy Research (NCAP) (now renamed as National Institute of Agricultural Economics and Policy Research or NIAP, New Delhi, India) and served it with distinction till 2005.

He was selected as the National Director of National Agricultural Innovation Project (NAIP) in 2005, which was supported by the World Bank and implemented by ICAR, and served with great distinction in that capacity till superannuation in September 2009. He helped in establishing 192 subprojects under various components across all parts of India, involving public-sector research organizations, general universities, government departments, private sector, and NGOs in a research consortia mode. The project impact was assessed by an independent agency and rated as a great success in terms of project design, delivery, and societal impact.

Utilizing his long experience and expertise, Dr. Mruthyunjaya continued his engagement with the profession as a consultant to the Ministry of Rural Development, Government of India (GOI), to monitor and evaluate its Rashtriya Krishi Vikas Yojana (RKVY) during 2009–11, Ministry of Agriculture, GOI, by preparing the report on "Status of Agriculture in India (2012)"; "Report on Status of Agriculture in Madhya Pradesh 2012" the Government of

Madhya Pradesh in 2013; "Agricultural Policy for Haryana State" in 2013; a report on "Augmentation of Revenue in SAUs of Rajasthan State," Government of Rajasthan in 2013; and "Vision and Strategic Plan for Krishi Gobesena Foundation (KGF)," Government of Bangladesh, in 2014. He worked the International Food Policy Research Institute (IFPRI), New Delhi, for developing the Priority Setting, Monitoring, and Evaluation (PME) Manual and state agricultural universities (SAUs) for ICAR in 2015; APAARI Vision 2030 for APAARI, Asia, in 2015; development of Vision 2030 for University of Agricultural and Horticultural Sciences (UAHS), Shimoga, Karnataka; Rice Strategy for India for IRRI; and Manual on Transparent Indicators of Assessment of Performance of Krishi Vigyan Kendras (KVKs) for IFPRI.

Dr. Mruthyunjaya is currently aiding the growth of social sciences in general and the discipline of Agricultural Economics in particular by serving in various capacities at the Agricultural Economics Research Association (AERA), New Delhi, since its establishment in 1988. As s consultant to IFPRI, he supported the Regional Strategic Analysis and Knowledge Support System (ReSAKKS), Asia for strengthening policy research networking with research institutions and the AERAs of India, Nepal, Bangladesh, and Sri Lanka. He has widely traveled abroad and visited more than 25 countries in various capacities.

Dr. J.K. Ladha

Dr. J.K. Ladha has devoted more than 32 years to the aspects of sustainable management of agriculture and natural resources for increasing food security and environmental quality in developing countries. He is an expert in soil fertility and plant nutrition, serving in different positions since 1980. Currently, he is a Principal Scientist and an Adjunct Senior Scientist at Columbia University, New York, NY, United States, and an Associate in the Agricultural Experiment Station at the University of California-Davis, Davis, CA. Dr. Ladha provided leadership to the Cereal Systems Initiative for South Asia and the Rice-Wheat Consortium Project that aims to sustainably enhance crop productivity. He was a "Frosty" Hill Fellow at Cornell University, Ithaca, NY (July 2007 to June 2008), and an Adjunct Professor of Soil Science at the University of the Philippines, Diliman, Philippines (1990–2004). He was born and raised in Gwalior, Madhya Pradesh, India, and earned his PhD degree from Banaras University, Varanasi, Uttar Pradesh, in 1976.

Dr. Ladha is recognized internationally as an authority on sustainable resource management for increasing food security and environmental quality. He has made immense contributions to international agriculture through his research, training, and extension activities in several Asian countries (such as Bangladesh, India, Nepal, Pakistan, Philippines, and Thailand) on problems across national and regional boundaries. Dr. Ladha is one of those unique scientists who has demonstrated success in conducting both basic and applied research. He has had an opportunity to pursue the full spectrum of basic, strategic, and applied research to find insights and develop technologies to solve farmers' problems.

He has published widely on issues related to sustainable and conservation agriculture.

The impact of Dr. Ladha's work is evident from his exceptionally high h-index for citations (Google Scholar, 69; Web of Science, 51; Scopus, 50). He served on the editorial boards of several international journals, including as the Regional Editor of Biology and Fertility of Soils. He has been involved with several international advisory/scientific review panels. He supervised 35 master's and doctoral students from a dozen countries.

He is a fellow of the American Association for the Advancement of Science (AAAS), American Society of Agronomy (ASA), the Soil Science Society of America (SSA), the Crop Science Society of America (CSSA), and NAAS, and is an associate member of the Philippine Council of Agricultural Research (PARC). He is a recipient of several awards and honors, notably, the Third World Academy of Sciences Agriculture Prize 2015, International Crop Science Award 2015, International Service in Agronomy Award 2011, International Soil Science Award 2010, and International Plant Nutrition Institute Science Award 2009. In 2000 and 2004, the Consultative Group on International Agricultural Research (CGIAR) awarded the Chairman's Excellence in Science Award for Outstanding Scientific Partnership and the prestigious King Baudoin Award for Outstanding Research to the Rice-Wheat Consortium in which Dr. J.K. Ladha was the key scientist and IRRI's coordinator.

Sampriti Baruah

Sampriti Baruah is a Lee Foundation Scholar pursuing her PhD degree in the Social Sciences Division of IRRI. Prior to joining IRRI as a scholar in 2015, she was also employed with IRRI as an Associate Scientist, Social Sciences, from 2012 to 2015 in New Delhi, India.

Before moving into this uniquely blended domain of research, agriculture, and development, she had 4 years of work experience from 2007 to 2010 as a development specialist in two other organizations, Gujarat State Disaster Management Authority (GSDMA), Gandhinagar, Gujarat, India, and The Friends of Women's World Banking (FWWB), Ahmedabad, Gujarat.

She holds an MA degree in Social Development and Sustainable Livelihoods from the Graduate Institute of International Development and Applied Economics (GIIDAE), University of Reading, Reading, United Kingdom, and another MA degree in Social Work with specialization in Social Welfare Administration from the Tata Institute of Social Science (TISS), Mumbai, Maharashtra, India. She also holds a BA degree in Political Science from Lady Shri Ram College, University of Delhi, New Delhi, India.

Dr. Elumalai Kannan

Dr. Kannan has a master's degree in Agricultural Economics from Chaudhary Charan Singh Haryana Agricultural University, Hisar, Haryana, India, and a PhD degree in Economics from Jawaharlal Nehru University (JNU), New Delhi. His areas of research interests include agricultural economics, international trade, futures market, NRM, institutions, and rural development.

He has worked at various reputed Indian institutions, such as NCAP; International Maize and Wheat Improvement Centre, New Delhi; National Council of Applied Economic Research, Madras School of Economics, Chennai, Tamil Nadu; and ISEC. Presently, he is working as an Associate Professor at the Centre for the Study of Regional Development (CSRD), JNU. Dr. Kannan served as a consultant to various international organizations, such as the World Bank and Food and Agriculture Organization (FAO). He was a visiting scholar at the University of Saskatchewan, Saskatoon, SK, Canada.

During recent years, he has undertaken important research studies on inclusive agricultural growth, stagnation in productivity of important crops in India, agrarian change and farm sector distress, pre- and postharvest losses, agricultural market reforms, impact evaluation of National Agriculture Development Programme, impact evaluation of Soil Micronutrients Programme, comparative advantage in the export of agroprocessed products, and India's textile and textile products.

Dr. A.V. Manjunatha

A.V. Manjunatha obtained his MSc degree in Agricultural Economics from UAS and an International MSc degree in Rural Development from the University of Ghent, Ghent, Belgium. He holds a doctoral degree in Agricultural Economics from Justus Liebig University Giessen, Giessen, Germany. He has worked in several national and international projects and has been involved in drafting policy documents for the state and central governments. He has published research articles in high-ranking journals, conferences, and leading national newspapers. His main research areas include NRM, agricultural water management, agricultural markets, impact analysis of developmental programs, technology adoption, international trade, and energy and climate change in agriculture. Currently, he is serving as an Assistant Professor in the Agricultural Development and Rural Transformation Centre of the Institute for Social and Economic Change (ISEC), Bengaluru, Karnataka, India.

Foreword

Rice is the most important food crop in India, accounting for more than one-fifth of the total gross cropped area. It is grown in various agroecological environments of the country and is spread across all states. It is also a staple food for the majority of India's billion-plus population and, on average, contributes more than one-fourth of the total calorie intake. The Indian rice sector has undergone remarkable transformation in the past 5 decades, with rice production almost tripling and making India a rice-surplus country. India is now one of the top exporters of rice in the world market. The credit for India's success goes to the advent of the Green Revolution with the development of high-yielding semi-dwarf varieties, improved management practices, and the commitment of the Indian government through support programs for farmers and investment in developing infrastructure. In the past 5 decades rice scientists have developed more than 1000 improved varieties and better crop management practices to counter biotic and abiotic threats and to enhance productivity growth to stay ahead of the rising demand. Government programs over the years involved market support in terms of minimum support price for rice farmers and input subsidies that covered seed, fertilizer, water, electricity, and machinery, among other inputs.

Despite these resounding successes, the Indian rice sector faces many emerging challenges, such as the rising cost of subsidies, depleting groundwater level in the major rice bowls (including northwest and southern India), and the growing threat of climate change. In addition, the Indian rice sector is affected by the ongoing changes in the economy through urbanization, rural outmigration, rising wage rates, and competition from other sectors of the economy. These changes are reflected in rice farming with mechanization and innovative models of land consolidation for operational efficiency without changing land titles. Even with all these challenges and transformations, India needs to produce more rice in the future to meet both domestic and export demands.

Given the emerging challenges and the ongoing structural transformation, this volume aims to provide direction on sustained growth in rice productivity; future investments in rice research, technology development, and transfer; value chain improvement; and policy reforms for sustainable development of the sector. This is also an opportune time to bring out this volume, as it is the 50th Anniversary of IR8, which changed the face of food security in India and other Asian countries.

This volume contains an array of articles prepared by chosen experts who have thoroughly examined the current status, emerging challenges and opportunities, and plausible solutions based on rigorous analysis for sustained growth of the rice sector in India. The various chapters in this publication extensively cover the existing and emerging opportunities to overcome the accumulated and emerging constraints besides tackling climate change, economic shocks, and the energy crisis. On the whole, an attempt has been made to provide a futuristic rice outlook for India. The main topics covered in different sections include unexploited technologies in genetic engineering, innovations in credit, inputs, processing, marketing, trade, insurance, agroadvisory services, and alliances with the private sector. The editorial board, led by Samarendu Mohanty (Chair), P.G. Chengappa, Mruthyunjaya, J.K. Ladha, and Sampriti Baruah, has worked hard to bring together the eminent experts to produce this timely publication. We believe that the publication is extremely valuable for policymakers, researchers, development and extension agents, and students engaged in different aspects of transforming India's rice sector and making it sustainable in the long term.

Matthew Morell
Director General, IRRI

T. Mohapatra
Secretary, DARE, and
Director General, ICAR

Acknowledgments

The Editorial Committee is deeply grateful to the International Rice Research Institute (IRRI) and the Indian Council of Agricultural Research (ICAR) for providing financial support and encouragement in the publication of this book. We would also like to acknowledge the institutional support provided by the Institute of Social and Economic Change (ISEC) to this initiative. The administrative staff of ISEC did an outstanding job in organizing the lead authors' meeting on the ISEC campus in December 2015. Our sincere appreciation goes to all lead authors, coauthors, and external reviewers for their contributions to this book.

Each chapter was reviewed by an external reviewer who stimulated us to work better and harder. Our team of reviewers consisted of Bisalaih Siddanaik, David Johnson, David Speilman, Derek Byerlee, Eric Wailes, Joseph Balagtas, Karachepone Ninan, Martin Gummert, Paolo Ficarelli, Ramachandra Shrikrishna Deshpande, Ranjita Pushkar, Remy Bitoun, Ajay Panchbhai, Reiner Wassmann, Sulaiman Rasheed, Ashok Misra, and Suresh Babu.

We would like to thank Bill Hardy for editing the chapters and Mary Grace Salabsabin for assisting in the editorial process. Finally, we thank Elsevier for partnering with us in the publication of this book.

Samarendu Mohanty
Chair, Editorial Committee

Overview

Rice is the most important food crop in India with 44 million ha of harvested area and more than 150 million tons of paddy production. In 2011–12, rice accounted for 22.3% of the total gross cropped area (195.2 million ha) and constituted 25.8% of total crops, for which small farmers produced more than 60% (Birthal et al., 2011). In the past 50 years, Indian rice production (paddy equivalent) has nearly tripled from 58 million tons in the late 1960s to more than 150 million tons in the past few years. This is largely due to the introduction of high-yielding varieties as part of the Green Revolution technology package. The greater availability of rice led to a rise in per capita annual consumption from 70 kg in the early 1970s to more than 80 kg in the early 1990s.

The success of Indian rice production started with the simultaneous introduction of high-yielding varieties in the mid-1960s as part of the Green Revolution technology package and the implementation of a suite of policies that included a minimum support price and input subsidies (for fertilizer, seed, water, electricity, and machinery, among other inputs) for farmers. These policies played a key role in improving India's food security by keeping arable land for rice cultivation, which would otherwise have been shifted to more profitable crops.

To continue the development and release of improved varieties and management practices, policy changes were backed up by a strong institutional setup and rice production programs. At the all India level, the Central Rice Research Institute (CRRI, which became the National Rice Research Institute, NRRI, in 2014) and the Directorate of Rice Research (which became the Indian Institute of Rice Research, IIRR, in 2014), in collaboration with the International Rice Research Institute (IRRI), have contributed significantly to the development of improved varieties and management practices. The International Network for Genetic Evaluation of Rice (INGER) and the All India Coordinated Rice Improvement Project (AICRIP) network have been responsible for the multilocation testing of improved varieties over the years. The AICRIP has so far been responsible for releasing 1084 rice varieties, including 72 hybrids, for various rice ecosystems through both central and state variety release authorities.

EMERGING CHALLENGES

India's total rice needs are projected to rise despite a reduction in rice consumption as a result of rising income and urbanization. According to Bhandari et al. (Chapter 5), Indian rice consumption will grow from 106 million tons in 2015 to 112 million tons by 2020 and to 122 million tons by 2030. The 2050 vision document published by the CRRI estimated Indian rice needs to continually rise after 2030, reaching 137 million tons by 2050.

The good news is that India will be able to meet its future rice demand if yield growth continues at the current rate and if rice area remains at the current level. However, the rapidly declining water table in key rice-growing states in northwest and southern India, urbanization, and competition from other profitable crops are likely to eat into rice area in the future. Therefore, the future challenge will be to not only "produce more with less water and less land but also with less labor and agro-chemical use."

In addition to resource scarcity, the country will have to face the effects of climate change on rice production. According to Rudrasamy et al. (Chapter 4), the impacts of climate change will differ for different rice-growing regions. The existing literature suggests that the rising temperature will decrease crop yield, whereas the increase in carbon dioxide (CO_2) concentration in the atmosphere will increase yield. The estimated net effect on yield ranges from a small increase to a large decline (up to 35%) in the long term. Rudrasamy et al. (Chapter 4) further suggest that the uncertainties in these associations are due to uncertainties in climate change projections; future technology growth; the availability of inputs, such as irrigation water; changes in crop management; and changes in genotype.

Farm-Level Structural Transformation

Rice farming is also transforming as a result of rural outmigration, aging farmers, and rising nonfarm opportunities that create acute labor shortages and rapid rises in farm wage rates. Rural wages in India increased by 35% between 2005–06 and 2012–13 (Wiggins and Keats, 2014). The rise in wage rates and labor scarcity are forcing farmers to explore the possibility of replacing labor-intensive activities, such as land preparation, transplanting, harvesting, and threshing, with appropriate mechanization to lower the fast-rising cost of production (Mohanty et al., 2015). Small-scale mechanization and custom-hiring machine arrangements are fast evolving as viable solutions for smallholder rice farmers in Asia. The ongoing transformation is also changing the role of women in rice farming (Bhandari et al., Chapter 5). With the outmigration of male family members, more and more women are taking over the role of farm managers and decision-makers (Mohanty et al., 2015), and they now contribute 46% of the total labor requirements.

MAJOR THEMES OF THE INDIAN RICE STRATEGY

Taking into consideration the ongoing structural transformation of India's rice farming systems and several emerging issues, including climate change, this book aims to bring together an array of eminent scientists and industry experts to develop and enrich the future rice-sector strategy in India. The chapters analyze the current and future scenario in light of global, regional, and national trends and priorities. A national strategy for rice can be all-inclusive when guided by insights and parameters drawn from various stakeholders within the nation.

This book has four themes. The first theme performs a detailed analysis of the significance of rice in the Indian economy and environment. This theme includes chapters on the changing temporal and spatial patterns of output, emerging challenges and opportunities that assure rice food security, the crop's ecological footprint and the impacts of climate change, and transforming the structure of the rice value chain to develop an internationally competitive system. According to the authors, rice will continue to be the most important crop in India in terms of both food security and contribution to the Indian economy. At the same time, rice is one of the most important contributors to greenhouse gas (GHG) emissions through the continuous flooding of rice fields and the intensive use of water, fertilizer, and pesticides. Despite the tremendous success of the Indian rice sector in the past 5 decades, this sector faces a series of emerging uncertainties and challenges, including the rapidly depleting water table in the traditional rice bowls of northwest and southern India. In addition, the sector faces a number of socioeconomic changes, such as rural outmigration, rising off-farm economic opportunities, rising income, and urbanization.

The second theme concentrates on technological advances and R&D policies. Chapters under this theme span topics on rice breeding, the status of entities in the rice seed system, opportunities for improving rice production in eastern India, postharvest management, the potential for improving value addition of rice and its by-products, facets of R&D investments, current technological innovations, and intellectual property issues. The authors discuss the recent advances in rice breeding and crop management practices that have the potential to overcome some of the emerging constraints described earlier.

The adoption of improved varieties and sustainable management practices would not be possible without an efficient seed system and proper market links. Therefore, the third theme draws attention to the evolving input and output markets and their distribution system by examining institutional innovations through experiences, export competitiveness, and the role of rice in India's food security.

The fourth theme discusses innovative platforms for increased impact, focusing on extension policy reforms; advanced approaches that integrate knowledge, technology, and markets within the extension framework; and future strategies of NARES in the development of India's rice sector. The extension system in India played a key role in conducting field demonstrations of high-yielding varieties of seeds and ensured timely delivery of inputs to farmers (Babu, Chapter 15). However, in light of the ongoing structural transformation of rice farming due to rural outmigration, mechanization, and feminization, it is necessary to review and develop new models of the extension system to efficiently reach farmers using advanced technology and know-how. This is particularly important for those farmers in remote areas.

What Needs to Be Done?

The scholars in this book extensively discuss the current and emerging opportunities that, if harnessed well, may not only overcome the emerging constraints attributed to climate change, economic shocks, and the energy crisis, but may also result in a more sustainable and commercial sector in the future. The opportunities discussed in different chapters include the unexploited existing and in-the-pipeline genetic engineering; crop management; and postharvest and value-addition technological advances. Several of these opportunities have already proven successful, but need to be scaled up and out with institutional and service innovations in production, credit, inputs, processing and marketing, trade, insurance, and agroadvisory services.

The various findings in this book can be summarized under five broad recommendations:

1. *Raising rice productivity and grain quality and intensifying production in eastern India*

   The productivity growth of paddy decelerated from 2.8% per annum during 1970–89 to 1.4% per annum during 1990–2013. One way to reverse this trend would be to close the yield gap among different rice-growing regions and ecosystems. These differences ranged from 4% in West Bengal to 41% in Assam, and from 20%–30% in irrigated ecosystems to 30%–50% in favorable shallow lowlands, and to more than 50% in unfavorable environments. Furthermore, out of the 342 main rice-growing districts, nearly 28% (mostly irrigated ecosystems) had high productivity, contributing ~56% of area with yield of more than 2.5 t/ha; 38% (mostly rainfed lowland) had medium productivity, contributing ~25% of area with yield of 1.5–2.5 t/ha; and 34% (mostly in rainfed upland) had the lowest productivity, ~ 16% of area with yield of less than 1.5 t/ha. This indicates a missed opportunity in one sense, but also scope to improve productivity in the future.

   Eastern India, which has the majority of rainfed rice ecosystems, could be prioritized to intensify rice production. This would relieve pressure on natural resource degradation that hinders these areas from meeting the expanding food demand in the traditional irrigated northern rice bowl, reduce poverty in the eastern region, and use abundantly available water in the region to serve as the future food basket. The low-productivity districts of rainfed areas in Eastern Uttar Pradesh, Madhya Pradesh, Bihar, Odisha, Jharkhand, and Chhattisgarh face serious technological, service, supply, and marketing constraints. It will be important to produce basmati rice in the northwestern states, high-grade rice in the southern states, and medium-grade rice in the northern and eastern states to utilize their comparative advantage. The need to diversify rice-based cropping systems toward high-value crops is highlighted. In addition, it is also noted that the relative shares of rice, coarse grains, and pulses are likely to decrease in line with dietary requirements; to avoid the adverse effects of continuous rice cultivation; and most importantly to avoid a drop in profitability and income from rice farming.

The application of frontier science and technologies in breeding, production practices, and postharvest management should be prioritized to raise productivity, improve quality, and reduce the environmental footprint. The breeding program should focus on agronomic and trait development for biotic and abiotic stress environments using modern biotechnological and molecular tools. The authors recommend targeted varietal development by integrating marker-assisted selection (MAS) in the breeding program, developing haploid induction systems in indica rice, nutrient-use efficiency for identifying genes for efficient nutrient uptake, biological nitrogen fixation, RNAi and genome-editing technologies for the improvement of selected traits, and GM technology.

There is also a need to implement improved varieties alongside integrated crop management systems that include mechanization, nutrient management, pest management, and water management. The introduction of ICT-based decision tools for nutrient, water, and weed management can also help to bridge the yield gap and decrease the risks associated with farming. Future technology developments could be further strengthened by taking advantage of the recent emergence of the private sector in rice R&D through innovative public–private partnership models.

R&D initiatives to design a suitable combine, particularly under waterlogged field conditions; a mechanical bulk handling system for easy transportation; and standardization of a suitable solar tunnel dryer or solar hot-air dryer for field-level drying are important with regards to harvest and postharvest aspects. On-farm mechanical dryers using biomass as fuel, mechanical grading, smaller-capacity sheller-cum-paddy separators, technology for producing rice milk from traditional rice with a longer shelf-life, and commercially generating electricity from rice husk are also critical in reducing postharvest loss and improving grain quality.

2. *Enhancing profitability and income of rice farmers*
This involves development and promotion of the rice value chain through institutional innovations and sustainable rice exports. To enhance the profitability and income of rice farmers, simply increasing yield will not be enough. It is well reported in this book that the decline in the average value of output and the increase in average paid-out costs have led to a drop in farm business income. This calls for rice business to be run differently. One way to achieve this is to pursue a value chain approach that facilitates linking small farmers with domestic and export markets. The current value chain is complex, consisting of producers, input suppliers, service providers, millers, wholesalers, processors, retailers, and consumers. Several institutional innovations are suggested to develop strong horizontal and vertical integration in production, marketing, and other points on the value chain. Horizontal integration of rice farming involves the replacement of labor-intensive activities, such as

land preparation, transplanting, harvesting, and threshing, with appropriate mechanization to lower the cost of production. Small-scale farm mechanization and custom-hiring arrangements for machinery are fast evolving as viable solutions for smallholder rice farmers in India. To offset the nonviability of mechanization due to small farm size, farmers are adopting various models of virtual land consolidation, such as small farmers' clubs, input producer companies (seed production, machinery rental, and water sharing), land-lease banks, and public–private partnerships for contract farming.

In the past the group approach has failed due to a lack of trust among members, unequal access to benefits, low incentives for participation, structural rigidities in the governance of some institutions, and financial irregularities. The success of these aggregation models could be significantly improved by undertaking various training programs to improve the organization and management skills, business orientation, and trust of members. In addition, linking these groups with input suppliers, traders, agribusiness companies, and group-based service provision could improve their survivability. Supportive policies and a legal environment with strict enforcement ability could encourage farmers to participate in these models and prove critical for their sustained success.

The bottom-line price received by farmers depends on India's competitiveness in the export market. In addition to traditional competitors, such as Thailand and Vietnam, India will also have to face emerging exporters (including the low-cost rice producers Myanmar and Cambodia) to keep its market share in the future. To remain competitive in the export market, India needs to reduce its production costs and improve grain quality by investing in state-of-the-art milling technology, proper packaging and labeling, and investing in an effective marketing strategy to establish generic branding for Indian-origin rice.

3. *Promoting sustainability of natural resources and climate resilience in rice farming*

Intensive rice monoculture has resulted in serious depletion and degradation of natural resources, including soil and water, as exemplified by poor drainage; soil salinity; nitrate and pesticide residues in soil and water resources; a decline in soil micronutrients; resistance to pesticides, disease, and nematode buildup; overextraction of groundwater; and the emission of GHGs, such as methane, carbon dioxide, and nitrous oxide, as a result of continuous flooding and the imbalanced use of nitrogenous fertilizers in rice fields. Apart from contributing to climate change through GHG emissions, rice production is itself affected by climate change through yield (quantity and stability) variability due to changes in temperature, rainfall, pest and disease outbreaks, and weed infestation.

Producing more rice per unit of land and water while simultaneously lowering GHG emissions is a major challenge. Suggestions to improve the management of natural resources and reduce GHG emissions include improving efficiency in the use of nutrients, water, and energy; improving crop and water management practices, such as adjusting the time of transplanting; scheduling of input application; using direct-seeded rice; using urea supergranules; incorporating alternate wetting and drying; amending soil and properly managing organic inputs; aerobic cultivation; zero tillage; and incorporation of rice straw.

One of the most visible impacts of climate change is the increased frequency of extreme weather events, such as floods and droughts. The development of varieties that are tolerant of submergence, drought, and salinity could go a long way to mitigate the negative impacts of climate change on yield variability. In recent years IRRI has developed many climate-resilient varieties that are tolerant of flood, drought, and salinity and has disseminated them in Asian rice-growing countries. More climate-resilient varieties, such as those with multiple stress tolerance, are in the pipeline. If successful, these varieties could have an even bigger impact on offsetting the effects of climate change compared with varieties with single-stress tolerance. In the future, efforts should be made to develop and release varieties with multiple-stress tolerance that could be effective in mitigating the negative impact of climate change.

4. *Increasing investments in rice research, capacity building, and extension and advisory services*
A significant push should be made to increase investments in rice science, which will lead to the utilization of advanced tools and technology options. This also includes increased investment in the capacity development of young rice scientists to attain excellence in advanced technologies using a repeat of the 1990s model on rice biotechnology by the Rockefeller Foundation. The success of AICRIP should be further strengthened by linking varietal development with testing and release. This will help to meet the growing need for high-yielding, stress-tolerant, and quality rice seeds at cheaper prices. It is therefore critical to further consolidate and enhance the success of AICRIP's partnership in rice R&D. The government should fully use the private sector and NGOs as vehicles to increase the reach of their extension programs by using digital technologies. In other words, rice extension and advisory service systems should integrate farmers and other value chain members with knowledge, technology, and markets using digital technologies. This will lead to innovative capacity development of everyone involved in the rice value chain.

5. *Policy reforms in the seed sector, IPR, and the food distribution system*
The Indian seed-provisioning system is extremely complex, encompassing various entities involved in the seed value chain, from R&D to seed retailers who deliver the seed to farmers. Nearly 120 national agricultural R&D centers, CGIAR centers, and 70 state agricultural universities are currently involved in varietal development programs for Indian farmers. According to the authors, the production of rice breeder seed has increased threefold

over the past 10 years, while certified seed has increased fourfold over the past 15 years. With more and more farmers preferring to use new seeds every year (as indicated by the rising seed replacement rate), the government should take various measures to develop modern infrastructure for storage, milling, and marketing of seeds.

As part of policy reforms, the public-sector financial institutions should provide financial support to the private seed sector and seed cooperatives at a low interest rate. The policy and regulatory framework should enable farmers to obtain due compensation from seed companies in the event of crop failure due to poor seed quality, enforce adherence to scientific procedures during the evaluation of new varieties and hybrids before commercialization, and enforce compulsory labeling of seeds with the prescribed information.

Protecting intellectual property rights is yet another policy measure that balances the needs and demands of different stakeholders in the rice system. It has to attain this balance without adversely affecting the public interest and welfare, while also facilitating innovation, providing access to farmers and plant breeders, and ensuring that quality seeds are available to farmers at affordable prices. There is a need for further empirical research (patent landscaping and policy issues) to permit development of a comprehensive policy framework to address contentious issues in defining and demarcating areas eligible for producing basmati rice. Furthermore, this research should also explore the scope for the use of geographical indication for promoting traditional rice varieties and the role of IP in emerging technologies, such as genome editing.

In the context of the National Food Security Act, the current food distribution system should be reformed to attain greater transparency. The reach of the food distribution system should be strengthened by allowing village organizations and women's self-help groups to manage ration shops, and nutritious and healthier rice should be made available through the food distribution system.

REFERENCES

Birthal, P.S., Joshi, P.K., Narayanan, A.V., 2011. Agricultural Diversification in India: Trends, Contribution to Growth and Small Farmer Participation. International Crops Research Institute for the Semi-Arid Tropics, Patancheru, India.

Mohanty, S., Bhandari, H., Mohapatra, V., Baruah, S., 2015. The ongoing transformation of rice farming in Asia. Rice Today 14 (3).

Wiggins, S., Keats, S., 2014. Real Wages in Asia. Overseas Development Institute Report. Overseas Development Institute, London.

Chapter 1

Rice Food Security in India: Emerging Challenges and Opportunities

Samarendu Mohanty, Takashi Yamano
International Rice Research Institute, Los Baños, Philippines

INTRODUCTION

Rice is a staple food for the majority of India's billion plus population and a source of livelihood for millions more households. Apart from its economic and strategic importance, rice is deeply ingrained in India's rich tradition and culture. India has the largest land area under rice cultivation but falls behind China in volume of production. In the past 50 years, Indian rice production has nearly tripled following the introduction of semidwarf modern varieties as part of the Green Revolution technology package. Release of the high-yielding semidwarf variety IR8 by the International Rice Research Institute (IRRI) in the late 1960s marked the beginning of the Green Revolution in India. Since then, production has kept up with population growth, with a steady increase in per capita production throughout the Green Revolution era in the 1970s and 1980s, before flattening out in the 1990s and finally declining in the 21st century (Fig. 1.1). However, per capita consumption declined at a faster rate than per capita production, making India a rice-surplus country. The decrease in rice consumption from the early 1990s coincided with economic reforms and trade liberalization, resulting in higher economic growth and diversification from rice to higher-value products. Despite a decline in per capita consumption, rice continues to be the single largest contributor to total calorie intake with a 28% share (FAOSTAT, accessed on December 7, 2015).

THE RISING COST OF SUCCESS

There is no such thing as a free lunch. India's resounding success in expanding rice production and improving the lives of millions of poor people has been achieved at a huge cost. Groundwater levels in the rice-growing belts of

The Future Rice Strategy for India. http://dx.doi.org/10.1016/B978-0-12-805374-4.00001-4

1

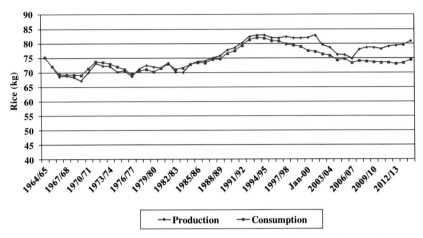

FIGURE 1.1 Indian rice production versus consumption on per capita basis (5-year moving average). *(USDA and FAO.)*

northwest and southern India have been declining at an alarming rate. As shown in the block-level groundwater stress map published by Shah (2013), the groundwater level is already exploited in the majority of blocks in the northwest (Punjab, Haryana, and Western Uttar Pradesh) and the south (Tamil Nadu, Andhra Pradesh, and Telengana). According to an article published in *National Geographic,*[a] in a study by the International Water Management Institute published in the *Journal of Water Resource Management*, the water table in Tamil Nadu is dropping at an average of 1.4 m/year. Farmers in the state are pumping out water at a rate of 8% more than the rate is being replenished. In the case of northwest India, the groundwater level is declining by 4 cm/year according to a study by Rodell et al. (2009) published in *Nature*. However, many other studies estimate the groundwater depletion in Punjab and Haryana to be much greater (Kumar et al., 2007; Pandey, 2014). Overall, water withdrawal for agricultural use in India has increased by 70% since the 1970s (Fig. 1.2).

India's subsidy for electricity, diesel, and shallow tubewells has worsened the excessive withdrawal of groundwater. In addition, India spends billions of dollars supporting its agricultural sector. The suite of policies it initiated in the mid-1960s, that is, a minimum support price and input subsidies for farmers (for fertilizer, seed, water, electricity, and machinery, etc.) and a food subsidy for the poor, remain in place. Without any doubt, these policies played a key role in improving India's food security by keeping arable land for rice cultivation, which would otherwise have been shifted to more profitable crops. However, over the years, these policies have become a serious burden to the Indian exchequer.

---

a. http://voices.nationalgeographic.com/2015/02/03/indias-food-security-threatened-by-groundwater-depletion/.

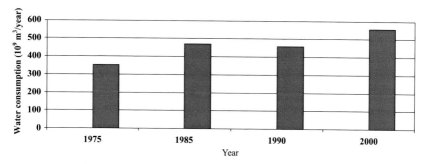

FIGURE 1.2 Indian agricultural water withdrawal. *(Zeigler, R., Mohanty, S., 2010. Support for international agricultural research: current status and future challenges. New Biotechnol. 27, 565–572.)*

FUTURE CHALLENGES

As India moves into the future it is safe to assume that higher incomes will bring about diversification of the food basket from cereal staples to more high-value products, with a continuing downward trend of per capita rice consumption for people from all economic spectra. At the same time, it is also safe to assume that the rate of diversification will be much slower than what has been witnessed in other East Asian countries during their development process. Thus total domestic rice consumption may still rise in the next three decades, with the World Bank predicting that by 2050 India's population will rise by another 340 million people from its current population of 1.28 billion. According to a 2050 vision document published by the Central Rice Research Institute (CRRI) in 2013 (CRRI, 2013), total rice consumption is estimated to reach 121.2 million tons by 2030, 129.6 million tons by 2040, and 137.3 million tons by 2050. In another study, Seck et al. (2012) estimated that total domestic rice consumption in India will rise by 14 million tons between 2010 and 2035. These consumption projections are less than what has been witnessed in the past because of the slowdown in population growth and food diversification. India's food subsidy program, which provides highly subsidized rice and wheat for poor households, is also likely to keep total consumption strong in the coming years.

If the current growth continues, and rice area remains at the existing level, India is likely to produce the additional rice needed to meet this demand. However, several emerging uncertainties such as growing water shortages, competition for rice land from nonagricultural uses and other cash crops, increasing frequency of extreme weather, and emerging pest outbreaks could derail this outcome. Depending on the extent of these problems, production growth in the future may fall well below the baseline projections.

A rapidly depleting water table in many northern and southern states is also a matter of concern for future productivity growth. Five major rice surplus states, Punjab, Haryana, Western Uttar Pradesh, Tamil Nadu, and old Andhra Pradesh, where groundwater depletion is a major issue, account for around 37% of India's rice production (in 2012–13). These states contribute a significant amount to the country's central pool, which is critical for India's food security and also to the functioning of India's food subsidy program, with which millions below the poverty line are provided with highly subsidized grains. Apart from water scarcity, the growing demand for land from urbanization, industrialization, and other cash crops is likely to cause a decline in rice area, which has been close to the all-time high of 45 million ha. According to the CRRI 2050 vision document, rice area may decline by 6–7 million ha by 2050, a decline of around 15% in the next 35 years. In other words, India will need to produce 137 million tons of rice on 37 million ha of land in 2050 compared with the current production of 105 million tons of rice on 43 million ha. Yield will have to increase by 50% in the next three decades to keep India food secure.

The target seems daunting considering the suite of emerging challenges, including the alarming rate of groundwater depletion in the major rice bowls of northwest and southern India. Although these regions contributed significantly to India's food security in the past three decades, it is widely believed that the key to India's future rice food security lies in eastern India (Adhya et al., 2008). Barah (2005) also argued in favor of a likely shift in rice production from the well-endowed irrigated areas of the northwest to the rainfed areas of eastern India.

According to Barah (2005), the shift is justified considering the recent change in policy, which is designed to abolish cereal land in the northwest in favor of crop diversification. According to a Haryana state agricultural policy document (http://agriharyana.nic.in/Agriculture%20Policy/English%20Haryana_State_Agriculture_Policy_Draft.pdf), the state is already concerned about degrading natural resources, particularly soil and water. In response, the state has already banned summer (dry season) rice. Various other options to reduce groundwater use in the wet season such as direct-seeded rice, aerobic basmati rice, and the promotion of maize cultivation are currently being explored. Similarly in Punjab, the state is mulling over restricting paddy cultivation to 1.6 million ha in areas where the groundwater level has not been adversely affected, compared with current paddy cultivation of 2.8 million ha (Agriculture Policy of Punjab, 2013, http://punjab.gov.in/documents/10191/20775/Agriculture+policy+of+punjab.pdf/9db4456f-55c5-4b55-882a-adf5811b2a53). The state would like to replace the 1.2 million ha of paddy land where the groundwater level is declining with other less water-intensive crops such as maize, cotton, sugarcane, soybean, pulses, and groundnut. Policymakers also realize that any such shift from rice to other crops will require an assured price and marketing similar to what is available for paddy. In the Tamil Nadu agricultural strategy document (www.tn.gov.in/dear/Agriculture.pdf), the government recognizes that the groundwater level in nearly half of the 386 blocks is already overexploited. Of the remaining half, the groundwater level in 33 blocks is critical, in 67 blocks is semicritical, and in 11 blocks the groundwater is of poor quality. According to the Tamil Nadu strategy document, policymakers are focusing on the urgent need to align cropping patterns with water availability and improve the efficiency of the irrigation system for long-term sustainability.

ENSURING FUTURE RICE FOOD SECURITY: OPPORTUNITIES IN EASTERN INDIA

Eastern India was left out of the first Green Revolution to reduce pressure on the water-stressed areas of northwest and southern India; however the time is now right for intensifying rice production in this area. Eastern India, comprising seven states, Odisha, West Bengal, Bihar, Eastern Uttar Pradesh, Assam, Jharkhand, and Chhattisgarh, accounts for more than 55% of the total rice area while contributing less than 50% to total production. The acceleration of paddy productivity growth in the region could have a significant impact on poverty alleviation, as the region accounts for 32% of India's rural poor population (Shah, 2013). West Bengal is a good example of where a combination of improved irrigation infrastructure and the widespread adoption of high-yielding varieties in the 1980s, particularly early-maturing varieties that allow the cultivation of a second crop, has resulted in strong yield growth in the past three decades (Shah, 2013), coinciding with a rapid decline in rural poverty. Rural poverty in West Bengal declined from 54.85% in 1983–84 to 35.66% in 1993–94, compared with the decline from 44.48% to 35.97% for the Indian rural average. According to Chakraborty and Mukherjee (2009), the sharp decline in rural poverty in the 1980s can be mostly explained by the unprecedented growth in agriculture, particularly rice production. This could have happened through multiple pathways including a rise in real income, employment generation, a decline in food prices, and creation of rural demand for products and services (Schneider and Gugerty, 2011).

Thirtle et al. (2003) concluded that a 1% increase in agricultural productivity is estimated to produce a 0.6%–1.3% reduction in poverty. Assuming the lower bound effect of productivity on poverty, a 50% growth in the productivity of rice could reduce poverty by 31% in eastern India.

In the eastern states, rainfall in the wet season (kharif) is much higher than in the western Indian breadbasket, and most of the rice produced in this area is rainfed; indeed, rainfall is so high in much of eastern India during kharif that few crops other than rice can be grown successfully. Supplying a larger proportion of India's rice needs from eastern Indian rainfed production could reduce the damage caused by overexploitation of water resources in the west, and increase the incomes of smallholder farmers in the poorest region of the country.

Despite high kharif rainfall, paddy yield in the eastern regions is severely affected by frequent climatic stresses such as droughts and floods. Despite having one-quarter of the country's usable groundwater and less than 1% being overexploited (Shah, 2013), the region frequently faces drought in the wet season because of irregular monsoons. The majority of India's 13.6 million ha that are prone to drought are located in eastern India (Pandey and Bhandari, 2009). In addition, 6 million ha of rice land are prone to submergence and waterlogged conditions (Barah, 2005) and 6.7 million ha are affected by salinization (Jehangir et al., 2013).

Varietal Improvement

With an expected rise in the frequency of adverse weather events in the future, minimizing yield loss due to abiotic stresses should be a top priority for policy-makers looking to improve productivity in the region. In addition, development of irrigation infrastructure for the efficient use of groundwater will not only help the wet season crop to escape the monsoon deficit, but will also allow the culti-vation of boro (winter) rice in many eastern states where millions of hectares of paddy fields are left fallow in the dry season.

Chhattisgarh is a relatively new eastern state in India, formed in 2000 by partitioning of the southeastern districts of Madhya Pradesh. Rice outputs of the districts that comprise Chhattisgarh were aggregated and then compared with output following formation of the district. It was found that rice produc-tion doubled from 1970 to 2010, although growth was erratic. Fluctuations in production were due to drought: a recurring event in the state. Chhattisgarh is vulnerable to drought-related crop damage since its rice area is predominantly rainfed. A similar erratic pattern was observed in Odisha, where fluctuations in the state's rice production are not only due to the occurrence of drought in high-lying fields but also to other abiotic stresses such as submergence in low-lying fields and salinity in coastal areas. In 2010, rice yields in both Chhattisgarh and Odisha were lower than the Indian national average.

The IRRI, in collaboration with its national partners, has introduced flood-tolerant modern varieties of rice (Sub1) in India that allow the plant to survive up to 2 weeks under water. This period would be long enough to completely destroy nonsubmergence-tolerant modern varieties. According to the IRRI estimates, these Sub1 varieties have the potential to increase production by up to 4 million tons in India and Bangladesh. A randomized control trial conducted by Dar et al. (2013) in Odisha found that the average yield of Swarna-Sub1 was 45% higher than that of Swarna, the recurrent parental variety, following submergence. They found no difference in the yields of the two varieties under normal conditions. Farmers who belong to disadvantaged castes, such as scheduled caste (SC) and scheduled tribe (ST), tend to occupy low-lying lands that are prone to submer-gence, and they benefited more than farmers who belong to other caste groups that have less flood-prone land. Furthermore, Emerick et al. (2016) reported that the use of flood-tolerant varieties had positive effects on the adoption of improved crop management practices, cultivation area, fertilizer usage, and credit. The au-thors concluded that varieties that reduce risk by protecting production in poor years have the potential to increase agricultural productivity in normal years.

Stress-tolerant rice varieties can withstand abiotic stresses such as submer-gence, drought, and salinity. Since 2007, several stress-tolerant varieties have been developed and disseminated in South and Southeast Asia. Important sub-mergence-tolerant rice varieties include Swarna-Sub1, Sambha Mahsuri-Sub1, IR64-Sub1, and BR11-Sub1. Similarly, drought-tolerant varieties such as Sahb-hagi Dhan, IR64-Drt1, and DRR dhan 44 were released recently in South Asia. The distribution of stress-tolerant rice varieties in South Asia started in 2008 under the Stress-Tolerant Rice for Africa and South Asia (STRASA) project, which coordinates seed multiplication with local counterparts and distributes seeds through NGOs, private seed producers, and government agencies. Seed minikits and field demonstrations were also used to show the value of these varieties to farmers. Seed distribution of STRASA products expanded exponentially in India after the National Food Security Mission began distributing stress-tolerant rice varieties in 2010.

Although it is too early to measure the adoption of stress-tolerant varieties in eastern India, a large-scale survey of ~9000 households, conducted in 2014, found that Swarna-Sub1 is already the fifth most popular variety grown in the region at 367,000 ha (Table 1.1). As shown in Table 1.1, the recently released Sambha Mahsuri-Sub1 is also attracting the interest of farmers in Eastern Uttar

TABLE 1.1 Estimated Area Under Popular Rice Varieties in Eastern India

Variety name	2013 Kharif season		2014 Kharif season	
	Total area (1000 ha)	Proportion of total area of rice (%)	Total area (1000 ha)	Proportion of total area of rice (%)
Swarna	4,571.3	30.6	3,808.5	27.7
Pooja	539.5	3.6	998.2	7.3
Lalat	429.6	2.9	897.7	6.5
Moti	411.6	2.8	276.8	2.0
Mahsuri	397.6	2.7	1,208.3	8.8
Swarna-Sub1	376.5	2.5	367.2	2.7
Sambha Mahsuri	330.6	2.2	220.2	1.6
ARIZE 6444	317.3	2.1	680.8	4.9
Sarju-52	315.1	2.1	349.9	2.5
MTU1001 (Vijetha)	408.8	2.7	522.7	3.8
MTU1010	283.0	1.9	347.5	2.5
Sahbhagi Dhan	—	—	34.6	0.3
Samba-Sub1	—	—	29.7	0.2
Other hybrid	411.0	2.8	232.4	1.7
Other improved	1,628.0	10.9	1,358.2	9.9
Other traditional	1,878.8	12.6	622.3	4.5
Unknown	2,632.5	17.6	1,801.6	13.1
Total	14,648.2	100.0	13,757.6	100.0

Estimated by the authors using Rice Monitoring Survey data.

Pradesh and Bihar with nearly 30,000 ha farmed in kharif 2015. Assam, another eastern state where a large proportion of the rice area is flood prone, has not been included in the survey, but Swarna-Sub1 is rapidly spreading there.

Apart from replacing Swarna in flood-prone areas, Swarna-Sub1 is also allowing cultivation of new areas that are normally left fallow in the monsoon season. According to Mohanty and Behura (2014), in Amathpur village in Odisha, 80 ha of land located between two rivers, the Birupa and Brahmani, used to be left fallow during the kharif season because of frequent flooding. The farmers planted the area with mungbean once the floodwater receded. In 2014, for the first time, farmers planted Swarna-Sub1 in the area through the Bringing Green Revolution to Eastern India seed distribution program. Coincidentally, the area was flooded for 6–8 days because of heavy rainfall that preceded the cyclone and, as expected, Swarna-Sub1 recovered quite well from the flood. In Odisha, some farmers in nonflood-prone areas grow Swarna-Sub1 because of its lighter husk color, which is preferred for offerings to their deities during religious rites (Mohanty and Behura, 2014). The drought-tolerant rice variety Sahbhagi Dhan has also started spreading in the region with nearly 35,000 ha in 2014 (Table 1.1).

Expected Impacts of Flood- and Drought-Tolerant Varieties on Income and Poverty

Based on the performance of Swarna-Sub1 in the initial years following adoption, we estimate that in the next 5–10 years, as these varieties are adopted by farmers, 1.3–1.8 million tons of rice will be saved in four eastern Indian states (Odisha, Assam, West Bengal, and Bihar), which would otherwise have been lost to floods (Table 1.2). This translates into an additional USD 300–400 million in income for farmers in the flood-prone regions of these four states. This

TABLE 1.2 Expected Swarna-Sub1 Impacts in the Next 5–10 Years

		Adoption scenario[a]	
		75%	100%
State	Total rice production in kharif 2014–15	Saved production loss in 1000 tons (% of total production)	
Assam	4,841	55 (1.14)	73 (1.51)
Bihar	5,694	441 (7.74)	588 (10.32)
Odisha	7,807	282 (3.61)	376 (4.82)
West Bengal	15,189	628 (4.13)	837 (5.51)
Total	33,531	1,406 (4.19)	1,874 (5.59)

[a]*Refers to the percentage of average submerged area between 2001 and 2009 for Bihar, Odisha, and West Bengal, and the percentage of submerged area for 2015 in Assam.*

is calculated based on area affected by short-, medium-, and long-duration submergence in the past decade and the yield advantage of submergence-tolerant rice varieties under different durations of flooding from earlier studies (0.5 t/ha for short-duration, 1.0 t/ha for medium-duration, and 0.2 t/ha for long-duration submergence).

Similarly, according to Mottaleb et al. (2016), India would have an additional 5–10 million tons of rice annually by 2035 under different climate scenarios if a successful drought-tolerant variety were made available in 2015. As expected, the majority of the additional production would be in eastern India.

The IRRI is also in the process of developing multiple-stress-tolerant rice varieties (submergence + salinity or drought + submergence) that will be released in India in the next few years. These varieties are likely to have a big impact on minimizing production loss and improving the livelihood of poor farmers.

Expansion of irrigation infrastructure in the eastern region, which has a quarter of the country's usable groundwater resources and hardly any groundwater-stressed blocks (Shah, 2013), will also allow a second rice crop in the boro (dry) season when yield is greater because of higher solar radiation.

High-Yielding Varieties for the Irrigated Belt With Desired Quality Traits

Apart from developing single- and multiple-stress-tolerant rice varieties to reduce yield loss in extreme weather, rice production in eastern India can be significantly improved by closing the yield gap in the irrigated rice area. According to Tiwari (2002), the experimental rice yield in the irrigated belt of eastern India was around 6 t/ha and 30%–40% of the potential yield is yet to be tapped. In a recent study by Das (2012), paddy yield in irrigated land in Odisha was around 3 t/ha in kharif and 3.5 t/ha in summer, which is 50% below the potential yield.

One of the major reasons for the low productivity of paddy in the irrigated belt of eastern India is the cultivation of varieties of rice that are more than 20 years old. As shown in Table 1.1, the top varieties grown in the region, with the exception of Swarna-Sub1, were all released before 2000, particularly during 1980–99. These varieties account for more than 60% of the total area of modern varieties. According to Das (2012), the strategy for eastern India's irrigated belt should be the development and dissemination of high-yielding varieties with desired quality traits.

Integrated Crop Management and Mechanization

In both the rainfed and irrigated belts, improving variety should be implemented alongside integrated crop management systems including mechanization, nutrient management, pest management, and water management. For example, yield gains obtained when using improved management systems have matched those attained when using a flood-tolerant variety of rice across a range of flooding depths, and the combination of improved variety and enhanced management systems more than doubled yields at flood depths of >35 cm (Sarangi et al., 2015). Similarly, good responses to crop management systems have been seen in saline areas where improvement in plant density and crop nutrition can provide worthwhile returns (Singh et al., 2016). Furthermore, changing cropping patterns and systems by combining management systems, variety, and additional crops such as pulses and oilseeds has the potential to raise the overall system productivity by several-fold over the coming years.

Rural outmigration, aging farmers, rising wage rates, and labor shortages are already transforming rice farming in the major rice-growing regions of Asia, including eastern India. Rice farmers are replacing labor-intensive activities such as transplanting, harvesting, and threshing with appropriate mechanization to lower the fast-rising cost of production. So far, the extent of mechanization in eastern India is much lower than in the intensive production regions of northwest and southern India. However, rising wage rates in eastern India are accelerating the pace of mechanization. In Odisha for example, the number of mechanical transplanters and combine harvesters grew from 19 in 2007–08 to 1,008 in 2013–14 (Fig. 1.3). During the same period, the number of tractors/power tillers grew from 4,069 to 20,678.

Many innovative models have started to emerge to offset the unviability of mechanization for smallholder farmers. Custom-hiring arrangements for machines are fast evolving as viable solutions for smallholder rice farmers in eastern India. In addition, apart from male service providers, many women and youths are finding it attractive to enter the service provider business for land preparation, transplanting, and harvesting. These service providers cover several hundred kilometers in a season, taking advantage of the different planting

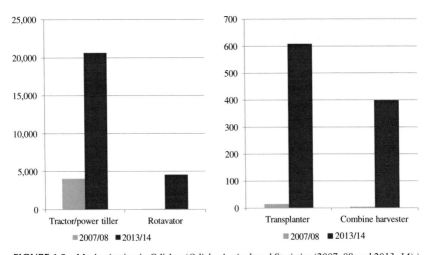

FIGURE 1.3 Mechanization in Odisha. *(Odisha Agricultural Statistics (2007–08 and 2013–14).)*

and harvesting periods between different regions. In India, the service providers start in Punjab and Haryana in the northwest as they harvest their crop first. They then move southeast through Uttar Pradesh, Bihar, Odisha, and West Bengal, covering nearly 1000 km in a season.

Small farmers are now renting additional land (available because of rural outmigration) to achieve economy of scale through the use of mechanical transplanters, combine harvesters, dryers, threshers, etc. Several forms of land consolidation models have been piloted in many parts of Asia. The "Small Farmers, Large Field" model in Vietnam, which allows small farmers to benefit from economies of scale by pooling their small farms into large fields of 50–500 ha to lower the per unit cost of using farm machinery, such as combine harvesters, is becoming popular among small farmers. Various forms of the "Small Farmers, Large Field" model are now being implemented in Vietnam and other neighboring countries. In the case of Thailand, an industrial rice farming scheme introduced by the Suphanburi Rice Millers Association in collaboration with the Suphanburi Rice Research Center has convinced farmers to grow one variety with synchronized planting and harvesting on around 400 ha. The idea is to lower the harvesting cost (USD 100 per ha) by 20%–30% by providing service providers with a bigger contract for custom harvesting. Eastern India is likely to implement some of these land-pooling models in the near future.

CONCLUDING REMARKS

India has come a long way in the past five decades: from a food-deficit country to one of the largest exporters of rice in the world. Transformation of the rice sector through improved productivity has supported overall economic growth in the country, thus contributing to rural poverty alleviation and a reduction in infant mortality and the undernourished population. On the flip side, India's resounding success in achieving food security and improving the lives of millions of poor people has come at a cost, with an alarming rate of decline in groundwater levels in the rice-growing belts of northwest and southern India. The affected states of Punjab, Haryana, and Tamil Nadu are mulling over strategies to arrest the decline in groundwater levels through improved crop management practices, restricting rice planting in sensitive areas through crop diversification, and even an outright ban on rice planting.

Although per capita consumption has stabilized for now, and is likely to decline in the coming years, total rice consumption is still expected to rise to 137 million tons by 2050 with the addition of another 340 million people to India's current population base of 1.28 billion. Taking the emerging uncertainties into account makes this target challenging, and a 50% increase in yield in the next three decades is required to keep India food secure.

Intensifying rice production in eastern India for future food security could have two benefits: lowering pressure on water-stressed areas of northwest and southern India and alleviating rural poverty in eastern India. The futuristic technological solution for the region should take into account the ongoing transformation of rice farming in eastern India, and these changes should be overlaid with the issue of sustainability of natural resources such as land, water, and environment to find an optimum solution for the intensification of the rice sector.

In both the rainfed and irrigated belts of eastern India, improved varieties of rice should be implemented alongside integrated crop management systems including mechanization, nutrient management, pest management, and water management. In the rainfed rice-growing belts of eastern India where the frequency of extreme weather such as floods and droughts is on the rise, the development and dissemination of rice varieties with abiotic stress tolerance, along with improved management practices, could have a significant impact on reducing yield loss and improving the livelihoods of millions of poor and marginal farmers and landless laborers. The strategy for eastern India's irrigated belt should be the development and dissemination of improved varieties with desired quality traits alongside sustainable management practices.

ACKNOWLEDGMENTS

The authors thank Dr. A.M. Ismail, Dr. D. Johnson, Dr. R.K. Singh, Dr. A. Kumar, Dr. J.K. Ladha, and Dr. J. Balagtas for some excellent suggestions.

REFERENCES

Adhya, T.O., Singh, N., Swain, P., Ghosh, A., 2008. Rice in Eastern India: causes for low productivity and available options. J. Rice Res. 2 (1), 1–5.

Barah, B.C., 2005. Dynamic of rice economy in India: emerging scenario and policy options. Occasional Paper Series 47, Department of Economic Analysis and Research, National Bank for Agriculture and Rural Development.

Chakraborty, A., Mukherjee, S., 2009. MDG-based poverty reduction strategy for West Bengal. Occasional Paper 20, Institute of Development Studies Kolkota, Calcutta University.

CRRI (Central Rice Research Institute), 2013. Vision 2050.

Dar, M.H., de Janvry, A., Emerick, K., Raitzer, D., Sadhoulet, E., 2013. Flood-tolerant rice reduces yield variability and raises expected yield, differentially benefiting socially disadvantaged groups. Sci. Rep. 3 (3315), 1–8.

Das, S.R., 2012. Rice in Odisha. IRRI Technical Bulletin No. 16. Los Baños (Philippines): International Rice Research Institute. 31 p.

Emerick, K., de Janvry, A., Sadoulet, E., Dar, M.H., 2016. Technological innovations, downside risk, and the modernization of agriculture. Am. Econ. Rev. 106, 1537–1561.

Jehangir, I.A., Khan, M.H., Bhat, Z.A., 2013. Strategies of increasing crop production and productivity in problem soils. Int. J. Forest Soil Erosion 3 (2), 73–78.

Kumar, V., Kumar, D., Kumar, P., 2007. Declining water table scenario in Haryana: a review. Water Energy Int. 64 (2), 32–34.

Mohanty, S., Behura, D., 2014. Swarna-Sub1: Odisha's food for a goddess. Rice Today 13 (1), 40–41.

Mottaleb, A.M., Rejesus, R., Mohanty, S., Murty, M.V.R., Li, T., 2016. Benefits of the development and dissemination of climate-smart rice: ex-ante impact assessment of drought-tolerant rice in South Asia. Mitig. Adapt. Strat. Glob. Chang., 1–23.

Pandey, R., 2014. Groundwater irrigation in Punjab: some issues and way forward. Working Paper No. 2014-140. National Institute of Public Finance and Policy, New Delhi.

Pandey, S., Bhandari, H., 2009. Drought, coping mechanisms and poverty. Occasional Paper, International Fund for Agricultural Development.

Rodell, M., Velicogna, M.I., Famiglietti, J.S., 2009. Satellite-based estimates of groundwater depletion in India. Nature 460999-1002 (www.nature.com/nature/journal/vaop/ncurrent/full/nature08238.html#B1)

Sarangi, S.K., Maji, B., Singh, S., Sharma, D.K., Burman, D., Mandal, S., Singh, U.S., Ismail, A.M., Haefele, S.M., 2015. Using improved variety and management enhances rice productivity in stagnant flood-affected tropical coastal zones. Field Crops Res. 174, 61–70.

Schneider, K., Gugerty, M.K., 2011. Agricultural productivity and poverty reduction: linkages and pathways. Evans School of Policy Analysis and Research 1 (1), 58–74.

Seck, P.A., Digna, A., Mohanty, S., Wopereis, M., 2012. Crops that feed the world 7: rice. Food Secur. 4, 7–24.

Shah, T., 2013. Water-energy nexus in the Eastern Gangetic Plains: old issues and new options. Moving from water problems to water solutions: research needs assessment for the Eastern Gangetic Plains, Proceedings of the International Workshop held at the National Agricultural Science Complex, Indian Council of Agricultural Research, New Delhi, India.

Singh, Y.P., Mishra, V.K., Singh, S., Sharma, D.K., Singh, D., Singh, U.S., Singh, R.K., Haefele, S.M., Ismail, A.M., 2016. Productivity of sodic soils can be enhanced through the use of salt tolerant rice varieties and proper agronomic practices. Field Crops Res. 190, 82–90.

Thirtle, C., Lin, L., Piesse, J., 2003. The impact of research-led agricultural productivity growth on poverty reduction in Africa, Asia and Latin America. World Dev. 31 (12), 1959–1975.

Tiwari, K.N., 2002. Rice production and nutrient management in India. Better Crops Int. 16, 18–22, special supplement.

FURTHER READING

Government of India, 2014. Agricultural Statistics at a Glance 2014. Available from: http://eands.dacnet.nic.in/PDF/Agricultural-Statistics-At-Glance2014.pdf

Chapter 2

Rice in the Public Distribution System

Sudha Narayanan

Indira Gandhi Institute for Developmental Research, Mumbai, Maharashtra, India

Gold is wealth, rice is wealth as well, but without rice, life has no sustenance
Turn to rice for your hunger as you would to water for your thirst

INTRODUCTION

Rice is a lifeline for a large proportion of the Indian population. In many communities across India, rice is considered wealth itself and is often accorded the same status as water, as the previously mentioned proverbs suggest. It is estimated that rice forms part of the consumption basket of 85% of all rural households and 88% of all urban households in India; paddy is grown on an estimated 59.9 million operational holdings.[a]

Although the per capita availability of rice has been stable for the past 25 years, primarily due to increased yields, per capita rice consumption in rural households has declined from ~83 kg in 1987–88 to 80 kg in 1999–2000 and more rapidly to 73 kg in 2009–10. Per capita rice consumption declined in urban households from 64 kg in 1987–88 to 62 kg in 1999–2000 and to 55 kg in 2009–10 (NCAER, 2014). This trend is part of a larger shift away from cereals to high-value commodities, such as dairy and fruits and vegetables, implying a shift toward higher-quality diets (Kumar et al., 2016; Kumar and Joshi, 2013). However, rice still remains an important part of the state's strategy for food security. In 2014–15, the Indian government procured 321.65 lakh tons of rice, one-third of the total rice produced in the country, and allocated as much as 87.7% of this (281.97 lakh tons) to various welfare schemes. This volume was roughly 1.5 times the allocation of wheat, making rice the single most important commodity in the nation's strategy for achieving food security. Despite years of consistent and robust economic growth, malnutrition in India

---

a. Consumption data are based on a 30-day recall period from the National Sample Survey (2011–12) and paddy cultivation data are from the Agricultural Census (2010–11), Government of India.

The Future Rice Strategy for India. http://dx.doi.org/10.1016/B978-0-12-805374-4.00002-6

is a persistent concern. Large-scale surveys show that significant improvements have been made in lowering stunting rates (a reported 48%–39% reduction in child stunting between 2005–06 and 2013–14); however, despite these gains, a substantial proportion of the population remains vulnerable. India has consistently fared poorly in indicators of hunger as reflected by the bbex (von Grebmer et al., 2015). The role of the state is therefore pertinent in India's struggle to eliminate hunger and malnutrition.

This chapter focuses on the role of rice for food security in India.[b] This work is designed to complement other chapters that deal with rice procurement and consumption by focusing on the place of rice in the nation's efforts to protect food security. Owing to the limitations of data availability, it is hard to isolate and distinguish the role of rice from that of wheat across different schemes. This chapter therefore deals more generally with food-related schemes administered by the nation. The following section traces the historical evolution of these schemes and outlines recent developments, culminating in the National Food Security Act. Many have questioned the virtue of large schemes that subsidize food, viewing these as leaky systems that are not cost-effective and hence impose a burden on the exchequer. The subsequent section therefore assesses the extent to which the major schemes have reached intended beneficiaries and offers a review of implementation, food security, and nutritional impacts of these programs. The proceeding section then assesses recent developments and provides thoughts on the architecture of a food security system that appears to veer toward cash transfers. The final section concludes with a discussion of the way forward, making a case for reforming rather than abandoning the current system.

FOOD SCHEMES IN INDIA

Historical Perspective and Recent Developments

India has a long history of food-focused schemes aimed at addressing food and nutritional security, and rice has been a key component of a majority of these. The keystone of India's food security strategy is the public distribution system (PDS)—the distribution of rice and wheat (mainly) at subsidized prices through ration shops—which originated from the need to address urban food scarcity after the Second World War. Food-price subsidies in India evolved in three broad phases: From the 1960s to the early 1990s, the PDS tackled food

---

b. Food security is characterized as "a situation … when all people, at all times, have physical, social and economic access to sufficient, safe and nutritious food that meets their dietary needs and food preferences for an active and healthy life" (FAO, 2002, www.fao.org/docrep/005/y4671e/y4671e06. htm-fn31). This understanding of food security incorporates the idea that access to food includes not just physical availability and affordability, but also requires that individuals do not face social barriers in feeding themselves. Food security implies nutritional security and further acknowledges that, in its attainment, it supports the actualization of individual capabilities. It is also important to note that individuals are the focus, although household-level or community-level food security is an appropriate concern.

scarcity by focusing largely on urban areas. The revamped public distribution system (RPDS, 1992–1997) sought to improve the program's reach to poorer rural areas. Later, in 1997, the targeted public distribution system (TPDS) modified the scheme by distinguishing consumers above the official state-specific poverty line (APL) from those below the poverty line (BPL) in both rural and urban areas. In most states the extent of the subsidy, and sometimes the allocations themselves, varies between the two categories. In some states the PDS functions as a universal program in which poor (BPL households) and nonpoor (APL households) are not distinguished. Many states provide additional subsidies of food grains, allowing consumers to purchase these at prices lower than in other states. Even the basket of commodities provided under the PDS varies between different states, with some offering just rice or wheat, sugar, and kerosene, and others offering fortified flour, iodized salt, and pulses.

Established in 1975, the Integrated Child Development Services (ICDS) scheme was designed to supplement nutrition for pregnant and lactating mothers and children in preschool, provide nutrition education to adolescent girls and women, and run preschool activities for children. The midday meal (MDM) scheme, served in designated categories of schools,[c] seeks to provide one hot meal a day for children during school days. In contrast to the ICDS scheme, which supplements nutritional intake, MDM schemes are intended to provide an entire meal. MDM schemes were systematically introduced in Tamil Nadu's primary schools as early as 1962–63, and by 1990–91 several other states had a similar program in place. In August 1995, the National Program of Nutritional Support to Primary Education provided free food grains for distribution. Although transportation was the central government's responsibility, all other costs and arrangements were taken care of by state governments. Several states adopted this system until orders were issued by the Supreme Court in 2001 mandating that hot cooked meals be served in schools. All of these programs include rice as an important component, although not exclusively. Aside from these key programs, other programs also include the distribution of rice. For example, programs targeted at adolescent girls, programs to support meals provided in state-run hostels for marginalized communities, grain provided to the elderly, and so on.

The National Food Security Act

In 2013, many of these schemes were brought together as legally backed entitlements under one umbrella, the National Food Security Act (NFSA) (see Table 2.1 for the list of schemes under the NFSA). However, the origins of

---

c. Children studying in primary and upper primary classes in government schools, government-aided schools, schools under local administration, Education Guarantee Schemes, and Alternative and Innovative Education (AIE) centers; Madarsa and Maqtabs supported under Sarva Shiksha Abhiyan (SSA), Education Guarantee Scheme (EGS), and National Child Labour Project (NCLP) Schools run by the Ministry of Labour are eligible for midday meals.

TABLE 2.1 Entitlements under NFSA, MGNREGA, and Related Programs

Target group	Scheme	Act	Entitlements	Eligibility	Type of assistance
Pregnant and lactating mothers	Janani Suraksha Yojana	—	INR 1,400 (700) in LPS (HPS) in rural areas, INR 1,000 (600) in urban areas for the mother, and INR 600 (200) for ASHA workers. Conditionality: delivery in government health centers or accredited private institutions	All SC and ST women in both LPS and HPS, delivered in a government health center or accredited private institution. In low-performance states, all pregnant women in HPS states; BPL pregnant women, aged 19 years and above	Cash transfer with conditionality
	Maternity entitlements	NFSA (Chapter 2)	Not less than INR 6,000 (IGMSY implemented on a pilot basis, expected to be universalized under the NFSA)	All pregnant women and lactating mothers in regular employment with the central or state government, public–sector undertakings, or those who are in receipt of similar benefits under any law for the time being in force	Cash transfer with eligibility criteria
	ICDS	NFSA	Take-home rations of 600 Cal containing 18–20 g of protein. Available during pregnancy and until 6 months after childbirth	Identified by the anganwadi	In-kind transfers
Preschool children	ICDS	NFSA	Take-home rations of 500 Cal containing 12–15 g of protein for children aged 6 months to 3 years. For 3–6 year olds, a morning snack and hot cooked meal are provided. 500 Cal containing 12–15 g of protein. If malnourished, then take-home rations include an additional 800 Cal containing 20–25 g of protein	Those attending anganwadi. State differences in the implementation occur	In-kind transfers

School-going children	School meals	NFSA	Hot cooked meals. At present, a midday meal provides an energy content of 450 Cal and protein content of 12 g at primary stage and an energy content of 700 Cal and 20 g of protein at upper primary stage	All enrolled children up to Grade 8 in government schools	In-kind transfers
Adults and children	PDS	NFSA	5 kg per person per month for priority households identified by state governments, 35 kg per household per month for AAY households, excluded households have no entitlements. The price is INR 3/kg for rice, INR 2/kg for wheat, and INR 1/kg for coarse grains	Priority households as defined by the state government, AAY households as defined by the state government (based on central government guidelines), and excluded households as defined by state governments (25% and 50% of the rural and urban population)	In-kind transfer at subsidized prices
Senior citizens	PDS	NFSA	Annapurna entitlements 10 kg/month food grains	Above 60 years of age	In-kind transfer
	Pensions	Schedule III in the NFSA	For 60–79 year olds, INR 200 per month; 80 years and above, INR 500/ month. For widows, and disabled, INR 300/month.	State implementation varies significantly in both coverage and amounts	Cash transfer with eligibility criteria

Notes: For midday meals, government schools means government schools, government-aided schools, schools run by local bodies, education guarantee schools, alternate and innovative education centers, madarsas and maqtabs supported under the sarva shiksha abhiyan, and national child labor project schools. Pensions are paid under Indira Gandhi national old age pension scheme, Indira Gandhi national widow pension scheme, and Indira Gandhi national disability pension scheme.
Source: Narayanan, S., Gerber, N., 2015. Social safety nets for food and nutritional security in India. Working Paper WP-2015-31, Indira Gandhi Institute of Development Research.

the NFSA go back to 2001. This was a time when the threat of severe hunger loomed large in many parts of the country, despite the government having accumulated stocks of ~50 million tons of grains (rice, wheat, and coarse cereals) (Srinivasan and Narayanan, 2008). In April 2001, the People's Union for Civil Liberties (PUCL), an active civil society group in the northern Indian state of Rajasthan, submitted a writ petition to the Supreme Court of India demanding that the country's food stocks be used without delay to protect people from hunger and starvation. This petition led to a prolonged "public interest litigation" (PUCL versus Union of India and Others, writ petition [civil] 196 of 2001). Supreme Court hearings have been held at regular intervals since then and significant interim orders have been issued by the court regarding the scope and implementation of eight food-related schemes run by the Government of India (Hertel, 2015; Srinivasan and Narayanan, 2008, for discussions on the Right to Food campaign). After protracted debates and discussions, which included explicit disapproval by at least one expert committee on account of the fiscal commitments entailed by these programs (discussed later), the NFSA was finally passed in 2013.

The NFSA was to form the backbone of a set of social safety nets for the poor. Envisioned as a comprehensive legislative framework for protecting an individual's right to food, it furthers the vision expressed in the Constitution of India to the right to life. It was originally conceived as a system of interventions following a life-cycle approach, whereby at every stage of an individual's life, the state would enable access to a safety net to ensure food security. This brought into its fold a whole range of interventions over the life cycle of an individual, from prenatal to senescence: child nutrition programs, maternity benefits, social security pensions, and other entitlements that would further food security. However, the bill became substantially restricted in its scope and vision over time. Only a few of the programs originally envisioned currently form part of the NFSA, although several other food-related schemes continue in a programmatic mode. The main difference between programs that are part of the NFSA and those that are not is that the former are associated with legal entitlements, defined in terms of quantities of rice and other commodities for specific beneficiaries, and are in theory actionable (Table 2.1; Narayanan and Gerber, 2015).

Despite a unifying national legislation, the cluster of programs that fall under the ambit of the NFSA have diverse goals, targeting specific beneficiaries and addressing different aspects that influence nutrition and health outcomes. Significant interstate differences regarding entitlement and implementation reflect the differentiated histories of these programs in different states, with several states providing benefits well beyond those prescribed in the NFSA. Although 2 years have passed since the NFSA was promulgated, the extent to which states have implemented the act, and their efforts to implement it, remain highly variable. In many states, the programs that constitute the NFSA continue as they did before.

SCALE OF THE PROGRAM: ALLOCATION, EXPENDITURE, AND REACH

The programs that form the NFSA have a massive reach. The MDM scheme is perhaps the largest school meal program in the world, the ICDS scheme reaches out to more children than any other single program, and the PDS is perhaps the largest subsidized food-grain distribution scheme in the world. In 2013–14, an estimated 104.5 million children benefited from hot cooked nutritious food in 1.16 million schools across the country (http://mdm.nic.in/). One hundred two million children and mothers were also reported to benefit from supplementary nutrition as part of the ICDS scheme, serviced by 1.26 million centers. The PDS distributes as much as 44.6 million tons of grains, including 25.5 million tons of rice, over half the volume of the world rice trade. Indeed, of the total allocations of rice to various welfare schemes, the PDS accounted for the lion's share (90%) in 2014–15 (Fig. 2.1).

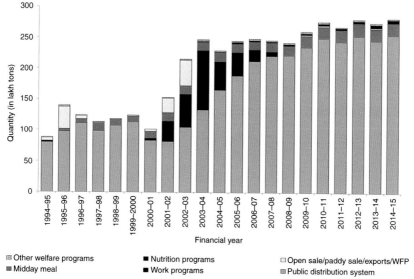

FIGURE 2.1   **Rice allocation to various welfare schemes.** Department of Food and Public Distribution website-foodgrain bulletins, http://dfpd.nic.in/foodGrain-archive.htm. Annual reports, http://dfpd.nic.in/annual-report.htm; Food Corporation of India website, http://fci.gov.in/imports.php?view=23; Indiastat.com; data for the years 1995–96 to 2001–02 and 2014–15 are provisional. Data for 2015–16 are provisional up to June 2015; allocation under the Annapurna scheme for 2002–03 is made on the basis of food grains (rice and wheat combined); PDS data for the years 2002–03 and 2004–05 are provisional; the graph does not include offtake data for relief and defense; food subsidy comprises offtake data from the PDS, RPDS, BPL, APL, Antyodaya, Annapurna, and the poorest district schemes; Work programs comprise offtake data on Sampoorna Grameen Rozgar Yojana, Jawahar Rozgar Yojana, and food for work program; Nutrition programs comprise offtake data on wheat-based nutrition program, nutrition program for adolescent girls, and the Rajiv Gandhi scheme for empowerment of adolescent girls; other welfare programs include hostels/welfare institutes and village grain banks.

BENEFITS AND IMPACT

Who Benefits?

In general, it appears that almost all food-related schemes in which rice is an important component are inclusive, in the sense that the poor and marginalized participate more than the nonpoor. Take the PDS for instance. It is often claimed that there is considerable mistargeting and that errors of inclusion and exclusion undermine the intent of the program. Although this is true, data from the National Service Scheme (NSS, obtained during the 66th and 68th rounds) suggest that the poor are nevertheless far more likely to access the PDS with a considerable proportion of the rich opting out (Narayanan and Gerber, 2015). This also holds true for reliance on the PDS, in terms of the proportion of consumed and purchased goods that are sourced from the PDS (Fig. 2.2), which decreases progressively as landholding and monthly per capita expenditure levels rise. Fig. 2.2 plots a local polynomial regression for the share of rice sourced from the PDS, in terms of total consumption and purchase of rice for those who use the PDS. The horizontal axis has the log monthly per capita expenditure. A similar exercise for whether or not households use the PDS shows that a larger proportion of poor households access the PDS than those who are nonpoor (data not shown). This holds true when looking at patterns over a range of landholding sizes or over monthly per capita expenditure levels. A similar pattern is observed for the MDM scheme, as the rural poor tend to send their children

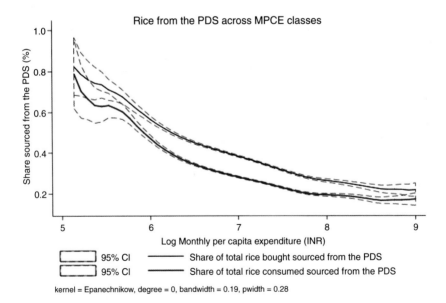

FIGURE 2.2   **Share of PDS purchases in total rice purchase and consumption by expenditure classes.** (*Computed from NSS data.*)

to the government school system and are therefore more likely to benefit from the MDM served in these schools (data not presented here). Overall, 55.8% of households with at least one child of primary school age were accessing MDMs in 2011–12. Participation rates in the MDM scheme at the district level show a strong correlation with enrollment in government schools. Given that the poor, who are also more likely to be undernourished, are more likely to send their children to government school than the nonpoor, the MDM serves an important function. Those households belonging to marginalized social groups are also more likely to access the PDS and these groups tend to rely more on PDS purchases of rice than those belonging to less backward communities (Table 2.2). Furthermore, those who have access to land or to regular salaries or wages are

TABLE 2.2 Rice Purchases From the PDS by Social and Occupational Groups

	Proportion accessing the PDS (%)	Share of the PDS in total rice consumption among those who use the PDS (%)	Share of the PDS in total rice purchases among those who use the PDS (%)
All	51.8	28	36.9
Social group			
Scheduled tribes	62.0	34.4	48.4
Scheduled castes	61.0	33.7	39.1
Other backward castes	51.0	28.0	37.1
Others	41.0	19.9	28.5
Occupation			
Self-employed in agriculture	42.0	22.2	42.9
Self-employed not in agriculture	49.0	24.7	28.5
Regular wage/salary earning	42.0	20.1	23.3
Casual labor in agriculture	71.0	39.4	43.1
Casual labor not in agriculture	59.0	31.1	33.9
Others	45.0	33.9	36.2

Source: Computed from the National Sample Surveys for rural areas.

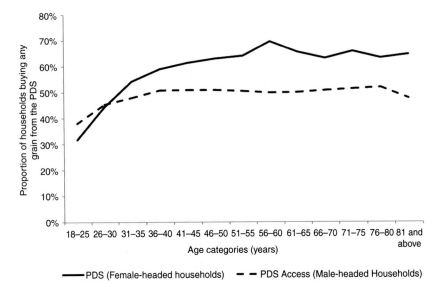

FIGURE 2.3   **Age and gender of heads of rural households accessing the PDS for rice and wheat.** *(Narayanan, S., Gerber, N., 2015. Social safety nets for food and nutritional security in India. Working Paper WP-2015-31, Indira Gandhi Institute of Development Research.)*

much less likely to access the PDS or depend on it. The PDS plays a significant role in the food security of households that depend on casual labor on- or off-farm (Table 2.2).

Other important dimensions of marginalization are gender and age. Fig. 2.3 offers an insight into the participation rates of individuals by gender across different age categories. The figure suggests that female-headed households and families with older heads depend on the PDS more than male-headed and younger households. These patterns are broadly similar for the proportion accessing the PDS, or the extent of reliance on the PDS, for total cereal consumption and purchases. Other research suggests that the PDS is of much greater importance to female-headed households and backward classes as it contributes a greater share of total calorie intake and also supplies more calories than other households (Ray, 2007). Overall, the beneficiaries of the major food-related schemes are the poor and the marginalized and, given that these groups are far more likely to be vulnerable to hunger and food insecurity, these schemes potentially form a critical safety net.

The Impact on Beneficiaries

We need to evaluate these rice-dominated schemes based on the definition of food security, that is, availability, access, absorption, and stability, and this is especially important at a time in which the implementation of these schemes is highly variable. Unfortunately, few studies systematically assess the contribution of rice in these food schemes, or the contribution of food schemes themselves to the food security of the beneficiaries. Perhaps the greatest benefit of the PDS is the psychological security of having access to grain at a time in which markets are underdeveloped and food availability uncertain. This is evident not only from anecdotal evidence but also from systematic surveys; for example, in the PEEP survey from 10 states, respondents repeatedly articulate that they are secure in the knowledge that they will not go hungry with the PDS (Drèze and Khera, 2014). Existing data on the incidence of hunger, that is, the proportion of households that report not having two square meals a day in the reference period, suggest that this declined significantly (14.7%–1.3%) from 1983–84 to 2009–10, although the data are considered somewhat unreliable (Kumar et al., 2012). Although this decline is attributable to many factors, including infrastructure improvements, market access, income increases, etc., it is plausible that food-related schemes have also contributed to this (Kumar et al., 2012). This is especially true when one considers the steep increase in allocations of rice (and wheat) in welfare schemes between 2000 and 2010.

Quantitative surveys, which estimate the impact of the PDS, tend to be narrowly focused on specific metrics, such as consumption expenditure, etc. A review of this literature suggests that virtually all studies indicate a positive impact of the PDS on nutritional outcomes (Narayanan and Gerber, 2015; see Tables 2.3 and 2.4). These include a decline in poverty rates and poverty gaps, as well as an increase in calorie intake and diet diversity. The impact on anthropometric measurements is far more limited, as one would expect, since these are sensitive to confounding factors, including sanitation, water, and healthcare. The MDM scheme has been shown to improve not only nutritional intake but also enrollment rates and attendance (Jayaraman, 2009; Narayanan and Gerber 2015; see Tables 2.3 and 2.4). There is even less evidence on the impact of the ICDS scheme. Although some studies show that participation in the ICDS has reduced stunting, others show that this is true only in specific areas, where perhaps complementary inputs are also present (Narayanan and Gerber, 2015; see Tables 2.3 and 2.4). Little research exists for other rice-based food security schemes.

CRITIQUES OF THE CURRENT SYSTEM

The fundamental critiques leveled at the current system are the implied presence of the state in grain markets, and the costs involved in procurement, storage, and distribution as part of what is widely seen as a flawed mechanism, (i.e., the PDS). The expert committee appointed by the government to review the draft NFSB, prepared by the National Advisory Council, observed that "massive procurement of food grains and a very large distribution network entailing a substantial step up in subsidy" was a problem. Their estimates suggested that the proposed PDS would require stocks of 54–74

TABLE 2.3 List of Selected Studies Examining Health and Nutrition Outcomes of the PDS.

Study	Location	Data set and method	Program	Dimension	Statistical significance (0,+,++)	Economic significance (−,−,0,+,++)
Himanshu and Sen (2013a,b)	All India	NSS 1993–94, 2004–05, 2009–10 Part I: Descriptive	MDM and PDS rations	Poverty reduction		MDM: + PDS: +
		Part II: Regression		Calorie intake	PDS: ++	PDS: ++
Rahman (2015)	KBK region, Odisha	NSS 2004–05, 2011–12 Difference-in-difference	PDS	Dietary diversity, macronutrient intake	++	++
Kochar (2005)	Major wheat-consuming states	NSS 1993, 1999–2000 OLS and IV regression analysis	PDS	Calorie intake	++	+
Chatterjee (2013)	Koraput, Odisha	Revisited HUNGaMA sample Probit model	PDS	Child anthropometry (wasting, stunting, underweight)	0	0
Jha et al. (2011)	Andhra Pradesh, Maharashtra, and Rajasthan	Primary survey IV regression	PDS and NREGA	Micronutrient intake	PDS: ++ NREGA: +	++ +
Kaul (2014)	Eight states with rice as the dominant staple (AP, AS, CG, JH, KA, KL, OD, and WB)	NSS 2002–08	PDS	Calorie intake, calories from different food groups, and cereal consumption	++	++

Study	Region	Data and method	Program	Outcome		
Kaushal and Muchomba (2013)	20 states	NSS 1993–94, 1999–2000, 2004–05 OLS and IV regression	PDS	Calorie intake	0	0
Khera, 2011a,b	Rajasthan	Primary survey Multivariate regression	PDS	Cereal consumption	0	0
				Wheat consumption	++	+
Ray (2007)	All India	NSS 1987–88, 1993–94, 1999–2000, 2001–02 Descriptive	PDS	Calorie consumption		++
Krishnamurthy et al. (2014)	Chhattisgarh	NSS 1999–2000, 2004–05 Difference-in-differences	PDS	Calorie consumption	++	++
Svedberg (2012)	All India	NSS 2007 Descriptive	PDS	Rice and wheat consumption	0	–
Tarozzi (2005)	Andhra Pradesh	NFHS 1992–93 OLS and difference-in-difference	PDS	Child anthropometry (weight-for-age)	Males: 0	+
					Females: 0	–
Drèze and Khera (2013)	All India	NSS 2009–10 Quantitative	PDS	Poverty	++	++

Note: For statistical significance, ++ or — refers to significance at the 1% level for positive impact and negative impact, respectively, + and − for significance up to the 10% level, and 0 if the relationship is not significantly different from 0. Economic significance is a loosely defined term that interprets the authors' own articulation of whether the findings show an impact that is significantly large. AP, Andhra Pradesh; AS, Assam; CG, Chhatisgarh; JH, Jharkhand; KA, Karnataka; KL, Kerela; OD, Odisha; WB, West Bengal.

Source: Extracted from Narayanan, S., Gerber, N., 2015. Social safety nets for food and nutritional security in India. Working Paper WP-2015-31, Indira Gandhi Institute of Development Research.

TABLE 2.4 List of Recent Studies Examining the Effectiveness and Outcomes of the ICDS and MDM Schemes

Study	Location	Data Set and Method	Program	Dimension	Summary of Results
Jain (2015)	All India	NFHS (DHS)	ICDS	Stunting	Children (0–2 years in rural India) who received daily supplementary feeding for a year were found to be 1 cm (0.4 z-score) taller than those who did not
Singh et al. (2014)	Andhra Pradesh	Young lives Instrumental variable	MDM	Stunting, underweight	The MDM scheme increased weight-for-age by 0.60 standard deviations and increased height-for-age by 0.27 standard deviations; the latter was not statistically significant
Afridi (2010)	Madhya Pradesh	Primary survey Difference-in-differences	MDM	Calorie consumption of children	A cost of 3 cents per child per school day reduced the daily protein deficiency of a primary school student by 100%, the calorie deficiency by almost 30%, and the daily iron deficiency by nearly 10%

million tons of grain and, at current economic costs of operations, outlays on the order of INR 90,000 crore.[d] Many writers believe the outlays on these two programs to be too great a burden on the exchequer and a waste of resources on account of leakages (Gulati et al., 2012), while others have contested these claims (Abreu et al., 2014; Sinha, 2013). The fiscal allocations

---

d. The proposal discussed here was drafted by the National Advisory Council and sent to an Expert Committee on the National Food Security Bill on October 27, 2010. More recent estimates of costs under different scenarios are available in Kozicka et al. (2015).

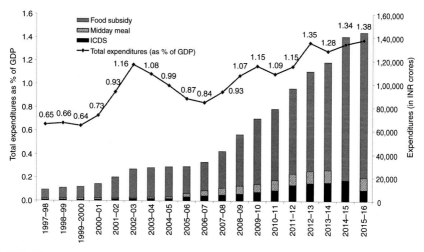

FIGURE 2.4 **Fiscal allocations.**

for these programs come mainly from the central government, although states can also allocate to these programs. Fig. 2.4 shows the historical split and progression of fiscal allocations to the main programs of the NFSA. In 2013–14, central government allocations to the ICDS, MDM scheme, and PDS accounted for just 1.3% of the gross domestic product (GDP)[e,f]. These budgetary allocations represent a very small proportion of the country's GDP. This is especially evident when compared with the percentage of GDP spent on social protection by low-income countries (3.4%), upper-middle-income countries (4%), China (5.4%), and high-income countries (10.2%), for which Japan stands out (19%). The fact that these food security schemes cover as many households as they do with such a small fraction of the GDP is noteworthy.

Other bones of contention include the method of support and the ability to reach intended beneficiaries. For example, some have asked whether subsidized food distribution is the best way for the PDS to achieve its goals or whether it can be replaced by cash transfers or food coupons (Kapur, 2011; Kapur et al., 2008; Kotwal et al., 2011). Others believe that India's policy should conform more with the WTO agreement on agriculture since the NFSA relies heavily on public procurement of grains (Hoda and Gulati, 2013). At

e. There are suggestions that one should look at the ratio of these expenditures to fiscal revenue or as a share of total public expenditures, but expenditures as a proportion of GDP provides a yardstick of comparison across countries. If one were to include state allocations, the number would be higher as it would include all the programs that are part of the NFSA or include other safety nets such as the Mahatma Gandhi National Rural Employment Guarantee Act (MGNREGA). The number cited here is meant to be a ballpark estimate rather than a precise estimate.

f. See also Himanshu and Sen (2013b) on the issue of fiscal allocations.

one extreme, economists advocate replacing the PDS in its entirety with cash transfers (Banerjee, 2011; Kapur et al., 2008; Somanathan, 2011) along with a supplementary nutrition component for children in the IDCS (Mehrotra, 2010). Others have suggested the implementation of cash transfers without dismantling the PDS, or moving to a system of food coupons (Basu, 2011; Kotwal et al., 2011).

An important concern in this context is the issue of "leakages" and implementation performance. The overall scale of leakages in the PDS is too high but recent improvements across states, especially in the poorer regions, represents a positive development. Khera (2010, 2011a,b) documented trends in the "diversion ratios in different states," that is, the quantity that is reported in government budgets as offtake for distribution to consumers under the PDS, which is not reflected in the NSS consumption data as PDS sourced. This is expressed as a percentage of the reported offtake and thus a higher mismatch would imply greater diversion and vice versa. Whereas the estimated diversion ratio was ~54% in 2004–05, it declined to 41% in 2009–10. It ranged from as low as 7% in Tamil Nadu to 85%–95% in Bihar, Jharkhand, Assam, and Rajasthan. The diversion rates declined in almost every state, with big improvements in some, even in states hitherto considered poor performers (Drèze and Khera, 2013; Khera, 2011a). An extension of this study in 2011–12 suggested that even states such as Bihar, long considered as chronically poor performers, have shown impressive improvements (Rahman, 2014).

Many estimates have also been put forth to prove that the PDS is not cost-effective. An early estimate suggested that it takes INR 3.65 to transfer INR 1 to the consumer; other recent estimates range from INR 3.42 to 9.53 (Jha and Ramaswami, 2010; Svedberg, 2012), with some of these assuming that food subsidies going to the nonpoor are also diversions. Although these estimates highlight the need for a cost-effective system, they neglect the fact that the PDS is only one side of the coin and that the costs involved in setting up this system involve substantial transfers to (rice) farmers as price support. When this is taken into account, the cost-benefit ratio changes substantially. Furthermore, the transaction costs of transferring INR 1 of benefits to both consumers and farmers was 41 paise (100 paise = INR 1) in 2010–11 and 39 paise in 2011–12 (Narayanan, S., Deuss, A., 2016. Estimating Food Subsidies in India, unpublished). Progress with the ICDS and MDM schemes is less well documented. The evidence that exists suggests large implementation differences across states (FOCUS, HunGAMA). However, one report cites that, with expenditures one-seventh those of the United States, India's MDM scheme in schools feeds 4 times as many.[g]

---

g. http://pulitzercenter.org/reporting/power-lunch-india-mid-day-meal-program. This does not imply efficacy per se, or discuss the impact, it merely suggests that per beneficiary costs are lower in India than in the United States.

TOWARD A NEW ARCHITECTURE: DEBATES AND CHALLENGES

Discussions on the merits and demerits of the existing food security support system continue. However, in 2014 the government initiated a cash-transfer program in lieu of in-kind distribution of grain in select union territories in India, in what is widely seen as a precursor to replacing in-kind transfers with cash. The central government reduced allocations to the ICDS in the union budget of 2015, the implicit justification being that the state share of resources was now augmented, enabling spending on these programs. The 2014–15 economic survey of the Government of India articulated a vision for cash transfers aimed at "wiping every tear from every eye" (page 21, Vol. 1 of the economic survey, Government of India, 2015). It stated that the combination of unique identity numbers, bank accounts, and mobile technology "offers exciting possibilities to effectively target public resources to those who need them most" and suggested that "success in this area will allow prices to be liberated to perform their role of efficiently allocating resources and boosting long-run growth" (page 21, Vol. 1, economic survey, Government of India, 2015). The government had framed guidelines for a cash-transfer program as of September 2015.

In theory, cash transfers are cost-effective since they have lower transaction costs and avoid the problem of having to procure, store, transport, and distribute commodities. They also offer beneficiaries freedom to direct the cash to particular household needs. In the context of food, for instance, this could imply a more diverse diet or better quality grain. Cash is also deemed to have multiplier effects that could potentially support local market development. In India, most proponents of cash transfers as a replacement for the PDS see them as a cost-effective alternative that is less prone to leakage or corruption.

However, how far can these expectations hold in reality and can cash transfers serve the central goal of food security? This is open to question. The canvas of empirical evidence suggests that cash transfers might be appropriate for some goals but not for others, and their efficacy is highly context dependent (Gentilini, 2016; earlier reviews available in Narayanan, 2011). There is now fairly consistent international evidence that, in the context of food and nutrition, a combination of food and cash might outperform either of these alone. Although cash transfers involve lower transaction costs, the impacts on nutrition are equivocal (see Gentilini, 2016 and Narayanan, 2011 for a review of earlier evidence). Recent evidence also suggests that cash transfers alone may have limited impact and it is with intensive communication of nutrition-focused messages that significant benefits are achieved.[h]

Cash transfers are often presumed to give people freedom of choice. Although this might be true in urban areas, effective choice is limited in most rural areas. In many parts of rural India, food commodities are nontradables, for which the transaction costs are too high for private traders to take food commodities into

---

h. https://www.ifpri.org/blog/cash-food-or-vouchers

other regions or for producers in a region to sell outside their markets. Under these conditions, rather than cash transfers promoting market development, they are likely to engender localized inflation in food commodities, bestowing existing traders with substantial power. Under inflationary conditions, in-kind transfers or commodity vouchers denominated in quantity offer beneficiaries the best protection. For instance, Sri Lanka's experience suggests that unindexed food stamps left the poor with a lower consumption of food compared with the traditional subsidies (Edirisinghe, 1987). Sabates-Wheeler and Devereux (2010) found that food transfers or "cash plus food" packages in the Ethiopian Productive Safety Net Program were superior to cash transfers alone. Since inflation quickly eroded the quantities cash could buy, the alternative enabled higher income growth, livestock accumulation, and self-reported food security. Even if cash transfers can in theory incorporate inflation, the valuation of food security entitlements to suit local inflationary conditions is extraordinarily difficult.

A recent survey of the PDS, and people's perceptions of cash versus in-kind transfers, provides useful insights into these issues. The survey found that in states where the PDS functions reasonably well, that is, where households obtain their entitlements, an overwhelming proportion of the respondents are in its favor and are averse to a cash-transfer system (Khera, 2011b). Overall, more than two-thirds preferred food and less than one-fifth preferred cash, with the others either having a conditional preference for one or the other or no clear preference at all. The greatest support for cash transfers was in states where the PDS does not function well, with people suggesting that they would be happy with an equivalent amount of money. In Bihar, for instance, only one-fifth preferred food and more than half preferred cash; only 18% of all respondents in Bihar obtained their full entitlement from the PDS. Interestingly, in states where the PDS did work well, those who stated that they preferred cash often suggested that the PDS was essential for those who were poorer. Even where markets were accessible, there were general apprehensions, for example, that traders might raise prices if the PDS were closed. The fungibility of cash implies that beneficiaries might use it in ways that undermine particular goals of the transfer. Although those who preferred a cash transfer mentioned that they would buy better quality grain, a significant proportion of those who opposed it feared that the money would be frittered away or diverted to alcohol. Indeed, the pay day–dry day in Delhi and Kerala, until recently, reflected the importance of this issue. A cash transfer to women of the household reduces this problem to some extent, but it could also provoke conflict within the household.

Another issue is whether or not the preconditions to implement these transfers exist. Difficult experiences for people with the National Rural Employment Guarantee Act (NREGA) wage payments through banks appear to make the option of cash transfers into bank accounts unattractive, especially where the banking infrastructure leaves a lot to be desired. "On October 2, a study commissioned by the Andhra Pradesh government, conducted by the Society for Social Audit, Accountability, and Transparency, found that over half of those who did not lift their rations in May 2015 reported reasons related to the use of Aadhaar: fingerprint recognition failed, Aadhaar numbers mismatched, or there was a malfunction in the E-PoS machine."[i] One reason for the unpopularity of piloted cash programs is the constraints posed by banking. Many respondents also suggested that a cash transfer would essentially make it harder for them to obtain food grains since they would now need to make two trips, one to the bank and another to the market, thus significantly increasing the time spent. The local bank was often said to be too far away, overcrowded, or difficult to deal with. The success of banking clients has been variable at best. People also expressed concern that a cash transfer (they had experienced this with other programs) would involve "cuts" they would have to pay out to various actors involved. The notion that cash transfers would reduce corruption and leakage and reduce the need for monitoring and supervision is at odds with this view.

The political ramifications of implementing cash transfers are also important. Economists have made arguments in the popular press touting cash transfers as a political winner (Somanathan, 2011). In many contexts within India, nothing could be further from the truth. The portrayal of cheap food aid as mere vote-bank politics is not wholly correct either (Kotwal et al., 2011). In states such as Tamil Nadu, the implementation of food-related schemes is an important issue and the demands made by voters on elected governments are often enlightened and sophisticated. For example, when the AIADMK lost the parliamentary elections in 2004, the newly elected chief minister (from the AIADMK) immediately reinstated eggs in midday meals and the ICDS. One survey respondent in a village in Tamil Nadu said: *"Votu pottu mutai vangittom"* (We got eggs back into the scheme with our votes!)[j] Similarly, when the state government tried to switch to a "targeted" PDS in 1997 following central government directives, it was forced to backtrack in just four days (CIRCUS, 2006). As an observer recently put it, "No elected government (in Tamil Nadu) will survive for even a day if they tamper with the PDS." Indeed, there are indications that the political process is increasingly incorporative of public demands for food security in a number of states.

REFORMS

According to critics, the current food security system needs overhauling to have a nimble, cost-effective system that delivers. However, what we know about alternative systems of support is that the contextual challenges need careful attention, without which systems such as cash transfers are bound to fail. Evidence from elsewhere in the world indicates that cash transfers too can engender

---

i. http://indianexpress.com/article/business/business-others/lpg-subsidy-transfer-centres-savings-not-more-than-rs-143-cr-while-it-claims-rs-12700-cr/

j. This is from a survey focused on the ICDS called the FOCUS Survey conducted in 2004, the results of which are reported in CIRCUS (2006).

corruption, be prone to elite capture, and be held hostage by nepotism. To avoid these problems, sophisticated tracking and monitoring systems are required (Devereux and Vincent, 2010), which increases the cost of administering cash transfers.

One shared feature in states where the PDS functions effectively has been the use of IT-based transparency measures, starting with a simple computerized record keeping system of the entire supply chain. Combined with GPS tracking of delivery trucks and SMS-based transmission of information to users, checks and balances exist that make diversion of food grains to the open market very difficult. In Tamil Nadu, consumers can obtain their stocks position via SMS, and in Chhattisgarh they are told the timing of the arrival of supplies to the fair price shop (Khera, 2011b). Tamil Nadu and Chhattisgarh also have functional grievance redressal mechanisms. Chhattisgarh has a system for tracking the entire chain from farmer to consumer in its local procurement operations (Dhand et al., 2010), and has been experimenting with portability of entitlements, implying that consumers can purchase from a ration shop of their choice. This would presumably weed out ration shopkeepers who siphon away food grain or cheat. The use of smart cards at PDS outlets in Andhra Pradesh, PDS coupons in Rajasthan, and Tamil Nadu's PDS pilot automation are known to have been effective in curtailing leakages in the "last mile," although more rigorous research is required to understand the efficacy of these latter measures. Other PDS innovations, including locating ration shops in supermarkets (as in Karnataka) and allowing village organizations and women's self-help groups (in Andhra Pradesh, Telengana, and Chhattisgarh, among several other states) to run ration shops, can go a long way to streamlining delivery systems.

A larger concern with a move to cash transfers is what this might mean for the strong link between price support and procurement support offered to farmers, and that tinkering with the form of support to consumers might unravel a system that offers critical support to farmers. This is especially relevant in the current context in which rice procurement increasingly comes from states such as Odisha and Chhattisgarh, where poor farmers supply most of the rice that is channeled into various welfare schemes. A system to deliver food hinges on an efficient procurement mechanism for food grains, and any reform of food security would have to factor in the consequences for poor rice farmers across the country.

Perhaps the greatest advantage of in-kind transfers of rice lies in the potential to use these as platforms to channel the distribution of nutrient-rich commodities. One criticism of the PDS and other food-related schemes is that they focus on rice and wheat often at the expense of crops and commodities that offer more nutritious alternatives, such as millets. Given the reach of these schemes, in-kind distribution of millets and even fortified food or nutrient-rich rice has a far greater possibility of success than cash transfers, even if accompanied by information campaigns. The potential of the PDS, the ICDS, and MDM scheme remains vastly underexplored. If India is to go beyond solving hunger and tackle food security in the complete sense of the term, it would have to use the key position rice occupies and go beyond it.

ACKNOWLEDGMENTS

Thanks are due to Krushna Ranaware for research assistance and to Dr. Mruthyunjaya, Dr. Chengappa, Dr. Ramesh Chand, staff of IRRI, and participants in seminars in Bangalore and Mumbai for their comments and suggestions. Any errors and omissions in the chapter are mine.

REFERENCES

Abreu, D., Bardhan, P., Ghatak, M., Kotwal, A., Mookherjee, D., Ray, D., 2014. Wrong numbers: attack on NREGA is misleading. Times of India, November 9, 2014.

Afridi, F., 2010. Child welfare programs and child nutrition: evidence from a mandated school meal program in India. J. Dev. Econ. 92 (2), 152–165.

Banerjee, A., 2011. A platter of choices. Indian Express, March 28, 2011. www.indianexpress.com/news/a-platter-of-choices/768124/

Basu, K., 2011. Food management in India. Working Paper, Ministry of Finance. New Delhi: Government of India.

Chatterjee, M., 2013. How effective is the public distribution system in India? Food Security and Child Undernutrition in Odisha, India. MPhil dissertation. University of Oxford, Queen Elizabeth House. Unpublished.

CIRCUS, 2006. FOCUS: Focus on children under six. Available from: www.righttofoodindia.org/data/rtf06focusreportabridged.pdf

Devereux, S., Vincent, K., 2010. Using technology to deliver social protection: exploring opportunities and risks. Dev. Pract. 20 (3), 367–379.

Dhand, V.K., Srivastav, D.K., Somasekhar, A.K., Rajeev, J., 2010. Computerization of paddy procurement and public distribution system in Chhattisgarh In: E-governance in Practice. www.csi-sigegov.org/egovernance_pdf/26_216-223.pdf

Drèze, J., Khera, R., 2013. Rural poverty and the public distribution system. Econ. Polit. Wkly. 47 (45–46), 55–60.

Drèze, J., Khera, R., 2014. Water for the leeward India. Outlook, March 24, 2014. www.outlookindia.com/article/water-for-the-leeward-india/289801

Edirisinghe, N., 1987. The Food Stamp Scheme in Sri Lanka: Costs, Benefits, and Options for Modification. International Food Policy Research Institute, Washington, DC.

Gentilini, U., 2016. Revisiting the "cash versus food" debate: new evidence for an old puzzle? World Bank Res. Obs. 31 (1), 135–167.

Government of India, 2015. Economic survey of India 2014–15, Volume 1, Ministry of Finance. New Delhi: Government of India.

Gulati, A., Gujral, J., Nandakumar, T., Jain, S., Anand, S., Rath, S., Joshi, P., 2012. National food security bill challenges and options. Discussion Paper No. 2, Commission for Agricultural Costs and Prices, Department of Agriculture & Cooperation, Ministry of Agriculture, Government of India, New Delhi.

Hertel, S., 2015. Hungry for justice: social mobilization on the right to food in India. Dev. Change 46 (1), 72–94.

Himanshu, Sen, A., 2013a. In-kind food transfers-I. Econ. Polit. Wkly. 48 (45–46), 45–55.

Himanshu, Sen, A., 2013b. In-kind food transfers-II. Econ. Polit. Wkly. 48 (47), 60–73.

Hoda, A., Gulati, A., 2013. India's agricultural trade policy and sustainable development goals. ICTSD Program on Agricultural Trade and Sustainable Development, Issue Paper No. 49, International Centre for Trade and Sustainable Development, Geneva, Switzerland.

Jain, M., 2015. India's struggle against malnutrition: is the ICDS program the answer? World Dev. 67, 72–89.

Jayaraman, R., 2009. The impact of school lunches on school enrolment: evidence from an exogenous policy change in India. Available from: http://client.norc.org/jole/soleweb/9001.pdf, https://www.econstor.eu/bitstream/10419/39926/1/15_jayaraman.pdf

Jha, S., Ramaswami, B., 2010. How can food subsidies work better? Answers from India and the Philippines. Economics Working Paper 221. Manila (Philippines): Asian Development Bank.

Jha, R., Bhattacharyya, S., Gaiha, R., 2011. Social safety nets and nutrient deprivation: an analysis of the national rural employment guarantee program and the public distribution system in India. J. Asian Econ. 22, 189–201.

Kapur, D., 2011. The shift to cash transfers: running better but on the wrong road? Econ. Polit. Wkly. 46 (21), 80–85.

Kapur, D., Mukhopadhyay, P., Subramanian, A., 2008. The case for direct cash transfers to the poor. Econ. Polit. Wkly. 43 (15), 37–43.

Kaul, T., 2014. Household responses to food subsidies: evidence from India. PhD thesis. University of Maryland, College Park. Unpublished.

Kaushal, N., Muchomba, F., 2013. How consumer price subsidies affect nutrition. National Bureau of Economics Research Working Paper 19404.

Khera, R., 2010. India's public distribution system: utilization and impact. J. Dev. Stud. 47, 1–23.

Khera, R., 2011a. Trends in diversion of grain from the public distribution system. Econ. Polit. Wkly. 46 (21), 106–114.

Khera, R., 2011b. Revival of the public distribution system: evidence and explanations. Econ. Polit. Wkly. 46 (44–45), 36–50.

Kochar, A., 2005. Can targeted food programs improve nutrition? An empirisat analysis of India's public distribution system. Econ. Dev. Cult. Change 54 (1), 203–235.

Kotwal, A., Murugkar, M., Ramaswamy, B., 2011. PDS forever? Econ. Polit. Wkly. 46 (21), 72–76.

Kozicka, M., Kalkuhl, M., Saini, S., Brockhaus, J., 2015. Modelling Indian wheat and rice sector policies. ZEF-Discussion Papers on Development Policy No. 197, Center for Development Research, Bonn, Germany.

Krishnamurthy, P., Pathania, V.S., Tandon, S., 2014. Food price subsidies and nutrition: evidence from state reforms to India's public distribution system. UC Berkeley Public Law Research Paper 2345675.

Kumar, P., Joshi, P.K., 2013. Household consumption pattern and nutritional security among poor rural households: impact of the MGNREGA. Agric. Econ. Res. Rev. 26 (1), 73–82.

Kumar, A., Bantilan, M.C.S., Kumar, P., Kumar, S., Jee, S., 2012. Food security in India: trends, patterns, and determinants. Indian J. Agric. Econ. 67 (3), 445–462.

Kumar, A., Parappurathu, S., Bantilan, M.C.S., Joshi, P.K., 2016. Public distribution in India: implications for poverty and food security. Available from: http://vdsa.icrisat.ac.in/Include/Mini-Symposium/12.pdf

Mehrotra, S., 2010. Introducing conditional cash transfers in India: a proposal for five CCTs. Planning Commission. New Delhi: Government of India. Available from: http://www.indiaenvironmentportal.org.in/files/cash-transfer-CCTS%20for%20India%20-%20Mehrotra.pdf

Narayanan, S., 2011, 41–48. A case for reframing the cash transfer debate in India. Econ. Polit. Wkly. 46 (21) (21), 41–48.

Narayanan, S., Gerber, N., 2015. Social safety nets for food and nutritional security in India. Working Paper WP-2015-31, Indira Gandhi Institute of Development Research.

NCAER, 2014. An analysis of changing food consumption pattern in India. A research paper prepared under the project Agricultural Outlook and Situation Analysis Reports.

Rahman, A., 2014. Revival of rural public distribution system: expansion and outreach. Econ. Polit. Wkly. 49 (20), 62–68.

Rahman, A., 2015. Universal food security program and nutritional intake: evidence from the hunger prone KBK districts in Odisha. Working Paper WP-2015-015. Indira Gandhi Institute of Development Research, Mumbai.

Ray, R., 2007. Changes in food consumption and the implications for food security and undernourishment: India in the 1990s. Dev. Change 38 (2), 321–343.

Sabates-Wheeler, R., Devereux, S., 2010. Cash transfers and high food prices: explaining outcomes on Ethiopia's productive safety net program. Food Policy 35 (4), 274–285.

Singh, A., Park, A., Dercon, S., 2014. School meals as a safety net: an evaluation of the midday meal scheme in India. Econ. Dev. Cult. Change 62 (2), 275–306.

Sinha, D., 2013. Cost of implementing the national food security act. Econ. Polit. Wkly. 48 (39), 31–34.

Somanathan, E., 2011. Money where the mouth is. Hindustan Times, March 23, 2011. Available from: http://www.im4change.org/latest-news-updates/money-where-the-mouth-is-by-e-somanathan-6764.html

Srinivasan, V., Narayanan, S., 2008. Food policy and social movements: reflections on the right to food campaign in India. In: Pinstrup-Andersen, P., Cheng, F. (Eds.), Case Studies on Food Policy and the Role of Governments in Developing Countries. Cornell University Press, Ithaca, NY, United States.

Svedberg, P., 2012. Reforming or replacing the public distribution system with cash transfers. Econ. Polit. Wkly. 47 (7), 53–62.

Tarozzi, A., 2005. The Indian public distribution system as provider of food security: evidence from child nutrition in Andhra Pradesh. Eur. Econ. Rev. 49, 1305–1330.

von Grebmer, K., Bernstein, J., de Waal, A., Prasai, N., Yin, S., Yohannes, Y., 2015 Global hunger index: armed conflict and the challenge of hunger. Bonn, Germany; Washington, D.C.; and Dublin, Ireland: Welthungerhilfe; International Food Policy Research Institute (IFPRI), and Concern Worldwide. Available from: http://dx.doi.org/10.2499/9780896299641

FURTHER READING

Drèze, J., Khera, R., 2011. PDS leakages: the plot thickens. The Hindu, p. 12.

Government of India, 2010. Report of the expert committee on national food security bill, Planning Commission. New Delhi: Government of India.

The HUNGaMA survey report, 2011. HUNGaMA Fighting Hunger and Malnutrition. Available from: www.naandi.org/wp-content/uploads/2013/12/HUNGaMA-Survey-2011-The-Report.pdf

Chapter 3

Spatial and Temporal Patterns of Rice Production and Productivity

Elumalai Kannan*, Ambika Paliwal**, Adam Sparks[†]
*Jawaharlal Nehru University (JNU), New Delhi, India; **International Rice Research Institute (IRRI), Los Baños, Philippines; [†]Centre for Crop Health, University of Southern Queensland, Queensland, Australia

BACKGROUND

Since the advent of Green Revolution technology in the late 1960s, India's rice production has nearly tripled, increasing from 37 million tons in 1968–69 to 104 million tons in 2012–13. During this period, rice area expanded to include non-traditional areas, such as Punjab, and Punjab now registers the highest paddy productivity (4 t/ha), closely followed by Tamil Nadu (3 t/ha) (GOI, 2014). However, concerns arose about the slowdown in the yield of major crops during the 1990s, with various studies showing that total factor productivity (TFP) growth of rice had decelerated in the major producing states (Chand et al., 2011; Evenson et al., 1999; Kumar et al., 2004, 2008; Murgai, 2001). Researchers and policy planners also expressed concern that incomes from crop cultivation had declined and that agriculture was nonremunerative. These concerns are consistent with findings from the Indian Government (GOI, 2005), which suggested that 27% of farmers did not like farming as it was not profitable and 40% would quit if they could choose another occupation.

Although there has been a growth revival in the agricultural sector since the mid-2000s (Chand et al., 2015), improvement in the yields of major crops (such as rice) has not been as remarkable as in the 1980s. In fact, higher variability has been shown in the output of various crops since the 1990s (Chand et al., 2015; Sen, 2016) and also their sources of growth (Birthal et al., 2014). Some studies have shown that diversification (an area shift) and output price were the important drivers of production growth in recent years (Birthal et al., 2014; Joshi et al., 2007). However, this level of growth may not be sustainable in the long term unless TFP growth is improved through innovation, as well as input efficiency. Researchers have not given adequate attention to analyzing the sources of productivity growth. This chapter analyzes the spatial and temporal patterns of rice production and productivity growth in India.

The Future Rice Strategy for India. http://dx.doi.org/10.1016/B978-0-12-805374-4.00003-8

39

This chapter is organized as follows: The first section discusses the distribution of rice in different agroecological zones (AEZs) in India. Rice cultivation in different ecosystems is provided in the second section, while the third presents various stresses in rice production. Temporal changes in rice production and productivity are presented in the fourth section. The fifth analyzes regional variation in cost and income from paddy cultivation. Technical efficiency and TFP growth are analyzed in the sixth section, and concluding remarks are made in the final section.

RICE IN THE AGROECOLOGICAL ZONES OF INDIA

Owing to its vast geographic extent and diverse topography, India's ecosystems vary from mountains to riverine deltas and from high-altitude forests to low-lying areas of coast. This leads to heterogeneous climatic conditions and different agroecologies. Therefore the entire country has been divided into 20 AEZs (Gajbhiye and Mandal, 2000). Agroecological zoning defines zones on the basis of a combination of soil, landform, and climatic characteristics (FAO, 1996). Fig. 3.1 shows the rice areas of India spread across different AEZs. Among these zones, the Bengal basin and Assam plain and the Eastern Plateau–Eastern Ghats collectively account for 31% of the cultivated rice area (Table 3.1). Five AEZs account for more than 60% of the total rice area in India. The Bengal basin–Assam Plain ecoregion covers major parts of Assam, West Bengal, and some parts of Jharkhand. The Eastern Plateau–Eastern Ghats ecoregion encompasses major parts of Odisha and some parts of Jharkhand and West Bengal. The Northern Plain ecoregion comprises Punjab, Haryana, Uttar Pradesh, and Bihar. The Eastern Plains ecoregion consists of northeastern Uttar Pradesh and northern Bihar. The Central Highlands cover Uttar Pradesh, parts of Rajasthan, Gujarat, and Madhya Pradesh.

RICE ECOSYSTEMS IN INDIA

In India, rice is cultivated under varied soil and climatic conditions. Significant variations exist in soil and water management, cultivation practices, and the type of rice grown. It is grown in contrasting geographic locations from below sea level in southern India up to 3000 m in northern India, and with no standing water in uplands to cultivation in deepwater in the lowlands of eastern India. The major rice ecosystems have been categorized into irrigated, rainfed lowland, upland, and deepwater (Fig. 3.2). In northern India, agriculture is predominantly irrigated, eastern India has rainfed lowland and deepwater ecosystems, while northwestern India has irrigated and upland ecosystems.

Irrigated Ecosystem

This is the most widespread ecosystem in India. Around 24 million ha of rice are irrigated, accounting for 57% of the total rice area (GOI, 2014). Irrigated rice is found in Punjab, Haryana, Uttar Pradesh, Bihar, Odisha, and parts of

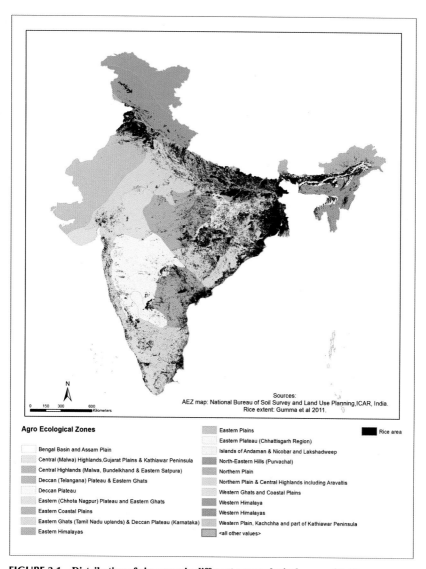

Agro Ecological Zones

Bengal Basin and Assam Plain
Central (Malwa) Highlands,Gujarat Plains & Kathiawar Peninsula
Central Highlands (Malwa, Bundelkhand & Eastern Satpura)
Deccan (Telangana) Plateau & Eastern Ghats
Deccan Plateau
Eastern (Chhota Nagpur) Plateau and Eastern Ghats
Eastern Coastal Plains
Eastern Ghats (Tamil Nadu uplands) & Deccan Plateau (Karnataka)
Eastern Himalayas

Eastern Plains
Eastern Plateau (Chhattisgarh Region)
Islands of Andaman & Nicobar and Lakshadweep
North-Eastern Hills (Purvachal)
Northern Plain
Northern Plain & Central Highlands including Aravallis
Western Ghats and Coastal Plains
Western Himalaya
Western Himalayas
Western Plain, Kachchha and part of Kathiawar Peninsula
<all other values>

Rice area

Sources:
AEZ map: National Bureau of Soil Survey and Land Use Planning,ICAR, India.
Rice extent: Gumma et al 2011.

0 150 300 600 Kilometers

FIGURE 3.1   Distribution of rice areas in different agroecological zones of India.

Andhra Pradesh, Tamil Nadu, and Karnataka. Rice is grown in rotation with other crops, such as wheat, maize, lentils, and mustard. Because of the presence of irrigation facilities, rice productivity is highest in this ecosystem. However, the main yield constraint in this ecosystem is drainage, which leads to salinity, sodicity, and inefficient nutrient use. The Green Revolution had a big impact on irrigated ecosystems.

TABLE 3.1 Distribution of Rice Area by Different AEZs in India

S. no.	Agroecological zones	Region	Geographic coverage	Contribution to total rice area (%)
1	Bengal Basin and Assam Plain	Hot subhumid (moist) to humid (inclusion of perhumid) ecoregions with alluvium-derived soils	Major parts of Assam, West Bengal, and some parts of Jharkhand	15.6
2	Eastern (Chhota Nagpur) Plateau and Eastern Ghats	Hot subhumid ecoregion with red and lateritic soils	Major parts of Odisha and some parts of Jharkhand and West Bengal	15.4
3	Northern Plain	Hot subhumid (dry) ecoregion with alluvium-derived soils	Punjab, Haryana, Uttar Pradesh, and Bihar	10.1
4	Eastern Plains	Hot subhumid (moist) ecoregion with alluvium-derived soils	Northeastern Uttar Pradesh and northern Bihar	10.0
5	Northern Plain and Central Highlands including Aravallis	Hot semiarid ecoregion with alluvium-derived soils	Uttar Pradesh, parts of Rajasthan, Gujarat, and Madhya Pradesh	9.7
6	Eastern Plateau (Chhattisgarh Region)	Hot subhumid ecoregion with red and yellow soils	Parts of Chhattisgarh and Jharkhand	8.7
7	Eastern Coastal Plains	Hot subhumid to semiarid ecoregion with coastal alluvium-derived soils	Parts of West Bengal, Odisha, Andhra Pradesh, and Tamil Nadu	6.3
8	Deccan (Telangana) Plateau and Eastern Ghats	Hot semiarid ecoregion with red and black soils	Telangana and the Eastern Ghats of Andhra Pradesh	4.9
9	Central Highlands (Malwa, Bundelkhand, and Eastern Satpura)	Hot subhumid ecoregion with red and black soils	Malwa and Bundelkhand plateaus of Madhya Pradesh	4.1
10	Eastern Ghats (Tamil Nadu uplands) and Deccan Plateau (Karnataka)	Hot semiarid ecoregion with red loamy soils	Tamil Nadu and parts of Karnataka	3.3

11	Western Ghats and Coastal Plains	Hot humid (perhumid) ecoregion with red lateritic and alluvium-derived soils	Coastal parts of Maharashtra, Karnataka, and Kerala	2.9
12	Deccan Plateau	Hot semiarid ecoregion with shallow and medium (with inclusion of deep) black soils	Major part of Maharashtra and some parts of Karnataka and Andhra Pradesh	2.4
13	Northeastern Hills (Purvachal)	Warm perhumid ecoregion with red and lateritic soils	Nagaland, Meghalaya, Manipur, Mizoram, and Tripura	1.7
14	Western Plain, Kachchha, and part of Kathiawar Peninsula	Hot arid ecoregion with desert and saline soil	Rajasthan, Gujarat, and some parts of Punjab	1.5
15	Western Himalayas	Warm subhumid to humid with inclusion of perhumid ecoregion with brown forest and podzolic soils	Jammu and Kashmir, Himachal Pradesh, and Uttarakhand	1.3
16	Central (Malwa) Highlands, Gujarat Plains, and Kathiawar Peninsula	Hot semiarid ecoregion with medium and deep black soils	Central Highlands, Gujarat Plains, parts of Madhya Pradesh, and Rajasthan	1.0
17	Eastern Himalayas	Warm perhumid ecoregion with brown and red hill soils	Arunachal Pradesh, Darjeeling, and Sikkim	0.6
18	Deccan Plateau	Hot arid ecoregion with red and black soil	Major parts of Karnataka and Andhra Pradesh	0.4
19	Western Himalayas	Cold arid ecoregion with shallow skeletal soils	Ladakh and northern Kashmir	0.0
20	Islands of Andaman and Nicobar, and Lakshadweep	Hot humid to perhumid island ecoregion with red loamy and sandy soils	Andaman and Nicobar Islands and Lakshadweep	0.0

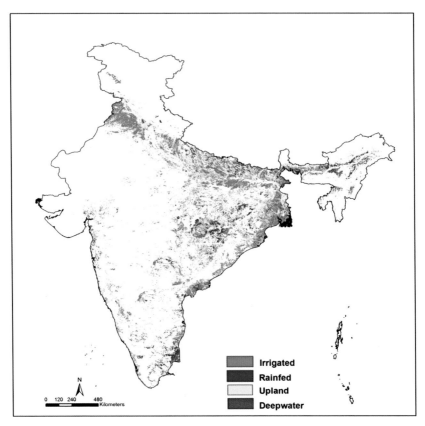

FIGURE 3.2 **The rice ecosystems of India.** *(Gumma, M.K., Nelson, A., Thenkabail, P.S., Singh, A.N., 2011. Mapping rice areas of South Asia using MODIS multitemporal data. J. Appl. Remote Sensing 5, 053547.)*

Rainfed Lowland Ecosystem

Approximately 35% of the total rice area (∼14 million ha) is classified as rainfed lowland. In this ecosystem, rice is grown in the deltas and lowland coastal areas. Dependency of this ecosystem on rain makes it prone to drought, submergence, and salinity. Prevailing impoverished socioeconomic conditions, unpredictable rains, and lack of suitable crop management interventions are the causes of the low productivity in this ecosystem. Rice is usually transplanted and grown in bunded fields to retain water.

Upland Ecosystem

The area under this rice ecosystem is 8% of the total rice area. Upland ecosystems can be found in West Bengal, Jharkhand, Madhya Pradesh, Odisha, and the northeastern hill region. Upland rice is usually directly seeded and productivity is low, being sufficient only to meet local demand. This ecosystem faces many challenges, such as insect pests, rodents, nematodes, and poor soil fertility. Upland rice is mostly nutrient deficient. The upland ecosystem can be both irrigated and rainfed.

Deepwater Ecosystem

This ecosystem is characterized by extreme flooding in the wet season and uncertain yields. It accounts for 0.8% of the total rice area and can be found in parts of West Bengal, Assam, and Andhra Pradesh. In this ecosystem, rice can withstand temporary submergence for up to 10 days. Salinity and cold intolerance at the seedling stage are the major constraints of this ecosystem.

STRESSES IN RICE PRODUCTION

Rice farmers face many abiotic and biotic stresses that adversely affect the growth and productivity of the rice crop. Environmental factors such as extreme temperatures, lack of optimal water supply, problem soils, and pests can cause stress to the rice crop. Rice stresses are generally classified as abiotic and biotic. The nature and distribution of common stresses are described later.

Abiotic Stresses

Rice is mainly grown in the monsoon season in India. It can thrive in waterlogged soil, sustain submergence to an extent, and can moderately tolerate salinity, but is very sensitive to drought. Abiotic stresses for rice are discussed briefly here.

Drought

Because rice is a water-intensive plant, water deficit can severely limit crop production. Drought can be defined by five approaches: meteorological, hydrological, agricultural, socioeconomic, and ecological. In India, drought is a regular phenomenon and is defined in terms of deviation of rainfall from the normal (GOI, 2009). An area is said to be affected by drought if it experiences 25% less rainfall than normal. If the rainfall is less than 50% of normal, then the drought is severe. The negative effects of drought can usually be avoided in irrigated environments with an appropriate water supply; however, rainfed systems tend to be badly affected. Among all Indian states, the rice crop in Uttar Pradesh is most prone to drought, with 20% of its total rice crop affected (Fig. 3.3). This is followed by Chhattisgarh (10%), Andhra Pradesh (9%), and Odisha (9%).

Submergence

Submergence affects crops in low-lying areas. Submergence can be defined on the basis of duration, depth, and frequency. Flooding causes submergence and damage to rice crops. Two types of environment cause submergence:

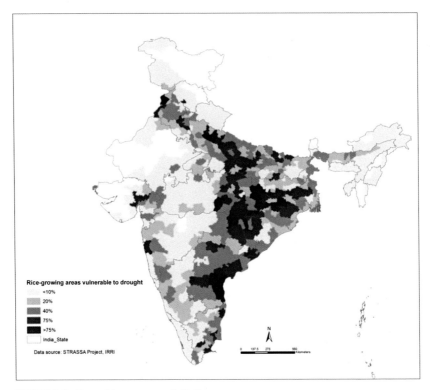

FIGURE 3.3   Drought-prone areas in India.

flash flooding and deepwater. Flash flood submergence is defined by water levels rising rapidly and plants remaining submerged for 1–2 weeks. Deepwater submergence is defined by water depths greater than 100 cm persisting for months. Among all the states, West Bengal is the most prone to submergence (Fig. 3.4). Out of the total submerged rice crop area, 20% is in West Bengal, 18% in Odisha, 16% in Assam, and 15% in Andhra Pradesh (Fig. 3.4).

Salinity/Alkalinity

This problem occurs in some irrigated and rainfed lowland rice ecosystems (Fig. 3.5). Irrigation and seawater intrusion are common causes of salinity in rainfed lowland ecosystems. Rice is a salt-sensitive crop. Its threshold for salinity is 4 dS/m, beyond which it negatively affects plant growth. The salt sensitivity of rice varies at different stages of growth, with highest sensitivity seen at the seedling and reproductive stages. Uttar Pradesh has the most salt-affected land (40%), followed by Rajasthan (31%) and Tamil Nadu (7%) (Wasteland Atlas of India, 2011).

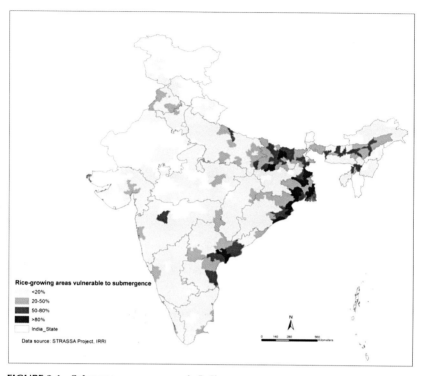

FIGURE 3.4   Submergence-prone areas in India.

Biotic Stresses

Diseases occur naturally as a part of any plant ecosystem. In managed cropping systems, the environment created by the farmer influences disease development. Some diseases are favored by cool and wet weather while others are favored by warm and wet weather. In India, many diseases affect rice and lead to yield loss. Four of the major diseases found in tropical and subtropical rice-growing areas are bacterial blight, caused by *Xanthomonas oryzae* pv. *oryzae*; brown spot, caused by *Cochliobolus miyabeanus*; leaf blast, caused by *Magnaporthe oryzae*; and sheath blight, caused by *Rhizoctonia solani,* all of which can reduce yield and grain quality. The relative risk of these diseases was measured using the EPIRICE model (Hijmans et al., 2015; Savary et al., 2012) along with inter-polated global summary of the day data at 15 arc-min for the 2001–08 period. The statistical software R (R Core Team, 2015) package raster (Hijmans, 2015) was used to extract the resulting area under the disease progress curves by state using FAO's global administrative unit layers and this was classified according to relative risk. Results were compared with district-level maps that had been vetted by expert Indian scientists and found to be suitable. Fig. 3.6 shows the spread of potential risk areas for each of these stresses in rice-growing areas of India.

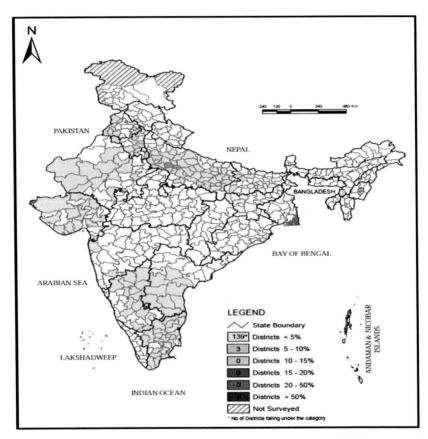

FIGURE 3.5   Salinity-affected land in India. *(Wasteland Atlas of India, 2011. Department of Land Resource-National Remote Sensing Centre Report.)*

Bacterial Blight

This is one of the most serious diseases. It is caused by *X. oryzae* and leads to yellowing and drying of leaves. Bacterial blight mainly develops in irrigated and rainfed lowland ecosystems that contain weeds, and is spread by strong winds and heavy rains. This disease leads to both yield loss and poor-quality grains. It is severe in eastern, northeastern, and southeastern parts of India, affecting almost all of West Bengal. It also intensely affects Uttar Pradesh, Assam, Odisha, Bihar, and Andhra Pradesh. Bacterial blight can be controlled by weeding, good drainage, and provision of nitrogen.

Sheath Blight

This fungal disease is caused by *R. solani*. It causes a reduction in leaf area, which leads to a further reduction in the canopies of rice plants. Tillers become

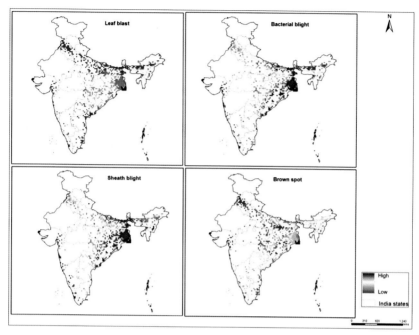

FIGURE 3.6   Severity of important biotic stresses that affect rice plants in India. *(STRASA.)*

infected, resulting in yield loss. Sheath blight occurs at high temperatures (28–32°C) and high humidity (85%–100%). Like bacterial blight, this disease severely affects eastern, northeastern, and southeastern parts of India. It severely affects West Bengal, Uttar Pradesh, Assam, Odisha, Bihar, and Andhra Pradesh. It can be managed by controlling weeds and using fungicide.

Brown Spot

This fungal disease kills the whole leaf. It occurs in nutrient-deficient soils and results in unfilled grains, therefore adversely affecting both the quality and quantity of rice. Rainfed rice is often affected by brown spot disease and thus Odisha is highly affected. West Bengal and Tamil Nadu are also moderately affected by this disease.

Leaf Blast

This disease is caused by *M. oryzae* and it affects all parts of the rice plant aboveground. This disease tends to be most severe in rainfed rice areas and upland areas where the cooler temperatures and greater daytime/nighttime temperature differences cause dew formation on the leaf. Far northeastern India and mountainous regions of the southwest and far north are at highest risk. Leaf blast can be devastating; to control it, crops should be planted after the start of the rainy season and proper fertilizer should be used. The recommendation is to not overfertilize with nitrogen and split its applications to help reduce the risk.

These stresses cause considerable yield loss and result in loss of income for cultivators. According to Kannan (2014), rice yield loss due to biotic stresses was estimated to range from 2.9% in Uttar Pradesh to 16.2% in Karnataka. Yield loss was as high as 15.1% in West Bengal, 9.1% in Tamil Nadu, 8.0% in Punjab, and 7.8% in Assam. Although yield loss varied across states, it remained relatively high among the major rice-producing states.

TEMPORAL CHANGES IN RICE PRODUCTION AND PRODUCTIVITY

Rice accounts for one-fifth of the total cropped area and one-sixth of the total value of output from the agricultural sector (GOI, 2014). Among cereals, rice constitutes about 40% of total food-grain production and its production performance influences the overall trend in food-grain production. In India, 90% of rice is grown during the kharif (rainy) season from June to October, and the remaining 10% is grown during the rabi (postrainy) season from November to April. The extent of rice grown in the kharif season[a] is given in Fig. 3.7. Uttar Pradesh accounts for ~14% of the total rice area in the kharif season and West Bengal accounts for ~10%. Odisha, Bihar, and Andhra Pradesh also contribute to the rice area during this season. In the rabi season, West Bengal and Andhra Pradesh contribute 3% and 2% of the total rice area, respectively.

Analysis of temporal changes in paddy area, production, and yield provides very useful results. In general, India's crop sector registered impressive growth during the 1980s because of the diffusion of Green Revolution technology in all regions. This included those regions that did not gain during the initial years following the introduction of new technologies (Bhalla and Singh, 2001). High-yielding varieties (HYVs) of rice were initially introduced in canal-irrigated regions and later extended to other regions, particularly groundwater-irrigated areas. HYVs of rice were suitable for cultivation only in irrigated regions, and therefore they performed poorly in rainfed conditions. As a result, the potential for rainfed areas to contribute to increased rice production remained underexploited (Barah, 2005). Notwithstanding, paddy production registered an annual growth of 4.3% from 1980–81 to 1989–90 (Table 3.2), with a yield growth of 3.4% positively contributing to the overall production of rice during this period. However, with area growth more or less stagnant, the fall in yield during 1990–91 to 1999–2000 led to a decline in total rice production. The 1990s did

---

a. Gumma et al. (2011) mapped the extent of rice in South Asia by using 500-m MODIS composite images. The map accuracy based on field observations was 80% and the estimated rice area corroborated well with subnational statistics at the district and state levels ($R^2 = 0.95$).

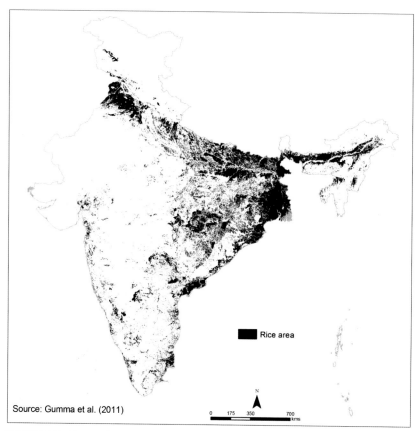

FIGURE 3.7   **Extent of rice in the kharif season in India.**

TABLE 3.2 Average Annual Growth in Area, Production, and Yield of Rice in India (%)

Period	Area	Production	Yield
1980–81 to 1989–90	0.65	4.32	3.35
1990–91 to 1999–2000	0.70	2.09	1.36
2000–01 to 2012–13	−0.40	1.80	1.99

Source: GOI (Government of India), 2014. Agricultural statistics at a glance. Ministry of Agriculture, Government of India, Oxford University Press, New Delhi.

not bode well for the agricultural sector as a whole. This occurred because of a shift in policy direction toward industrialization through economic reforms, with a thrust on reallocation of resources at the macro level, and a consequent decline in public investment in agriculture. Although rice yield revived from 2000–01 to 2012–13, the decrease in area led to a fall in production.

At the state level, growth performance of paddy appears to be slightly irregular. Some parts of eastern Indian states fall under the AEZ of the Bengal basin–Assam plain, and although this accounts for a considerably large rice area, rice productivity in this zone was low. Smaller, irrigated northern and southern areas had higher productivity. Among all states, Punjab and Tamil Nadu had the highest rice yield per hectare with 3.9 and 3.2 t/ha, respectively (GOI, 2014), and Punjab and West Bengal were best for rice production (Fig. 3.8).

In terms of average annual growth rates, all major states showed positive production growth during the 1990s (Table 3.3). The highest annual increase in rice production was registered in Bihar, mainly due to an impressive increase in yield. Despite a negative yield in Haryana, an increase in cultivation area led to a production growth rate of 4.64%. Karnataka also registered a relatively high production growth rate owing to improvements in both area and yield. West Bengal recorded appreciable yield growth (2.65%) in the 1990s; however this declined to 1.67% during the 2000–01 to 2012–13 period. A fall in yield and negative growth in area meant a production growth of only 1% for this state. Except for Andhra Pradesh, all states showed a decline in paddy area during the 2000s compared with the 1990s. Tamil Nadu registered negative growth in area, production, and yield, with the rate of decline in yield seeming to be faster than the area decline. However, some eastern states, such as Assam, Bihar, and Odisha managed to show impressive growth in yield during this recent period.

Analysis of growth performance at the state level masks the existence of disparity in rice production within states. Progressive and lagging regions are historically being left out of the development process because of a policy bias toward regions endowed with natural resources. Table 3.4 shows state districts that have been classified based on rice productivity groups. All districts in Punjab registered rice yield surpassing 2.5 t/ha while most districts in Tamil Nadu and Haryana fell under the high-productivity category. Interdistrict variation in rice productivity was observed to be highest in Madhya Pradesh, Bihar, and Odisha. Of 44 rice-growing districts in Madhya Pradesh, 40 districts fell within the low-productivity category with average yields of less than 1.5 t/ha. These districts account for 98.7% of the total paddy area and 96.9% of total production in the state. This suggests that there is potential to improve rice production through the adoption of new technologies and provision of proper incentives to farmers in these districts of the Central highlands (Malwa, Bundelkhand, and eastern Satpura) and Eastern plateau (Chhattisgarh region). Furthermore, most of these districts also fall under the rainfed rice ecosystem, which requires appropriate production techniques.

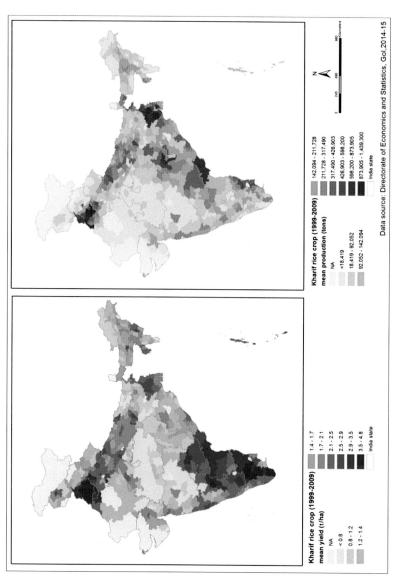

FIGURE 3.8   Average rice crop yield (t/ha) and average rice production (tons) (1999–2009).

TABLE 3.3 Average Annual Growth in Area, Production, and Yield of Rice by States (%)

State	1990–91 to 1999–2000			2000–01 to 2012–13		
	Area	Production	Yield	Area	Production	Yield
Andhra Pradesh	0.48	2.35	1.38	0.71	2.64	1.68
Assam	0.73	2.04	1.23	−0.38	2.73	2.92
Bihar	−0.71	11.59	12.10	−2.57	8.21	8.79
Haryana	5.82	4.64	−0.58	1.07	3.60	2.77
Karnataka	2.47	5.11	2.54	−0.46	2.26	1.74
Madhya Pradesh	0.51	2.16	1.70	0.47	7.28	6.71
Odisha	0.51	2.02	1.27	−0.99	8.35	8.63
Punjab	2.93	3.47	0.56	0.71	2.16	1.44
Tamil Nadu	2.02	4.17	1.90	−1.82	−0.61	−0.12
Uttar Pradesh	0.92	3.15	2.23	0.09	2.21	1.50
West Bengal	0.65	3.25	2.65	−0.71	1.00	1.67

Source: GOI (Government of India), 2014. Agricultural statistics at a glance. Ministry of Agriculture, Government of India, Oxford University Press, New Delhi.

In Bihar, 28 districts fell within the low-productivity category. These districts constituted 74.1% of paddy area and 59.4% of total production. Similarly, 22 districts within Odisha and 14 within Assam also registered productivity under 1.5 t/ha. These low-productivity districts are in the rainfed ecosystem and they lack appropriate production technologies. Since farmers in this ecosystem face uncertain production conditions, and market discrimination, they do not have the incentive to make adequate investments in farm production technologies. These conditions lead to low input–low output production systems, which tend to perpetuate poverty among farmers in rainfed areas (Barah, 2005).

The use of HYVs of rice alongside chemical fertilizers in irrigated conditions contributed to a significant increase in rice production over time. The adoption of modern rice varieties progressed much faster in irrigated regions because of their suitability and higher returns. However, the level of adoption of HYVs varies markedly across states as seen in the data obtained for the major rice-growing states. Andhra Pradesh and Punjab registered 100% of rice area under HYVs (Fig. 3.9), while uptake was 87% and 79% in Karnataka and Odisha, respectively. Conversely, the adoption of HYVs was less than 50% of the area under paddy in Madhya Pradesh, Haryana, and Bihar. Low adoption of modern varieties is one of the reasons for the low productivity of rice in these states. The promotion of suitable HYVs with proper institutional mechanisms for the supply of inputs and marketing facilities encourages farmers to adopt

TABLE 3.4 Classification of Rice-Growing Districts Based on Productivity Groups: 2002–03 to 2006–07

State	High productivity (>2.5 t/ha)			Medium productivity (1.5–2.5 t/ha)			Low productivity (<1.5 t/ha)		
	Number of districts	Area share (%)	Production share (%)	Number of districts	Area share (%)	Production share (%)	Number of districts	Area share (%)	Production share (%)
Andhra Pradesh	14	78	83	8	22	17	—	—	—
Assam	—	—	—	12	45	5	14	55	47
Bihar	1	6	11	9	20	30	28	74	59
Haryana	12	78	83	7	22	17	—	—	—
Karnataka	11	37	48	12	52	47	4	11	5
Madhya Pradesh	—	—	—	4	1	3	40	99	97
Odisha	—	—	—	8	26	32	22	74	68
Punjab	19	100	100	—	—	—	—	—	—
Tamil Nadu	25	72	80	4	29	20	—	—	—
Uttar Pradesh	5	5	6	56	91	91	9	5	3
West Bengal	8	49	55	10	51	45	—	—	—

Source: Computed from GOI (Government of India), 2009. Rice in India during tenth plan. Directorate of Rice Development (Patna), Ministry of Agriculture, Government of India.

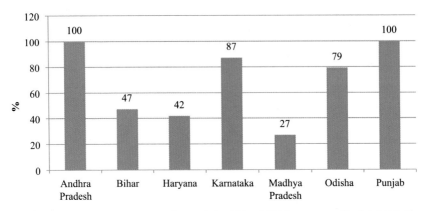

FIGURE 3.9 Area under high-yielding varieties (%). *(GOI (Government of India), 2014. Agricultural statistics at a glance. Ministry of Agriculture, Government of India, Oxford University Press, New Delhi.)*

improved cultivation practices, which in turn is found to have a positive effect on improving their livelihoods (Prahladachar, 1983).

Various studies also expressed concerns about the gap in yield between farmers' fields and those of experiment stations. Singh (2012) estimated the yield gap between improved varieties of rice in farmers' fields and control varieties in demonstration plots. As per the estimate of this study, the yield gap was lowest in West Bengal (4%) and highest in Assam (41%). The yield gap in Uttar Pradesh was as high as 33%, and in Tamil Nadu, a high-productivity state, the yield gap was surprisingly estimated at 27%. It was 18% in Karnataka, 15% in Andhra Pradesh, and 13% in Bihar (relatively low). This analysis infers that rice production could be enhanced by bridging the yield gap through effective extension services for the transfer of technology and public investment in irrigation infrastructure.

REGIONAL VARIATIONS IN COST AND INCOME

This section presents regional variations in real cost and income from rice cultivation. To compute per hectare farm business income, both input and output data were deflated by relevant price deflators at 2004–05 prices. Farm business income was calculated as the difference between paid-out cost and gross value of output. For analytical purposes, paid-out cost is widely used to track changes in the welfare of farmers. It includes all actual expenses in cash and kind incurred by cultivators, and rent paid for leased-in land. A weighted state-level income series was constructed by using the area share of crops in total cropped area as weight.

The trends in average real cost, output, and income are presented in Table 3.5. Rice cultivation is highly profitable in Punjab, with average farm business income of INR 24,168/ha in the 2000–01 to 2012–13 period. Although paid-out

TABLE 3.5 Growth in Real Cost, Output, and Income From Paddy Cultivation

Period	Average value (INR/ha)			Annual growth rate (%)		
	Paid-out cost	Gross value of output	Farm business income	Paid-out cost	Gross value of output	Farm business income
Andhra Pradesh						
1990–91 to 1999–2000	13,972	27,975	14,003	0.79	1.69	2.81
2000–01 to 2012–13	16,824	34,581	17,757	1.09	1.85	2.62
Assam						
1990–91 to 1999–2000	4,260	12,754	8,494	2.10	1.92	1.82
2000–01 to 2012–13	6,411	13,332	6,921	1.50	0.12	−1.52
Bihar						
1990–91 to 1999–2000	5,709	12,975	7,266	2.48	2.02	1.62
2000–01 to 2012–13	7,639	13,856	6,218	0.91	1.68	2.43
Haryana						
1990–91 to 1999–2000	11,477	27,874	16,397	0.13	−2.06	−3.58
2000–01 to 2012–13	14,650	38,234	23,585	−0.85	2.16	4.13
Karnataka						
1990–91 to 1999–2000	11,716	29,590	17,874	4.39	3.07	2.21
2000–01 to 2012–13	17,177	34,168	16,991	−1.62	0.66	3.14
Madhya Pradesh						
1990–91 to 1999–2000	5,675	11,856	6,181	2.77	2.11	1.34

(continued)

TABLE 3.5 Growth in Real Cost, Output, and Income from Paddy Cultivation (*cont.*)

Period	Average value (INR/ha)			Annual growth rate (%)		
	Paid-out cost	Gross value of output	Farm business income	Paid-out cost	Gross value of output	Farm business income
2000–01 to 2012–13	6,594	13,996	7,402	1.50	7.94	15.86
Odisha						
1990–91 to 1999–2000	6,496	15,011	8,515	1.33	−0.59	−2.08
2000–01 to 2012–13	9,357	17,513	8,156	1.10	1.80	2.61
Punjab						
1990–91 to 1999–2000	12,874	28,724	15,850	−0.48	0.55	1.43
2000–01 to 2012–13	15,364	39,533	24,168	−0.56	1.23	2.50
Uttar Pradesh						
1990–91 to 1999–2000	7,042	16,556	9,514	−0.59	1.81	3.92
2000–01 to 2012–13	9,703	21,052	11,348	1.49	3.79	5.85
West Bengal						
1990–91 to 1999–2000	11,408	24,927	13,519	−0.28	−0.94	−1.51
2000–01 to 2012–13	13,144	22,901	9,757	1.69	2.36	3.21
Overall (weighted)						
1990–91 to 1999–2000	8,455	19,028	10,573	0.77	0.71	0.67
2000–01 to 2012–13	10,891	22,411	11,520	1.07	2.85	4.61

Source: GOI (Government of India), various years. Cost of cultivation of principal crops in India. Ministry of Agriculture, Government of India, New Delhi.

cost increased marginally between 1990–91 to 1999–2000 and 2000–01 to 2012–13, a more than proportionate increase in the value of output led to higher growth in farm business income. Haryana registered the second highest average income from rice cultivation at INR 23,585/ha during the most recent period. In Andhra Pradesh, average income increased from INR 14,003 to INR 17,757 per ha between 1990–91 to 1999–2000 and 2000–01 to 2012–13. Both the gross value of output and paid-out cost registered a relatively high growth rate during 2000–01 to 2012–13 compared with 1990–91 to 1999–2000. Although average farm business income increased in absolute terms, its growth actually decelerated during the 2000s vis-à-vis the 1990s. Uttar Pradesh and Madhya Pradesh showed increased average income from paddy cultivation over time.

In Bihar, the annual growth in per hectare income was higher during the 2000s than in the 1990s. However, in absolute terms, average farm business income fell from INR 7266 to INR 6218 between 1990–91 to 1999–2000 and 2000–01 to 2012–13. A rise in the level of paid-out cost led to a decline in average income per hectare during the most recent period. Average income from rice cultivation also declined in Assam, Karnataka, Odisha, and West Bengal during the 2000s vis-à-vis the 1990s. West Bengal registered a 28% fall in income between the 1990s and 2000s. A decline in the average value of output, and an increase in the average paid-out cost, led to a fall in farm business income during the recent period. In Assam, average per hectare income from rice cultivation fell by 19% during the 2000s vis-à-vis the 1990s. The growth in paid-out cost was much higher than the value of output, which caused a decline in income during the recent period. Although the decline in average income is only marginal in Karnataka and Odisha, the rising paid-out cost is worrisome. Weighted average income of all states increased from INR 10,571 to 11,520 per hectare between 1990–91 and 1999–2000 and 2000–01 to 2012–13. Although average paid-out cost increased over time, a more than proportionate increase in the value of output led to an increase in farm business income. In fact, the value of output registered impressive growth of 2.85% per annum and paid-out cost recorded growth of 1.07% during the 2000–01 to 2012–13 period.

Table 3.6 provides an analysis of the growth trend for input cost in rice cultivation. With the exception of animal labor and irrigation, most other inputs have increased since 1990. The use of seeds increased in all states except Assam. Growth in the use of seeds, perhaps hybrids and other improved varieties, was by far the highest in Punjab (3.19%), but was relatively high in Haryana and Uttar Pradesh as well. For fertilizers, most states showed positive growth from 1990–91 to 2012–13. The negative growth in the use of fertilizers in Andhra Pradesh and Punjab is negligible. All states registered positive growth in human labor use with the highest growth being recorded in Odisha and the lowest in Punjab and Andhra Pradesh.

Regarding agricultural machinery, agriculturally progressive states such as Punjab, Haryana, Andhra Pradesh, and Uttar Pradesh have all shown negative growth rates. These states have actually achieved a higher level of capital

TABLE 3.6 Growth in Major Input Use in Paddy Cultivation (%): 1990–91 to 2012–13

State	Seed	Fertilizer and manure	Human labor	Animal labor	Machine labor	Insecticides	Irrigation
Andhra Pradesh	1.23	−0.02	1.67	−5.29	−1.94	8.33	−6.44
Assam	−3.05	9.41	2.56	2.10	5.67	7.88	48.95
Bihar	0.89	2.44	2.59	−6.25	1.88	28.28	10.87
Haryana	2.62	0.58	2.10	−9.83	−2.82	5.19	−1.69
Karnataka	1.99	2.26	2.29	−0.93	2.88	3.94	−3.83
Madhya Pradesh	1.33	2.15	2.51	−1.45	3.79	15.28	−1.91
Odisha	0.10	2.65	6.41	1.90	5.65	6.80	−0.39
Punjab	3.19	−0.08	1.57	−4.25	−4.59	7.05	−4.74
Uttar Pradesh	2.51	3.39	2.58	−2.79	−3.34	4.67	4.30
West Bengal	0.72	2.12	2.48	−1.66	−0.03	5.92	−0.14

Source: GOI (Government of India). Various years. Cost of cultivation of principal crops in India. Ministry of Agriculture, Government of India, New Delhi.

intensification and farmers spend more on replacement cost for the initial capital investment in an attempt to maintain the same level of efficiency (Singh, 2009). It is clear from the analysis that the use of inputs in rice cultivation is increasing and this poses two key questions: Is input intensification driving rice output growth? How sustainable is this growth in the long run? TFP growth is an important measure of sustainable output growth and hence it is important to analyze the performance of TFP growth in the rice sector.

TECHNICAL EFFICIENCY AND TFP GROWTH IN RICE

Analytical Method

Data envelopment analysis (DEA) has been employed to decompose the sources of productivity growth into technical change (innovation) and technical efficiency change (catching-up effect). Empirically, TFP is calculated as the ratio of aggregate output to aggregate input. In India, most studies used the Tornqvist index to construct output and input indices by using quantity and price information (Chand et al., 2011; Evenson et al., 1999; Kumar et al., 2004, 2008; Kumar and Mruthynjaya, 1992; Mukherjee and Yoshimi, 2003). This index is preferred for its niceties of theoretical properties as established by Diewert (1976, 1978). However, this index is considered to be a descriptive measure of productivity change that does not allow for decomposition of productivity growth (Färe et al., 1994) and neither does it require knowledge of underlying production technology (Ray, 2004, p. 274). Furthermore, estimation of TFP growth using this method is inappropriately interpreted as technical change (Mahadevan, 2002). However, the distance function-based Malmquist productivity index decomposes productivity growth into technical efficiency change (catch-up) and technical change (innovation) between two time periods. The Malmquist productivity index was originally introduced by Caves et al. (1982) and its use in empirical studies was popularized by Färe et al. (1994).

Following Färe et al. (1994), the output-oriented Malmquist productivity index can be written in the following form:

$$M_0(x^{t+1}, x^t, y^t)$$
$$= \frac{D_0^{t+1}(x^{t+1}, y^{t+1})}{D_0^t(x^t, y^t)} \times \left[\left(\frac{D_0^t(x^{t+1}, y^{t+1})}{D_0^{t+1}(x^{t+1}, y^{t+1})} \right) \times \left(\frac{D_0^t(x^t, y^t)}{D_0^{t+1}(x^t, y^t)} \right) \right]^{1/2}$$

The component outside the square bracket is the ratio of technical efficiency in period t to technical efficiency in period $t + 1$. This efficiency change component indicates how far the observed production is getting closer to or farther from the frontier. The expression inside the bracket indicates a shift in technology frontier (technical change) between period t and $t + 1$. It is measured as the geometric mean of a shift in technology between two periods evaluated at input levels x^t and x^{t+1}. A value of the efficiency change component greater than 1 indicates that the production unit is catching up to the frontier in period $t + 1$, compared with period t. An improvement in technical change provides evidence of innovation between two periods and a value of technical change greater than 1 shows technical progress. In other words, TFP growth can be written as:

$$\text{TFP growth} = \text{Technical efficiency change} \times \text{Technical change}$$
$$\text{(catching-up effect)} \qquad \text{(frontier shift effect)}$$

Following Färe et al. (1994), technical efficiency change can be decomposed into pure technical efficiency change (estimated under variable returns to scale) and scale efficiency change. This can be specified as:

$$\text{Technical efficiency change} = \text{Pure technical efficiency change}$$
$$\times \text{Scale efficiency change}$$

Overall efficiency change is the product of pure technical efficiency change and scale efficiency change. A major advantage of using DEA is the identification of sources of growth for appropriate policy suggestions to improve productivity growth in India's rice sector. Rice yield (quintals/ha) was captured as the output variable. Input variables included seed (kg/ha), fertilizer (kg nutrients/ha), manure (quintals/ha), human labor (person-hours/ha), animal labor (pair-hours/ha), insecticide, and irrigation.

Analysis of Technical Efficiency and TFP Growth

The average rice TFP growth from 1991–92 to 2012–13 was 3.28% (Table 3.7). Most of the growth in TFP was due to growth in technical change. This implies that farmers adopted high levels of new technologies during this period. Little improvement in technical efficiency change was observed; however a decrease in pure technical efficiency change was seen. Subperiod analysis also showed

TABLE 3.7 Technical Efficiency and TFP Growth in Rice (%)

Period	TFP growth	Technical change	Technical efficiency change	Pure technical efficiency change	Scale efficiency change
1991–92 to 1995–96	2.27	1.73	0.53	0.00	0.53
1996–97 to 2000–01	1.84	3.73	−1.79	−0.36	−1.45
2001–02 to 2005–06	4.78	2.97	1.72	0.39	1.36
2006–07 to 2012–13	3.98	4.23	−0.24	−0.14	−0.07
1991–92 to 2012–13	3.28	3.26	0.02	−0.04	0.08

a deterioration in pure technical efficiency gains during 1996–97 to 2000–01 and 2006–07 to 2012–13. This implies that there were no efficiency gains because of improper application of technologies/combination of inputs in rice production.

The highest growth (4.78%) in TFP was observed between 2001–02 and 2005–06, of which 2.97% was attributable to technical change and 1.72% to technical efficiency change. Both the frontier effect (technical change) and catching-up effect, or movement toward the best frontier (technical efficiency change), helped to achieve higher TFP growth. Improvement in technical efficiency change was mainly driven by scale efficiency change (1.36%) and pure technical efficiency change (0.39%). TFP growth was lowest (1.84%) between 1996–97 and 2000–01. Although the frontier effect showed impressive growth of 3.73%, negative growth in technical efficiency change pulled down the overall productivity improvements. Deterioration in pure technical efficiency and scale efficiency was observed during this subperiod. TFP growth during the most recent period (2006–07 to 2012–13) was as high as 3.98% and it was largely driven by growth in technical change. These results broadly indicate that technical change was the main driver of TFP growth in rice alongside little improvement in technical efficiency.

The trend in technical efficiency change, technical change, and TFP index for the entire period of analysis is presented in Fig. 3.10. The trend in technical efficiency change remained more or less constant except during the mid-1990s and mid-2000s. Furthermore, the trend in TFP change has closely mirrored that of technical change. In fact, the increase in technical change seems to have pushed the trend in overall productivity during recent years.

Analysis of spatial and temporal patterns in TFP growth (and its components) provides more interesting results (Table 3.8). With the exception of Assam, all states showed positive TFP growth from 1991–92 to 2012–13. No improvement

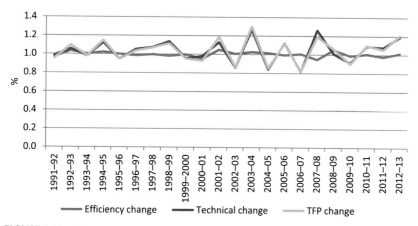

FIGURE 3.10   Trend in technical efficiency, technical change, and TFP index.

TABLE 3.8 Technical Efficiency and TFP Growth Across States (%)

State	1991–92 to 1999–2000			2000–01 to 2012–13			1991–92 to 2012–13		
	Efficiency change	Technical change	TFP growth	Efficiency change	Technical change	TFP growth	Efficiency change	Technical change	TFP growth
Andhra Pradesh	1.19	2.48	3.70	0.10	6.12	6.23	0.54	4.61	5.19
Assam	0.00	−8.09	−8.09	0.00	−1.59	−1.59	0.00	−4.30	−4.30
Bihar	0.00	4.17	4.17	0.00	1.82	1.82	0.00	2.78	2.78
Haryana	0.00	6.42	6.42	0.00	2.15	2.15	0.00	3.88	3.88
Karnataka	0.00	1.30	1.30	−0.52	4.10	3.57	−0.31	2.95	2.64
Madhya Pradesh	−0.66	−0.83	−1.49	0.45	3.98	4.47	−0.01	1.98	1.99
Odisha	−3.23	2.84	−0.47	2.45	2.71	5.22	0.09	2.77	2.86
Punjab	0.00	11.33	11.33	0.00	1.98	1.98	0.00	5.71	5.71
Uttar Pradesh	0.00	4.43	4.43	0.00	1.72	1.72	0.00	2.82	2.82
West Bengal	1.13	−1.35	−0.26	−0.97	1.83	0.85	−0.12	0.52	0.39

in efficiency change was observed and technological regression was seen in Assam. Recent efforts to bring technological progress to eastern India (including Assam) do not seem to have affected TFP growth. With the exception of Assam, TFP growth was largely attributed to the frontier shift resulting from the adoption of modern technologies by rice-cultivating farmers. The highest TFP growth was observed in Punjab (5.71%), and was almost entirely driven by technical change (5.71%) as no technical efficiency gains were observed in this state. The second highest growth in TFP was recorded in Andhra Pradesh (5.19%), which was mainly attributed to technical change (4.61%) and to a slight extent by growth in technical efficiency (0.54%). TFP growth in Haryana, Odisha, Uttar Pradesh, Bihar, and Karnataka was 3.88%, 2.86%, 2.82%, 2.78%, and 2.64%, respectively. TFP growth in West Bengal was as low as 0.39% from 1991–92 to 2012–13. Although the frontier shift effect was positive, movement of paddy farmers toward a fully efficient frontier had deteriorated in this state.

Subperiod analysis of TFP growth provides mixed results. With no improvement in technical efficiency change, Assam registered negative growth in TFP during the 1990s and 2000s. Madhya Pradesh experienced technical inefficiency as well as technological regression during the 1990s, which resulted in a negative TFP growth of 1.49%. However, during the 2000s there was a turnaround in TFP growth (4.47%) with a significant improvement in technical change (3.98%) and technical efficiency change (0.45%). In West Bengal, the negative frontier effect outweighed the positive catching-up effect, leading to negative TFP growth during the 1990s. However, a reversal in their effects with dominance of the frontier effect (technological progress) led to positive TFP growth during the 2000s. Similarly, Odisha registered a spectacular performance in TFP growth during the 2000s vis-à-vis the 1990s, with both technical efficiency and technical change contributing to overall productivity improvements. In the case of Andhra Pradesh, it was interesting to observe that technical change became more dominant than technical efficiency change during the 2000s. Overall, analysis revealed that technical change due to the adoption of improved technologies had largely contributed to TFP growth. There was little improvement in technical efficiency change in most rice-producing states.

CONCLUSIONS

This chapter analyzed the spatial and temporal changes in rice production and productivity growth in India. The distance function-based Malmquist productivity index was used to decompose TFP growth into technical change and technical efficiency change. In India, rice is grown in different AEZs under varying ecosystems. Rice production and productivity showed marked variations across AEZs. Spatial analysis showed that, out of the 20 AEZs, 5 zones accounted for more than 60% of the total rice area. Unfortunately, these five zones have a large number of low-productivity districts and rice is mostly cultivated in rainfed ecosystems, which seem to lack appropriate rice production technologies. On the other hand, all districts in Punjab, and most rice-growing districts in Tamil Nadu and Haryana, registered rice productivity above 2.5 t/ha. Interdistrict variation in rice productivity was highest in Madhya Pradesh, Bihar, Odisha, and Assam. These low-productivity districts are in rainfed ecosystems and lack appropriate production technologies. There is huge potential for the improvement of rice production through the adoption of suitable modern varieties and institutional mechanisms for the supply of inputs and provision of marketing facilities.

Among cereals, the sheer dominance of paddy output influences the overall trend in food-grain production. HYV technology contributed to rice production increasing from 40 million tons in 1968–69 to 104 million tons in 2012–13. With more or less stagnant area growth, rice yield more than doubled from 1.1 to 2.5 t/ha between these periods and seems to have significantly contributed to the increase in production. The adoption of HYVs varied markedly across states, with 100% area coverage in Andhra Pradesh and Punjab but less than 50% in Madhya Pradesh, Haryana, and Bihar. Low adoption of modern varieties was one of the reasons these states recorded low rice productivity. Furthermore, the yield gap was lowest (4%) in West Bengal and highest (41%) in Assam. This implies that great potential exists to enhance rice production by bridging the yield gap through effective extension services for the transfer of technology and public investment in irrigation infrastructure.

At the national level, weighted average income from rice cultivation increased from INR 10,571 to 11,520 per hectare from 1990–91 to 1999–2000 and 2000–01 to 2012–13. Although the average paid-out cost increased over time, a more than proportionate increase in the output value led to an increase in farm business income. Analysis of input cost growth revealed that, with the exception of animal labor and irrigation, the use of most other inputs including seed, fertilizer, pesticide, and machine labor had increased over time. With respect to productivity analysis, average TFP growth for rice was estimated at 3.28% from 1991–92 to 2012–13. Most of the growth in TFP was attributed to technical change. There was little improvement in technical efficiency change, but growth was negative in pure technical efficiency change. With the exception of Assam, all states showed positive growth in TFP from 1991–92 to 2012–13. Assam showed a worsening of the frontier shift effect. Punjab registered the highest TFP growth (5.71%), closely followed by Andhra Pradesh (5.19%). However, with the exception of Andhra Pradesh, all states showed zero technical efficiency gains, which means that a higher amount of rice is produced by using a higher quantity of inputs per hectare. Overall, analysis revealed that technical change was the main driver of TFP growth in rice with little improvement in technical efficiency. Rice farmers face many abiotic and biotic stresses that tended to adversely affect the growth and productivity of the rice crop. There is a need to manage biotic and abiotic stresses effectively through scientific approaches, which will help to reduce the output loss and improve efficiency in input use.

REFERENCES

Barah, B.C., 2005. Dynamic of rice economy in India: emerging scenario and policy options. Occasional Paper 47. National Bank for Agriculture and Rural Development (NABARD), Mumbai.

Bhalla, G.S., Singh, G., 2001. Indian Agriculture: Four Decades of Development. Sage Publications, New Delhi.

Birthal, P.S., Joshi, P.K., Negi, D.S., Agarwal, S., 2014. Changing sources of growth in Indian agriculture: implications for regional priorities for accelerating agricultural growth. Discussion Paper 01325. Washington, DC, International Food Policy Research Institute.

Caves, D.W., Christensen, L.R., Diewert, W.E., 1982. The economic theory of index numbers and the measurement of input, output, and productivity. Econometrica 50 (6), 1393–1414.

Chand, P., Kumar, P., Kumar, S., 2011. Total factor productivity and contribution of research investment to agricultural growth in India. Policy Paper 25. National Centre for Agricultural Economics and Policy Research, New Delhi.

Chand, R., Saxena, R., Rana, S., 2015. Estimates and analysis of farm income in India, 1983–84 to 2011–12. Econ. Polit. Wkly. 50 (22).

Diewert, W.E., 1976. Exact and superlative index numbers. J. Econom. 4, 115–145.

Diewert, W.E., 1978. Superlative index numbers and consistency in aggregation. Econometrica 46 (4), 883–900.

Evenson, R.E., Pray, C.E., Rosegrant, M.W., 1999. Agricultural Research and Productivity Growth in India. Research Report 109. International Food Policy Research Institute, Washington, DC.

FAO, 1996. Agro-ecological Zoning Guidelines. FAO Soils Bulletin 73. FAO, Rome, Italy.

Färe, R., Grosskopf, S., Norris, M., Zhang, Z., 1994. Productivity growth, technical progress, and efficiency change in industrialized countries. Am. Econ. Rev. 84 (1), 66–83.

Gajbhiye, K.S., Mandal, C., 2000. Agro-ecological zones, their soil resource and cropping systems, In: Cropping Systems: Status of Farm Mechanization in India, pp. 1–32, Indian Agricultural Statistics Research Institute, New Delhi.

GOI (Government of India), 2005. Situation Assessment Survey of Farmers, Some Aspects of Farming, Report No. 496(59/33/3), Ministry of Statistics and Programme Implementation, Government of India.

GOI (Government of India), 2009. Rice in India during tenth plan. Directorate of Rice Development (Patna), Ministry of Agriculture, Government of India.

GOI (Government of India), 2014. Agricultural statistics at a glance. Ministry of Agriculture, Government of India, Oxford University Press, New Delhi.

Gumma, M.K., Nelson, A., Thenkabail, P.S., Singh, A.N., 2011. Mapping rice areas of South Asia using MODIS multitemporal data. J. Appl. Remote Sensing 5, 053547.

Hijmans, R.J., 2015. Raster: geographic data analysis and modeling. R package version 2.4-20. Available from: http://CRAN.R-project.org/package=raster

Hijmans, R.J., Aunario, J., Sparks, A., 2015. Cropsim: functions for use in the dynamic and mechanistic simulation of crop (plant) growth and development, and of plant diseases; and to implement these functions in a number of models. R package version 0.2.0-5. Available from: https://r-forge.r-project.org/projects/cropsim/

Joshi, P.K., Gulati, A., Birthal, P.S., 2007. Agricultural diversification in India: status, nature, and pattern. In: Joshi, P.K., Gulati, A., Cummings, Jr., R. (Eds.), Agricultural Diversification and Smallholders in South Asia. Academic Foundation, New Delhi, pp. 219–242.

Kannan, E., 2014. Assessment of Pre- and Post-Harvest Losses of Important Crops in India. Research Report IX/ADRTC/153A. Institute for Social and Economic Change (ISEC), Bangalore, India.

Kumar, P., Kumar, A., Mittal, S., 2004. Total factor productivity of crop sector in the Indo-Gangetic plain of India: sustainability issues revisited. Indian Econ. Rev. 34 (1), 169–201.

Kumar, P., Mittal, S., Hossain, M., 2008. Agricultural growth accounting and total factor productivity in South Asia: a review and policy implications. Agric. Econ. Res. Rev. 21, 145–172.

Mahadevan, R., 2002. A DEA approach to understanding the productivity growth of Malaysia's manufacturing industries. Asia Pac. J. Manag. 19, 587–600.

Mukherjee, A.N., Yoshimi, K., 2003. Productivity growth in Indian agriculture: is there evidence of convergence across states? Agric. Econ. 29 (1), 43–53.

Mruthynjaya, Kumar P., 1992. Measurement and analysis of total factor productivity growth in wheat. Indian J. Agric. Econ. 47 (3), 451–458.

Murgai, R., 2001. The Green Revolution and the productivity paradox: evidence from Indian Punjab. Agric. Econ. 25 (2–3), 199–209.

Prahladachar, M., 1983. Income distribution effects of the Green Revolution in India: a review of empirical evidence. World Dev. 11 (11), 927–944.

R Core Team, 2015. R: a language and environment for statistical computing. R Foundation for Statistical Computing, Vienna, Austria. Available from: https://www.R-project.org/

Ray, S.C., 2004. Data Envelopment Analysis: Theory and Techniques for Economics and Operations Research. Cambridge University Press, Cambridge, United Kingdom.

Savary, S., Nelson, A., Willocquet, L., Pangga, I., Aunario, J., 2012. Modelling and mapping potential epidemics of rice diseases globally. Crop Prot. 34, 6–17.

Sen, A., 2016. Some reflections on agrarian prospects. Econ. Polit. Wkly. 51 (8), 12–15.

Singh, K., 2009. Agrarian crisis in Punjab: high indebtedness, low returns, and farmers' suicides. In: Reddy, N., Mishra, S. (Eds.), Agrarian Crisis in India. Oxford University Press, New Delhi, pp. 261–284.

Singh, M., 2012. Yield gap analysis in rice, wheat, and pulses in India. Ann. Agric. Res. New Series 33 (4), 247–254.

Wasteland Atlas of India, 2011. Department of Land Resource, National Remote Sensing Centre Report.

FURTHER READING

Minhas, B.S., Vaidyanathan, A., 1965. Growth of crop output in India, 1951–54 to 1958–61: an analysis by component elements. J. Indian Soc. Agric. Stat. 17 (2), 230–252.

Chapter 4

Ecological Footprints of and Climate Change Impact on Rice Production in India

Rudrasamy Balasubramanian, Venkatachalam Saravanakumar, Kovilpillai Boomiraj
Tamil Nadu Agricultural University, Coimbatore, Tamil Nadu, India

INTRODUCTION

The increasing use of external chemical inputs and diminishing returns to major inputs, stagnation in technological development, dwindling of the agricultural water supply, environmental pollution, and climate change are emerging as major challenges for sustaining future agricultural productivity and food security. Land-augmenting agricultural intensification has made massive contributions to ensure food and nutritional security for millions of poor people in India, while saving vast areas of forest from being converted for agricultural use; however, intensive agriculture has attendant environmental problems. For example, the excessive use of natural resources, coupled with climate change, has led to a plateauing of yield in Green Revolution crops, such as rice and wheat, and some crops have seen negative yield growth (Saharawat et al., 2012). The application of both fertilizer and water played a key role in the success of Green Revolution technologies; however, the excessive and unbalanced use of fertilizer and the extensive and intensive use of water are now common problems. The excessive use of both fertilizer and water has caused serious damage to soil and water, two of the most important natural resources that support agriculture and human livelihood, as borne out by the studies cited in the ensuing sections of this chapter. Problems such as deterioration in soil quality as exemplified by poor drainage, soil salinity, nitrate and pesticide residues in soil and water resources, nutrient mining, depletion of soil micronutrients and organic carbon, herbicide tolerance of weeds, pest resistance to pesticides, disease and nematode buildup, and overextraction of groundwater have received considerable research attention. These problems are especially important and extremely serious in intensive rice and rice–wheat monocultures, with far-reaching consequences for future food

The Future Rice Strategy for India. http://dx.doi.org/10.1016/B978-0-12-805374-4.00004-X

security. Emission of greenhouse gases (GHGs) from rice fields is another key issue for rice cultivation, and this has also received substantial research attention. Continuous flooding of rice fields, along with the intensive use of nitrogenous fertilizer, has generated large quantities of GHGs, such as methane (CH_4), carbon dioxide (CO_2), and nitrous oxide (N_2O). Apart from contributing to climate change through GHG emissions, rice production is itself affected by climate change. The task of increasing rice productivity is not as easy as expected due to rapid changes in climatic and environmental factors, including soil and water, the two crucial components for rice production (Banerjee and Sanyal, n.d.). In light of the critical environmental issues facing rice ecosystems, this chapter attempts to provide a synthesis and overview of the ecological impacts of rice production, GHG emissions from rice production, and the impact of climate change on rice production in India. A plausible set of recommendations is also provided to ensure sustainable rice production in this increasingly challenging environment.

AGROCHEMICALS AND SUSTAINABILITY OF RICE ECOSYSTEMS

Fertilizer Use and Soil Health in Rice Ecosystems

Cultivating high-yielding varieties of rice is a chemical-intensive enterprise with heavy dependence on both fertilizer and pesticide. Increased rice yields have been largely attributed to an increased use of fertilizer coupled with an assured water supply. It has been estimated that the production of 1 ton of rice removes about 15 kg of nitrogen, 4 kg of phosphorus, and 24 kg of potash from the soil (Hegde, 1992). Hence, heavy nutrient removal by high-yielding rice has led to the impoverishment of soil fertility and a decline in crop production (Nambiar et al., 1992). Long-term degradation of rice production environments was noticed as early as the 1990s, as indicated by the stagnation in yield and declining partial factor productivity, especially for fertilizer (Cassman et al., 1995; Cassman and Pingali, 1993; Pingali, 1992; Pingali et al., 1990). Long-term (1972–82) fertilizer experiments conducted in four Indian states (Punjab, Uttar Pradesh, Odisha, and Andhra Pradesh) found that yields were declining in two locations and remaining stagnant in two (Nambiar and Ghosh, 1984). A follow-up study of these four locations revealed that mean rice yields declined in all locations over a 16-year period, and this remained true for all treatments (Yadav, 1998). Long-term yield trials conducted in Pantnagar showed that rice yields declined by 2.8% per year (Nambiar, 1994). In addition to groundwater depletion, the resource-intensive rice–wheat cropping system has resulted in the mining of macro- and micronutrients from the soil and an increase in soil salinization in the Indo-Gangetic Plains (IGP) (Ladha et al., 2000; Tiwari, 2002; Tiwari et al., 2009). These findings corroborate the fatigue of the Green Revolution with severe stress on soil and water ecosystems.

The heavy usage of chemical fertilizers, especially nitrogenous fertilizers, coexists with the low efficiency of fertilizer nitrogen usage, especially in tropical regions where rice is the principal crop (Prasad, 2013). Indirect nitrogen budget estimates show that up to 50% of applied nitrogen may be lost via nitrification–denitrification, ammonia volatilization, and nitrate leaching in irrigated porous soils under wetland rice. The nitrogen-use efficiency of rice is only 30% of applied nitrogen; therefore, application of high doses of fertilizer nitrogen coupled with inefficient management of irrigation water for shallow-rooted crops may lead to nitrate leaching and groundwater pollution (Milkha and Bijay-Singh, 1997). Nitrate pollution of groundwater is common in agriculturally intensive states, such as Punjab, Andhra Pradesh, and Maharashtra, where fertilizer application rates are high and the rice crop has a nitrogen-use efficiency of only 28%–34% (Rao and Puttanna, 2010). In Punjab, groundwater nitrate content increased by ~2 mg/L between 1982 and 1988, alongside an increase in N fertilizer use from 56 kg/ha to 188 kg/ha, most of which was used in the rice–wheat cropping system (Bijay-Singh et al., 1991). This is further corroborated by the fact that nitrogen leaching from Punjab rice fields was estimated to be on the order of 27.3%–30.7% of total applied nitrogen (Kumar et al., 2006). Using remote sensing and GIS techniques, Chhabra et al. (2010) estimated nitrogen loss through leaching in kharif[a] and rabi season rice in the IGP, establishing that 35% and 40% of the applied nitrogenous fertilizer was lost in kharif and rabi seasons, respectively. This study provides evidence of the magnitude of the nutrient leaching problem as well as providing insights into the role of nitrogenous fertilizer as a nonpoint source pollutant from agriculture. The high leaching rate of nitrogen in the sandy soils of Punjab and Haryana has led to groundwater pollution. Approximately 3% (26,796 and 10,588 ha) of the area in six districts of Haryana was estimated to fall within moderately high (7.5–10 mg/L) and high (>10 mg/L) risk categories in terms of groundwater nitrate concentration (Chandna et al., 2011).

Overextraction of groundwater in coastal rice ecosystems has resulted in seawater intrusion and salinity buildup. This is in addition to the inland salinity caused by improper water management, waterlogging, and poor drainage facilities in many canal-irrigated rice production environments (Ramasamy et al., 2005). The falling water table, and resulting seawater intrusion, has led to a deterioration in groundwater quality in southwestern and central districts of Punjab (Hira, 2009). Waterlogging caused by poor drainage and a rise in the groundwater table are posing serious challenges to the sustainability of rice-based cropping systems. A study in Haryana revealed that two-thirds of the rise in the level of the water table was attributable to rice. This is emerging as a threat to rice–wheat cultivation and hence a reduction in rice area is mandatory to mitigate the rise in the water table (Singh et al., 2012). Basmati rice regions in India and Pakistan have some of the highest salinity problems in the world

---

a. Kharif and rabi are the two main cropping seasons in India, based on monsoon. Kharif rice is sown during May–August and harvested during September–January. Rabi season rice is sown during December–January and harvested during April–May. The wide sowing and harvesting window for the kharif season is due to the wide variation in the date of onset of the southwest monsoon across different agro-climatic regions of the country.

(Maredia and Pingali, 2001). Complications of intensive rice cultivation include a decline in the level of the water table, increasing problems of salinity and alkalinity, declining soil fertility because of the depletion of macro- and micronutrients (N, P, and Zn), deterioration of physicochemical soil conditions because of heavy tillage and puddling, and imbalanced use of fertilizer (Singh, 2000). Policies that favored excessive water and fertilizer use included price subsidies for inputs, such as electricity, canal water, and fertilizer, and price support for many agricultural commodities, including rice (World Bank, 2007).

Pesticide Use in Rice Ecosystems

Large quantities of pesticides and fungicides are used in modern rice production environments because of the wide array of pests and diseases that attack the crop at various stages of growth. The use of chemical pesticides in the major rice-growing regions of India poses serious health hazards for the rural poor as farmers are led to overuse pesticides, often considering them as productive rather than protective inputs (Mencher, 1991). High production levels of rice could not be maintained because many regions often witnessed outbreaks of rice brown planthopper and rice stem borer, which caused crop failure. The irrational use of pesticides has led to serious problems for crops such as rice, for which the costs of pesticide use have exceeded the benefits (Rola and Pingali, 1993). In some cases, such as with the rice stem borer, continuous use of a single pesticide has resulted in pesticide resistance (Su et al., 2014). Recent research demonstrated that broad-spectrum insecticides killed natural enemies of the rice stem borer, thus disrupting the natural balance in unsprayed rice fields and leading to these outbreaks. Pesticide use in rice is estimated to account for ~17% of total pesticide use in the country (Subbaiah, 2006). Pesticide residues that cause serious damage to soil and aquatic ecosystems have been reported in many countries, including India. Apart from causing damage to beneficial soil-borne microbes and other living organisms (i.e., arthropods and earthworms), pesticide-related health damage to humans and livestock has also been reported. An economic study on pesticide use in Tamil Nadu found that the average environmental impact quotient[b] (EIQ) field rating for rice was highest in the *thaladi* season at 16.01, followed by *samba* and *kuruvai*[c] season rice, with values of 13.01 and 10.63, respectively (Ashok and Kennedy, 2004).

Excessive use of pesticides, along with a lack of precautionary measures, has resulted in several acute and chronic health complications among pesticide

---

b. Pesticides have a multitude of environmental impacts. The EIQ of pesticides compresses different environmental impact information into a single metric. EIQ field use rating has been developed to account for different formulations and different use patterns in farmers' fields. The higher the value of the EIQ field rating, the greater the environmental impact of pesticides.

c. Kuruvai, samba, and thaladi are the three main rice seasons of Tamil Nadu. Kuruvai rice is sown during June–July, samba rice during August, and thaladi (or late samba) is sown during September–October.

applicators. This is aside from the presence of pesticide residues in food, fodder, soil, and aquatic ecosystems. Hospital expenditure for these farmers was in the INR 450–3780 range, with a mean cost of INR 1536. The duration of hospitalization ranged from 1 day to 1 week and almost a quarter of the daily earnings of pesticide applicators are lost because of the health complications arising out of pesticide exposure (Devi, 2007). It was found that the value of the rice crop lost to pests is invariably lower than the pesticide-related health costs. If the pesticide-related human health and environmental damage costs were taken into account, nonuse of pesticides in rice cultivation would be the best option (Pingali et al., 1997). Another important dimension to the issue of pesticide use in rice is insecticide resistance. Brown planthopper's resistance to neonicotinoids has been observed in many Asian countries, including India (Harris, 2006). The emergence of pesticide resistance will lead to overapplication of more toxic chemicals to manage pest outbreaks with both private-economic and social-environmental costs.

The Water Footprint of Rice Production

Rice accounts for ~45% of the total global cropped area under irrigation and around 57% of all harvested rice comes from irrigated fields. Furthermore, rice cultivation accounts for ~85% of total irrigation water use. In Asia, ~84% of the water withdrawn from surface or underground sources is used for agriculture, mostly for flooded rice irrigation (Thiyagarajan and Gujja, 2013). India is one of the largest rice producers in the world, but per capita water availability steadily decreased from 5831 to 1518 m^3 per annum between 1950 and 2010 and it is projected to further decrease to 1341 m^3 and 1140 m^3 per annum by 2025 and 2050, respectively (Ministry of Water Resources, Government of India). India has the largest rice production water footprint, amounting to 432.9 billion m^3 per annum during 2000–04 (Chapagain and Hoekstra, 2011). As rice is mostly identified with irrigated agriculture, the spread of high-yielding varieties of rice was highly correlated with access to irrigation (Pingali, 2012). Econometric analysis of rice yield revealed that access to both public and private irrigation increases yield by 15%, while access to private irrigation alone leads to a 9% increase in yield (Jin et al., 2012). Consequently, the increasing use of water for rice cultivation has become a common phenomenon in India as elsewhere.

The water-intensive nature of rice has led to the twin problems of excessive groundwater depletion and waterlogging and salinity. Groundwater over-extraction has become a widespread phenomenon due to intensive rice farming with groundwater (Killebrew and Wolff, 2010). In states such as Punjab and Telengana, where groundwater irrigation plays a critical role in rice cultivation, the fall in the water table and drying up of wells are likely to emerge as major challenges to future agriculture (Fishman et al., 2011). As noted by the World Bank report (World Bank, 1999), the groundwater table has dropped by 3 m in most parts of Punjab due to the overexploitation of water for Green Revolution agriculture. In Punjab, extensive and intensive cultivation of rice, early planting of rice (especially in May), and a regular supply of electricity cause excessive evapotranspiration loss and a decline in the water table (Hira, 2009; Katya, 2015; Singh, 2013a). Major rice-growing states such as Punjab, Haryana, and Tamil Nadu have exploited groundwater heavily with their groundwater development (groundwater draft as a percentage of net groundwater availability) reaching 170%, 127%, and 80%, respectively, during 2011 (CGWB, 2014). The water table has declined by 3–10 m in agriculturally intensive rice–wheat cropping districts of Haryana, since 60%–65% of irrigation water is procured from groundwater sources (Singh, 2000). Both demand-side factors (i.e., the huge water requirement for rice and price support for rice to make rice farming a profitable enterprise) and supply-side factors (i.e., government support to groundwater development through subsidized electricity, cheaper credit for well drilling, absence of regulatory mechanisms for groundwater extraction, etc.) have contributed to groundwater depletion. Given the serious stress on and attrition of India's water resources, future policies for sustainable rice cultivation in India need to be carefully crafted.

Conventional rice cultivation requires approximately 1500 mm of water, although this may vary depending on the soil and agroclimatic factors. The actual amount of water applied by farmers varies because of the differences in water management knowledge, climatic conditions, source and availability of water, and soil characteristics. Despite the significance of estimating the water requirement for rice, estimation of farm-level water use is fraught with serious challenges, such as continuous flooding of rice fields, field-to-field irrigation in most of the surface irrigation command areas, and the unlined canal system that makes flow measurements difficult. In recent times, remote-sensing techniques have become some of the most economical methods for estimating actual farm-level water use over large areas (Bastiaanssen et al., 2003; Nayak, 2006). In crop production, water footprint analysis is an important indicator of water-use efficiency, and it is helpful to assess the relative performance of different rice production environments in terms of their water-use efficiency. The water footprint is defined as the quantity of water required to produce 1 kg of rice. Water footprint estimates from different studies reveal wide inter- and intraregional variations (Table 4.1). Estimates from Tamil Nadu and the IGP showed comparatively low intraregional variation in the water footprint of rice cultivation. However, in Andhra Pradesh, Adusumilli and Laxmi (2010) estimated a water footprint of 8851 L/kg of rice under conventional methods and 3554 L/kg of rice under the system of rice intensification (SRI), while Murthy et al. (2006) reported a water footprint of 1754 L/kg under conventional methods and 488 L/kg of rice under the SRI. A similar observation is made in West Bengal, where the water footprint varied from 2662 L/kg (Ghosh et al., 2014) to 8973 L/kg (Datta and Mahato, 2012) of rice. Better water control leads to better water-use efficiency and a lower water footprint. For example, the water footprint for nonbasmati rice was estimated to be 1850 L/kg of rice in groundwater-irrigated

TABLE 4.1 Estimates of the Water Footprint of Rice Production Under Different Rice Production Environments

S. no.	Study	Region	Water required (L) per kg of rice			Remarks
			Traditional method	SRI	Other	
1	Adusumilli and Laxmi (2010)	Andhra Pradesh	8851	3554		
2	Rao et al. (2013)	Andhra Pradesh	2183–2433	1492–1538		
3	Raju and Sreenivas (2008)	Andhra Pradesh		1221		
4	Murthy et al. (2006)	Andhra Pradesh	1754	488		72% savings in water under SRI
5	Duttarganvil et al. (2014)	Andhra Pradesh	2700–4400	1590–2812		
6	Suryavanshi et al. (2013)	IGP	3831	2809		
7	Singh (2013b)	IGP	2841	1873		
8	Saha et al. (2014)	IGP	3121	2355	2078	Aerobic rice
9	Jain et al. (2013)	IGP	3846	2632	2564	Modified SRI
10	Bhushan et al. (2007)	IGP	2325–2702	—	2222	Direct-seeded rice
11	Dass et al. (2016)	IGP	3195	2732		
12	Srivastava et al. (2015)	IGP (Punjab)	1846 (nonbasmati), 3557 (basmati)			Exclusively groundwater
13	Mahajan et al. (2009)	IGP (Punjab)	1470–2000			
14	Mahajan et al. (2011)	IGP (Punjab)	2500	—	2439	
15	Thakur et al. (2011)	Odisha	2778	1463		

(continued)

TABLE 4.1 Estimates of the Water Footprint of Rice Production under Different Rice Production Environments (cont.)

S. no.	Study	Region	Water required (L) per kg of rice			Remarks
			Traditional method	SRI	Other	
16	Thakur et al. (2013)	Odisha	3030	1754		
17	Mishra et al. (2012)	Odisha	1887–5000			
18	Mitra et al. (2014)	West Bengal	2778	1667		
19	Datta and Mahato (2012)	West Bengal	8973	3375		
20	Ghosh et al. (2014)	West Bengal	2662–3259	1834	1409	
21	Vijayakumar et al. (2006)	Tamil Nadu		1639–2024		
22	Satyanarayana et al. (2007)	Tamil Nadu	2272–2564	1149–1389		
23	Senthilkumar et al. (2012)	Tamil Nadu	1960–3333			
24	Geethalakshmi et al. (2011)	Tamil Nadu	2778	1724		
25	Kumar et al. (2010)	Multilocation trials in major rice-growing regions of India (2004–07)	2083	1471		
26	Kumar et al. (2013a)	Multilocation trials in major rice-growing regions of India (2008–10)	2277	1368–1736		Wet season–kharif (2008–09)
		Multilocation trials in major rice-growing regions of India (2008–10)	3776	2531–3099		Dry season–kharif (2008–09)
		Multilocation trials in major rice-growing regions of India (2008–10)	2779	1414–1520		Dry season–rabi (2008–09)
		Multilocation trials in major rice-growing regions of India (2008–10)	2507	1263–1362		Dry season–rabi (2009–10)
27	Singh et al. (2010)	India	2222			
28	Chapagain and Hoekstra (2011)	India	2020			

rice in Punjab (Srivastava et al., 2015), which is lower than the water footprint estimated from other studies in the IGP. Groundwater is under the control of farmers; therefore, this demand-based water application facilitates better control over water application than canal water, which is supply-based. The intraregional differences between studies could be because rice is cultivated under a wide range of agroclimatic conditions and water management regimes. Rice grown under the SRI (with alternate wetting and drying) has a lower water footprint than continuously flooded rice. Similarly, direct-seeded rice requires less water than transplanted rice. Canal irrigation systems, under which rice is a predominant crop, are characterized by wide differences in water conveyance, water distribution, and control systems: completely concrete-lined versus unlined canal networks, continuous versus intermittent release of water in the canals, and canal systems with varying degrees of supplementation of well water. Differences in the method and precision of water measurement also play an important role in determining differences in water-use efficiency and water footprint estimates.

Regional/state averages (Table 4.2) reveal that West Bengal had the highest water footprint of 4904 L/kg of rice, followed by Andhra Pradesh with 4116 L/kg. Both of these estimates are much higher than those of other states and the national average, probably due to the very high estimates (close to 9000 L/kg) from one study. Estimates for Tamil Nadu and the multilocational estimates across India are very close to each other, at 2600 L/kg of rice. Chapagain and Hoekstra (2011) estimated a water footprint of 2020 L/kg of rice in India, which is second only to Pakistan among the 13 major rice-producing countries addressed in their study. This implies a huge need for water savings in

TABLE 4.2 Summary of the Water Footprint of Rice in Different Regions/States Across India

State/region	Water required in L/kg of rice		Water savings under SRI over traditional method (%)
	Traditional method	SRI	
Andhra Pradesh	4116	1796	56.37
Tamil Nadu	2614	1608	38.49
IGP	2825	2480	12.21
Odisha	3084	1609	47.83
West Bengal	4904	2292	53.26
India (multilocation trials)	2607	1724	33.87
National average	3150	1953	38.80

IGP, Indo-Gangetic Plains.
Average values calculated from Table 4.1.

rice cultivation in India so as to put India's rice water footprint on a par with that of the other major rice-growing economies of the world. The average water footprint of India's rice production is estimated to be 3150 L/kg of rice under conventional rice cultivation and 1953 L/kg under SRI cultivation (Table 4.2). The differences in water footprint estimates across locations could be attributed to differences in climatic and soil conditions, sources and availability of water, and differences in the methods of estimating water used. Since the water footprint is estimated in terms of water consumed per unit of output, yield variability due to other factors, such as varietal differences, technology and input use, etc., would also be important in affecting water footprint estimates. In recent years, the introduction of new aerobic rice technology and the SRI have made it possible to obtain higher yields alongside 30%–40% savings in water. Water savings under the SRI vary from 12% in the IGP to 56% in Andhra Pradesh, with a national average of 39% (Table 4.2).

Strategies to Reduce the Water Footprint of Rice

Several strategies and water management techniques are available to reduce the water footprint of rice cultivation (Table 4.3). Laser leveling of rice fields has the largest impact on water savings with more than a 22% reduction in water use compared with the baseline. Timed irrigation with a tensiometer and the SRI method have the potential to save water by one-fifth of the baseline water requirement. Late planting of rice has also been recommended to reduce the water requirement. In fact, in 2009, both the Punjab and Haryana state governments introduced legislative measures that prohibited transplanting of paddy before the second week of June, with the aim of reducing the irrigation water requirement. To ensure compliance with the law, agricultural electricity supply

TABLE 4.3 Water-Saving Potential of Individual Interventions in Rice Cultivation (Water Savings are not Cumulative)

Proposed intervention	Reduction in water requirement (mm)
Laser leveling	410 (22.28)
Delayed transplanting by 1 month	210 (11.41)
Timed irrigation with tensiometer	370 (20.11)
Short-duration rice varieties	300 (16.30)
System of rice intensification	370 (20.11)
Baseline water requirement for rice = 1840 mm	

Note: Numbers in parentheses are percentage savings in water compared with the baseline water requirement.
Source: World Bank, 2010. Deep wells and prudence: towards pragmatic action for addressing groundwater overexploitation in India. Washington, DC: The World Bank.

and procurement dates were also deferred in accordance with the late planting of rice. Delayed planting of rice could reduce consumptive water demand by at least 200 mm per crop (Table 4.3) without compromising rice yields (World Bank, 2010; Humphreys et al., 2010). Chahal et al. (2007) provided experimental evidence of the water-saving impact of delayed rice planting in Punjab, where the mean water requirement to produce 1 kg of rice declined from 5495 to 3226 L with a steady increase in yield from 5745 to 7428 kg/ha when the planting date was postponed from May 1 to July 1. Scarcity does not often lead to efficiency, especially for groundwater use. For example, Varghese et al. (2013) found that the increase in groundwater scarcity on southern Indian rice farms has led to inefficient use of water. Hence, external coercion such as delayed planting, conjunctive use, and supply augmentation assumes greater significance. Furthermore, direct-sown rice under zero tillage and best management practices with residue retention have the potential to save water by 30%–50%, with equal or greater yield than conventional transplanted rice under farmers management practices (Gathala et al., 2013). Water control measures that could be implemented in canal systems such as the turn system of irrigation/intermittent release of water, improving water conveyance efficiency at the system level, and volumetric pricing of water rather than area-based water tariff could also play an important role in economizing water use in rice cultivation. Enabling farmers to have better control over water management through demand-based application instead of the current practice of supply-based water application will facilitate the economical use of water, thus reducing the water footprint. Synchronized planting, community nurseries, and zero-tillage practices are some of the options to reduce the water footprint through improved water control measures.

GREENHOUSE GAS EMISSIONS FROM RICE FIELDS

Agriculture is both a source and a sink of GHGs. Rice production is not only affected by global climate change but it also contributes to global warming through the release of GHGs such as CH_4, CO_2, and N_2O (Matthews and Wassmann, 2003). GHG emissions from rice fields are a major challenge since rice cultivation is closely associated with food security in the most populous countries, such as India. In the following section, we provide an overview of the various estimates of GHG emissions from rice fields from a wide range of studies. In 2007, agricultural GHG emissions in India were 17% of total GHG emissions, and N_2O was estimated to be responsible for 13% of all Indian agricultural GHG emissions (MoEF, 2010). Out of the total agricultural emissions (334 million tons CO_2 equivalent), rice cultivation accounted for more than one-fifth of total GHG emissions (20.9%), while on-field burning of crop residues accounted for only 2% of total agricultural emissions (INCCA, 2010). Wide regional variations in CH_4 emissions were noted, likely due to differences in the methods of rice cultivation, specifically the type of water management regime.

Among the major rice-growing states, Uttar Pradesh (which has the largest area under rice) had the lowest CH_4 emissions with only 4 kg of CH_4/ha, while Andhra Pradesh, Bihar, and West Bengal recorded the highest CH_4 emissions, amounting to ~100 kg/ha (Bhatia et al., 2013). The reason for the high CH_4 emission rates is because these states are dominated by continuously flooded rice cultivation.

Extrapolating emission factors for the United States and Europe, Ahuja (1990) made an early estimate of annual CH_4 emissions from Indian rice paddies of 37.5 million tons. With sustained and systematic indigenous research, CH_4 emission estimates have been rationalized. Multilocational experiments across the major rice-growing regions in India revised the estimate to 4 million tons of CH_4 from Indian rice fields during the base year 1991. Emission estimates vary widely with the methodology adopted and assumptions made about the importance of different factors affecting CH_4 emissions. Based on field measurements made up to 1990, CH_4 emissions of ~3.0 million t/year were estimated (Mitra, 1992; Parashar et al., 1991). Bhatia et al. (2004) estimated the CH_4 emissions from Indian rice fields to be ~2.9 million t/year during 1994–95 with Andhra Pradesh, West Bengal, and Tamil Nadu topping the list of major CH_4 producers. Parashar et al. (1996) further revised the emissions to be 4.0 million t/year with a range of 2.7–5.4 million t/year. Matthews et al. (2000) used the MERES model to simulate CH_4 emissions from rice paddies in India and estimated a value of 2.14 million t/year. Gupta et al. (2002), using average emission factors for all paddy water regimes, estimated CH_4 emissions to be 5.0 million t/year, whereas Yan et al. (2003), using region-specific emission factors, estimated India's CH_4 emissions to be 5.9 million t/year.

Table 4.4 provides a comprehensive summary of GHG emissions from Indian rice fields. Nationally, total CH_4 emissions from rice fields ranged from 1.10 to 4.09 million tons under continuously flooded rice. Of the seven studies that estimated national emissions, only one study (Pathak et al., 2005) reported emissions for rice cultivation with midseason drainage, which was 0.125 million tons per annum. Of the 18 studies that addressed regional CH_4 emissions, most focused on the IGP and only one was from southern India. A wide variation in CH_4 emissions was observed across these 18 studies, likely due to the differences in production systems as well as the interventions made by the researchers. Under the continuously flooded conventional method of rice cultivation, CH_4 emissions varied from 28.10 kg/ha (Pathak et al., 2003a) to as high as 388.28 kg/ha (Tyagi et al., 2010), with both figures obtained from field measurements in the IGP. The lowest emissions of CH_4 observed under conventional rice cultivation with intermittent drying were 14.6 kg/ha in the IGP (Pathak et al., 2003a), whereas the highest CH_4 emissions with intermittent drying were 229.35 kg/ha (Tyagi et al., 2010).

Very few studies estimated emissions of N_2O and CO_2 from rice fields. Total N_2O emissions from rice fields in the whole of India ranged from ~0.02 million tons (Bhatia et al., 2012) to 0.19 million tons (Bhatia et al., 2007).

TABLE 4.4 Summary of Various Studies on Estimates of GHG Emissions From Indian Rice Fields

S. no.	Study	Region/situation	Unit	CH_4	N_2O	CO_2 eq.	GWP	Remarks on methodology
1	Bhatia et al. (2004)	India (42.25 million ha)	Million tons	2.900	0.08			For the years 1979–2006
2	Gupta et al. (2008)	India and statewise IPCC 1996	Million tons	3.62 ± 1.00 to 4.09 ± 1.19				
3	Bhatia et al. (2013)	Multiple aeration (9.91 million ha)	Million tons	0.149				
		Single aeration (8.192 million ha)	Million tons	0.553				
		Continuous flooding (6.747 million ha)		1.140				
		Deepwater (1.35 million ha)		0.257				
		Rainfed (drought prone) (9.02 million ha)		0.574				
		Rainfed (flood prone) (3.686 million ha)		0.700				
		India total emissions for 43.86 million ha		3.370	0.138		148.94	
4	Pathak et al. (2005)	India (42.25 million ha), continuous flooding	Million tons	1.100	0.045	21.16–60.96	130.93–272.83	
		India (42.25 million ha), midseason drainage	Million tons	0.125	0.055	16.66–48.80	91.73–211.80	

(continued)

TABLE 4.4 Summary of Various Studies on Estimates of GHG Emissions from Indian Rice Fields (*cont.*)

S. no.	Study	Region/situation	Unit	CH_4	N_2O	CO_2 eq.	GWP	Remarks on methodology
5	Bhatia et al. (2012)	India (42.21 million ha)	Million tons	2.070	0.02	72.9	88.5	Infocrop model with validation sites in Pantnagar, Delhi, Karnal, 24 South Parganas, Cuttack, Chennai, and Trivandrum
6	Bhatia et al. (2007)	India (42.21 million ha)	Million tons	2.070	0.19	72.90		Infocrop simulation model
7	Matthews et al. (2000)	India	Million tons	2.140				Crop simulation model and GIS technique
8	Gupta et al. (2002)	Rainfed (flood prone)	kg/ha	190.0 ± 60.0				All estimates are averages over the 1991–99 period from studies conducted in major rice-growing states of India, including West Bengal, Odisha, Tamil Nadu, Kerala, Haryana, Uttar Pradesh, Andhra Pradesh, and Delhi. Total emissions for India are the average of emissions across water regimes.
		Rainfed (drought prone)		91.2				
		Continuously flooded		292.0				
		Intermittently flooded, single aeration		91.2				
		Intermittently flooded, multiple aeration		36.0				
		Deepwater		190.0 ± 60.0				
		India total emissions (for 42.23 million ha of rice area)	Million tons	3.61 ± 1.40				

#	Reference	Treatment					Notes
9	Garg et al. (2001)	Continuously flooded	kg/ha	251.0 ± 84.0			Quoted from Parashar et al. (1997)
		Intermittently flooded, single aeration		60.0 ± 15.0			
		Intermittently flooded, multiple aeration		13.6 ± 5.7			
10	Suryavanshi et al. (2013)	Continuously flooded	kg/ha	32.33			IGP, field measurements
		SRI		19.93			
		Double transplanted		29.30			
11	Bhatia et al. (2005)	100% chemical NPK	kg/ha	35.90	0.569	1,578	IGP, field measurements
		100% organic source		57.10	0.537	2,081	
12	Tirol-Padre et al. (2016)	Conventional tillage	kg/ha	12.8 ± 4.4	1.4 ± 0.7		IGP, metaanalysis means from field measurements
		Zero tillage, direct seeded	kg/ha	5.6 ± 4.7			
13	Pathak et al. (2003a)	Continuously flooded (urea)	kg/ha	28.1 ± 1.8			IGP, field measurements
		Intermittent drying (urea)		14.6 ± 3.5			
		Continuously flooded (urea + FYM)		45.4 ± 0.4			
		Intermittent drying (urea + FYM)		27.7 ± 0.5			
14	Pathak and Wassmann (2007)	With urea + FYM and without nitrification inhibitor	kg/ha	92.0	0.33		IGP (Haryana), field measurements
		With urea only and with nitrification inhibitor		47.0	0.27		

(continued)

TABLE 4.4 Summary of Various Studies on Estimates of GHG Emissions from Indian Rice Fields (cont.)

S. no.	Study	Region/situation	Unit	CH$_4$	N$_2$O	CO$_2$ eq.	GWP	Remarks on methodology
15	Malla et al. (2005)	Urea only	kg/ha	27.0	0.76		938	IGP, field measurements
		Urea with nitrification inhibitor		23.4–28.4	0.50–0.68		754–841	
16	Ghosh et al. (2003)	Urea only	kg/ha	37.3 ± 5.1	0.17 ± 0.008			IGP, field measurements
		Urea with nitrification inhibitor (DCD)		29.0 ± 2.7	0.08 ± 0.011			
		Ammonium sulfate		33.0 ± 2.1	0.15 ± 0.008			
		Ammonium sulfate with nitrification inhibitor		28.7 ± 3.4	0.082 ± 0.0067			
17	Tyagi et al. (2010)	Continuous flooding	kg/ha	388.28			8,153.88	IGP, field measurements
		Tillering-stage drainage		352.92			7,411.32	
		Midseason drainage		245.77			5,161.07	
		Multiple drainage		229.35			4,816.25	
18	Pathak et al. (2002)	Continuous flooding	kg/ha		0.534			IGP, field measurements
		Intermittent drying			0.623			
		Continuous flooding (with nitrification inhibitor, DCD)			0.483			
		Intermittent drying (with nitrification inhibitor, DCD)			0.540			

#	Reference	Treatment	Units	Value	Location/Notes
19	Majumdar et al. (2000)	Urea only		0.060	IGP, field measurements
		Urea + DCD		0.050	
		Neem-coated urea		0.053	
20	Jain et al. (2013)	Conventional	kg/ha	22.59	IGP, field measurements
		SRI		8.81	
		Modified SRI		8.16	
21	Khosa et al. (2011)	Continuous flooding	kg/ha	55.7	IGP (Punjab), field measurements
		Intermittent flooding		34.5	
22	Adhya et al. (2000)	Urea N	kg/ha	42.30	Eastern India (Odisha), field measurements, 1996 wet season
		Sesbania + urea N		131.97	
		Compost + urea N		65.44	
		Azolla + urea N		67.71	
		Continuously flooded		18.61	Eastern India (Odisha), field measurements, 1997 dry season
		Continuously flooded + rice straw		36.18	
		Alternately flooded		15.73	
		Alternately flooded + rice straw		27.14	
23	Nayak et al. (2006)	Wet season, early planting	kg/ha	409.0	Eastern India (Odisha), field measurements
		Wet season, late planting		309.0	

(continued)

TABLE 4.4 Summary of Various Studies on Estimates of GHG Emissions from Indian Rice Fields (*cont.*)

S. no.	Study	Region/situation	Unit	CH_4	N_2O	CO_2 eq.	GWP	Remarks on methodology
		Dry season, early planting		431.0				
		Dry season, late planting		329.0				
24	Bhattacharyya et al. (2012)	Control	kg/ha	69.7	0.23	1,100.3	5,862	Eastern India (Odisha), field measurements
		Urea		92.6	1.00	1,447.7	8,084	
		Urea + rice straw		115.4	0.84	1,680.6	9,418	
		Rice straw + green manure		122.7	0.72	1,858.5	10,188	
25	Datta et al. (2013)	N fertilizer (dry and wet seasons)	kg/ha	80.27 to 451.27				Eastern India (Odisha), field measurements
		NPK fertilizer (dry and wet seasons)		34.60–233.66				
26	Khan and Zargar (2013)		kg/ha	39.32				Kashmir, field measurements
27	Rajkishore et al. (2013)	Conventional	kg/ha	44.6–55.5				Tamil Nadu, field measurements

IGP, Indo-Gangetic Plains.

Regional-level studies have estimated N_2O emissions to be less than 1 kg/ha. Through metaanalysis, Tirol-Padre et al. (2016) estimated N_2O emissions to be 0.5% of total nitrogen applied for transplanted rice. Their field experiments in Haryana estimated emissions to be 1.66–4.04 kg/ha. In spite of their small quantity, N_2O emissions had a higher global warming potential (GWP) than CH_4 under both transplanted puddled rice and direct-sown rice with zero tillage. While the GWP of CH_4 emissions has more than halved from 320 kg of CO_2 equivalent/ha under transplanted rice to 140 kg of CO_2 equivalent/ha under direct-sown rice with zero tillage, the GWP of N_2O emissions stood steady at 375 kg of CO_2 equivalent/ha under both methods of rice cultivation. Only two studies estimated total CO_2 emissions from rice fields in India, with reported total emissions of 16.66 million tons under midseason drainage and 72.90 million tons under continuous flooding. In accordance with the wide variation in the estimates of GHG emissions from rice fields, GWP also showed a huge variation from 88.5 million tons (Bhatia et al., 2012) to 272.83 million tons of CO_2 equivalent (Pathak et al., 2005). The estimated GWP of rice in Odisha varied from 5,862 kg in control plots to 10,188 kg of CO_2 equivalent/ha in rice fields applied with rice straw and green manure (Bhattacharyya et al., 2012). The highest CH_4 emissions were from irrigated, continuously flooded rice (34% of total methane emissions from rice fields), followed by rainfed, flood-prone rice (21%). Rainfed drought-prone, single-aeration, deepwater, and irrigated multiple-aeration rice ecosystems contributed 17%, 16%, 8%, and 4% of CH_4, respectively (Bhatia et al., 2013). Substituting 50% of N with FYM increased CH_4 emissions in the rice–wheat system by 172% compared with the application of the entire amount of N through urea. Incorporation of a combination of rice straw and green manure resulted in the maximum GWP, whereas the application of rice straw with inorganic fertilizer was the most effective strategy for the sequestration of soil organic carbon (1.39 t/ha), with higher grain yield in the rice–rice cropping system (Bhattacharyya et al., 2012). The wide variations in GHG emission estimates across studies could be due to the geographic agroclimatic differences among the experimental locations; the type of experiments; methods of measurement (modeling versus field measurements); the various types of interventions introduced into the experiments, such as quantities of manure, fertilizer, and water; single or multiple aeration; conventional versus zero-tillage cultivation; etc. While some studies followed a cropping systems approach, others measured emissions from monocrop rice.

GHG Emissions and Environmental Pollution from Burning of Rice Straw

Wide variations exist within studies that estimated the quantity of crop residues burned on fields in India. The total amount of residue generated in 2008–09 was 620 million tons, of which ~15.9% was burned on-farm (rice straw contributed 40% of the total residue burned). On-farm burning of 98.4 million tons of crop residue led to the emission of 141.15 million tons of CO_2, which alone accounted for 91.6% of total emissions (Jain et al., 2014). Burning of crop residues emitted 0.25 million tons of CH_4 and 0.007 million tons of N_2O in 2007, of which rice straw constituted 39%. Out of 97.12 million tons of rice straw produced in India, Gadde et al. (2009a) estimated that 13.92 million tons were burned, whereas Streets et al. (2003) estimated the quantity of rice straw subjected to open field burning to be 84 million tons. Intensive rice–wheat cropping systems implemented in Punjab, Haryana, and parts of Uttar Pradesh are the main contributors of total rice straw undergoing open field burning in India. Open field burning of rice straw in India has resulted in the emission of 16.26 million tons of CO_2, 13,359 tons of CH_4, and 779 tons of N_2O (Gadde et al., 2009a). The total GHG emissions potential from this burning has been estimated to be 566,165 tons of CO_2 equivalent per annum, assuming 100% collection efficiency of straw, although the GHG emission contribution from open field burning of rice straw accounts for only 0.05% of total GHG emissions from India (Gadde et al., 2009b). Large-scale burning of rice residues in Punjab, Haryana, and western Uttar Pradesh is a matter of serious concern not only for GHG emissions but also in terms of pollution, hazard to health, and nutrient loss (Pathak et al., 2006). In Punjab, rice straw burning produced 0.02 million tons of NO_2 and 3000 tons of CH_4 during October 2005 (Badarinath and Chand Kiran, 2006). The use of combine harvesters is the single most dominant factor that promotes rice residue burning while innovative farmers with access to technical knowledge are less likely to burn the rice straw (Gupta, 2012). The use of balers for bundling rice residue and alternative usages for rice straw, such as for biochar and bioenergy production and as animal feed and organic manure, are some of the options that will reduce the residue burning problem. Sowing wheat in rice stubble using a Happy Seeder, which facilitates the use of rice straw as mulch, is another option to reduce residue burning. However, the low collection efficiency of rice straw due to its wide geographic distribution, low thermal energy, and high cost of collection are the major hindrances for turning rice straw into alternative energy generation.

Strategies to Mitigate GHG Emissions From Agriculture

Increasing rice output per unit of land and water while lowering GHG emissions is a major challenge for both technocrats and policymakers. Technological and management interventions should take into consideration the trade-off between increasing rice yield and reducing GHG emissions. It has been shown that increased use of water and nitrogen can increase rice yields but with a concomitant increase in GHGs (Bhatia et al., 2010). Hence, efforts to mitige GHGs in agriculture should carefully focus on both scientific and policy solutions. Emphasis should be placed on improving efficiency in the use of nutrients, water, and energy to achieve reduced GHG emissions. Similarly, crop and water management practices such as adjusting the time of transplanting, scheduling of input application, direct seeding of rice, use of urea supergranules, alternate wetting and drying, soil amendments, and proper management of organic inputs can be manipulated to mitigate GHG emissions from agriculture. The addition of organic manure and green manure has increased methane emissions (Bhattacharyya et al., 2012), whereas the SRI method of rice cultivation has significantly reduced methane emissions (Jain et al., 2013; Rajkishore et al., 2013; Suryavanshi et al., 2013). Application of nitrification inhibitors such as dicyandiamide (DCD) has resulted in a significant reduction in GHG emissions (Ghosh et al., 2003; Pathak et al., 2002; Pathak and Wassmann, 2007). Compared with conventional transplanted rice, direct-sown rice with zero tillage coupled with best management practices has shown huge potential to save water and reduce GHG emissions. Furthermore, water regimes play an important role in determining the quantity of CH_4 and N_2O emissions (Gathala et al., 2013; Tirol-Padre et al., 2016). Multiple drainage (intermittent flooding) of rice fields has resulted in a significant reduction in CH_4 emissions (Gupta et al., 2002; Pathak et al., 2002, 2003a, 2005).

Bringing ~25% of rice area under the SRI or aerobic cultivation would reduce CH_4 emissions by ~0.062 million tons (12%) in northeastern India (Das et al., 2012). Application of DCD along with urea reduced CH_4 emissions to 70% of those seen with urea treatment alone. Although DCD seems to have the potential to mitigate CH_4 emissions, its cost is a bottleneck to resource-poor farmers, as there was no yield advantage of applying DCD along with urea. Intermittent wetting and drying of rice soil is a potential technology for decreasing methane emissions, provided that the nitrogen is applied within one day of irrigation so that nitrous oxide emissions can be mitigated. Therefore, water management and nutrient application strategies need to be carefully implemented to reduce GHG emissions without compromising yield (Pathak et al., 2003a). Late planting of rice in eastern India could reduce methane emissions without significantly compromising rice yield, and this is especially suitable for rainfed rice (Nayak et al., 2006). The selection of suitable rice cultivars, application of sulfate-containing fertilizers, and the use of fermented organic manure with a low C/N ratio are also recommended for reducing CH_4 emissions from rice fields (Singh et al., 2003). Water management and organic amendments are found to be the key factors affecting CH_4 emissions from rice fields, while climatic variability is the least important. Apart from mid-season drainage and organic amendments in the rice-growing season, draining the rice field in the preceding fallow season, incorporating rice straw in the previous season, or using composting could mitigate methane emissions (Yan et al., 2005). Experimental results from Haryana reveal that shifting from the conventional rice cultivation method to zero-tillage direct-sown rice could reduce methane emissions by more than 55% and total GWP by 23%. The quantity and timing of the application of nitrogeneous fertilizers, rather than methods of cultivation, had a significant effect on N_2O emissions (Tirol-Padre et al., 2016).

IMPACT OF CLIMATE CHANGE ON RICE

Rice is both a heat- and water-loving plant, requiring high temperatures and an adequate moisture supply. Recent studies on crop responses to climate change reveal that a doubling of CO_2 causes an increase in biomass and grain yield of C_3 crops; however, rising temperatures might reduce yield because of spikelet sterility (Kim et al., 1996; Oh-e et al., 2004). Based on data from 73 meteorological stations in India, mean temperature has increased 0.4°C over the past 100 years (Hingane et al., 1985). The Intergovernmental Panel on Climate Change (IPCC) has projected that rainfall in India will increase 10%–12% by the end of the 21st century with more frequent and heavy rainfall days, while the mean annual temperature will rise by 3–6°C (IPCC, 2014). Climate change reduces rice yield through a decrease in crop duration, spikelet sterility, grain number, and grain-filling duration, while rainfall affects rice yields through changes in moisture availability, waterlogging, and occasional physical damage to the crop through cyclones, storms, etc. Pests, disease outbreaks, and weed infestation are other mechanisms through which both precipitation and temperature affect rice yields. Increased CO_2 concentration in the atmosphere causes a CO_2 fertilization effect on crops leading to an increase in crop yields. Temperature is found to be a risk-increasing input that increases the variability of production of major crops, including rice (Ranganathan, 2009). In the following sections, we provide a broad overview of the impact of climate change on rice yield across India, generated by both crop simulation and econometric models, each of which covers a number of studies.

Impact of Climate Change on Rice Yield: Crop Simulation Models

In an early study, Mohandass et al. (1995) used the ORYZA1 model to simulate Indian rice production under current and future climates. They predicted an increase in rice production under the general circulation model scenarios used. This was mainly due to an increase in the yield of main-season crops for which the CO_2 fertilization effect will more than compensate for yield loss due to increased temperatures. Although large decreases were predicted for many second-season crops (due to the high temperatures encountered), the relatively low contribution of second-season rice to total rice produced meant that the overall effect on rice production was small. Through crop simulation models (CERES-Rice and ORYZA1N), it was found that an increase in temperature of 1–2°C, without a concomitant increase in CO_2, resulted in a 3%–17% decrease in grain yield (Aggarwal and Mall, 2002). In general, as temperature increased, rice yields in eastern and western India were lowly affected, rice yields in northern India were moderately affected, and rice yields in southern India were severely affected. Grain yield increased in all regions as CO_2 concentration increased, with a doubling of CO_2 concentration resulting in a 12%–21% increase in yield. The beneficial effect of 450 ppm of CO_2 was nullified by an increase in temperature of 1.9–2.0°C in northern and eastern regions and of 0.9–1.0°C in southern and western regions. A 28%–35% increase in rice yield was obtained as atmospheric CO_2 concentration doubled.

Pathak et al. (2003b) found that the negative trends in solar radiation and an increase in minimum temperature are likely to result in a declining yield potential of rice in the IGP of India. A decrease in solar radiation by 1.7 MJ/m^2 per day resulted in a 27% reduction in rice yield. Using ORYZA1 and Infocrop rice simulation models at the current CO_2 level of 380 ppm, Krishnan et al. (2007) predicted an average yield loss of 7.20% and 6.66%, respectively, for every 1.8°C increase in temperature. However, increasing CO_2 concentration up to 700 ppm led to average yield increases of 30.73% and 56.37% in the ORYZA1 and Infocrop models, respectively. Zacharias et al. (2014) examined climate change scenarios from 2070 to 2100 for several key locations in India, analyzing their impacts on rice using InfoCrop-Rice and the regional climate model PRE-CIS. The PRECIS model projects an increase in temperature over most parts of India, especially in the IGP. An increase in rainfall would reduce rice yields by 24%–35% under different scenarios in IGP regions. Using the Infocrop simulation model along with climate projections from PRECIS, Kumar et al. (2013a,b) quantified climate change[d] impacts on irrigated and rainfed rice. Climate change is likely to reduce irrigated rice yields by ~4% in 2020 (2010–39), ~7% in 2050 (2040–69), and ~10% in 2080 (2070–99). Rainfed rice yields in India are likely to decrease by ~6% in the 2020 scenario; however, in the 2050 and 2080, scenarios rice yield is projected to decrease only marginally (<2.5%). In the states of Punjab, Haryana, and Rajasthan, the yield loss will be 6%–8% in 2020 and 15%–17% in 2050, whereas the estimated loss in rice yield in the central Indian states of Maharashtra and Madhya Pradesh will be 5%.

The impact of climate change [as projected by the Indian Network for Climate Change Assessment (INCCA, 2010)] revealed an approximate 10% loss in the yield of irrigated rice in the majority of the Indian coastal districts, with the exception of some coastal districts of Maharashtra and Odisha. For rainfed rice, yields are projected to increase up to 15% in the eastern coastal districts, but decrease by up to 20% on the west coast. Kumar et al. (2011) used Infocrop and PRECIS to assess the impact of climate change on major crops in ecologically sensitive areas, such as the Western Ghats, coastal districts, and northeastern states. These models revealed mixed impacts on rice yields. Under the climate change scenario, yields of irrigated rice will decline by 10% in the majority of coastal districts by 2030. However, in some coastal districts of Maharashtra, northern Andhra Pradesh, and Odisha, irrigated rice yields are projected to marginally increase (<5%). Rainfed rice yields are projected to increase by up to 15% in many of the districts on the east coast, but decline by up to 20% on the west coast. Estimates using the Decision Support System

---

d. The temperature and rainfall data from MIROC3.2.HI A1b and B1 for 2020, 2050, and 2080 scenarios, and the PRECIS A1b, A2, and B2 for 2020 and 2080 scenarios were used for the analysis.

for Agro-technology Transfer (DSSAT), with the climate outputs of regional climate model PRECIS, revealed a 356 kg/ha loss in rice yield in the Cauvery Delta region of Tamil Nadu by 2100 (Geethalakshmi et al., 2011). Climate change is expected to exacerbate current stresses on water resources and freshwater availability in Central, South, East, and Southeast Asia, particularly in large river basins. As rice production depends heavily on water availability, and irrigated lowlands account for 55% of the total area of harvested rice and produce 2 to 3 times the crop yield of rice grown under nonirrigated conditions (IRRI, 2002), the impact of climate change on water resources will have serious consequences for rice cultivation.

Climate Change Impact on Rice: Estimates From Econometric Models

Estimates from a panel regression model indicated that there will be a loss in Indian rice yield of 4.5%–9.0% by 2039 (Guiteras, 2009). Using a similar model, Pattanayak and Kumar (2013) estimated that India will lose an average of 4.4 million tons of rice annually, or experience a cumulative loss of 172 million tons over a 39-year period. Some studies have attempted to examine the impact of climate change, atmospheric brown clouds (ABCs), and GHG emissions on rice yields. For example, Auffhammer et al. (2006) estimated the impact of climate change and ABCs on rice harvests using time series data in agricultural and meteorological models. Adverse climate change due to ABCs and GHGs contributed to the slowdown in rice productivity growth that India has seen during the past 2 decades. The simultaneous reduction of ABCs and GHGs would have increased the mean annual rice harvest by 6.18% during 1966–84 and by 14.4% during 1985–98. Similarly, Burney and Ramanathan (2014) focused on the combined effects of climate change and the direct effects of short-lived climate pollutants, such as tropospheric ozone and black carbon, on rice yields in India from 1980 to 2010. The projected rice yields in 2010 were 5%–15% lower due to climate change and air pollution.

Auffhammer et al. (2012) used fixed effects regression models to estimate the impact of climate variables, and Monte Carlo simulations to examine the future impacts of climate change on *kharif* rice in India. The study confirmed the negative impact of climate variables on rice yield, with drought having a much greater effect than extreme rainfall. Rice yield would have been 1.7% lower on average due to climate change, especially because of the increased frequency of drought since 1960. Saravanakumar (2015) estimated the sensitivity of rice yield to climate change in the southern Indian state of Tamil Nadu using a panel regression model. These regression estimates were then used to identify yield sensitivities in the future based on projected climate scenarios using the Regional Climate Model (RegCM4). As temperature and rainfall increase, rice yield initially increases up to a threshold level, and then decreases. Projections suggest that there will be a decrease of 283 kg/ha of rice by 2100. This represents a 10% decline in rice yield by the end of the 21st century, relative to current yield. Birthal et al. (2014) found that an increase in maximum temperature had an adverse effect on rice yield but rainfall had a positive effect; however, this could not counterbalance the negative effect of temperature. The projections of climate change impacts toward 2100 have suggested that rice yield will be lower by 15% vis-à-vis current yield.

Table 4.5 provides a classified summary of climate change studies based on the magnitude of impact. These quantitative estimates still have uncertainties associated with them, largely due to uncertainties in climate change projections, future technology growth, availability of inputs (i.e., water for irrigation), changes in crop management, and genotype. These projections nevertheless provide a direction of likely change in crop productivity in future climate change scenarios. These results have important implications for the allocation of resources, including agricultural land, among the different varieties of rice and for devising appropriate climate zone-specific adaptation policies to reduce rice-yield variability and ensure food security in developing countries, such as India. The estimates of climate change impact on rice yields vary widely across regions because of interregional differences in current and projected climate scenario, solar radiation, CO_2 fertilization effect, and the irrigation regime under which rice is cultivated. The variations are also due to differences in the models (crop simulation versus econometric models) as well as the modeling approach. For example, different studies used different types of crop simulation models (ORYZA1, CERES-Rice, Infocrop, and DSSAT with PRECIS), and econometric models rely on both panel data and cross section data, which are likely to produce different sets of results. Although the inputs for crop simulation models mostly come from experimental data, econometric models rely heavily on farm survey data. Hence, the projected yield loss in rice shows wide variation across regions as well as across studies. Though a few studies project the yield loss to surpass 30%, the majority of the studies predict a yield loss of 10%–20%. IRRI's research estimated that, by 2050, yield loss in rice would be 10%–15% and rice prices would increase by 32%–37% (http://irri.org/news/hot-topics/rice-and-climate-change). However, from the perspective of evolving technological changes, specific climate change adaptation technologies in the form of varieties tolerant of biophysical stresses, and short-duration varieties suitable for direct sowing, determining the exact magnitude of climate change impact is fraught with serious challenges. Nonetheless, these estimates provide a framework for prioritizing research and development activities on adaptation strategies across different rice production environments. Research on adaptation to climate change is already under way through the ClimaRice program and the National Initiative for Climate-Resilient Agriculture. Efforts are under way to develop C_4 rice that will have greater photosynthetic efficiency. Proper seed and seedbed management, direct seeding, and the evolution of healthier, taller, and more flood-tolerant rice varieties are some of the options for mitigating flood damage.

TABLE 4.5 Summary of Estimates of Climate Change Impact on Rice in India

Range of impact (yield reduction)	Study	Region	Impact
≤10%	Saseendran et al. (1999)	Kerala	For every 1°C increment in temperature, the decline in yield is ~6%
	Guiteras (2009)	India	4.5%–9.0% reduction in rice yield by 2039
	Kumar et al. (2011)	Western Ghats and northeastern states of India	Yield of irrigated rice will decline by up to 10% by 2030 in the majority of coastal districts. In some coastal districts of Maharashtra, northern Andhra Pradesh, and Odisha, irrigated rice yields are projected to marginally increase (<5%)
	Kumar et al. (2013b)	India	Irrigated rice yields to decrease by ~4% in 2020 (2010–39), by ~7% in 2050 (2040–69), and by ~10% in 2080 (2070–99) in different parts of India
	Srivastava et al. (2009)	Kerala and Tamil Nadu	In Kerala and Tamil Nadu, a continual decrease in yield was observed at higher temperatures, with higher yield reduction in Tamil Nadu
	INCCA (2010)	Coastal districts of India	Loss in yields of irrigated rice of about 10% in a majority of the coastal districts of India
	Saravanakumar (2015)	Tamil Nadu	There will be a yield loss in rice of 283 kg/ha per decade (i.e., 10% of current yield) in 2100 due to an increase in temperature of 3.26°C and rainfall (9% increase in baseline rainfall)
10%–20%	Aggarwal and Mall (2002)	India	An increase in temperature by 1–2°C without any increase in CO_2 resulted in a 3%–17% decrease in grain yield in different regions of India
	Auffhammer et al. (2006)	India	Simultaneous reduction in ABCs and GHGs would have increased mean annual rice harvest by 6.18% during 1966–84 and 14.4% during 1985–98
	Krishnan et al. (2007)	Eastern India	An increase in temperature by +4.8°C would reduce rice yield from 7.63%–15.86% under different climate scenarios

	INCCA (2010)	Coastal districts of India	For rainfed rice, yields are projected to increase up to 15% in the districts of the east coast, but decline by up to 20% in the west coast
	Kumar et al. (2011)	Coastal districts of India	Rainfed rice yields are projected to increase up to 15% in many of the districts in the east coast, but decline by up to 20% in the west coast
	Geethalakshmi et al. (2011)	Tamil Nadu	The projected loss in rice yield will be 356 kg/ha (i.e., 13% of current yield) for rice variety ADT 43 over the Cauvery Delta Zone of Tamil Nadu by 2100
	Kumar et al. (2013a,b)	India	In northern India (Punjab, Haryana, and Rajasthan), the yield loss will be 15%–17% in 2050 scenarios
	Burney and Ramanathan (2014)	India	Rice yields in 2010 were up to 5%–15% lower because of climate change and air pollution
	Birthal et al. (2014)	India	The projections of climate impacts toward 2100 have suggested that rice yield will be lower by 15%
21%–30%	Pathak et al. (2003b)	IGP	Rice yield decreased by 27% due to decreased solar radiation by 1.7 MJ/m² per day
	Kavikumar and Parikh (1998)	India	Rice yields will fall by 15%–25% if the temperature rises by 4°C
> 30%	Zacharias et al. (2014)	IGP	An increase in temperature and rainfall would likely reduce rice yields from 24% to 35% under climate change scenarios
	Krishnan et al. (2007)	Eastern India	An increase in CO_2 concentration up to 700 ppm led to average yield increases of 30.73% by ORYZA1 and 56.37% by Infocrop rice models
	Pattanayak and Kumar (2013)	India	Annual average rice yield loss of 4.4 million t/year in India

CONCLUSIONS

Academics and policymakers recognize that rice ecosystems are facing serious and multidimensional challenges. The dwindling water table; pollution of soil and water ecosystems; pesticide residues in the atmosphere and food chains, and their harmful effects on beneficial flora and fauna and human health; GHG emissions from rice fields; and the negative impacts of climate change and ABCs on rice production, are some of the key challenges that will continue to harm intensive rice production ecosystems long into the future. This chapter has been a modest attempt to provide a broad-based review of studies on the water footprint of rice production, GHG emissions from rice paddies, and the effects of climate change on rice yield. These three seemingly distinct problems are intricately interwoven and hence their implications for rice ecosystems, rice production, and water resources are addressed through a review of a wide range of studies across India. Barring a few studies on climate change impact, most of the studies covered are field experiments and/or modeling exercises with attendant limitations for wider generalization. Nevertheless, they provide a broad picture of the impending crisis facing rice agriculture vis-à-vis climate change, water scarcity, and GHG emissions. Solutions to these problems are multidimensional, often characterized by synergies as well as by trade-offs. For example, organic manures could be eco-friendly, soil enriching, and healthy, but they make matters worse when it comes to GHG emissions. Similarly, alternate wetting and drying could save water and reduce methane emissions but at the same time it will increase N_2O emissions. A recent study by Sapkota et al. (2015) highlights the trade-offs involved in GHG emissions control, climate change adaptation, and enhanced productivity in the rice–wheat cropping system. This study, conducted in the IGP, found that although the zero-tillage based rice–wheat system with residue retention and precision nitrogen management reduced CH_4 emissions, this benefit is counterbalanced by higher N_2O production compared with conventional cultivation practices. Hence, the complexity of the problems encompassing GHG emissions, climate change adaptation, water footprint, and nutrient management in rice production needs intensive research and policy attention. Technological progress leading to yield improvement or input savings for a given level of yield will ensure a win-win situation of increased yield and profitability along with a reduced water and ecological footprint. Replacing the conventional practice of intensive tillage in rice–wheat systems with conservation agriculture and best management practices is found to be an effective strategy to enhance the yield and economic performance of intensive rice–wheat systems. Water productivity was also highest under conservation agriculture with best management practices (Ranjan et al., 2014). These technologies could also play an important role in mitigating the impact of climate change. Best management practices incorporating fertilizer management and integrated pest control also offer several indirect benefits, such as improved soil and water quality, improved human health, and reduced chemical residues in the food chain. Being common property resources, water resources and pest susceptibility to pesticides need to be carefully managed. From an intergenerational perspective, even soil fertility is a common property resource. However, the short-term profit-maximizing goal of farmers with excessive, irrational, and imbalanced use of chemicals is in conflict with the long-term societal goal of ensuring sustainable production systems with minimal impact on ecosystems and climate. Hence, policies and incentives, especially those affecting the environment and climate, should be carefully crafted with a long-term vision.

The impacts of climate change will not be felt evenly across all regions and may not all be negative. Climate change is likely to exhibit three types of impacts on rice: (1) regions that are adversely affected by climate change can gain in net productivity with adaptation, (2) regions that are adversely affected will still remain vulnerable despite adaptation gains, and (3) rainfed regions (with currently low rainfall) that are likely to gain due to an increase in rainfall can further benefit by adaptation. Regions falling in the vulnerable category, even after suggested adaptation to climate change, will require more intensive, specific, and innovative adaptation options. Adaptation to climate change can not only minimize negative impacts but can also substantially increase productivity. As a result, current varieties may be able to perform better under increased input and efficiency regimes, offsetting yield loss in the near future; however, in the long run, developing suitable varieties coupled with improved and efficient crop husbandry will become essential. For irrigated rice, genotypic and agronomic improvements will become crucial. For rainfed conditions, improved management such as zero tillage, direct-sown rice, and additional fertilizer with proper nutrient management strategies will be needed in the near future. Laser leveling and timed irrigation with a tensiometer are other important techniques for reducing the water requirement for rice.

Policies and programs that will promote balanced and precision use of fertilizers, enhanced water- and energy-use efficiency, and the adoption of water- and energy-saving technologies are necessary. A combination of residue management strategies such as the use of manure, organic mulch, and feedstock in energy generation and biochar production will help prevent GHG emissions that would otherwise be generated from residue burning. Direct-seeded rice with zero tillage and the SRI hold significant potential to reduce both the water footprint and GHG emissions. The application of rice straw as an organic amendment with inorganic fertilizer for rice has been found to be the best option for organic carbon sequestration in soils with higher rice yield (Bhattacharyya et al., 2012). Simple but effective policy measures, such as mandatory late planting of rice in Punjab, need to be evolved to protect ecosystems supporting rice cultivation and to lower GHG emissions from rice cultivation. Environmentally damaging input subsidies such as for electricity, water, and fertilizer; price distortions; and other direct/indirect support to rice agriculture need to be critically revisited so they can be replaced with suitable environment-friendly subsidies or cash support programs to save rice production from ecosystem crisis. Policy support for the adaptation of rice cultivation to the impacts of climate change, mitigation of GHG emissions, and water-saving technologies are the needs of the time, instead of conventional input-based subsidies, such as fertilizer and electricity subsidies. Incentives and subsidies that promote water savings, reduced use of chemical fertilizers and pesticides, and systems of cultivation that reduce GHG emissions are necessary. The seemingly isolated problems of GHG emissions, water footprint, and climate change impacts on rice are all closely connected to each other from the perspective of seeking solutions. Best management practices covering fertilizer and water management, reduced/zero tillage and modified transplanting techniques, and evolving new stress-tolerant varieties will help to overcome many of the critical problems facing rice cultivation in the future.

REFERENCES

Adhya, T.K., Bharati, K., Mohanty, S.R., Ramakrishnan, B., Rao, V.R., Sethunathan, N., Wassmann, R., 2000. Methane emission from rice fields at Cuttack, India. Nutr. Cycl. Agroecosys. 58, 95–105.

Adusumilli, R., Laxmi, S.B., 2010. Potential of the system of rice intensification for systemic improvement in rice production and water use: the case of Andhra Pradesh. India. Paddy Water Environ. 9 (1), 89–97.

Aggarwal, P.K., Mall, R.K., 2002. Climate change and rice yields in diverse agro-environments of India: effect of uncertainties in scenarios and crop models on impact assessment. Clim. Change 52 (3), 331–343.

Ahuja, D.R., 1990. Estimating Regional Anthropogenic Emissions of Greenhouse Gases. US-EPA Climate Change Technical Series, Washington, DC.

Ashok, K.R., Kennedy, J.S., 2004. Pesticide Use in Agriculture and its Environmental Implications. Final Report of the Research Project Submitted to the Indian Council of Agricultural Research (New Delhi). Department of Agricultural Economics, Tamil Nadu Agricultural University, Coimbatore.

Auffhammer, M., Ramanathan, V., Vincent, J.R., 2006. Integrated model shows that atmospheric brown clouds and greenhouse gases have reduced rice harvests in India. Proc. Natl. Acad. Sci. USA 103, 19668–19672.

Auffhammer, M., Ramanathan, V., Vincent, J.R., 2012. Climate change, the monsoon, and rice yield in India. Clim. Change 111, 411.

Badarinath, K.V.S., Chand Kiran, T.R., 2006. Agriculture crop residue burning in the Indo-Gangetic plains: a study using IRS-P6 AWiFS satellite data. Curr. Sci. 91 (8), 1085–1089.

Banerjee, H., Sanyal, S.K., n.d. Emerging soil pollution problems in rice and their amelioration. Rice Knowledge Management Portal (RKMP), Directorate of Rice Research, Rajendranagar, Hyderabad.

Bastiaanssen, W.G.M., Ahmed, M.-u.-D., Tahir, Z., 2003. Upscaling water productivity in irrigated agriculture using remote-sensing and GIS technologies. In: Kijne, J. et al. (Eds.), Water Productivity in Agriculture: Limits and Opportunities for Improvement. Comprehensive Assessment of Water Management in Agriculture. CABI Publishing in Association with the International Water Management Institute, Wallingford, UK.

Bhatia, A., Aggarwal, P.K., Pathak, H., 2007. Simulating greenhouse gas emissions from Indian rice fields using the InfoCrop Model. Int. Rice Res. Notes. 32, 38–40.

Bhatia, A., Pathak, H., Aggarwal, P.K., 2004. Inventory of methane and nitrous oxide emissions from agricultural soils of India and their global warming potential. Curr. Sci. 87 (3), 317–324.

Bhatia, A., Pathak, H., Jain, N., Singh, P.K., Singh, A.K., 2005. Global warming potential of manure amended soils under rice–wheat system in the Indo-Gangetic plains. Atmos. Environ. 39, 6976–6984.

Bhatia, A., Pathak, H., Aggarwal, P.K., Jain, N., 2010. Trade-off between productivity enhancement and global warming potential of rice and wheat in India. Nutr. Cycl. Agroecosys. 86 (3), 413–424.

Bhatia, A., Aggarwal, P.K., Jain, N., Pathak, H., 2012. Greenhouse gas emission from rice and wheat-growing areas in India: spatial analysis and upscaling. Greenh. Gases 2, 115–125.

Bhatia, A., Jain, N., Pathak, H., 2013. Methane and nitrous oxide emissions from Indian rice paddies, agricultural soils, and crop residue burning. Greenh. Gases 3, 196–211.

Bhattacharyya, P., Roy, K.S., Neogi, S., Adhya, T.K., Rao, K.S., Manna, M.C., 2012. Effects of rice straw and nitrogen fertilization on greenhouse gas emissions and carbon storage in tropical flooded soil planted with rice. Soil Till. Res. 124, 119–130.

Bhushan, L., Ladha, J.K., Gupta, R.K., Singh, S., Tirol-Padre, A., Saharawat, Y.S., Gathala, M., Pathak, H., 2007. Saving of water and labor in a rice–wheat system with no-tillage and direct seeding technologies. Agron. J. 99, 1288–1296.

Bijay-Singh, Sadana, U.S., Arora, B.R., 1991. Nitrate pollution of groundwater from nitrogen fertilizers and animal wastes in the Punjab, India. Indian J. Environ. Health 33, 57–67.

Birthal, P.S., Khan, T.M., Negi, D.S., Agarwal, S., 2014. Impact of climate change on yields of major food crops in India: implications for food security. Agric. Econ. Res. Rev. 27 (2), 145–155.

Burney, J., Ramanathan, V., 2014. Recent climate and air pollution impacts on Indian agriculture. Proc. Natl. Acad. Sci. USA 111 (46), 16319–16324.

Cassman, K.G., Pingali, P.L., 1993. Extrapolating trends from long-term experiments to farmers' fields: the case of irrigated rice systems in Asia. In: Barnett, V., Payne, R., Steiner, R. (Eds.), Agricultural Sustainability: Economic, Environmental, and Statistical Considerations. John Wiley and Sons Ltd., New York.

Cassman, K.G., De Datta, S.K., Olk, D.C., Alcantara, J., Samson, M., Descalsota, J., Dizon, M., 1995. Yield decline and the nitrogen economy of long-term experiments on continuous, irrigated rice systems in the tropics. In: Lal, R., Stewart, B.A. (Eds.), Soil Management: Experimental Basis for Sustainability and Environmental Quality. Lewis Publication, Boca Raton, FL, pp. 181–222.

CGWB (Central Groundwater Board), 2014. Dynamic groundwater resources of India (as of March 31, 2011), Ministry of Water Resources, River Development and Ganga Rejuvenation, Government of India.

Chahal, G.B.S., Sood, A., Jalota, S.K., Choudhury, B.U., Sharma, P.K., 2007. Yield, evapotranspiration, and water productivity of rice (*Oryza sativa* L.)–wheat (*Triticum aestivum* L.) system in Punjab (India) as influenced by transplanting date of rice and weather parameters. Agr. Water Manage. 88, 14–22.

Chandna, P., Khurana, M.L., Ladha, J.K., Punia, M., Mehla, R.S., Gupta, R., 2011. Spatial and seasonal distribution of nitrate-N in groundwater beneath the rice–wheat cropping system of India: a geospatial analysis. Environ. Monit. Assess. 178 (1), 545–562.

Chapagain, A.K., Hoekstra, A.Y., 2011. The blue, green, and grey water footprint of rice from production and consumption perspectives. Ecol. Econ. 70, 749–758.

Chhabra, A., Manjunath, K.R., Panigrahy, S., 2010. Non-point source pollution in Indian agriculture: estimation of nitrogen losses from rice crop using remote sensing and GIS. Int. J. Appl. Earth Obs. Geoinf. 12, 190–200.

Das, A., Patel, D.P., Ramkrushna, G.I., Munda, G.C., Ngachan, S.V., Choudhury, B.U., Mohapatra, K.P., Rajkhowa, D.J., Kumar, R., Panwar, A.S., 2012. Improved rice production technology: for resource conservation and climate resilience. Farmers' Guide. Extension Bulletin No. 78. ICAR Research Complex for NEH Region, Umiam, Meghalaya.

Dass, A., Chandra, S., Choudhary, A.K., Singh, G., Sudhishri, S., 2016. Influence of field re-ponding pattern and plant spacing on rice root–shoot characteristics, yield, and water productivity of two modern cultivars under SRI management in Indian Mollisols. Paddy Water Environ. 14 (1), 45–59.

Datta, D., Mahato, B.C., 2012. Productivity improvement and water saving under the system of rice intensification (SRI): a study in eastern India. Paper presented at the IWA Conference on Water, Climate, and Energy, Dublin, May 13–18, 2012.

Datta, A., Santra, S.C., Adhya, T.K., 2013. Effect of inorganic fertilizers (N, P, K) on methane emission from tropical rice field of India. Atmos. Environ. 66, 123–130.

Devi, P.I., 2007. Pesticide use in the rice bowl of Kerala: health costs and policy options. SANDEE Working Paper No. 20–07. South Asian Network for Development and Environmental Economics (SANDEE), Kathmandu, Nepal.

Duttarganvil, S., Tirupataiah, K., Reddy, K.Y., Sandhyrani, K., et al., 2014. Yield and water productivity of rice under different cultivation practices and irrigation regimes. Paper presented at the International Symposium on Integrated Water Resources Management (IWRM–2014), February 19–21, 2014, CWRDM, Kozhikode, Kerala, India.

Fishman, R.M., Siegfried, T., Raj, P., Modi, V., Lall, U., 2011. Overextraction from shallow bedrock versus deep alluvial aquifers: reliability versus sustainability considerations for India's groundwater irrigation. Water Resour. Res. 47, W00L05.

Gadde, B., Bonnet, S., Menke, C., Garivait, S., 2009a. Air pollutant emissions from rice straw open field burning in India, Thailand, and the Philippines. Environ. Pollut. 157, 1554–1558.

Gadde, B., Menke, C., Wassman, R., 2009b. Rice straw as a renewable energy source in India, Thailand, and the Philippines: overall potential and limitations for energy contribution and greenhouse gas mitigation. Biomass Bioenergy 33, 1532–1546.

Garg, A., Bhattacharya, S., Shukla, P.R., Dadhwal, V.K., 2001. Regional and sectoral assessment of greenhouse gas emissions in India, 2001. Atmos. Environ. 35, 2679–2695.

Gathala, M.K., Kumar, V., Sharma, P.C., Saharawat, Y.S., Jat, H.S., Singh, M., Kumar, A., Jat, M.L., Humphreys, E., Sharma, D.K., Sharma, S., Ladha, J.K., 2013. Optimizing intensive cereal-based cropping systems addressing current and future drivers of agricultural change in the northwestern Indo-Gangetic plains of India. Agric. Ecosyst. Environ. 177, 85–97.

Geethalakshmi, V., Lakshmanan, A., Rajalakshmi, D., Jagannathan, R., Gummidi Sridhar, Ramaraj, A.P., Bhuvaneswari, K., Gurusamy, L., Anbhazhagan, R., 2011. Climate change impact assessment and adaptation strategies to sustain rice production in Cauvery basin of Tamil Nadu. Curr. Sci. 101 (3), 342–347.

Ghosh, S., Majumdar, D., Jain, M.C., 2003. Methane and nitrous oxide emissions from an irrigated rice of North India. Chemosphere 51, 181–195.

Ghosh, R.K., Sentharagai, S., Shamurailatpam, D., 2014. SRI: a methodology for substantially raising rice productivity by using farmers' improved thinking and practice with farmers' available resources. J. Crop Weed 10 (2), 4–9.

Guiteras, R., 2009. The Impact of Climate Change on Indian Agriculture. Manuscript, Department of Economics. University of Maryland, College Park, Maryland, USA.

Gupta, R., 2012. Causes of emissions from agricultural residue burning in north-west India: evaluation of a technology policy response. SANDEE Working Paper No. 66–12. Kathmandu: South Asian Network for Development and Environmental Economics.

Gupta, P.K., Sharma, C., Bhattacharya, S., Mitra, A.P., 2002. Scientific basis for establishing country greenhouse gas estimates for rice-based agriculture: an Indian case study. Nutr. Cycl. Agroecosys. 64, 19–31.

Gupta, P.K., Gupta, V., Sharma, C., Das, S.N., Purkait, N., Adhya, T.K., Pathak, H., Ramesh, R., Baruah, K.K., Venkatratnam, L., Singh, G., Iyer, C.S.P., 2008. Development of methane emission factors for Indian paddy fields and estimation of national methane budget. Chemosphere 74 (4), 590–598.

Harris, R., 2006. Monitoring of neonicotinoid resistance in *Nilaparvata lugens* and subsequent management strategies in Asia Pacific. In: Proceedings of the International Workshop on Ecology and Management of Rice Planthoppers, May 16–19, 2006, Hangzhou, China, Zhejiang University.

Hegde, D.M., 1992. Cropping system research highlights. Coordinator's report, presented during XX Workshop on AICRP on Rice at the Tamil Nadu Agricultural University, July 6–8, Coimbatore.

Hingane, L.S., Rupakumar, K., Ramana Murthy, B.V., 1985. Long-term trends of surface air temperature in India. Int. J. Climatol. 5, 521–528.

Hira, G.S., 2009. Water management in northern states and the food security of India. J. Crop Improv. 23 (2), 136–157.

Humphreys, E., Kukal, S.S., Christen, E.W., Hira, G.S., Singh, B., Yadav, S., Sharma, R.K., 2010. Halting the groundwater decline in north-west India: which crop technologies will be winners? Adv. Agron. 109, 155–217.

INCCA (Indian Network for Climate Change Assessment), 2010. Climate change and India: A 4X4 assessment. A sectoral and regional analysis for 2030s. INCCA REPORT #2. Indian Network for Climate Change Assessment, Ministry of Environment & Forests, Government of India.

IPCC, 2014. Summary for Policymakers. In: Field, C.B. et al., (Eds.), Climate Change 2014 Impacts, Adaptation, Vulnerability. Part A: Global, Sectoral Aspects. Contribution of Working Group II to the Fifth Assessment Report of the Intergovernmental Panel on Climate Change. Cambridge University Press, Cambridge (UK), pp. 1–32.

IRRI (International Rice Research Institute), 2002. Rice Almanac: Source Book for the Most Important Economic Activity on Earth. CABI Publishing, Oxon, United Kingdom.

Jain, N., Dubey, R., Dubey, D.S., Singh, J., Khanna, M., Pathak, H., Bhatia, A., 2013. Mitigation of greenhouse gas emission with system of rice intensification in the Indo-Gangetic Plains. Paddy Water Environ. 12 (3), 355–363.

Jain, N., Bhatia, A., Pathak, H., 2014. Emission of air pollutants from crop residue burning in India. Aerosol Air Qual. Res. 14, 422–430.

Jin, S., Jansen, H.G.P., Muraoka, R., Yu, W., 2012. The impact of irrigation on agricultural productivity: evidence from India. Paper presented at the International Association of Agricultural Economists (IAAE) Triennial Conference, Foz do Iguaçu, August 18–24.

Katya, V., 2015. Overuse of water by Indian farmers threatens supply. State of the Planet. Earth Institute, Columbia University. Available from: http://blogs.ei.columbia.edu/2015/04/14/overuse-of-water-by-indian-farmers-threatens-supply/

Kavikumar, K.S., Parikh, J., 1998. Climate change impacts on Indian agriculture: the Ricardian approach. In : Dinar, A., Mendelsohn, R., Evenson, R., Parikh, J., Sanghi, A., Kumar, K., McKinsey, J., Lonergan, S. (Eds.), Measuring the Impact of Climate Change on Indian Agriculture. Technical Paper No. 402. Washington, DC: The World Bank.

Khan, M.A., Zargar, M.Y., 2013. Experimental quantification of greenhouse gas (CH_4) emissions from fertilizer amended rice field soils of Kashmir Himalayan Valley. J. Environ. Eng. Ecol. Sci.

Khosa, M.K., Sidhu, B.S., Benbi, D.K., 2011. Methane emission from rice fields in relation to management of irrigation water. J. Environ. Biol. 32, 169–172.

Killebrew, K., Wolff, H., 2010. Environmental impacts of agricultural technologies. EPAR Brief No. 65. Evans School Policy Analysis and Research, University of Washington, Seattle, Wash., USA.

Kim, H.Y., Horie, T., Nakagawa, H., Wada, K., 1996. Effects of elevated CO_2 concentration and high temperature on growth and yield of rice. II. The effect on yield and its components of Akihikari rice. Jpn. J. Crop Sci. 65, 644–651.

Krishnan, P., Swain, D.K., Chandra Bhaskar, B., Nayak, S.K., Dash, R.N., 2007. Impact of elevated CO_2 and temperature on rice yield and methods of adaptation as evaluated by crop simulation studies. Agric. Ecosyst. Environ. 122, 233–242.

Kumar, M., Singh, K.G., Shakya, S.K., 2006. Modeling of fertilizer nitrogen in irrigated rice fields. Hydrol. J. 28 (3–4), 73–88.

Kumar, R.M., Rao, P., Somasekhar, N., Surekha, K., Padmavathi, C.H., Prasad, M.S., Babu, V.R., Rao, L.V.S., Latha, P.C., Sreedevi, B., Ravichandran, S., Ramprasad, A.S., Muthuraman, P., Gopalakrishnan, S., Goud, V.V., Viraktamath, B.C., 2010. Research experiences on system of rice intensification and future directions. J. Rice Res. 2 (2), 61–71.

Kumar, S.N., Aggarwal, P.K., Rani, S., Jain, S., Saxena, R., Chauhan, N., 2011. Impact of climate change on crop productivity in Western Ghats, coastal, and northeastern regions of India. Curr. Sci. 101 (3), 33–42.

Kumar, S.N., Aggarwal, P.K., Saxena, R., Rani, S., Jain, S., Chauhan, N., 2013b. An assessment of regional vulnerability of rice to climate change in India. Clim. Change 118 (2–3), 683–699.

Kumar, R.M., Rao, P., Somasekhar, N., Surekha, K., Padmavathi, C.H., Prasad, M.S., Babu, V.R., Rao, L.V.S., Latha, P.C., Sreedevi, B., Ravichandran, S., Ramprasad, A.S., Muthuraman, P., Gopalakrishnan, S., Goud, V.V., Viraktamath, B.C., 2013a. SRI: a method for sustainable intensification of rice production with enhanced water productivity. Agrotechnology S11 (009.), 1–6.

Ladha, J.K., Fischer, K.S., Hossain, M., Hobbs, P.R., Hardy, B., (Eds.) 2000. Improving the productivity and sustainability of rice–wheat systems of the Indo-Gangetic Plains: a synthesis of NARS-IRRI partnership research. Discussion Paper No. 40. Los Baños (Philippines): International Rice Research Institute.

Mahajan, G., Bharaj, T.S., Timsina, J., 2009. Yield and water productivity of rice as affected by time of transplanting in Punjab, India. Agr. Water Manage. 96 (3), 525–532.

Mahajan, G., Timsina, J., Kuldeep-Singh, 2011. Performance and water-use efficiency of rice relative to establishment methods in northwestern Indo-Gangetic plains. J. Crop Improv. 25 (5), 597–617.

Majumdar, D., Kumar, S., Pathak, H., Jain, M.C., Kumar, U., 2000. Reducing nitrous oxide emission from an irrigated rice field of North India with nitrification inhibitors. Agric. Ecosyst. Environ. 81, 163–169.

Malla, G., Bhatia, A., Pathak, H., Prasad, S., Jain, N., Singh, J., 2005. Mitigating nitrous oxide and methane emissions from soil in rice–wheat system of the Indo-Gangetic plain with nitrification and urease inhibitors. Chemosphere 58, 141–147.

Maredia, M., Pingali, P., 2001. Environmental Impacts of Productivity-Enhancing Crop Research: A Critical Review. TAC Secretariat. Food and Agriculture Organization of the United Nations, Rome.

Matthews, R., Wassmann, R., 2003. Modelling the impacts of climate change and methane emission reductions on rice production: a review. Eur. J. Agron. 19, 573–598.

Matthews, R.B., Wassmann, R., Knox, J., Buendia, L.V., 2000. Using a crop/soil simulation model and GIS techniques to assess methane emissions from rice fields in Asia. IV. Upscaling to national levels. Nutr. Cycl. Agroecosys. 58, 201–217.

Mencher, J.P., 1991. Agricultural labor and pesticides in rice growing regions of India: some health considerations. Econ. Polit. Wkly. 26 (39), 2263–2268.

Milkha, S.A., Bijay-Singh, 1997. Nitrogen losses and fertilizer N use efficiency in irrigated porous soils. Nutr. Cycl. Agroecosys. 47, 197–212.

Mishra, A., Verma, H.C., Kannan, K., 2012. Improving existing practices of water delivery in a run-of-the-river based canal system for better water use efficiency. Irrig. Drain. 61 (3), 330–340.

Mitra, A.P. (Eds.), 1992. Greenhouse gas emission in India: 1991 Methane Campaign, Centre on Global Change, New Delhi: National Physical Laboratory Scientific Report No. 2.

Mitra, B., Mookherjee, S., Biswas, S., Mukhopadhyay, P., 2014. Potential water saving through system of rice intensification (SRI) in *Terai* region of West Bengal. India. Int. J. Bio-resour. Stress Manage. 4 (3), 449–451.

MoEF (Ministry of Environment and Forests), 2010. India: greenhouse gas emissions 2007. Ministry of Environment and Forests, Government of India.

Mohandass, S., Kareem, A.A., Ranganthan, T.B., Jayaraman, S., 1995. Rice production in India under current and future climates. In: Matthews, R.B., Kropff, M.J., Bachelet, D., Vann Laar., H.H. (Eds.), Modeling the Impact of Climate Change on Rice Production in Asia. CAB International, Wallingford (UK), pp. 165–181.

Murthy, K.M.D., Reddy, C.V., Rao, A.U., Zaheruddeen, S.M., 2006. Water management and varietal response of rice under system of rice intensification (SRI) in Godavari delta of Andhra Pradesh. In: Proceedings of National Symposium on System of Rice Intensification (SRI): Present Status and Future Prospects, November 17–18, 2006, ANGRAU, Rajendranagar, Hyderabad, India.

Nambiar, K.K.M., 1994. Yield sustainability of rice–rice and rice–wheat systems under long-term fertilizer use. Indian Society of Agronomy, New Delhi, India.

Nambiar, K.K.M., Ghosh, A.B., 1984. Highlights of research of a long-term fertilizer experiment in India (1971–1982). LTFE Research Bulletin No. 1. Indian Agricultural Research Institute, New Delhi, India.

Nambiar, K.K.M., Soni, P.N., Vats, M.R., Sehgal, K., Mehta, D.K., 1992. Annual Reports, 1987–88 and 1988–89. All India Coordinated Project on Long-Term Fertilizer Experiments. Indian Council of Agricultural Research, New Delhi.

Nayak, A.B., 2006. An analysis using LISS III data for estimating water demand for rice cropping in parts of Hirakud command area, Orissa, India. Unpublished thesis submitted to Indian Institute of Remote Sensing, Dehradun, India, and International Institute for Geo-Information Science and Earth Observation, Enschede, Netherlands.

Nayak, D.R., Adhya, T.K., Babu, Y.J., Datta, A., Ramakrishnan, B., Rao, V.R., 2006. Methane emission from a flooded field of Eastern India as influenced by planting date and age of rice (*Oryza sativa* L.) seedlings. Agric. Ecosyst. Environ. 115, 79–87.

Oh-e, I., Saitoh, K., Kuroda, T., 2004. Effects of rising temperature on growth, yield, and dry-matter production of rice grown in the paddy field, Proceedings of the Fourth International Crop Science Congress, Brisbane, Australia, Sept 26–Oct 1, 2004. Available from: www.cropscience. org.au/icsc2004/poster/2/7/1/661_ohe.htm

Parashar, D., Rai, C.J., Gupta, P.K., Singh, N., 1991. Parameters affecting methane emission from paddy fields. Indian J. Radio Space Phys. 20, 12–17.

Parashar, D.C., Mitra, A.P., Gupta, P.K., Rai, J., Sharma, R.C., Singh, N., Koul, S., Ray, H.S., Das, S.N., Parida, K.M., Rao, S.B., Kanungo, S.P., Ramasami, T., Nair, B.U., Swamy, M., Singh, G., Gupta, S.K., Singh, A.R., Saikia, B.K., Barua, A.K.S., Pathak, M.G., Iyer, C.S.P., Gopalakrishnan, M., Sane, P.V., Singh, S.N., Banerjee, R., Sethunathan, N., Adhya, T.K., Rao, V.R., Palit, P., Saha, A.K., Purkait, N.N., Chaturvedi, G.S., Sen, S.P., Sen, M., Sarkar, B., Banik, A., Subbaraya, B.H., Lal, S., Venkatramani, S., Lal, G., Chaudhary, A., Sinha, S.K., 1996. Methane budget from paddy fields in India. Chemosphere 33 (4), 737–757.

Parashar, D.C., Gupta, K.P., Bhataachaya, S., 1997. Methane emissions from paddy fields in India: an update. Indian J. Radio Space Phys. 26, 237–243.

Pathak, H., Wassmann, R., 2007. Introducing greenhouse gas mitigation as a development objective in rice-based agriculture. I. Generation of technical coefficients. Agric. Syst. 94, 807–825.

Pathak, H., Bhatia, A., Prasad, S., Singh, S., Kumar, S., Jain, M.C., Kumar, U., 2002. Emission of nitrous oxide from rice–wheat systems of Indo-Gangetic plains of India. Environ. Monit. Assess. 77, 163–178.

Pathak, H., Prasad, S., Bhatia, A., Singh, S., Kumar, S., Singh, J., Jain, M.C., 2003a. Methane emission from rice–wheat cropping system in the Indo-Gangetic plain in relation to irrigation, farmyard manure, and dicyandiamide application. Agric. Ecosyst. Environ. 97, 309–316.

Pathak, H., Ladha, J.K., Aggarwal, P.K., Peng, S., Das, S., Singh, Y., Singh, B., Kamra, S.K., Mishra, B., Sastri, A.S.R.A.S., Aggarwal, H.P., Das, D.K., Gupta, R.K., Mall, R.K., 2003b. Trends of climatic potential and on-farm yields of rice and wheat in the Indo-Gangetic plains. Field Crops Res. 80, 223–234.

Pathak, H., Li, C., Wassmann, R., 2005. Greenhouse gas emissions from Indian rice fields: calibration and upscaling using the DNDC model. Biogeosciences 1, 1–11.

Pathak, H., Singh, R., Bhatia, A., Jain, N., 2006. Recycling of rice straw to improve wheat yield and soil fertility and reduce atmospheric pollution. Paddy Water Environ. 4 (2), 111–117.

Pattanayak, A., Kumar, K.S.K., 2013. Weather sensitivity of rice yield: evidence from India. Working Paper 81/2013. Madras School of Economics, Chennai.

Pingali, P.L., 1992. Diversifying Asian rice farming systems: a deterministic paradigm. In: Barghouti, S., Garbux, L., Umali, D. (Eds.), Trends in Agricultural Diversification: Regional Perspectives. Paper No. 180. Washington, DC: The World Bank. pp. 107–126.

Pingali, P., 2012. Green revolution: impacts, limits, and the path ahead. Proc. Natl. Acad. Sci. USA 109 (31), 12302–12308.

Pingali, P.L., Moya, PF, Velasco, LE. 1990. The post-green revolution blues in Asian rice production: the diminished gap between experiment station and farmer yields. Social Sciences Division Paper No. 90–01 (January). International Rice Research Institute, Los Baños, Philippines.

Pingali, P.L., Hossain, M., Gerpacio, R.V., 1997. Asian rice bowls: the returning crisis? CAB International, Wallingford (UK) in association with International Rice Research Institute, Manila (Philippines).

Prasad, R., 2013. Fertilizer nitrogen, food security, health, and the environment. Proc. Indian Natl. Sci. Acad. Special Issue, Part B 79 (4), 997–1010.

Rajkishore, S.K., Doraisamy, P., Subramanian, K.S., Maheswari, M., 2013. Methane emission patterns and their associated soil microflora with SRI and conventional systems of rice cultivation in Tamil Nadu, India. Taiwan Water Conserv. 61 (4), 126–134.

Raju, R.A., Sreenivas, C., 2008. Agronomic evaluation of rice intensification methods in Godavari delta. Oryza 45, 280–283.

Ramasamy, C., Selvaraj, K.N., Anoop, G., 2005. Impact assessment of salinity on rice production in Tamil Nadu: food security issues and investment priorities. CARDS Series 15. Final Report of the NATP Project of the Department of Agricultural Economics, TNAU, Coimbatore, sponsored by the Indian Council of Agricultural Research, New Delhi.

Ranganathan, C.R., 2009. Quantifying the impact of climatic change on yield and yield variability of major crops and optimal land allocation for maximizing food production in different agroclimatic zones of Tamil Nadu, India: an econometric approach, Working Paper, Research Institute for Humanity and Nature, Kyoto University, Kyoto, Japan.

Ranjan, L., Sharma, S., Idris, M., Singh, A.K., Singh, S.S., Bhatt, B.P., Saharawat, Y., Humphreys, E., Ladha, J.K., 2014. Integration of conservation agriculture with best management practices for improving system performance of the rice–wheat rotation in the Eastern Indo-Gangetic plains of India. Agric. Ecosyst. Environ. 195, 68–82.

Rao, E.V.S.P., Puttanna, K., 2010. Nitrates, agriculture, and environment. Curr. Sci. 79 (9), 1163–1168.

Rao, A.U., Ramana, A.V., Sridhar, T.V., 2013. Performance of system of rice intensification (SRI) in Godavari delta of Andhra Pradesh. Ann. Agr. Res. New Series. 34 (2), 118–121.

Rola, A.C., Pingali, P.L., 1993. Pesticides, Rice Productivity and Farmers' Health: An Economic Assessment. International Rice Research Institute and World Resources Institute, Manila, Philippines.

Saha, S., Singh, Y.V.G., Gaind, S., 2014. Water productivity and nutrient status of rice soil in response to cultivation techniques and nitrogen fertilization. Paddy Water Environ. 13 (4), 443–453.

Saharawat, Y.S., Ladha, J.K., Pathak, H., Gathala, M., Chaudhary, M., Jat, M.L., 2012. Simulation of resource-conserving technologies on productivity, income, and greenhouse gas emission in rice–wheat system. J. Soil Sci. Environ. Manage. 3 (1), 9–22.

Sapkota, T.B., Jat, M.L., Shankar, V., Singh, L.K., Rai, M., Grewal, M.S., Stirling, C.M., 2015. Tillage, residue, and nitrogen management effects on methane and nitrous oxide emission from rice–wheat system of Indian Northwest Indo-Gangetic Plains. J. Integr. Environ. Sci. 12 (S1), 31–46.

Saravanakumar, V., 2015. Impact of climate change on yield of major food crops in Tamil Nadu, India. Working Paper No. 91–15. South Asian Network for Development and Environmental Economics (SANDEE), Kathmandu, Nepal.

Saseendran, S.A., Singh, K.K., Rathore, L.S., Singh, S.V., Sinha, S.K., 1999. Effects of climate change on rice production in the tropical humid climate of Kerala, India. Clim. Change 12, 1–20.

Satyanarayana, A., Thiyagarajan, T.M., Uphoff, N., 2007. Opportunities for water saving with higher yield from the system of rice intensification. Irrig. Sci. 25, 99–115.

Senthilkumar, K., Bindraban, P.S., de Ridder, N., Thiyagarajan, T.M., Giller, K.E., 2012. Impact of policies designed to enhance efficiency of water and nutrients on farm households varying in resource endowments in south India. NJAS Wagen. J. Life Sci. 59, 41–52.

Singh, R.B., 2000. Environmental consequences of agricultural development: a case study from the green revolution state of Haryana, India. Agric. Ecosyst. Environ. 82 (1–3), 97–103.

Singh, J., 2013a. Depleting water resources of Indian Punjab agriculture and policy options: a lesson for high potential areas. Global J. Sci. Frontier Res. Agric. Veter. 13 (4), 17–23.

Singh, Y.V., 2013b. Crop and water productivity as influenced by rice cultivation methods under organic and inorganic sources of nutrient supply. Paddy Water Environ. 11, 531–542.

Singh, S.N., Verma, A., Tyagi, L., 2003. Investigating options for attenuating methane emission from Indian rice fields. Environ. Int. 29, 547–553.

Singh, R., Kundu, D.K., Bandyopadhyay, K.K., 2010. Enhancing agricultural productivity through enhanced water use efficiency. J. Agric. Physics 10, 1–15.

Singh, A., Panda, S.N., Flugel, W.-A., Krause, P., 2012. Waterlogging and farmland salinisation: causes and remedial measures in an irrigated semiarid region of India. Irrig. Drain. 61 (3), 357–365.

Srivastava, S.K., Meena Rani, H.C., Bandyopadhyay, S., Hegde, V.S., Jayaraman, V., 2009. Climate risk assessment of rice ecosystems in India. J. South Asia Disaster Stud. 2 (1), 155–166.

Srivastava, S.K., Chand, R., Raju, S.S., Jain, R., Kingsly, I., Sachdeva, J., Singh, J., Kaur, A.P., 2015. Unsustainable groundwater use in Punjab agriculture: insights from cost of cultivation survey. Indian J. Agric. Econ. 70 (3), 365–378.

Streets, D.G., Yarber, K.F., Woo, J.-H., Carmichael, G.R., 2003. Biomass burning in Asia: annual and seasonal estimates and atmospheric emissions. Global Biogeochem. Cycles 17 (4), 1099–1118.

Su, J., Zhang, Z., Wu, M., Gao, C., 2014. Changes in insecticide resistance of the rice striped stem borer (Lepidoptera: Crambidae). J. Econ. Entomol. 107 (1), 333–341.

Subbaiah, S.V., 2006. Several options being tapped. The Hindu Survey Indian Agric., 50–54.

Suryavanshi, P., Singh, Y.V., Prasanna, R., Bhatia, A., Shivay, Y.S., 2013. Pattern of methane emission and water productivity under different methods of rice crop establishment. Paddy Water Environ. 11, 321–329.

Thakur, A.K., Rath, S., Patil, D.U., Kumar, A., 2011. Effects on rice plant morphology and physiology of water and associated management practices of the system of rice intensification and their implications for crop performance. Paddy Water Environ. 9, 13–24.

Thakur, A.K., Mohanty, R.K., Patil, D.U., Kumar, A., 2013. Impact of water management on yield and water productivity with system of rice intensification (SRI) and conventional transplanting system in rice. Paddy Water Environ. 11, 13–24.

Thiyagarajan, T.M., Gujja, B., 2013. Transforming rice production with SRI knowledge and practice. National Consortium on SRI (NCS), New Delhi, and AgSri, Hyderabad, India. Available from: www.agsri.com/images/documents/sri/SRI_Book_Final_Version.pdf

Tirol-Padre, A., Rai, M., Kumar, V., Gathala, M., Sharma, P.C., Sharma, S., Nagar, R.K., Deshwal, S., Singh, L.K., Jat, H.S., Sharma, D.K., Wassmann, R., Ladha, J.K., 2016. Quantifying changes to the global warming potential of rice–wheat systems with the adoption of conservation agriculture in northwestern India. Agric. Ecosyst. Environ. 219, 125–137.

Tiwari, K.N., 2002. Balanced fertilization for food security. Fert. News 47, 113–122.

Tiwari, V.M., Wahr, J., Swenson, S., 2009. Dwindling groundwater resources in northern India, from satellite gravity observations. Geophys. Res. Lett. 36, L18401.

Tyagi, L., Kumari, B., Singh, S.N., 2010. Water management: a tool for methane mitigation from irrigated paddy fields. Sci. Total Environ. 408, 1085–1090.

Varghese, S.K., Veettil, P.C., Speelman, S., Buysse, J., Huylenbroeck, G.V., 2013. Estimating the causal effect of water scarcity on the groundwater use efficiency of rice farming in South India. Ecol. Econ. 86 (1), 55–64.

Vijayakumar, M., Ramesh, S., Chandrasekaran, B., Thiyagarajan, T.M., 2006. Effect of system of rice intensification (SRI) practices on yield attributes, yield, and water productivity of rice (*Oryza sativa* L.). Res. J. Agric. Biol. Sci. 2 (6), 236–242.

World Bank, 1999. The green revolution and the productivity paradox: evidence from the Indian Punjab. Policy Research Working Papers. Available from: http://dx.doi.org/10.1596/1813-9450-2234

World Bank, 2008. Agriculture for Development: World Development Report 2008. The World Bank, Washington, D.C.

World Bank, 2010. Deep Wells and Prudence: Towards Pragmatic Action for Addressing Groundwater Overexploitation in India. The World Bank, Washington, D.C.

Yadav, R.L., 1998. Factor productivity trends in a rice–wheat cropping system under long-term use of chemical fertilizers. Exp. Agric. 34, 1–18.

Yan, X., Ohara, T., Akimoto, H., 2003. Development of region-specific emission factors and estimation of methane emission from rice fields in the East, Southeast, and South Asian countries. Glob. Change Biol. 9, 237–254.

Yan, X., Yagi, K., Akiyama, H., Akimoto, H., 2005. Statistical analysis of the major variables controlling methane emission from rice fields. Glob. Change Biol. 11, 1131–1141.

Zacharias, M., Naresh Kumar, S., Singh, S.D., Swaroopa Rani, D.N., Aggarwal, P.K., 2014. Assessment of impacts of climate change on rice and wheat in the Indo-Gangetic plains. J. Agrometeorol. 16 (1), 9–17.

FURTHER READING

Kim, H.Y., Horie, T., Nakagawa, H., Wada, K., 2010. Effects of elevated CO_2 concentration and high temperature on growth and yield of rice. II. The effect on yield and its components of Akihikari rice. Jpn. J. Crop Sci. 65, 644–651.

Chapter 5

Structural Transformation of the Indian Rice Sector

Humnath Bhandari*, Praduman Kumar**, Parshuram Samal†
*International Rice Research Institute, Dhaka, Bangladesh; **Indian Agricultural Research Institute, ICAR, New Delhi, India; †National Rice Research Institute, Cuttack, Odisha, India

INTRODUCTION

Rice is a staple food for the majority of India's 1.31 billion population and a vital source of livelihood, directly or indirectly, for the majority of the 0.86 billion rural Indians. Rice farming is directly and indirectly linked to poverty and food security, as well as economic, social, cultural, political, and environmental aspects. Increasing productivity and profitability of the whole rice value chain is central for improving rural livelihoods, assuring an adequate supply of rice at affordable prices, and earning foreign exchange reserves from exports. The Green Revolution (GR) led to a dramatic increase in rice and wheat production, made famines a historical phenomenon, and significantly improved food security and rural livelihoods in India. Continuous improvement in the genetics of high-yielding varieties (HYV) of rice, as well as crop and farm management practices; policy support to improve irrigation facilities, market infrastructure, and the supply of chemical fertilizers and agricultural credit; subsidies on production inputs; and farmers' enthusiasm to adopt HYV were the major drivers of impressive cereal production growth and total factor productivity (TFP) over the past 4 decades (Hazell and Ramasamy, 1991; Janaiah et al., 2006). The transformation of the rice sector is startling, with India becoming the largest rice exporter in the world in 2011. The biggest successes for the Indian rice sector have been the improvement of TFP, development and adoption of climate-smart varieties, vertical integration of value chain members, and transformation of the basmati rice industry. The world basmati rice market was dominated by Pakistan until the early 1990s; however, the significant production increase due to improved technologies, farm management practices, upgrading of postharvest systems (drying, storage, milling, and processing), improved value addition (packaging and branding), and wider marketing brought success to the Indian basmati industry, making India the leader in the world basmati rice market today (Wani et al., 2015). Favorable

The Future Rice Strategy for India. http://dx.doi.org/10.1016/B978-0-12-805374-4.00005-1
107

trade policy and licensing procedures for agricultural exports, private sector efforts to promote rice exports, and the zeal of basmati rice exporters to establish themselves as reliable suppliers contributed to the success of the Indian basmati industry. Although the Indian rice sector has achieved astounding success in the past, the future is likely to see both enormous challenges and exciting opportunities arising from demand and supply sides.

Regarding the demand-side challenges, India will have the world's largest population in 2022, surpassing China (UNDESA, 2015), and is likely to grow by ~220 million during 2015–30. Approximately 22% of the country's population is poor[a] and lacks the purchasing capacity to obtain adequate food for a healthy and productive life, while 17% of the population is also undernourished.[b] Rice is the predominant item in the poor and undernourished population's food menu and is supplemented by vegetables, pulses, and meat. Cheap rice allows the poor to consume greater quantities of nutritious nonrice foods, and therefore the supply of adequate rice at affordable prices is pivotal for improving food and nutrition security for the poor. When the income of the poor rises, their dietary patterns will shift from coarse grain to rice, and from low- to medium-quality rice, thereby increasing the demand for both quantity and quality of rice. Rising incomes of the rich will reduce rice demand somewhat as diversification of the food basket increases, although it will increase the demand for higher-quality rice.

Regarding the supply-side challenges, a burgeoning number of constraints are threatening production, economic viability, and sustainability of future rice production. The availability of land and water for rice farming is decreasing due to population growth, urbanization, industrialization, and the diversification of rice-based cropping systems. Declining human fertility rate, growth of the nonfarm sector, and rural labor outmigration have led to increasing wage rates and a scarcity of rural laborers, an aging farming population, and feminization of agriculture, with access to production resources and services being inadequate and unequal. In addition, climate variability and change are exacerbating the problem by decreasing land availability and increasing production risk. Moreover, rice farming is under pressure to adopt sustainable production practices and reduce environmental footprints, such as land degradation, water depletion, air and water pollution, unbalanced nutrient cycles, greenhouse gas emissions, and biodiversity loss. Given the lower profit from rice vis-à-vis many other crops, diversification of the rice-based cropping system is desirable and should be promoted. However, other emerging supply-side constraints will threaten production, which in turn will make the staple food costlier and also debilitate exports.

On the optimistic side, rising incomes, a growing middle-income class population, urbanization, changing lifestyles and diet patterns, growing demand for

---

a. A poor refers to a person that is not able to earn enough income to meet the daily minimum dietary energy requirements.
b. Undernourishment refers to a person that is not able to obtain enough food to meet the daily minimum dietary energy requirements over a 1-year period.

processed and convenience food, higher awareness of health and nutrition, globalization, greater economic and trade liberalization, and a higher rice demand in the world market have created new market opportunities for the Indian rice sector. The public sector's prioritization of, and investment in, agricultural R&D; the private sector's increased investment in agribusiness firms; technological innovations; better access to innovative technologies and information; new models for providing agricultural services (extension and finance); better connections to markets; marketing through modern supply chain systems (i.e., supermarkets); and greater national and international integration along the supply chain have also generated opportunities for value addition and upgradation of the rice value chain.

These ongoing social and economic transformations in the wider Indian economy are posing new challenges and opportunities in the rice sector. The rice sector is feeling the heat and gradually undergoing structural transformations to meet new challenges and exploit new opportunities. We define structural transformation of the rice sector in terms of a fundamental change in form, function, or structure of rice production, consumption, marketing, and trade. Although the Indian economy is rapidly transforming, the rice sector is responding slowly. Past visions for rice sector development focused on public sector-led transformation for yield growth through improvements in seeds, agrochemicals, irrigation, technology transfer, and access to credit and information. This successfully transformed agriculture from subsistence to semi-subsistence and commercial farming. A new vision and strategy are required in the evolving context. The future vision for the rice sector must bolster transition from commercial farming to agribusiness with emphasis on value addition and efficiency gain throughout the value chain. The challenges are to craft and implement strategies, policies, and programs that lead to an inclusive, sustainable, resilient, and internationally competitive rice value chain, which improves the income and welfare of millions of rice farmers, consumers, and traders. The objectives of this chapter are manyfold: (1) to provide a broad overview of transformation in rice production, consumption, and trade; (2) to provide comprehensive information about the ongoing structural transformation in India's rice economy; (3) to estimate future rice demand and supply; (4) to analyze potential consequences, both intended and unintended, of ongoing socioeconomic changes in the rice sector; and (5) to draw conclusions to further transform the rice sector in terms of developing an efficient and internationally competitive rice value chain system. The scope of this chapter is broader than typical academic-style papers that concentrate on gathering narrowly focused data and testing specific hypotheses. The analytics within this chapter are descriptive and predictive in nature, providing comprehensive information and strategic views about the ongoing structural transformation in India's rice economy, and deriving conclusions to develop a vibrant and competitive rice sector. The lessons will be useful for rice scientists, policymakers, traders, the private sector, and farmers who are engaged at different steps of the rice value chain.

This chapter is organized in 11 sections to provide a comprehensive background, attributes of transformative changes, and future challenges and opportunities in the Indian rice sector. Section "Rice Production Performance" examines rice production performance, while section "Trends in Rice Exports" examines rice export performance, especially basmati rice. Section "Diversification of the Rice-Based Cropping System" examines diversification of the rice-based cropping system, while section "The Changing Economic Role of Rice" examines the importance of rice in cereal production, agricultural GDP, and overall economy. Section "Rice Demand and Supply" analyzes rice demand and supply. Section "Future Challenges for the Rice Sector" analyzes the challenges facing the rice value chain, focusing on land and water scarcity, labor crisis, climate change, and access to credit. Section "Changing Role of Women and Youth in the Rice Value Chain" analyzes the changing role of women and youth in the rice value chain, focusing especially on reducing gender gaps and attracting youth to agriculture. Section "Vertical Integration in the Rice Value Chain" investigates the benefits, opportunities, and potential of horizontal and vertical integration in the rice value chain. Section "International Competitiveness of the Indian Rice Sector" investigates international competitiveness of the Indian rice sector, while the final section concludes the chapter with suggestions to develop an economically vibrant and internationally competitive rice value chain.

RICE PRODUCTION PERFORMANCE

Rice is India's most important crop in terms of area, production, and consumption. It accounts for 36% of the total crop area and for 42% of food grain production. During 1970–2014, rice's share in food grains slightly increased in area, but remained virtually constant in production. The contribution of rice to food grain production varies enormously between states depending on the agroecology. The top three rice-producing states are West Bengal, Uttar Pradesh, and Andhra Pradesh, which together account for 40% of India's rice production.

Spearheaded by the GR, India has achieved an impressive growth in rice production in the past 4 decades. Rice production grew from 63 to 159 million tons (2.3% per year) between 1971–73 and 2012–14, exceeding the growth rate of both food grain production and the population. Yield growth contributed 83% to the production growth (Fig. 5.1). A higher growth rate in rice production than in population increased per capita paddy rice production from 113 to 128 kg, despite a 70% increase in population over the past 4 decades. This resulted in an adequate supply of rice at affordable prices, which was instrumental in increasing food and nutrition security, reducing poverty, and uplifting rural livelihoods. It also made the country a major player in the world rice market. The TFP growth for rice production was 0.7% per year during 1970–2005 and growth was highest in southern states and lowest in eastern states (Chand et al., 2012). However, a deceleration in the annual growth rate of rice production from 2.8% during 1970–89 to 1.4% during 1990–2014 is worrisome.

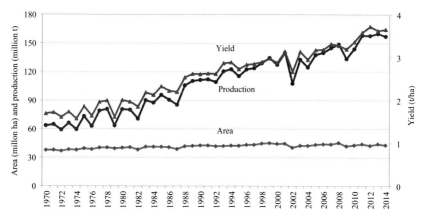

FIGURE 5.1   Trends in paddy rice area, production, and yield in India, 1970–2014. *(Data taken from United States Department of Agriculture (USDA), 2015. Production, Supply, and Distribution Online Database of Agriculture. Foreign Agricultural Services. USDA, Washington, DC.)*

The average paddy yield more than doubled from 1.7 to 3.6 t/ha over the past 4 decades (Fig. 5.2). Yield increased in all states; however, the magnitude of the increase and current yield levels vary considerably. The average yield varied widely across states, on the order of 1.1–3.2 t/ha in 1971–73 to 2.2–5.8 t/ha in 2011–13. Yields were considerably higher in northwestern states of Punjab and Haryana and southern states of Andhra Pradesh, Karnataka, Kerala, and Tamil Nadu, where irrigation infrastructure is well developed and adoption of HYV is high, while yields were considerably lower in eastern states. This shows there is great potential to increase rice production by reducing yield gaps in eastern India.

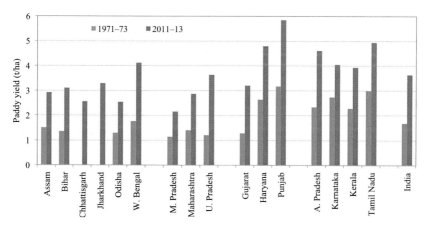

FIGURE 5.2   Trends in paddy rice yield in major rice-growing states in India, 1971–73 and 2011–13. *(Data taken from the Directorate of Economics and Statistics (DES), 2015. Estimates of area, production, and yield of foodgrains in India. DES, Department of Agriculture, Cooperation, and Farmers Welfare, Ministry of Agriculture and Farmers Welfare, GOI, New Delhi, India; INDIASTAT, 2015. INDIASTAT.COM: an online socioeconomic statistical information of India. Available from: www.indiastat.com)*

TRENDS IN RICE EXPORTS

India ranks second and first in the world in terms of rice production and exports, respectively. It accounts for 27% of the world's rice exports amounting to 42.5 million tons (FAO, 2015), and of this 11.5 million tons of rice, 33% is basmati and 67% nonbasmati [Agricultural and Processed Food Products Export Development Authority (APEDA), 2015]. Basmati occupies a prime position in the world rice market because of its premium grain quality in terms of appearance, aroma, and taste. Basmati is produced only in India (Indo-Gangetic region) and Pakistan, which have unique climatic and soil conditions. Of the 12.4 million tons of milled basmati produced globally in 2014–15, India accounted for 71% and Pakistan 29%. In 2013–14, the size of the world basmati exports was 4.5 million tons (valued at USD 5.8 billion), accounting for 11% of the total rice export value; India accounted for 84% of the total basmati exports and 43% of basmati rice is exported from India.

Trends show a significant upward slope for exports of both basmati and nonbasmati rice from India. Exports of rice grew 9.9% per year during 1990–2014, and the growth rate was significantly higher for basmati (11.0%) vis-à-vis nonbasmati (7.7%). Rice exports grew rapidly after 2000 and spectacularly after 2010, making India the leader in the world rice export market (by beating Thailand) after 2011. The Indian rice export market size increased significantly from 0.5 million tons in 1990 (valued at USD 0.25 billion) to 11.5 million tons in 2014 (USD 7.9 billion). The basmati market size increased significantly from 0.23 million tons in 1990 (USD 0.16 billion) to 3.8 million tons in 2014 (USD 4.67 billion). Likewise, the nonbasmati market size increased from 0.27 million tons (USD 0.09 billion) to 7.7 million tons (USD 3.26) billion over the same time period. Basmati exports are currently half those of nonbasmati exports in terms of quantity, but 43% higher in terms of value. The ~80% higher price of basmati (FOB USD 1245 per ton) compared with nonbasmati (FOB USD 695 per ton) resulted in a higher income from basmati. This shows a great potential to increase farm income through production and export of basmati rice.

The middle-class population (defined as earning PPP$ 10–100 per capita per day) will increase from ~525 million to 3.23 billion in the Asia–Pacific during 2009–30 (Kharas, 2010) and from ~50 to 475 million in India during 2010–30 (EY, 2013). A growing middle-class, rice-consuming population will increase the demand for high-quality rice in domestic and international markets. This creates great opportunities for India to earn foreign currency by exporting basmati and high-quality nonbasmati rice. Harnessing this potential, however, largely depends on the availability of surplus rice in the domestic market and price competitiveness in the international market. This requires achieving higher rice productivity and profitability, especially in the context of a growing population and supply-side constraints. Traditional basmati varieties have a relatively long growth period and low yield potential, resulting in lower system productivity. The improved pusa basmati rice varieties (i.e., Pusa Basmati-1121 and Pusa Basmati-1509)—which have a relatively shorter growth duration, lower water requirements, and higher yield potential—assure cropping system intensification, increased exports, and higher profits. Interventions to increase basmati production include improved yield potential of basmati varieties, increased availability of good-quality seeds, development of irrigation facilities, improved postharvest systems, improved marketing systems (including packaging and branding), and development of infrastructure for marketing logistics and exports. The rapidly emerging new markets in India are increasing consumers' awareness of brands, consumption of branded rice, and demand for better-quality rice with nutrition and health benefits. This is reflected in the shift toward packaged, branded rice products with better color, grain size, and postcooking attributes, such as aroma and taste. Future rice markets should internalize these transformations in consumer preferences. Rice packs should be launched in different sizes to cater to the needs of small-to-large families. In the face of new market opportunities, investment from large multinational food companies (such as REI Agro, KRBL, LT Foods, Kohinoor Foods, McCormick, and Ebro Foods) has increased manyfold in the last 5 years. Conductive policies and business environments can attract more foreign direct investment (FDI) in the rice industries.

DIVERSIFICATION OF RICE-BASED CROPPING SYSTEMS

Rice occupies the largest area (44 million ha) of all crops. Rice's share in the total gross cropped area marginally declined from 23% to 22% between 1990–92 and 2010–12. In the last 2 decades, India has witnessed some diversification from food grains to higher-value crops. The relative areas of rice, coarse grains, and pulses have decreased, while those of maize, wheat, fiber crops, sugarcane, and horticultural crops have increased. Continuous rice cultivation in the same field for a long period, intensive input use, and high varietal uniformity have all depleted soil organic matter and soil fertility, increased macro- and micronutrient deficiency, and increased insect and disease pressure. Andhra Pradesh, Punjab, and Tamil Nadu are all examples of this. Income from rice is lower than from cash crops (DES, 2015), and market opportunities for cash crops are increasing due to changing consumption patterns, better connections to markets, and globalization. Diversification away from rice farming is necessary to increase farmers' incomes. This will eventually happen and it warrants a two-pronged approach: first, to craft and implement appropriate policies to accelerate diversification of rice-based cropping systems, and second, to increase investment in rice R&D to raise rice productivity, which is essential to offset the decline in rice production due to diversification of rice-based cropping systems.

Farmers in states or districts with intensive input use and high yields would be interested in increasing profits from efficient allocation of inputs. Those farmers might be more interested in cultivating high-quality, higher-value rice varieties instead of further increasing the yield of average-quality rice varieties that require extra application of agrochemicals. The cultivation of higher-value rice varieties would increase profit and make rice farming economically attractive. The rice strategy should promote higher-quality rice varieties in agroecologically suitable areas with comparative advantages to grow such varieties.

THE CHANGING ECONOMIC ROLE OF RICE

Rice is still the most important food crop in terms of production, but its share in cereal production value, agricultural value, and total economic value has substantially declined over the last 4 decades. During 1970–2013, rice's share in cereal production value fell from 70% to 68%, from 24% to 17% in agricultural value, and from 10% to 3% in GDP (Fig. 5.3). The contribution of rice to cereal production is still large, but its contribution to total economic output is trivial and declining further due to structural transformation of the economy. Despite making only a small contribution to economic output, rice farming underpins food security, income, and employment for millions of Indians, particular the poor. Rice's cultural, social, and political importance is much larger than its economic value. The rice sector also supplies raw materials and other resources to manufacturing and service industries. Despite generating only a low income, farmers will continue to grow rice, at least the quantity required to ensure family food security and for social prestige. However, low incomes will ultimately induce diversification away from rice in the short term and migration from farming when better opportunities arise in the long term.

RICE DEMAND AND SUPPLY

Over the past 2 decades, food consumption patterns have depicted significant changes in food baskets, from coarse grains to superior grains (rice and wheat), and from grains to livestock and horticultural products (Joshi and Kumar, 2011; Kumar et al., 2007). This has considerable implications for future food demand, research priority setting, and resource allocation to achieve food and nutrition security. This section examines patterns of rice consumption and price and income elasticities of rice demand and makes projections for rice demand and supply (production) by the year 2030.

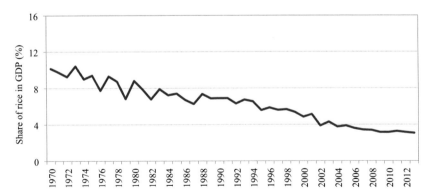

FIGURE 5.3   **The percentage share of rice in total GDP in India, 1970–2013.** *(Authors, estimation based on data taken from the Food and Agriculture Organization of the United Nations (FAO), 2015. FAOSTAT: FAO Online Statistical Database. FAO, Rome, Italy.)*

Consumption of Rice

Household-level data on dietary patterns and consumer expenditure, collected by the National Sample Survey Organization (NSSO), were used to study rice consumption trends. This covered the years 1983, 1993, 2004, and 2011, pertaining to the various major NSS rounds. Households across rural and urban areas were classified into three income groups based on the poverty line (PL). Households below the PL were defined as "poor," households between the PL and 150% of the PL were defined as "middle income," and those above 150% of the PL were classified as "high income." The consumption of rice per person showed an increasing trend across poor households, and a declining trend across nonpoor households (Table 5.1). Rice consumption was higher in rural households than in urban households in all the rounds. The declining trends have resulted from diet diversification, which could be because of economic growth, rising per capita income, and changing tastes and food preferences (Kumar et al., 2007).

TABLE 5.1 Trends in Annual Per Capita Consumption of Milled Rice in India, 1983–2011

	Milled rice consumption (kg)			
Year	Poor households	Middle-income households	High-income households	All households
Rural India				
1983	69.44	96.60	101.78	87.02
1993	69.22	78.32	79.90	76.00
2004	79.98	88.40	96.82	90.45
2011	82.16	85.36	86.83	85.45
Urban India				
1983	55.94	76.15	82.06	70.15
1993	59.46	68.08	72.85	67.59
2004	63.36	73.01	78.86	72.60
2011	67.94	73.93	71.61	71.50
All India				
1983	64.79	89.97	94.02	81.09
1993	65.79	75.05	76.94	72.92
2004	72.81	84.25	90.85	84.44
2011	76.72	81.89	80.24	80.10

Source: Computed from various NSS rounds.

Demand Elasticity

Demand elasticity was computed using the food characteristic demand system following Bouis and Haddad (1992). The income and own-price elasticities of rice demand at rural, urban, and national levels were computed and are presented in Table 5.2. These elasticities were used to project rice demand in India. The demand elasticities for rice have previously been found to be highly inelastic, close to zero, and even negative for high-income groups. The magnitude of elasticity declined with rising income in this study, but did not vary across rural and urban households. The own-price elasticity was found to be negative, as expected. The own-price elasticity was found to be lower for rich than poor households, and varied widely across income groups. An inflation in rice prices will affect poor households more than rich households.

Rice Demand by 2030

Total rice demand comprises direct and indirect demand. Direct demand consists of rice consumption at home and outside the home, while indirect demand includes its use as seed, feed, its industrial usage, and wastage. Following the methodology of Joshi and Kumar (2011), total rice demand (except for exports) was estimated by adding the direct and indirect demand. Rice demand has been projected under three GDP growth scenarios (current, low growth, and high growth) with and without deceleration in expenditure elasticities (Table 5.3).

Rice demand in 2015 is estimated to be ~106 million tons, growing to 112 million tons by 2020 and 122 million tons by 2030. During 2010–20, demand is projected to increase with an annual growth rate of 1.2%–1.3%. A deceleration in the growth rate of rice demand has been observed due to diversification from cereals to high-value commodities in the consumer's food basket. Rice demand will grow at a rate lower than the population growth rate; thus, in India, per capita rice consumption will continue to decline as has been observed in the past. A slight shift in consumption from rice, and a substantial shift from coarse cereals to wheat, is predicted (Joshi and Kumar, 2011). Easy availability of wheat through the public distribution system, under the food security program of the Government of India, is leading to an increase in wheat consumption even in traditionally rice-eating states of the country.

Supply of Rice

The area, TFP, supply elasticity, and input–output price environment are the major sources of supply growth of a commodity. The rice supply elasticity is 0.236, while input price elasticity is almost zero (Table 5.4). Following methodology from Kumar et al. (2010a), rice supply growth was estimated under four scenarios (presented in Table 5.5). The annual growth in rice supply is predicted to be 2.64% corresponding to the baseline scenario. In the scenario without TFP growth, rice supply is estimated to grow at 1.92% per year. As the possibility

TABLE 5.2 Expenditure and Own-Price Elasticities for Rice by Income Group in India

	Expenditure elasticity				Own-price elasticity			
	Poor households	Middle-income households	High-income households	All households	Poor households	Middle-income households	High-income households	All households
Rural India	0.157	0.041	−0.017	0.049	−0.463	−0.291	−0.163	−0.289
Urban India	0.130	0.015	−0.029	0.008	−0.477	−0.330	−0.222	−0.293
All India	0.146	0.028	−0.024	0.026	−0.469	−0.309	−0.200	−0.291

Source: Joshi, P.K., Kumar, P., 2011. Food Demand and Supply Projections for India: 2010–2030. Trade, Agricultural Policies, and Structural Changes in India's Agrifood System: Implications for National and Global Markets (TAPSIM) Project No. KBBE212617. IFPRI, New Delhi, India.

TABLE 5.3 Domestic Demand and Growth in Demand for Rice Under Different Scenarios in India, 2004–30

GDP growth	Domestic demand (million tons)						Demand growth (%)		
	2004	2010	2015	2020	2025	2030	2004–10	2010–20	2020–30
Scenario: constant income elasticities in the predicted period									
Current	90.0	98.7	105.8	111.8	116.9	122.4	1.54	1.26	0.91
Low	90.0	98.7	105.7	111.5	116.3	121.5	1.54	1.23	0.87
High	90.0	98.7	106.0	112.2	117.5	123.4	1.54	1.29	0.96
Scenario: 50% deceleration in expenditure elasticity during 2005–30									
Current	90.0	98.6	105.6	111.4	116.1	121.0	1.53	1.22	0.84
Low	90.0	98.6	105.5	111.1	115.7	120.5	1.53	1.20	0.82
High	90.0	98.6	105.7	111.6	116.5	121.6	1.53	1.24	0.86

Source: Joshi, P.K., Kumar, P., 2011. Food Demand and Supply Projections for India: 2010–2030. Trade, Agricultural Policies, and Structural Changes in India's Agrifood System: Implications for National and Global Markets (TAPSIM) Project No. KBBE212617. IFPRI, New Delhi, India.

TABLE 5.4 Supply Response Elasticities for Rice in India

Output price (P)	Input price				
	w/P	b/P	m/P	r/P	i/P
0.2357	−0.0017	−0.0004	0.0004	0.0001	0.0017

Notes: w = wage (INR/h), b = cost of animal labor (INR/h), m = cost of machine labor (INR/h), P = price of crop (INR/100 kg), r = cost of fertilizer (NPK) (INR/kg), and i = cost of irrigation (INR/ha).
Source: Kumar, P., Shinjo, P., Raju, S.S., Kumar, A., Rich, M., Msang, S., 2010a. Factor demand, output supply elasticities, and supply projections of major crops of India. Agric. Econ. Res. 23(1), 1–14.

of rice area expansion is limited, in the assumption of no area expansion, rice supply will grow at 2.26% per year. However, in a scenario without growth of TFP or rice area, rice supply would grow at only 1.54%. The average production during 2011–13 has been used as the base year domestic rice production and projected rice supply has been projected by 2030 and compared with demand projections (Table 5.5).

An assessment of crop demand–supply balance for four alternative scenarios provides valuable insights into possible rice self-sufficiency levels in India. Domestic rice production was estimated to be on the order of 103–107 million tons in 2015, 111–121 million tons in 2020, 120–138 million tons in 2025, and 130–157 million tons in 2030. Looking at past trends, India has produced surplus rice and has been exporting in small volumes and building its buffer stock. However, it is highly likely that in the future rice will be met with a marginal surplus under the scenarios with or without TFP growth and area response. Under the TFP growth scenario, India will not only be self-reliant, but could also export rice. Therefore, there is a need to strengthen efforts toward maintaining and accelerating the TFP growth to increase the productivity of rice. This emphasizes the need to strengthen the efforts for increasing rice production through technological changes and raising resource productivity in the less-developed areas.

TABLE 5.5 Projected Domestic Supply and Demand of Rice in India

Year	Supply under different scenarios (million tons)				Demand (million tons)
	S1	S2	S3	S4	
2015	106.5	104.3	105.3	103.1	105.8
2020	121.3	114.7	117.8	111.3	111.8
2025	138.2	126.1	131.7	120.1	116.9
2030	157.4	138.7	147.3	129.7	122.4

Notes: S1 = Baseline assumptions (i.e., current growth of input–output prices, area, and TFP).
S2 = Baseline assumptions without TFP growth.
S3 = Baseline assumptions without area growth.
S4 = Baseline assumption without TFP and area growth.

FUTURE CHALLENGES FOR THE RICE SECTOR

Natural Resource Scarcity

Small and steadily declining farm size, water scarcity, soil degradation, and growing competition for land and water are emerging natural resource related challenges for future rice production. The average operational landholding per household is relatively small in India and it decreased from 2.28 to 1.16 ha during 1971–2011. More than two-thirds of Indian farmers are smallholders, operating less than a 1-ha farm, and their share in the total number of farmers is increasing (Table 5.6). The average rice farm size is even smaller (0.92 ha) and this tiny parcel is fragmented into three to four plots. The competition for rice land is increasing from agricultural diversification, industrialization, urbanization, and residential use; climate change also adds to the problem. The small and fragmented farm size hinders mechanization, commercialization, and economies of scale, and the average farm size is only expected to decline in the short- to medium-term future. On the optimistic side, the trend of declining farm size has reached a tipping point in periurban areas and in economically advanced states, such as Punjab and Haryana, where agriculture is capital intensive and economic opportunities in nonfarm sectors are growing. For example, although farm size has declined by 25% over the past 2 decades at the all-India level, it has remained virtually constant in Punjab and increased by 10% in Haryana. In addition, growing economic opportunities in the nonagricultural sectors and large rural outmigration promoted land consolidation. Policy solutions are needed to facilitate real or virtual land consolidation. One solution is to liberalize and facilitate well-organized land rental markets that assure efficient allocation of agricultural lands. Other solutions are synchronized block, cooperative, group, and contract farming.

To overcome the problem of smallholder farming, several models of virtual land consolidation have emerged in different parts of Asia. The "small farmers–large field" model is becoming popular among small farmers in Vietnam. It

TABLE 5.6 Percentage Share of Farm Size Group in the Total Number of Operational Holdings in India

Year	Marginal	Small	Semimedium	Medium	Large	All
1971	51	19	15	11	4	100
1981	56	18	14	9	2	100
1991	59	19	13	7	2	100
2001	63	19	12	5	1	100
2011	67	18	10	4	1	100

Notes: Farm size group: marginal (0–1 ha), small (1–2 ha), semimedium (2–4 ha), medium (4–10 ha), and large (>10 ha).
Source: The Agriculture Census Division (ACD), various years.

allows small farmers to benefit from economies of scale by pooling their small farms into large fields of 50–500 ha. The aims are to lower per unit production cost using farm machinery (i.e., combine harvesters), lower marketing costs through one-stop sales, and reduce transaction costs. Similarly, an industrial rice-farming scheme introduced by the Suphanburi Rice Millers Association in collaboration with the Suphanburi Rice Research Center in Thailand has provided incentives for farmers to grow one variety with synchronized planting and harvesting time on around 400 ha. This will cut rice-harvesting costs by 20%–30% per hectare by providing service providers with a bigger contract for custom machine harvesting (Mohanty et al., 2015).

The growing scarcity, pollution, and competition for water will increase constraints for future rice production. For example, in Tamil Nadu, rice lands are being diverted to less water-intensive crops because of the growing water scarcity. During 2000–13, rice area in Tamil Nadu decreased by more than 15%. Recent technological innovations in soybeans, millets, and maize, which promise higher returns with less water use, are likely to encourage farmers in semiarid areas to substitute these crops for irrigated rice. The falling groundwater table, due to expansion of irrigated agriculture in Punjab and western Uttar Pradesh, has increased irrigation costs and pressure to improve water productivity. The future rice strategy should invest in water-saving technologies and practices, including rice varieties, as well as crop and farm management practices. Some viable options include the adoption of green super-rice varieties, short-duration and aerobic rice varieties, direct seeding crop establishment method, alternate wetting and drying irrigation, drip irrigation, precision irrigation, and community-based management of irrigation systems.

Agricultural Labor Scarcity

The declining human fertility rate, large rural labor outmigration, aging farming population, youth eschewing farming, and increased employment opportunities in the nonfarm sector have increased wage rates and scarcity of agricultural laborers. The annual growth rate of the agricultural labor force decreased from 1.6% to 1.2% between the 1980s and the 2000s, and it is projected to further dip to 0.8% in the 2010s (FAO, 2015). The scarcity of agricultural laborers, especially during the peak cropping season, is a common phenomenon, but it is most severe in areas dominated by medium and large farms, intensive cropping systems, and good rural–urban links. The average real wage rate of casual male agricultural labor increased by ~50% during 2000–13. The rising labor crisis increased production costs and also compelled farmers to reduce or even abandon the cultivated land. Rice farming is highly labor intensive in states with low mechanization, with labor accounting for more than 40% of the total production cost. The growing labor crisis will increase rice production costs further, which will in turn reduce rice production, profitability, and international competitiveness. Farmers have responded to the crisis by mechanizing labor-intensive activities, such as land preparation, transplanting, threshing, and harvesting (Fig. 5.4).

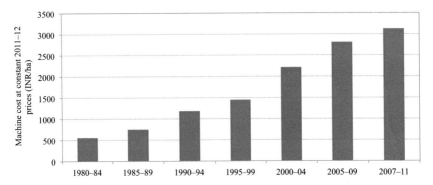

FIGURE 5.4   Trends in machine cost in rice farming in India, 1980–84 to 2007–11 (2011–12 prices). *(Data from The Directorate of Economics and Statistics (DES), 2015. Cost of cultivation of principal crops in India: an online database. DES, Department of Agriculture, Cooperation, and Farmers Welfare, Ministry of Agriculture and Farmers Welfare, Government of India, New Delhi, India.)*

As we anticipate an agricultural labor crisis in the future, states with labor-intensive rice farming should promote labor-saving technologies, such as mechanization, direct seeding crop establishment method, and conservation agriculture. In many cases, large-size agricultural machines may not be suitable for rice farming, where 85% of farmers operate less than 2-ha farms and holdings are fragmented into three to four scattered plots. Two-pronged approaches are warranted to increase mechanization among smallholders. First, promotion of appropriate-scale machines is required to ensure full capacity utilization and an improvement in cost effectiveness. Second, promotion of the machine rental markets through capacity building of young men and women entrepreneurs who can provide machine custom-hiring services for payment. Small-scale farm mechanization and machine custom-hiring services are fast evolving as viable solutions for smallholders. Many young men and women are involved in the service provider business. These service providers cover several hundred kilometers in a season, taking advantage of the differences in planting and harvesting period among different regions. The service providers start in Punjab and Haryana in the northwest, as they harvest their crop early, before they move toward the southeast through Uttar Pradesh, Bihar, Odisha, and West Bengal, covering nearly 1000 km in a season (Mohanty et al., 2015). Existing farm machines have been designed to consider the needs of men; in the context of feminization of agriculture, research and development programs should also promote mechanical technologies considering the needs of women farmers.

Climate Change and Environmental Footprints

Climate change is an emerging challenge for future rice production. It can reduce rice area and yield by changing the amount and pattern of precipitation, increasing the frequency and intensity of climate disasters, enhancing river erosion and salinity intrusion, and elevating sea levels. It also affects rice plant growth and yield through changes in air and water temperature and evapotranspiration. Some studies have shown that an increase of 2°C could decrease rice yield by ~0.75 t/ha in high-yielding areas of India (Sinha and Swaminathan, 1991), while other studies have predicted that climate change will reduce freshwater availability, especially in large river basins. Environmental concern is also expected to exert pressure on rice farming to lower the use of water and agrochemicals. Scientists have already developed abiotic stress–tolerant rice varieties (drought, flood, salinity, and high temperature) to minimize the impacts of climate change, and continuous technological innovations of climate-smart rice varieties would increase farmers' resilience to climate change. Although large investments are being made to develop climate-smart, average-quality rice varieties, future rice strategies should also develop climate-smart basmati and other higher-quality rice varieties. Water-saving technologies alongside water management practices, such as rainwater harvesting, multiuse of reservoirs, and integrated management at the watershed level, can help farmers cope with water scarcity. More investment could be provided for rice R&D for climate-smart varieties, irrigation water management, and crop insurance.

Agricultural Credit Availability

Capital is a critical input to commercial farming. Studies have shown that availability and access to adequate, timely, low-cost, and long-gestation period institutional credit can enable smallholder farmers to adopt improved technologies, transition from subsistence to commercial farming, and add value along the food value chain (Biradar, 2013; Kumar et al., 2010b). This in turn leads to increased production and productivity, superior food security, building assets, higher farm income, and better resilience to shocks. However, access to institutional credit is highly skewed toward large farms; for example, Indian farmers possessing less than 2 ha of land (85%) account for only 51% of the institutional credit, while farmers possessing more than 2 ha of land (15%) account for 49% of the institutional credit (Kumar et al., 2010b). The government has initiated several policy measures to improve farmers' accessibility to institutional credit. In 2004, a comprehensive credit policy was launched to double the flow of agricultural credit within 3 years (with reference to the base year 2003–04). This credit policy emphasized strengthening credit flow at the ground level through credit planning, adoption of region-specific strategies, and rationalization of lending policies and procedures. The Kisan Credit Card (KCC) scheme was launched in 1998–99 to provide farmers with adequate and timely support from the banking system in a flexible and cost-effective manner. By the end of 2014, ~51 million KCC were in operation. As a result of these transformations, institutional credit flow to agriculture increased remarkably and exceeded targets since 2003–04. The institutional credit flow to agriculture increased 10-fold over the last decade from INR 835 billion in 2003–04 to INR 8410 billion in 2014–15 (NABARD, 2014). In 2013–14, the institutional credit flow achievement was 104% of the target; however, evidence suggests that a large proportion of farmers, particularly smallholders, still rely on noninstitutional credit supplied by moneylenders, landlords, and traders, often at high interest rates relative to institutional credit (Rao and Jeromi, 2009). In 2012–13, noninstitutional credit accounted for 44% of the total rural credit flow in India (RBI, 2015). The transformation and implementation of various credit programs have significantly increased farmers' access to institutional agricultural credit, but the persistence of noninstitutional rural credit is a serious concern. A large inequity in access to institutional credit among different categories of farmers is another matter of concern (Kumar et al., 2010b). Credit reforms, especially the simplification of procedures; low interest rates; extension of repayment gestation periods; provision of credit without collateral; and application of ICT and other innovative tools for credit disbursement are needed for smallholders and disadvantaged farmers to have better access to institutional agricultural credit.

CHANGING ROLE OF WOMEN AND YOUTH IN THE RICE VALUE CHAIN

India has the world's second highest number of women working in the agricultural labor force. Women constitute one-third of India's agricultural labor force, with 60% of all "working" women finding employment in agriculture. Rice is the most labor-intensive crop among food grains and women are an indispensable part of rice farming. Women contribute to all activities throughout the rice value chain, although the extent of their contribution varies by farming operation and geographic location. Women contribute to rice farming primarily as paid and unpaid family laborers, and secondarily as farm managers. Gender roles and responsibilities are typically stipulated by several interrelated sociocultural (e.g., religion, social class, caste, ethnicity, age, marital status, and social norms and values), economic, environmental, and human biology factors. In Indian rice farming, women are primarily involved in transplanting, weeding, and harvesting, as well as postharvest operations, while men take the lead in land preparation, crop management, machine operation, threshing, marketing, and farm investment decisions. Typically, men take charge of heavy, physical, and major decision-making tasks. However, gender roles and responsibilities are dynamic and transform in response to socioeconomic and biophysical changes. For example, the increasing outmigration of male labor from rural areas has transformed the traditional division of labor in rice farming, with women not only providing labor inputs but also taking on managerial and decision-making roles on the farm, in the household, and in the community.

At the national aggregate level, on average, women contribute 46% of the total labor inputs required to cultivate 1 ha of rice farm in India, although the labor share is more than 60% in states such as Tamil Nadu and Uttar Pradesh where male outmigration is high. Women constitute half of the farming population and significantly contribute to rice farming, but gender stereotypes and socio-cultural restrictions often create barriers and inequalities for women. This can be seen in access to, and control over, resources, such as land, capital, credit, and agricultural inputs; technology, such as improved crop varieties and farm machinery; services, such as extension, training, information, and marketing; social networks; and household income. For example, women's share in the total number of farm holdings is only 13% in India, although it is gradually transforming toward increased land ownership by women (ACD, 2014). Improving women's access to technologies, resources, and services can improve agricultural productivity, enhance food and nutrition security, reduce poverty, and empower women, which is central to promoting inclusive and sustainable development (Grassi et al., 2015; Meinzen-Dick et al., 2010). Agricultural scientists and policymakers must acknowledge the increasing role of women in rice farming and cater to the needs of women farmers. The future rice strategy should strengthen women's access to technologies, resources, and services; target women in technology development and dissemination, as well as in capacity development interventions; target a higher participation of women in training and technology demonstration activities; and promote agricultural enterprises that benefit women.

India has the world's largest youth (aged 15–29 years) population amounting to approximately 355 million. An additional 1 million will be added every year between 2015 and 2030. This youth population is a valuable asset if properly utilized, but it becomes a burden if not developed. Failure to provide jobs for this large youth population will hamper economic growth, livelihoods, and social stability, as well as perpetuate the intergenerational cycle of poverty. Agriculture, being the major economic sector, will need to generate large numbers of jobs over the next 15 years. Through direct, backward, and forward linkages, agriculture has a great potential to create numerous jobs in input supply, seed production, provision of machine rental services, food production, marketing, processing, retailing, and research. Employing youth in agriculture will be a win–win situation for both agriculture and the whole economy because it will increase employment and productivity. Youth are productive resources with creative minds, who can transform the traditional rice value chain by adopting innovative technologies and practices, such as the use of improved agricultural technologies, mechanization, commercial farming, market-oriented production, application of ICT, and by adding value to primary products. However, today's rural youth are increasingly eschewing farming and most parents discourage their children from taking up the farming profession when alternatives exist. Agriculture is not attractive to the youth because it is arduous, high risk, capital intensive, unstable, less prestigious, less economically attractive, and less intellectually stimulating. Rural life is also boring because it lacks entertainment. Huge numbers of rural youths migrate to cities or abroad to work in nonfarm jobs with higher incomes, which allow them to maintain a decent livelihood. The younger generation will take up rice farming as a profession only if farming becomes economically rewarding, intellectually exciting, and socially prestigious. This requires transformation of the rice value chain. Policy and programs to attract and retain young men and women in rice farming should focus on improving access to land and capital, improving access to developing technologies, enabling access to services (i.e., research, extension, credit, and market) targeting young farmers, mechanization, commercial production, market development, and facilitation of collective production and marketing. Programs should also motivate and develop the capacity of young men and women through training, skills development, mentoring, and exposure. In addition, infrastructure, ICT-based knowledge management tools and services, and connections to urban areas should be developed so that youth living in rural areas can connect to urban life and do business with urban centers.

VERTICAL INTEGRATION IN THE RICE VALUE CHAIN

The Indian rice value chain is highly complex, spanning input suppliers (including manufacturers, wholesalers, and retailers), producers, large numbers of intermediaries (including collectors, traders, commission agents, and brokers), wholesalers, processors (including millers and food companies), retailers, and end consumers. The value chain encompasses huge diversity and variety at each stage. The traditional rice sector development approach, which largely focused on on-farm production systems, has been transformed with the focus now on a systems approach spanning the whole value chain, with a greater emphasis on horizontal and vertical integration.

Most Indian rice farmers are smallholders who operate less than 1-ha rice farms. A large portion of rice produced by smallholders is self-consumed and only a small surplus quantity is sold, mainly to local traders. Therefore, the rice passes through a long supply chain before reaching end consumers. Smallholders sell their products to traders individually, which results in "low volume, high cost" for traders; however, large farmers sell their products directly to large traders or millers, making the supply chain shorter. Rice farmers face several constraints, including small farm size, untimely availability of quality inputs, low mechanization, poor access to new technologies and finance, poor technical knowledge on improved farming practices, production risk, low and volatile output prices, and poor access to markets and information. Private agribusiness companies, such as input suppliers, wholesalers, organized retailers, and food companies, face different bottlenecks, including selling their inputs and services to farmers, securing sufficient quantities of good-quality products, high logistical costs (transportation, storage, and distribution), and logistic coordination. Traditional traders face bottlenecks related to storage, transportation, packaging, branding, quality, certification, and high logistical costs. Lack of appropriate postharvest facilities and large numbers of members cause product losses and inefficiency in the value chain.

The ongoing transformation in supply and demand sides has created opportunities to increase productivity of the rice value chain. On the supply side, higher public and private investment in R&D across the value chain as well as technological innovations in plant genomics and ICT have generated a continuous stream of innovations. The private sector is evolving to supply services, including extension and training, technology transfer, machine custom hiring, crop protection, soil testing, input supply, and digital finance. The growth of modern food retailers has also widened opportunities. They can source products directly from farmers and greatly improve efficiency in the supply chain by reducing wastage, increasing prices for producers, and reducing prices for consumers. The contract farming of basmati rice between Pepsi Foods Private Limited and farmers in Punjab is a case in point. On the demand side, population growth, economic growth, and growing concerns for health and nutrition are driving the demand for improved quantity, quality, and nutrition of rice. The growing middle-class and youth population, urbanization, dietary diversification, changing consumer preferences, health and wellness concerns, and the growing demand for ready-to-prepare and ready-to-eat foods have created new markets for different food products, including rice.

The changing drivers of food supply and demand have brought enormous potential for the rice value chain to transform and move away from where it is today. However, this potential cannot be realized under the business-as-usual scenario because smallholders lack the capacity to exploit the supply-side opportunities and to respond to the demand-side requirements. For example, smallholders are lacking in the knowledge, skills, and capacity to satisfy the quantity, quality, and continuous supply requirements of modern food retailers. Therefore, the solution lies in horizontal and vertical integration. Horizontal integration—such as organizing smallholders into a "club" or "group" and integrating their rice fields into one large field—benefits farmers in terms of labor sharing; economies of scale; market access; providing bargaining power with traders; access to land, finance, and other resources; access to technologies, such as improved varieties, storage, and farm machines, such as combine harvesters and dryers; reduced production costs, as private firms are willing to offer mechanization services at lower costs due to scale; collective marketing; and contract farming with agribusiness firms. Input firms benefit through reduced transaction costs, as they are dealing with only one farmers' club instead of many individual farmers. Output firms benefit from fewer transactions; an assured supply of high-quality rice, as farmers' clubs grow rice varieties based on market demand; and improved capacity utilization of large rice mills due to a higher supply of the same rice varieties. Milling of the same rice variety with uniform physical traits will result in an increased milling recovery rate, decreased percentage of broken grains, and enhanced quality. The farmers' club could also be a good organizational tool for coordinating input purchases, arranging collective acquisition and use of machine custom rental services, transferring technology, and attracting training programs for members. Despite several benefits, this group approach often fails because of a lack of trust among members, unequal access to benefits, low incentives to participate, and financial irregularities. Training to develop organizational capacity and business skills of the farmers' club, provision of incentives to work in groups, group-based service provision, and trust development among members would enhance horizontal linkages.

Strong vertical integration among rice value chain members brings several advantages, such as flow of products and information from upstream (farmers) to downstream (retailers); flow of orders, consumer preferences, and information from downstream to upstream; higher efficiency in the value chain; supply of produce to modern food retailers; assured markets for farmers and assured supplies for retailers; shortening of the supply chain, resulting in higher returns for farmers and lower prices for consumers; reduced postharvest losses; greater control over upstream and downstream activities; lower costs for companies; potential opportunities for agribusiness companies to sell their inputs and services to farmers; access to new technologies, knowledge, and skills; access to resources and services; promotion of innovations; and enhanced value addition. In strong formal vertical integration, some agribusiness firms provide certified seeds and a package of practices for cultivation to be assured of pure and quality products. Other firms invest in the relationship by prefinancing other inputs, such as seeds, fertilizers, and pesticides. Vertical integration facilitates and provides impetus for productivity improvement in the whole value chain.

Formal vertical integration is weak in the Indian rice value chain, but it is gradually improving, especially for premium-quality rice. An informal vertical integration, based on personal relationships, can be found between smallholder farmers and local traders at the village level. Information on seeds, proper use of fertilizers and pesticides, and crop production techniques flows from input suppliers and dealers to farmers on an ad hoc basis. At the same time, market intelligence and information about consumer preferences flow from traders to farmers, but true price discovery mechanisms are not transparent. Credit is extended between members, such as from big traders to small traders, from small traders to farmers, and from input suppliers to farmers. A future rice strategy that aims to improve efficiency in the value chain should foster horizontal and vertical integration in two ways: first, by establishing, nurturing, and strengthening farmer clubs; and second, by increasing market orientation of farmer clubs by linking them with input suppliers, traders, millers, and agribusiness firms. The current market underutilizes end markets as signals for product development and value chain upgrading. Policies and programs should establish transparent, incentive-based vertical integration of members in the value chain through mechanisms, such as promotional events, prepaid purchasing schemes, price premiums, and contract arrangements. For example, agribusiness firms establishing direct contact with farmer clubs; agribusiness firms providing inputs including seed, fertilizer, pesticides, and technical knowledge; and farmer clubs following production and postharvest practices recommended by firms, which would then buy the products on preagreed arrangements. This will generate a win–win situation for both farmer clubs and traders.

INTERNATIONAL COMPETITIVENESS OF THE INDIAN RICE SECTOR

Comparison of rice production, consumption, and trade with other top rice producers will help India to assess its competitiveness in the world rice market and permit the development of strategies to further increase competitiveness. The international competitiveness of the Indian rice sector was analyzed by comparing 13 indicators between India and the world's other top 5 rice producers, namely China, Indonesia, Bangladesh, Vietnam, and Thailand (Table 5.7). India's per capita income in real terms grew 5.1% per year during 1991–2013. Despite this impressive growth, India's per capita income (USD 1610) is still far below that of the other five countries, with the exception of Bangladesh. A relatively low income and high poverty rate (22% or 270 million) makes rice an important component of the Indian food menu, as it is a cheap source of dietary energy.

India's paddy yield (3.6 t/ha) is the second lowest among the major rice-producing countries, following Thailand. This large yield gap indicates a great potential to increase rice production in India. Despite a large growth in rice production in the past, per capita milled rice production (84 kg) in India is lowest among all the major rice-producing countries. However, relatively low per capita milled rice consumption (70 kg/year) made India the leading exporter in the world. As the population is expected to grow by 15 million every year for the next 15 years, India must increase rice productivity and production to supply adequate quantities of rice at affordable prices to domestic consumers, and enable it to retain the title of the leading player in the world rice market. Rice is the single largest source of dietary calorie intake in India, but its relative importance in the dietary energy supply is one of the lowest among the top rice-producing countries. A secured supply of nutritious rice at an affordable price is pivotal to improving nutrition security in India.

India's average rice farm size (0.92 ha) is the second highest among the five countries, following Thailand. Hence, Indian rice farmers are relatively better placed in terms of farm size and economies of scale than those in the other four countries. However, this farm size is still inadequate for viable mechanization, attracting private sector professional service providers, commercial farming, and generating large marketable surplus. The solution lies in the horizontal integration of smallholders to form a business model of "small farmers–large field," which will allow smallholders to benefit from economies of scale by organizing themselves and integrating their small rice areas into one large field. Horizontal and vertical integrations provide several benefits to smallholders (see section "Vertical Integration in the Rice Value Chain" for details).

Irrigation expansion can partially offset reduced production due to small farm size. The proportion of rice irrigated area in India is much lower than in China and Vietnam, although it is much higher than in Bangladesh and Thailand. Irrigation expansion will increase rice production and make the Indian rice sector more competitive by reducing production cost per unit quantity via higher yield. India's adoption of HYV rice is lower than China and Vietnam,

TABLE 5.7 Comparison of India's Rice Sector With the Other Top Five Rice-Producing Countries, 2015

Variable	Unit	Year	IND	PRC	INO	BAN	VIE	THA
Income per capita	USD	2013–14	1610	7380	3650	1080	1890	5410
Paddy yield	t/ha	2011–13	3.6	6.7	5.1	4.4	5.6	3.0
Rice production per capita	kg	2011–13	84	97	184	217	315	361
Rice consumption per capita	kg	2011–13	70	79	134	172	144	114
Rice's share in food calories	%	2007–09	30	26	48	70	53	44
Paddy farm size	ha/hh	1993–08	0.92	0.55	0.36	0.40	0.74	2.60
Paddy irrigated area	%	2009–11	59	94	60	47	88	26
Paddy area under HYV	%	2005–11	79	95	84	80	96	46
NPK use in paddy	kg/ha	2010–11	160	221	117	109	151	54
Labor use in rice farming	d/ha	2005–11	89	182	80	118	78	86
Paddy cost of production	USD/t	2010–14	239	287	277	259	156	212
Paddy farm harvest price	USD/t	2010–12	227	404	384	205	195	388
Rice net trade quantity	Mt	2012–14	10.41	–3.52	–1.04	–0.62	6.31	8.83

Notes: Paddy refers to rice measured in unmilled form, while rice refers to rice measured in milled form. Net trade refers to exports minus imports. BAN, Bangladesh; d, day; ha, hectare; HYV, high-yielding varieties; IND, India; INO, Indonesia; kg, kilogram; Mt, million tons; PRC, China; t, ton; THA, Thailand; VIE, Vietnam.
Source: Data were collected from many sources. Please contact the authors for detailed information.

although it is comparable with that of Bangladesh and Indonesia, and higher than Thailand. Higher adoption of HYV rice, especially premium quality, is a good option for increasing rice production and international competitiveness. India's chemical fertilizer use (in terms of NPK) in rice farming is one of the highest among the six countries. This suggests a need to improve the efficiency of fertilizer use to reduce production costs and environmental footprints of rice farming. The average labor use in Indian rice farming is ~90 person-days per hectare, which is significantly lower than in China and Bangladesh, but by and large comparable to Indonesia, Vietnam, and Thailand. The mechanization of rice farming is relatively low in many states of India. In the context of rising scarcity and wage rates of agricultural laborers, increased appropriate-scale mechanization can save labor inputs, reduce production costs, and make the Indian rice sector more competitive internationally.

Production cost is an important determinant of market price, comparative advantage, and international competitiveness of crops. Comparison of the average production cost for medium-quality grain paddy rice indicates that India's production cost (~USD 240 per ton) is lower than that of China, Indonesia, and Bangladesh, but higher than that of the leading rice exporters, namely Thailand and Vietnam. India must lower the production cost per ton to compete with Thailand and Vietnam. The options for reducing production costs lie in either increasing yield or improving the technical efficiency of input use. Horizontal integration of small farms to make a large operating field, optimum use of inputs, and mechanization are other potential options to help reduce production costs. The producer price of paddy in India was ~USD 230 per ton, which is significantly lower than in China, Indonesia, and Thailand, but slightly higher than in Bangladesh and Vietnam. When compared to the production cost, producer price is lower in India and Bangladesh, but higher in China, Indonesia, Vietnam, and Thailand. A high minimum support price for paddy offered by the government caused a relatively high producer price in China and Indonesia, which benefited producers. A relatively low production cost resulted in a low producer price in Vietnam and made its production competitive in the international market. The higher producer price in Thailand can be attributed to higher grain quality. A positive profit margin makes Vietnamese and Thai rice more competitive than Indian rice, which incurs a negative profit margin per ton.

India beat its counterpart Thailand to become the world's largest rice exporter in 2011, and it has maintained this leadership position ever since. Among the world's top rice producers, the average net traded (exports minus imports) quantity of milled rice in 2012–14 was highest in India (10.41 million tons), followed by Thailand (8.83 million tons) and Vietnam (6.31 million tons). Bangladesh, China, and Indonesia are net rice importers, whereas India can export rice.

The world rice market is highly competitive and new players, such as Cambodia and Myanmar, are entering the game. India cannot remain complacent and must increase its competitiveness to retain its leadership title in the world rice market. The strategy to improve competitiveness of the Indian rice sector should focus on five programs: (1) increasing marketable surplus of medium-grade rice; (2) increasing the supply of basmati and other premium-quality rice that can compete with upper-grade Thai rice and other premium-grade rice; (3) reducing production costs and market price; (4) adding value by adopting effective marketing strategies, including branding, packaging, and labeling of product contents; and (5) crafting and implementing conducive policies to attract private investment and promote exports. Continuous investment in R&D for both basmati and nonbasmati rice is necessary to increase productivity, profitability, and quality. The production strategy should be based on comparative advantage: basmati in northwestern states, high-grade rice in southern states, and medium-grade rice in northern and eastern states. A strong vertical integration in the rice value chain not only reduces costs, but also assures a secured supply of the desired quantity of high-quality products demanded in the market.

CONCLUSIONS

India's remarkable rice productivity growth in the past 4 decades increased food security, decreased poverty, improved livelihoods, and made the country a leader in the world rice market. However, the ongoing rapid social and economic transformations are posing new challenges and opportunities to the rice sector. On the demand side, population growth, a growing middle-class population, urbanization, changing food habits and diets, and awareness about health and nutrition effects of food are affecting the rice sector. On the supply side, land and water scarcity, rising labor shortage, rising input prices, aging farmers, feminization of agriculture, climate change, and pressure to reduce environmental footprints are affecting the rice sector. At the same time, access to innovate technologies and service provision, better information flow, globalization, economic and trade liberalization, vertical integration in the supply chain, marketing through modern food retailers, increased investment in postharvest and value addition, and increased investment of the private sector in agricultural R&D have opened up new opportunities.

The Indian rice sector is transforming in the right trajectory. Key transformative changes have seen a shift from subsistence to semisubsistence and commercial farming, from on-farm production to a value chain approach, from high volumes to higher-value and nutritious rice, and from northwest India to eastern India. Other transformative changes included intensification and diversification of rice-based cropping systems, expansion of irrigated rice, mechanization, application of ICT for agro-advisory service provision, increased access to institutional credit for smallholders, increased efficiency in input use, automation of the milling system, improved packaging and branding, and increased participation of modern food retailers in processing and marketing. Additional changes included firm-farm contract rice farming, horizontal and vertical intergradations in the rice value chain, higher investment of both the public and private sectors in rice technologies and business, focus on attracting the youth in agriculture, targeted programs for women farmers, and favorable policies and business enabling environment for private sector involvement and exports. Although these transformations are encouraging, the pace of change has been too slow. The rice sector must undergo rapid transformative changes to address new challenges and to exploit new opportunities. The rapid transformation is necessary to increase and sustain higher productivity, profitability, and efficiency of the rice value chain, which is central to ensure future food and nutrition security, contain the spike in rice prices, generate handsome incomes for farmers, attract youth into farming, increase competitiveness in the world rice market, and earn foreign exchange reserves.

Rice demand will increase in the next 15 years, but the demand growth rate will be lower than the population growth rate. Consequently, per capita rice consumption in India will steadily decelerate in the future. At the same time, rice supply will increase, but the magnitude will largely depend on TFP growth and area expansion. In the next 15 years, domestic rice supply will exceed domestic rice demand with or without TFP growth and area expansion. The surplus quantity will be small without TFP growth, but large with TFP growth. India could remain a large rice exporter with TFP growth and therefore efforts to increase rice production should target maintaining and accelerating TFP growth.

Basmati growers earn more revenue than nonbasmati growers. Rice farmers' livelihoods can be significantly improved by promoting the cultivation of premium-quality rice varieties complemented by good market systems. Growing incomes, a middle-class population, awareness about health and nutrition benefits of food, demand for high-quality and branded products, and increasing participation of supermarkets in food marketing have opened up new market opportunities for the rice industry, especially for premium-quality rice.

A relatively small farm size, low mechanization and irrigation, high production cost, low yield, low profit, and high price make Indian rice production less competitive vis-à-vis the major rice exporters, such as Thailand and Vietnam. Despite this, India has emerged as the world's largest rice exporter primarily because of the high quality of Indian rice. India has a great potential to dominate the world rice export market and earn a large amount of foreign currency by exporting rice, such as basmati. This requires increasing domestic marketable surplus and price competitiveness in the world rice market.

The following strategies are recommended to develop a vibrant, efficient, and internationally competitive Indian rice sector: (1) Policies and programs should be implemented that increase production of average-grade nonbasmati rice. This should be mediated through technological change and raising resource productivity in less-developed rice-growing eastern and northern states where the rainfed ecosystem is predominant, subsistence risk-averse farmers are relatively less receptive to improved technologies, average yield is low, and yield gaps are high. (2) Policies and programs should be implemented that increase production of basmati rice in northwestern states and high-grade nonbasmati rice in other states, especially southern states. This should occur through the enhancement of technological change and resource productivity. (3) Increased investment should be made in postharvest and value-addition steps to reduce postharvest losses, improve market systems, and enhance product processing, packaging, and branding. (4) Horizontal integration of smallholder producers and vertical integration of members in the supply chain should be strengthened. This will increase access to improved technologies and services, reduce cost per unit quantity, and assure a secured supply of the desired quantity and quality of rice. (5) Adequate investment in R&D should be made, particularly in the development and dissemination of improved technologies (i.e., climate-smart, nutrient-rich, and healthy rice varieties), which should be complemented by enhanced infrastructure, mechanization, credit, and agro-advisory services. (6) Policies that enhance production and exports should be crafted and implemented alongside the generation of businesses that will attract FDI from modern grocery retailers. (7) Policies and programs to increase the participation of women and youth in the rice value chain should be crafted and implemented. Together, these policies will allow India to remain as the leader in the world rice market by enhancing productivity and profitability of the rice value chain.

REFERENCES

The Agriculture Census Division (ACD), 2014. All India report on number and area of operation holdings. Agriculture Census 2010–11. ACD, Department of Agriculture, Cooperation and Farmers Welfare, Ministry of Agriculture, Government of India, New Delhi, India.

Agricultural and Processed Food Products Export Development Authority (APEDA), 2015. India export of agro food products: an online database. Ministry of Commerce and Industry, Government of India. Available from: http://agriexchange.apeda.gov.in/indexp/reportlist.aspx

Biradar, R.R., 2013. Trends and patterns of institutional credit flow for agriculture in India. J. Asia Business Stud. 7 (1), 44–56.

Bouis, H., Haddad, L., 1992. Are estimates of calorie-income elasticities too high? A recalibration of the plausible range. J. Dev. Econ. 39 (2), 333–364.

Chand, R., Kumar, P., Kumar, S., 2012. Total factor productivity and returns to public investment on agricultural research in India. Agric. Econ. Res. Rev. 25 (2), 181–194.

The Directorate of Economics and Statistics (DES), 2015b. Cost of cultivation of principal crops in India: an online database. DES, Department of Agriculture, Cooperation, and Farmers Welfare, Ministry of Agriculture and Farmers Welfare, Government of India, New Delhi, India.

Ernst and Young, 2013. Hitting the Sweet Spot: The Growth of the Middle Class in Emerging Markets. Ernst and Young Global Limited, New York, USA.

Food and Agriculture Organization of the United Nations (FAO), 2015. FAOSTAT: FAO Online Statistical Database. FAO, Rome, Italy. Available from: http://faostat3.fao.org/home/E.

Grassi, F., Landberg, J., Huyer, S., 2015. Running Out of Time: The Reduction of Women's Work Burden in Agricultural Production. FAO, Rome, Italy.

Hazell, P., Ramasamy, C., 1991. The Green Revolution Reconsidered: The Impact of High-Yielding Rice Varieties in South India. Johns Hopkins University Press, Baltimore.

Janaiah, A., Hossain, M., Otsuka, K., 2006. Productivity impact of the modern varieties of rice in India. Dev. Econ. 44 (2), 190–207.

Joshi, P.K., Kumar, P., 2011. Food Demand and Supply Projections for India: 2010–2030. Trade, Agricultural Policies, and Structural Changes in India's Agrifood System: Implications for National and Global Markets (TAPSIM) Project No. KBBE212617. IFPRI, New Delhi, India.

Kharas, H., 2010. The emerging middle class in developing countries. OECD Development Centre Working Paper No. 285. The Organization for Economic Cooperation and Development (OECD), Development Centre, Paris, France.

Kumar, P., Mruthyunjaya, Birthal, P.S., 2007. Changing consumption patterns in South Asia. In: Joshi, P.K., Gulati, Jr., A., Cummings, R. (Eds.), Agricultural Diversification and Smallholders in South Asia. Academic Foundation, Kolkata, India, pp. 151–187.

Kumar, P., Shinjo, P., Raju, S.S., Kumar, A., Rich, M., Msang, S., 2010a. Factor demand, output supply elasticities, and supply projections of major crops of India. Agric. Econ. Res. 23 (1), 1–14.

Kumar, A., Singh, K.M., Singh, S., 2010b. Institutional credit to agriculture sector in India: status, performance, and determinants. Agric. Econ. Res. 23, 253–264.

Meinzen-Dick, R., Quisumbing, A., Behrman, J., Biermayr-Jenzano, P., Wilde, V., Noordeloos, M., Ragasa, R., Beintema, N., 2010. Engendering Agricultural Research, Development, and Extension. International Food Policy Research Institute, Washington, DC.

Mohanty, S., Bhandari, H., Mohapatra, B., Baruah, S., 2015. The ongoing transformation in rice farming in Asia. Rice Today 14 (4), 37–39.

National Bank for Agriculture and Rural Development (NABARD), 2015. Annual Report 2014–15. NABARD, Bandra, Mumbai, India.

Rao, V.M., Jeromi, P.D., 2009. Modernizing Indian agriculture. In: Krishna, K.L., Kapila, U. (Eds.), Readings in Indian Agriculture and Industry. Academic Foundation, New Delhi, pp. 77–130.

The Reserve Bank of India (RBI), 2015. All India debt and investment survey. RBI, Government of India, New Delhi, India.

Sinha, S.K., Swaminathan, M.S., 1991. Deforestation, climate change and sustainable nutrition security: a case study of India. Clim. Change 19, 201–209.

United Nations, Department of Economic and Social Affairs (UNDESA), 2015. World Population Prospects: The 2015 Revision. Population Division, The United Nations, New York, NY.

Wani, S.A., Manhas, S.K., Kumar, P., 2015. Basmati rice in India: its production and export. A newsletter article. FNB News. Available from: www.fnbnews.com.

Chapter 6

Frontiers in Rice Breeding

Darshan S. Brar*, Kuldeep Singh*, Gurdev S. Khush**
*Punjab Agricultural University, Ludhiana, Punjab, India;
**University of California, Davis, Davis, CA, United States

Rice is a staple food for more than half of the world's population with over 3.5 billion people depending on rice for at least 20% of their daily calorie intake. Global rice demand is estimated to rise from 723 million tons in 2015 to 763 million tons in 2020 and to 852 million tons in 2035, an overall increase of 18% (129 million tons) in the next 20 years. Increasing rice yields from existing land remains the primary strategy for increasing production. Grain production is largely influenced by the yield potential of rice varieties and management practices. Therefore, improvement in the yield potential of rice is a major strategy for increasing world rice production. However, both genetic enhancement and management technologies are important to realize the yield potential of rice varieties. To produce 129 million tons of additional paddy rice by 2035, the yield potential of rice needs to increase from the present 10 to 12.3 t/ha under favorable irrigated conditions (Khush, 1995). This may sound difficult but the history of rice breeding is encouraging and is an inspiration for achieving future requirements.

World rice production increased from 256 million tons in 1960 to 680 million tons in 2010. This was achieved by increased productivity, primarily through the application of principles of Mendelian genetics and conventional plant breeding. Achieving the target of an additional 129 million tons in the next 20 years is more challenging given the continual threat to rice by major diseases, such as bacterial blight (BB), blast, tungro, and sheath blight; insects, such as the brown planthopper (BPH), stem borers, and gall midges; and abiotic stresses, such as drought, floods, salinity, cold, heat, etc., particularly in the context of global climate change. The major challenge at present is how to overcome these constraints and increase rice productivity in an ecofriendly way using fewer chemicals, less water, less land, and less labor (Khush, 2013). Producing enough rice for the world's population may be easy but doing it at an acceptable cost to the planet will depend on research in the cutting-edge science of genomics and implementation of this research from high-tech to low-tech farming practices. This calls for an "evergreen" revolution in agriculture, leading to the improvement of crop productivity, and sustainability without associated ecological harm and environmental degradation.

Recent advances in biotechnology, particularly in the fields of cellular and molecular biology and genomics, have opened new avenues to overcome some of the constraints that limit crop productivity, to develop improved germplasm with novel genetic properties, to accelerate crop breeding programs, and to understand the function of the genes that govern agronomic traits. Khush and Brar (2001) and Brar and Khush (2006, 2013) have reviewed advances in rice biotechnology and applied genomics. Developments in rice biotechnology include (1) the production of dense molecular maps and identification of a new generation of markers, such as SNPs that facilitate mapping of genes/QTLs and the precise characterization of genetic resources; (2) tagging and pyramiding of many important genes/QTLs for use in marker-assisted selection (MAS) to accelerate breeding programs for enhancing tolerance of biotic and abiotic stresses and improved quality traits; (3) map-based cloning of agronomically important genes that facilitate gene-based MAS and allele mining; (4) broadening the gene pool of rice through the transfer of novel genes from wild relatives and the characterization of alien introgression using molecular markers; (5) molecular characterization of pathogen populations for the deployment of specific resistance genes under temporal and spatial conditions; (6) the availability of a large set of new genomics resources, such as T-DNA insertion lines, deletion mutants, and YAC, BAC, and EST libraries for functional genomics; (7) high-throughput genotyping [SNP chips, genotyping-by-sequencing (GBS)] for MAS and gene chips for gene discovery and expression analysis; (8) high-throughput protocols for *Agrobacterium*-mediated transformation and large-scale production of transgenics with new genetic properties, including pyramiding of transgenes and; (9) the availability of cloned genes and genome sequences for future research in forward and reverse genetics.

Genome-wide selection, sequence-based MAS, and next-generation sequencing offer new opportunities to further accelerate breeding. Genomics approaches are important components of molecular breeding as they permit the development of designer crops with value-added traits. RNA interference (RNAi) technology is becoming increasingly important for enhancing pest resistance and improving the quality characteristics of crops as it allows gene expression to be modified or silenced. Genome editing technologies are expected to become useful tools in rice research, especially in instances when directed changes are required to suppress the expression of a gene, such as an antinutritional gene. New tools in biotechnology are also being used to engineer the photosynthetic system and to introduce C_4 pathway-related genes into C_3 rice. The identification and manipulation of endophytes offer new ways for enhancing biological nitrogen fixation. Some traits, such as nutrient-use efficiency, apomixis, and the identification of QTLs for enhancing heterosis, need biotechnological interventions. Integration of the abovementioned molecular approaches with conventional breeding is emphasized to accelerate breeding programs and to increase rice productivity and sustainability.

Using conventional genetic and cytogenetic techniques coupled with molecular marker technology, several specialized genetic stocks, such as recombinant inbred lines (RILs), near-isogenic lines (NILs), chromosome segment substitution lines (CSSLs), alien introgression lines (AILs), TILLING populations, and sequence-based characterized germplasm, are used for breeding and functional genomics.

CONVENTIONAL BREEDING APPROACHES

Traditional hybridization and empirical selection will remain a widely used strategy for developing crop varieties with a high yield potential. In this methodology, segregating populations, derived from crosses between two parents, are screened for desirable recombinants and selected lines are evaluated in replicated yield trials. This approach is therefore based on the variability created through hybridization between diverse parents and the subsequent selection of desirable offspring. Using this strategy, an approximate 1% per year increase in yield potential of cereals, including wheat, barley, and rice, has been observed, and this is a time-tested strategy. Conventional breeding using genetic resources for dwarfing genes, male sterility, heterotic groups, including donors for various ideotypes, early maturity, tolerance of major biotic and abiotic stresses, and quality characteristics has been used worldwide to develop high-yielding, pest-resistant varieties tolerant of the major abiotic stresses and possessing superior quality characteristics. Several success stories are known in which dwarfing genes, male sterility, heterotic gene pools, early maturity, different ideotypes, intersubspecific crosses (indica × japonica rice), genetic donors, and genes identified through classical Mendelian genetics for tolerance of major biotic stresses (diseases, insects) and abiotic stresses (drought, submergence, salinity, heat, cold, unfavorable soil conditions, etc.) and for superior quality characteristics have played a major role. As an example, more than 1000 rice varieties have been released in India as a result of conventional breeding. Similarly, lines bred by the International Rice Research Institute (IRRI) have been released as 864 rice varieties in 78 countries. Conventional rice breeding has, so far, relied on the following approaches and strategies.

Ideotype Breeding: The New Plant Type

Ideotype breeding, aimed at modifying plant architecture, is also a time-tested strategy to increase yield potential. Selection for short-statured cereals such as rice and wheat resulted in a doubling of their yield potential. Yield is a function of total biomass and harvest index (HI). Tall and traditional rice varieties could produce a biomass of 13–14 t/ha under most conditions, and their HI was around 0.3. The first semidwarf or short-statured variety (IR8) developed at IRRI had a combination of desirable traits including profuse tillering, dark green and erect leaves for good canopy architecture, and sturdy stems for lodging tolerance. It was responsive to nitrogenous fertilizer and could produce a biomass of ~20 t/ha. This plant type was widely accepted and most global rice breeding programs adopted this plant. To increase the yield potential of semidwarf rice further, IRRI scientists proposed a new plant type (NPT) in 1989 (IRRI, 1989) with characteristics such as reduced tillering (9–10 tillers for transplanted conditions), zero nonproductive tillers, 200–250 grains per panicle, dark green and erect leaves, a vigorous and deep root system, growth duration of 110–130 days, multiple disease and insect resistance, and a higher HI.

Breeding efforts to develop NPTs began in the early 1990s by crossing tropical japonica (TJ) varieties called *Bulu* rice with semidwarf indica varieties. The *Bulu* rice varieties from Indonesia are known to have low tillering ability, large panicles, sturdy stems, and a deep root system. *Bulu* varieties were initially crossed with a semidwarf japonica breeding line (Sheng Nung 89-366) from Shenyang Agricultural University in China. More than 2000 crosses were made and more than 100,000 pedigree lines were evaluated. Breeding lines with desirable ideotype traits were selected and more than 500 NPT-TJ-type lines were evaluated in yield trials. Some of these lines also showed good agronomic performance in high temperate areas where japonica-type grain quality is preferred. Three of these lines (Diancho-1, Diancho-2, and Diancho-3) were released as varieties in Yunnan Province in China (Khush, 1995). To improve the adaptability of these NPT-TJ lines for tropical conditions, and to improve their yield potential, they were crossed with elite indica lines, as well as varieties possessing disease and insect resistance and good grain quality. Several lines (designated as NPT-IJ types) were evaluated in yield trials, where many outyielded the best indica varieties (i.e., IR72) by as much as 1.0–1.5 t/ha. These lines have been continuously used in breeding programs and have contributed to increased genetic diversity through the introduction of japonica germplasm into otherwise purely indica breeding materials.

Similar to the NPT breeding program, Chinese scientists proposed another ideotype with attributes such as moderate tillering capacity, heavy panicles (5 g/panicle), slightly taller plant height (about 100 cm), and the top three leaves having characteristics including flag-leaf length of ~50 cm and length of ~55 cm for the second and third leaves. All three leaves should be above panicle height and should remain erect until maturity. Leaf angles of the flag leaf and second and third leaves should be 5, 10, and 20 degrees, respectively. The leaves should be narrow and V-shaped (2 cm when flattened), and should be thick and dark green. Leaf area index of the top three leaves should be ~6 and it should have a HI of 0.55 (Yuan, 2001). Rice breeders in China have developed many inbreds and hybrids of this ideotype.

Heterosis Breeding

Hybrid breeding exploits the increased vigor or heterosis (yield advantage of hybrids over varieties) of F_1 that is generally observed in outcrossing species.

In maize, major yield improvement has been associated with the introduction of F_1 hybrids on a commercial scale, so much so that the total maize area in the United States and Europe is planted as hybrids. The average yield advantage of maize hybrids over other varieties is approximately 15%–20%. Rice hybrids were introduced in China during the mid-1970s and are now planted on 13–14 million hectares or 50% of the total rice area in that country. The average yield advantage of the current rice hybrids over other varieties is 10%–15%. The limited adoption of rice hybrids in other countries is due to the low level of heterosis. Most breeding programs in the tropics have used indica germplasm in hybrid development, which has limited genetic diversity. In their super hybrid breeding program, Chinese scientists developed parental lines with variable amount of japonica alleles in an otherwise indica background. These hybrids have higher heterosis. In this context, the use of NPT-IJ-type lines in the breeding program for hybrid development should help to improve heterosis. In the United States, the two-line system is used in which male sterile female lines are tropical japonicas and restorers are indicas. These hybrids are reported to have as much as a 20%–25% yield advantage over inbred parents. Besides China, the public and private sectors in India, Bangladesh, Indonesia, and Vietnam have active hybrid programs; however, the area under hybrids has not increased significantly in these countries. As an example, hybrids in India occupy only 2.5 million hectares of the total 44 million hectares under rice. Breeding strategies should focus on enhancing heterosis coupled with superior grain quality.

With the exception of China and the United States, the area under hybrid rice has not surpassed 5% of the total rice area planted in any country. This is primarily because of the low heterosis in indica × indica crosses. Enough evidence is available in the literature to indicate that higher heterosis rates can be achieved with indica × japonica crosses, although this has not been capitalized on so far because of the sterility in F_1s, poor grain quality, and susceptibility of tropical japonicas to insect pests and diseases. Therefore, prebreeding of tropical japonicas for specific improvements of these traits following MAS can make them usable for developing high-yielding hybrids. Tropical japonicas have several yield-related traits, such as heavy panicles (more grains per panicle), higher grain weight, and sturdy stems, but they possess coarse grains and are susceptible to major diseases and insects. Hence, prebreeding of tropical japonicas for improvement of their grain quality and resistance to diseases and insect pests is an important breeding strategy for their use in exploiting heterosis. A breeding strategy for prebreeding of tropical japonicas appears in section "Prebreeding Tropical Japonicas to Enhance Heterosis." Prebreeding lines possessing pest resistance, good grain quality, and other desirable traits should be used in crosses with indica rice. The isolation of desirable recombinants possessing key target traits could enhance the yield potential of inbreds and also lead to the development of heterotic indica rice hybrids suitable for commercial cultivation.

Wide Hybridization

Wide hybridization involves hybridization between cultivated species and their wild relatives. Broadening the gene pool of a crop is an important plant breeding method as it can enhance tolerance of major biotic and abiotic stresses and improve the quality characteristics of the plant. Several examples demonstrate the successful transfer of useful genes from wild species to wheat, oats, rice, cotton, Brassica, tomato, and other crop plants. The genus *Oryza* to which cultivated rice (*Oryza sativa* $2n = 24$, AA) belongs has 24 species (having $2n = 24$, 48) representing different genomes (AA, BB, CC, BBCC, CCDD, EE, FF, GG, HH, JJ, and HHJJ). These wild species act as a reservoir of useful genes for resistance to diseases and insects, for tolerance of abiotic stresses, and for cytoplasmic diversification. Interspecific hybrids can be produced through direct crosses between rice and AA genome wild species, and wild species genes can be transferred following backcrossing with the recurrent rice parent. However, crosses between rice and all wild species other than AA genome require embryo rescue to produce hybrids and backcross progenies (AILs) with a stable chromosome number ($2n = 24$). Backcross progenies are evaluated in the field for yield and other target traits in replicated trials. Khush et al. (1977) transferred grassy stunt resistance from *O. nivara* into IR24. The first set of grassy stunt-resistant varieties (IR28, IR29, and IR30) was released in 1974, and many IR varieties now carry the gene for grassy stunt resistance. Resistance was identified in only one accession of *O. nivara* among the 6000 rice accessions tested. Since then, a number of agronomically important genes governing resistance to grassy stunt, tungro disease, BB, blast, and BPH; tolerance of acid sulfate soils; and new cytoplasm sources have been transferred into rice from several wild species, including *O. nivara* (AA), *O. rufipogon* (AA), *O. longistaminata* (AA), *O. minuta* (BBCC), *O. officinalis* (CC), *O. australiensis* (EE), and *O. brachyantha* (FF) (Brar and Khush, 1997). Some of the IRRI breeding lines from wide crosses (MTL 98, MTL 103, MTL 105, MTL 114, Matatag 9, AS 996, and NSIC Rc 112) have been released in different countries for commercial cultivation (Brar and Singh, 2011). Other varieties, such as Dhanrasi and Yan Duo, have been released in India and China, respectively.

A large number of varieties derived from crosses of *O. sativa* × *O. glaberrima*, commonly called NERICA rice, have been released for commercial cultivation in African countries by the Africa Rice Center. Several genes introgressed from wild species have been tagged with molecular markers and used in MAS for pyramiding and transfer into different varieties. These will be discussed in the next section.

BIOTECHNOLOGICAL APPROACHES TO AUGMENT CONVENTIONAL RICE BREEDING

The following biotechnological approaches should be integrated into conventional breeding for the genetic enhancement of germplasm for different target traits and varietal development:

- Molecular tagging and pyramiding of genes/QTLs for agronomic traits for use in MAS and accelerated breeding.
- Introgression of wild species alleles, pyramiding of cloned genes for yield component traits, and engineering of C_4 pathways into C_3 rice to enhance yield potential (of inbreds/hybrids).
- Introgression of novel genes/QTLs from wild species to broaden the gene pool of rice for tolerance of major biotic and abiotic stresses.
- Molecular characterization of pathogen populations for deployment of genes for resistance.
- Doubled-haploid technology for accelerated breeding.
- Genetic engineering to develop transgenics with new genetic properties.
- RNAi and genome editing technologies for modifying gene expression and silencing of genes in a targeted way to develop pest-resistant germplasm and improved quality characteristics.
- Functional genomics using different "Omics" platforms to understand the function of genes that govern agronomic traits and to manipulate such genes in rice breeding programs.
- Combining well-characterized and specific genes/QTLs for different target traits to develop designer rice.

MOLECULAR TAGGING AND PYRAMIDING OF GENES/QTLs AND MAS TO ACCELERATE BREEDING

The advent of various kinds of molecular markers (RFLP, SSR, and, more recently, SNPs) has led to the publication of highly saturated and comprehensive molecular genetic maps for rice. This has led to a major breakthrough in mapping genes/QTLs for various agronomic traits, such as tolerance of biotic and abiotic stresses, and different quality characteristics, including several yield components and traits. Several review papers on the molecular mapping of genes/QTLs that govern agronomic traits and MAS protocols for many traits have become available. Some of the genes/QTLs tagged are now routinely used in the transfer of genes into elite breeding lines (through MAS). As an example, several genes/QTLs have been tagged with molecular markers for resistance to BB, blast, BPH, and green leafhopper; tolerance of abiotic stresses, such as submergence, drought, salinity, and cold; fertility restoration, wide compatibility, aroma, etc. Multiple disease resistance genes in rice have been cloned using a map-based strategy (17 genes for blast resistance, 6 for BB resistance, and 4 for BPH), which has facilitated the use of gene-based MAS for developing resistant varieties. The MAS products serve as an important source for developing prebreeding products and for varietal development. In rice, *Xa* genes for resistance to BB and BPH have been transferred through MAS and varieties released for commercial cultivation (Table 6.1). The *Xa21* gene, originally transferred from *O. longistaminata,* has also been transferred into many elite inbreds and parental lines of hybrids by several institutes in India, the Philippines, Thailand,

TABLE 6.1 MAS Products Resistant to BB and BPH Released as Varieties in Rice

Variety	Variety type	Gene(s)	Country
BB resistance			
NSIC Rc 142 (Tubigan 7)	Inbred	*Xa4 + Xa21*	Philippines
NSIC Rc 154 (Tubigan 11)	Inbred	*Xa4 + Xa21*	Philippines
Improved Sambha Mahsuri	Inbred	*xa5 + xa13 + Xa21*	India
Improved Pusa Basmati 1	Inbred	*xa13 + Xa21*	India
Punjab Basmati 3	Inbred	*xa13 + Xa21*	India
Angke	Inbred	*Xa4 + xa5*	Indonesia
Conde	Inbred	*Xa4 + Xa7*	Indonesia
Xieyou 218, Zhongyou 218	Hybrid	*Xa21*	China
Ilyou 218, Zhongbai You 1	Hybrid	*Xa21*	China
Guodao 1, Guodao 3	Hybrid	*Xa4 + xa5 + xa13 + Xa21*	China
Neizyou, Ilyou 8006	Hybrid	*xa5 + xa13 + Xa21*	China
BPH resistance			
Suweon 523	Inbred	*Bph18*	Korea

BB, Bacterial blight; BPH, brown planthopper.
Source: Modified from Brar, D.S., Singh, K., 2011. *Oryza.* In: Kole, C. (Ed.), Wild Crop Relatives: Genomic and Breeding Resources: Cereals. Springer-Verlag, Berlin, pp. 321–365.

and China. One recent and successful example of MAS was observed in Punjab Basmati 3, a variety of rice developed and released at Punjab Agricultural University (PAU). Transfer of the dwarfing gene *sd1*, two BB resistance genes (*xa13* and *Xa21*), and background selection with markers linked to aroma and amylose content (waxy locus) resulted from the improvement of the traditional variety Basmati 386 (Bhatia et al., 2011; Singh et al., 2014). Genomic selection (GS) coupled with GBS is expected to play a larger role in plant breeding as the cost of marker technologies continues to decline in conjunction with the development of robust statistical methods. Lorenz et al. (2011) have described the usefulness and limitations of GS to accurately estimate breeding values, accelerate the breeding cycle, and introduce greater flexibility in phenotypic evaluation and selection. GS has great potential but needs to be tested in actual breeding programs.

TABLE 6.2 MAS Products Tolerant of Submergence With the *SUB1* Gene Released as Varieties in Rice

Breeding line	Year of release	Variety name	Country
IR05F102 (Swarna)	2009	Swarna-Sub1	India
	2009	INPARA-5	Indonesia
	2010	BRRI dhan51	Bangladesh
	2011	Swarna-Sub1	Nepal
	2011	Yemyoke Khan	Myanmar
IR07F102 (IR64)	2009	NSIC Rc 194	Philippines
	2009	INPARA-4	Indonesia
IR07F290 (BR11)	2010	BRRI dhan52	Bangladesh
IR09F436 (Ciherang)	2015	Binadhan 11	India
	2014	BINA dhan11	Bangladesh
IR07F101 (S. Mahsuri)	2011	Sambha Mahsuri-Sub1	Nepal
	2014	BINA dhan12	Bangladesh
	2014	Sambha Mahsuri-Sub1	India

MAS, Marker-assisted selection.
Source: http://strasa.irri.org/: Courtesy of Dr. U.S. Singh, IRRI, Manila, Philippines.

The *SUB1* gene for submergence tolerance has been transferred through MAS into megavarieties and many Sub1 versions of the megavarieties have been released in South and Southeast Asia (Table 6.2). More recently, QTLs for drought tolerance (*DRO1*, *DTY1.1*), salt tolerance (*SALTOL*), phosphorus starvation tolerance (*PSTOL1*), thermo-tolerance (*TT1*), and many more characters are being stacked through MAS. Several institutes in India are engaged in the stacking of major genes for biotic stresses and QTLs for abiotic stresses in elite varieties.

ENHANCING YIELD POTENTIAL OF INBREDS AND HYBRIDS (BREAKING THE YIELD PLATEAU)

During the last few decades, major progress has been made in increasing rice production. The adoption of semidwarf, high-yielding varieties coupled with improved production technologies ushered in the Green Revolution. Several reviews discuss the development of these high-yielding varieties through conventional breeding techniques and the way forward for increasing yield potential (Khush, 1995, 2001, 2005, 2013; Yuan, 2001). Plant physiologists have proposed increased photosynthetic efficiency and increased sink size as possible approaches to increase yield potential, but, to break the yield ceiling, emphasis should be given to a new breeding strategy.

Introgression of Novel Genes/QTLs From Wild Species

The yield of cereals (particularly wheat and rice) has not increased significantly since the release of semidwarf, high-yielding, fertilizer-responsive, and lodging-resistant varieties in the early 1970s. Plant breeders are now attempting to tap new genes in wild species to improve the yield potential of rice. Molecular markers have facilitated the tracking of alien introgression in segregating populations. Several QTLs were identified in the wild species *O. rufipogon* (among others) that were reported to increase yield in rice; yield-enhancing loci *yld1.1* and *yld2.1* were identified on chromosomes 1 and 2, respectively. Since then, several QTLs for yield component traits have been reported in wild species of rice (Imai et al., 2013). At PAU-Ludhiana, as many as 70 accessions from six different species (*O. glaberrima, O. barthii, O. nivara, O. rufipogon, O. longistaminata,* and *O. glumaepatula*) were crossed with two *O. sativa* cultivars. The F_1s were backcrossed 2 or 3 times to the recurrent parents, coupled with the selection of agronomically desirable plants, and followed by single seed descent from the BC_2F_2/BC_3F_2 generation. More than 2000 introgression lines were generated and evaluated in an augmented design over 3–4 years. Several lines with alien introgression for one or more yield component traits were identified and are being used in rice breeding programs, as well as for mapping of the alien introgressed QTL. One introgression line, which demonstrates BB resistance and improved yield over the recurrent parent, is being evaluated in farmers' fields for its potential to be released as a variety. Likewise, several AILs have been generated at the IIRR, Hyderabad, and many of these are being evaluated under coordinated network trials. Future research should focus on the integration of molecular marker technology for the identification and introgression of novel QTLs for productivity traits from wild species so as to enhance the yield potential of rice. GS coupled with GBS as a breeding method for the simultaneous identification, mapping, and transfer of desirable genes from wild to cultivated species is being investigated by a few groups in rice, wheat, and other crops. PAU has genotyped more than 400 accessions of *O. rufipogon* using the GBS approach. Based on population structure and diversity, backcross populations with about 35 diverse accessions are being used as training populations for the mapping and transfer of yield-related QTLs from *O. rufipogon* into cultivated rice.

Pyramiding of Cloned Genes/QTLs for Yield-Related Traits

Most genetic research to date has focused on identifying and pyramiding genes/ QTLs for tolerance of biotic and abiotic stresses. However, with the development of genomics and the availability of whole genome sequencing, it has now become easier to clone genes/QTLs even for yield-related traits. Map-based cloning strategies and specialized genetic stocks (i.e., NILs and CSSLs) have been used to clone genes/QTLs for key yield components, such as (1) number of grains per panicle, (2) grain size and weight, and (3) number of tillers per plant, including grain filling. More than 20 yield-related QTLs have been cloned to date and their biology is well understood (Ashikari and Matsuoka, 2006; Bai et al., 2012; Miura et al., 2011). As an example, for grain number alone, five genes (*Gn1*, *Dep3*, *Ghd7*, *DTH8*, and *Dep1*) located on chromosomes 1, 6, 7, 8, and 9, respectively, are cloned. The nature of the desirable alleles is also varied and encompasses loss of function (*dep1*), increased expression (*WFP*), and reduced expression (*Gn1a*). One major challenge with stacking these QTLs is not technological but biological, as it is not known how these genes will interact with each other when desirable alleles at all loci are pyramided. A second challenge to the pyramiding of these QTLs is the identification (polymorphism) of alleles present in the recipient line. Many times, the functional marker or the markers closely linked to the QTL/gene are not associated with the functional domains of the gene. Hence, before initiating pyramiding of QTLs for the same trait, it is advisable to sequence the allele in the recipient parent and compare it with the donor parent allele. If the two alleles are functionally different, then only the pyramiding should be continued, or else one may not find any superiority in the phenotype of the pyramided lines.

Prebreeding Tropical Japonicas to Enhance Heterosis

The suggested strategy to improve and use tropical japonicas to improve heterosis in rice should include:

1. Genotyping a larger set (~500 accessions) of TJ types following GBS.
2. Identifying a smaller set (~100 accessions) of diverse lines (core).
3. Crossing the core with the best available CMS lines and evaluating F_1 hybrids for yield and yield components.
4. Selecting highly heterotic hybrids and improving the TJ lines for grain quality and disease and insect resistance using MAS.
5. Ensuring that the improved TJ lines possess WC and Rf genes.
6. Using the improved TJ lines for the production of F_1 hybrids.

Engineering Photosynthetic Systems to Enhance Yield Potential: Transfer of C_4 Pathway Genes into C_3 Rice

Improving radiation-use efficiency (RUE) through the regulation of Rubisco, introducing C_4-like traits, such as CO_2-concentrating mechanisms, improving light interception, and improving photosynthesis at the whole-canopy level are all being explored as options to increase yield potential in rice. Efforts must focus on the identification and use of photosynthetically efficient germplasm, and the genes/QTLs for efficient mobilization and loading of photosynthates from source to sink. High-throughput protocols for the selection of photosynthetically efficient germplasm in breeding programs are needed. Ultimately, RUE depends on the surplus of photosynthesis over respiration. Little can be done to decrease respiration, whereas higher RUE, biomass, and yield require increased photosynthesis. The C_4 photosynthesis system is more efficient and offers several advantages over C_3 including (1) faster and more complete translocation of assimilates from leaves, (2) reduced photorespiration, (3) almost twice the efficiency in dry matter production per unit of water transpired, and (4) greater photosynthetic efficiency at high temperature. Some approaches for converting C_3 to C_4 involve (1) searching for genetic variability in the wild-species germplasm of C_3 plants for C_4 photosynthesis, (2) incorporating such variability into C_3 plants through wide hybridization, (3) identifying the genes responsible for compartmentalization, (4) identifying candidate genes for key components in C_4 plants, (5) using comparative genomics for the identification of chromosomal regions carrying key traits/genes for C_4-ness, and (6) using genetic engineering approaches to transfer C_4 characteristics/key enzymes from different species, such as maize and sorghum, into rice.

The main challenge is to make the photosynthetic pathway of C_3 resemble that of C_4 by reducing photorespiration and modifying leaf anatomy. There is a need to understand the complex interactions between canopy architecture and the biochemistry of the photosynthetic apparatus. C_4 is a long-term project requiring extensive experimentation and exploratory research on several basic components before C_3 plants can be successfully converted to C_4 for commercial use. IRRI has a collaborative project with several laboratories in Europe and the United States, which aims to alter the photosynthesis of rice from the C_3 to C_4 pathway by introducing cloned genes (from maize and other systems) that regulate the production of the enzymes responsible for C_4 synthesis. Converting C_3 plants to C_4 is a long-term (15–20 years) option, but C_4 rice could yield 25% more than the existing C_3 rice in addition to having higher nutrient- and water-use efficiency.

BROADENING THE GENE POOL OF CROPS THROUGH INTROGRESSION OF NOVEL GENES/QTLs FROM UNADAPTED GERMPLASM AND WILD SPECIES

Pyramiding of Genes/QTLs to Develop Superior Prebreeding Products and Varieties With Value-Added Traits

A number of genes/QTLs for resistance to diseases and insects, tolerance of drought, and other traits have been pyramided in rice. When used in combination, these genes show a wider spectrum of resistance than when used individually. Some important examples include pyramiding of four genes for resistance to BB in rice (*Xa4*, *Xa21*, *xa5*, and *xa13*), for rust resistance in wheat

(*Lr24, Lr28, Yr10,* and *Yr15*), and for leaf curl virus resistance in tomato (*Ty1, Ty2, Ty3,* and *Ty4*). These pyramided lines have become a useful source for developing prebreeding products with value-added traits. As an example, three BB resistance genes (*xa5, xa13,* and *Xa21*) were pyramided in one of the leading varieties of rice in Punjab (PR106) in 2002. This line could not be released as a variety although it has been used as one of the parents in five recently released varieties in Punjab (PR121, PR122, PR123, PR124, and Punjab Basmati 3).

Introgression of Novel Genes/QTLs From Wild Species

Wild species are important reservoirs of useful genes for resistance to diseases and insects and tolerance of abiotic stresses, as well as being a rich source of productivity traits (Brar and Khush, 1997, 2006; Brar and Singh, 2011). A number of genes have been transferred from wild species into rice, wheat, tomato, cotton, Brassica, and several other crops. As an example, of the 67 genes catalogued for leaf rust resistance in wheat, 32 are from alien sources; 22 of the 45 genes for stripe rust, 20 of the 40 genes for stem rust, 18 of the 30 genes for powdery mildew, and 8 of the 10 genes for cereal cyst nematode are from wild relatives of wheat. In rice, several genes for resistance to BB, blast, BPH, tungro disease, and grassy stunt virus have been transferred from different wild species of *Oryza*. Some of the IRRI breeding lines with introgression of wild species genes have been released as varieties. Of the 6000 *O. nivara* accessions tested in India, only one was found to be resistant to grassy stunt virus and the resistance gene (*Gs*) has subsequently been transferred into several rice varieties that are now grown in tropical countries (Khush et al., 1977). These rice varieties have held their resistance for the past 40 years.

Genes can be transferred through direct crosses of cultivated × wild species but require embryo rescue in several interspecific crosses. Molecular markers have facilitated fast-track introgression of segments from wild species and in mapping genes/QTLs introgressed from wild species. Several genes that confer tolerance of biotic and abiotic stresses have been transferred from wild species to crop plants and some varieties have been released. Genes for resistance to BPH were successfully transferred from *O. officinalis* and other wild species of rice to elite breeding lines. Similarly, genes for resistance to BB have been transferred from *O. minuta, O. rufipogon,* and *O. nivara*. Several agronomically important genes have been tagged with molecular markers and one of these genes (*Xa21*, originally transferred from *O. longistaminata*) has been transferred through MAS into many breeding lines and 12 varieties have been released. Other BB resistance genes (*Xa33* and *Xa38*, recently transferred from *O. nivara*) are being pyramided through MAS into a series of high-yielding varieties. QTLs introgressed for aluminum toxicity and drought tolerance have also been identified. Wild-species germplasm is being investigated for novel variations for traits such as C_4-ness and biological nitrogen fixation (BNF).

Future priorities include the identification and transfer of novel genes/QTLs from wild relatives for tolerance of major biotic and abiotic stresses, yield enhancement, quality characteristics, and for BNF. Some of the wild species of maize and pearl millet are well known and are rich sources of genes for apomixis. Future efforts are needed to introgress such genes into cultivated rice so as to develop apomictic hybrids, which has not been achieved to date.

Pyramiding of Small-Effect QTLs

For many diseases and insects, such as sheath blight (ShB), brown spot, and leaf folder, no major resistance genes have been identified in the cultivated or wild germplasm. However, germplasm lines with partial resistance are known for each of these biotic stresses. For ShB, partial resistance has been reported in several germplasm types including landraces and wild species. Near complete resistance can be achieved by pyramiding a minor-effect QTL for ShB and other such biotic stresses as has been achieved for Fusarium head blight in wheat (Buerstmayr et al., 2009). Emphasis should be given to the mapping of the QTL in diverse genetic backgrounds followed by pyramiding of the minor-effect QTL into elite varieties.

Alternative Dwarfing Genes

The dwarfing gene *sd1* is widespread in cultivated rice germplasm. However, 50 years of rigorous use of this single source has led to the erosion of many agronomically important genes, especially ones conferring abiotic stress tolerance and nutritional value (for both plant and human nutrition). The *sd1* locus is also unsuitable for deep sowing, a requirement for direct-seeded rice. Variation in plant height exists in *O. nivara*. Moreover, the genes for dwarfness in this species appear to be different from *sd1* because several *O. nivara* accessions are dwarfs, and derivatives from crosses with *O. nivara* are shorter than the *O. sativa* recipient lines, which carry *sd1* (Kuldeep Singh, personal communication). Variation in coleoptile length and depth emergence also exists in cultivated as well as wild-species germplasm; *O. nivara* and *O. glaberrima* show variations in these traits. Mapping and transfer of QTLs for seedling emergence and plant height from *O. nivara* and *O. glaberrima* will bring about new variability for rice. These genes are expected to improve rice productivity, especially in unfavorable environments that constitute more than 35% of the total area under rice cultivation.

Allele Mining for QTLs for Abiotic Stresses and Nutrient-Use Efficiency

In any diploid species, breeders will work with a maximum of two alleles at any given locus; however, many alternative alleles will also exist for this locus.

An excellent example has been described in wheat for the powdery mildew resistance gene *Pm3* where each functional allele, of which there were more than 18, showed a differential reaction to a set of powdery mildew pathotypes (Bhullar et al., 2010). In homozygous inbred lines or varieties, only one allele could be present at any given locus; however, multiple alleles for the same locus could be introduced into a single line through the transgenic approach demonstrated by the Beat Keller group for the *Pm3* locus in wheat (Brunner et al., 2010). In rice, several QTLs/genes have been cloned for submergence tolerance (*SUB1A*), drought tolerance (*DRO1*), phosphorus-use efficiency (*PSTOL1*), stem culm strength (*SCM2*), etc. It would be interesting to look for allelic variation at these loci and evaluate the performance of various functional haplotypes for specific stresses, with the aim of using the most efficient allele in breeding programs. Alternatively, a combination of alleles could be stacked through the transgenic approach and their efficiency evaluated.

Identification of germplasm that efficiently uses nutrients is constrained by the screening techniques and quantifiable phenotype. The breakthrough for developing varieties resistant to diseases was the development of efficient screening techniques by plant pathologists and their close interaction with breeders. Therefore, a well-knit network of crop physiologists, soil scientists, plant breeders, and molecular geneticists is now required to develop efficient screening techniques and quantifiable phenotypes for the mapping and cloning of genes/QTLs that can use N, P, K, and other nutrients more efficiently.

Improving Nutritional Quality

Almost 50% of the world population depends on rice as a major source of calories. In India, 48% of children below the age of five are suffering from protein malnutrition, which leads to stunting of growth. A large proportion of the population also suffers from iron (Fe) and zinc (Zn) deficiency. As rice is the staple food for the majority of the population, it is eaten in higher volumes than anything else; therefore, nutritionally enhanced rice could help to minimize nutrient malnutrition. At present, nutrient consistency (protein, Zn, and Fe) varies widely in both cultivated and wild species of *Oryza*. For example, *O. glumaepatula* has high grain Zn concentration while *O. meridionalis* has high protein content. QTLs for these traits need to be mapped and transferred from landraces and wild species into elite lines. *Triticum dicoccoides*, a wild species of wheat, was identified as having high grain protein content and this trait was transferred to cultivated bread wheat through a conventional backcross breeding approach. The QTL associated with increased grain protein content has been cloned and designated as *Gpc-B1* (Uauy et al., 2006). The clear demonstration that *Gpc-B1* is associated with an increase in grain protein content and the availability of functional markers mean that this gene is now being transferred into several elite varieties through MAS by several groups around the world. The same strategy needs to be tested in rice.

Improving Nutrient-Use Efficiency

Nitrogen (N), phosphorus (P), and potassium (K) are three key nutrients that are required in higher quantities. These three nutrients add to the major input cost of farmers, invariably more than seed cost. On average, the rice plant uses 30%–40% of applied N, meaning that there is a considerable loss of N due to denitrification and nitrification. This N leaches into the lower strata of the soil, leading to contamination of groundwater, which has become a health concern in several regions of the world. Rock phosphate, a source for P fertilizer, is concentrated in just five countries globally and is expected to become depleted within a century; the same is true for K. Thus, breeding for N, P, and K efficiency should be a priority for rice geneticists and breeders so as to save farmers' costs and the environment. A protein kinase gene conferring phosphorus starvation tolerance (*PSTOL1*) has been cloned from an upland rice variety (Kasalath) and it is invariably absent in varieties developed for lowland irrigated conditions. This gene is present in all *O. glaberrima* and *O. rufipogon* accessions. Allelic variations, if any, need to be identified from this germplasm and similar genes need to be identified. Trait development groups in breeding programs should emphasize the identification of QTLs/genes that are efficient in the use of N, P, and K so as to protect farmers, the environment, and countries importing NPK fertilizers.

MOLECULAR CHARACTERIZATION OF PATHOGEN POPULATIONS FOR TEMPORAL AND SPATIAL DEPLOYMENT OF GENES FOR RESISTANCE

Sustaining yield potential is as important as improving yield potential. Diseases and insects regularly cause 15%–20% yield losses. It is well known that pest populations, including insects, fungi, bacteria, viruses, and nematodes, are highly variable in morphological, physiological, and pathological (virulence) characteristics. Effective deployment of the genes could be achieved provided information on pathogen variability and virulence is known. Traditionally, differential sets are used for analyzing variability within pathogen/insect populations. With DNA sequencing technologies becoming cheaper and more high-throughput, the whole genome of several insects and other pathogenic populations has become available. In rice, for example, whole genome sequences of the BB pathogen (*Xanthomonas oryzae* pv. *oryzae*), blast fungus (*Pyricularia grisae),* and BPH (*Nilaparvata lugens*) are available and in-depth understanding of the *vir/avr* genes is available for these pathogens. Therefore, it is important to precisely characterize the pathogen population structure in different areas and hot spots so as to deploy relevant resistance genes into these areas based on the actual pathogen population and race/biotype of the pest. Molecular markers and DNA fingerprinting are important for characterizing pathogen populations.

In addition, isogenic lines carrying different genes/QTLs for resistance can be tested in different areas and environments to identify resistance genes for deployment. As an example, several resistance genes have been identified in rice: more than 60 for blast, 40 for BB, and 30 for BPH. Before genes can be deployed to appropriate locations, characterization of the pest population must be undertaken. Future research should focus on developing good molecular markers for the characterization of pest populations to strengthen the deployment of different genes for resistance.

ACCELERATING CROP BREEDING USING DOUBLED-HAPLOID TECHNOLOGY

Plant breeders are generating crosses among desired parents, selecting desirable recombinants in the segregating populations, and developing homozygous lines. Reaching near homozygosity takes five to six generations from F_1 onward, which is expensive and time-consuming. Haploids can be conveniently used to accelerate the breeding process and they have been produced in more than 250 plant species to date using various methods. Some of the more commonly used methods are anther culture, wide crosses resulting in selective chromosome elimination of one parent, and haploid inducer genes/genetic stocks. In general, haploids representing only one set of chromosomes are highly sterile and thus their chromosome number is doubled. Doubled haploids, produced from segregating populations, can be used to develop homozygous lines in the immediate (same) generation unlike conventional methods that take four to five or more generations to achieve homozygosity. Guha and Maheshwari from Delhi University were the first to produce haploids in *Datura* in 1966. Since then, anther culture has been used to produce haploids in a large number of crop species and several varieties have been released through haploid breeding. However, in many species (including rice), anther culture is genotype specific. For example, japonica rice is highly responsive to anther culture, while indica types are less responsive. The production of haploids through chromosome elimination in crosses of *Hordeum vulgare* × *H. bulbosum* has been successfully used to produce several varieties of barley. A similar system of haploid production from crosses of wheat × maize is being used successfully at the International Maize and Wheat Improvement Center in Mexico and PAU-Ludhiana. Several doubled-haploid lines have been produced in wheat and are now being tested for their agronomic performance (N.S. Bains, personal communication). The third method of using haploid inducer stocks includes haploid inducer stocks in maize and *Solanum phureja* stocks in potato. However, no such system is available in rice.

Future research in rice should therefore focus on enhancing the production of haploids in recalcitrant and less responsive genotypes, such as those of indica rice grown on larger areas in Asia. Some of the priority areas of research in rice haploidy breeding involve (1) the identification and transfer of genes/QTLs for high anther culturability from japonica into indica genotypes, (2) the search for a chromosome elimination method in rice involving distant crosses as in wheat and barley, (3) the identification of haploid inducer stocks as in maize, and (4) a novel genetic engineering approach, such as modification of the *CenH3* gene through genome editing or through induced mutations followed by TILLING.

GENETIC ENGINEERING APPROACHES TO DEVELOP TRANSGENICS WITH NEW GENETIC PROPERTIES

Introducing alien genes (from bacteria, viruses, fungi, and unrelated plants) into crop species allows plant breeders to achieve breeding objectives that were not considered possible before. Transgenics in rice were proposed as early as 1988 by three laboratories in Japan, the United States, and the United Kingdom. Before this time, polyethylene glycol, electroporation, and biolistic methods had been used. We are now in a time when new, high-throughput methods using *Agrobacterium*-mediated transformation have become available for several dicot as well as monocot species. Varieties of transgenic rice carrying genes for resistance to diseases and insect pests, tolerance of herbicides and abiotic stresses, such as drought and salinity, and with quality characteristics, such as beta-carotene (provitamin A Golden rice) and high grain iron content, have been produced (Table 6.3). GM crops now occupy more than 170 million hectares worldwide (www.isaaa.org/resources/publications/briefs/51/default.asp). In the United States, these GM crops are grown on 80% of the soybean area, 80% of cotton, 35% of maize, and 30% of canola, whereas in India transgenic cotton occupies more than 90% of the cotton area. Key traits exploited commercially through transgenic approaches are herbicide tolerance and insect resistance.

Conventional breeding approaches to develop stem borer resistance in rice have not met with much success. Several *Bt* genes—*cryI(A)*, *cryI(B)*, *cryI(C)*, and *cryI(2A)*—conferring resistance to stem borers have been transferred and pyramided to develop a wide spectrum of resistance to stem borers. Besides tolerance of biotic and abiotic stresses, improvement of the nutritional quality of rice is being pursued as a priority. Ye et al. (2000) first established a proof of concept and developed rice with provitamin A properties. Since then, several attempts have been made to enhance provitamin A. In Golden rice-2, *psy* from maize has been used in combination with *Erwinia uredovera* carotene desaturase (*crtl*), which led to as high as 25 μg/g carotenoid content accumulation in the rice grains. The carotenoid locus from the leading GR2 event has been introgressed by IRRI into megavarieties of rice (IR64, PSBRc18, and BR 29) using marker-aided backcrossing. Evaluations of transgenic Golden rice introgression lines in confined field tests have shown that these megavarieties are not as high yielding as the recurrent parents. This was found to be attributable to the transgene integrating within a hormone-producing gene. As a result, plants homozygous for the transgene have a nonfunctional native

TABLE 6.3 Some Examples of GM Rice Carrying Agronomically Important Genes

Trait	Transgene(s)
Herbicide resistance	Bar
Tolerence of tungro and stripe virus	Coat protein
Sheath blight resistance	Chitinase
Stem borer resistance	(Bt genes) cryI(A), cryI(B), cryI(C), cry(2A), CpTI, mwt 1B (winged bean trypsin inhibitor)
Resistance to brown planthopper	Gna
Bacterial blight resistance	Xa21
Bacterial streak resistance	Rxo1
Blast resistance	Afp, Gns1
Salt tolerance	Cod A, Gs2
Drought tolerance	OtsA, OtsB, OsTPS1, HvA1, HvCBF4, SNAC1, DREB IA, HRD
Heat tolerance	Hsp 101
Increased iron content	Ferritin
Provitamin A (Golden rice)	psy, crtI, Icy; psy (maize), crtI
Reduced amylose	Antisense Waxy (Wx) gene
Increased photosynthetic activity	PEP, PEPC, PPDK, PEPC + PPDK, NADP-ME
Nitrogen-use efficiency	OsNRT2.3b (overexpression)

Source: Modified from Brar, D.S., Khush, G.S., 2013. Biotechnological approaches for increasing productivity and sustainability of rice production. In: Bhullar, G.S., Bhullar N.K. (Eds.), Agricultural Sustainability: Progress and Prospects in Crop Research. Academic Press, London, pp. 151–175.

gene, which makes these plants shorter than the recurrent parent (A.K. Singh, personal communication).

Overexpression of native genes, especially transcription factors, through the transgenic approach has been reported in rice and other crops for a series of traits, including induction of tolerance of biotic and abiotic stress, nutrient-use efficiency, and improving yield component traits. One such recent example is overexpression of a pH-sensitive nitrate transporter gene in rice (Fan et al., 2016). Overexpression of the native gene *OsNRT2.3b* enhanced the pH-buffering capacity of the plant, thus increasing N, Fe, and P uptake. In field trials, increased expression of *OsNRT2.3b* was shown to increase grain yield and nitrogen-use efficiency (NUE) by 40%.

Future priorities in transgenic research should focus on the use of high-throughput transformation protocols, the production of marker-free transgenics, the use of organ-specific promoters, promoters for inducible gene expression, and pyramiding of transgenes. Emphasis should be given to identifying new transgenes, particularly for those traits for which conventional and other non-transgenic approaches are not successful, with emphasis on tolerance of abiotic stresses and improving nutritional value.

Although various laboratories have generated a multitude of transgenics for a series of traits, no commercial release has been made so far. It is suggested that all the available transgenic events be collected at key institutes within a country and be thoroughly evaluated in terms of agronomic performance and the target traits. Reliable events should be thoroughly characterized for the expression and stability of the target trait at the phenotypic and molecular level. The promising transgenic events should be evaluated as a priority in multilocation trials along with their biosafety. There is also a need for increased public awareness on the benefits and risks in adopting GM technology.

RNAi TECHNOLOGY FOR MODIFYING GENE EXPRESSION, SILENCING OF GENES TO DEVELOP PEST-RESISTANT GERMPLASM, AND IMPROVED QUALITY CHARACTERISTICS

RNA interference (RNAi) refers to the biological process in which an RNA molecule inhibits gene expression by causing degradation of the specific mRNA molecule. Two scientists, Andrew Fire and Craig Mello, were awarded the Nobel Prize in 2006 for their discovery of RNA interference–gene silencing by dsRNA. Different kinds of RNAs (dsRNA), hairpin RNAs, siRNAs, and microRNA (miRNA) regulate gene expression. RNAi technology offers breeders an opportunity to improve crops, produce pest-resistant germplasm, and improve the nutritional characteristics of breeding/crop products. Several promising products have been generated using RNAi; however, to date, only a few have been approved for commercial cultivation. For example, approval has been granted for the release of two apple varieties for nonbrowning of Arctic Apple, Arctic "Golden Delicious," and Arctic "Granny Smith." Other examples of products that have nearly reached commercialization through genome editing include nonbrowning low-acrylamide potato and nonbrowning mushrooms, which are both under field evaluation in Canada and the United States.

Both RNAi and genome editing hold great promise for developing resistance against sheath blight, stem borer, and leaf folder, for which limited genetic variability is available in the rice germplasm. It may also be of benefit for antinutritional genes whose functions could be marred either through RNAi or through genome editing.

FUNCTIONAL GENOMICS OF AGRONOMICALLY IMPORTANT GENES

During the last decade, major advances have been made in genome sequencing of crop plants. A large number of genetic resources for functional genomics have also become available, including T-DNA insertion lines, RNAi, and

TILLING. Some 100,000 T-DNA-tagged insertion lines carrying a large number of well-characterized T-DNA inserts, Tos17 retrotransposon insertional mutants with 30,000 lines carrying more than 250,000 Tos17 mutants, and more than 50,000 deletion mutants produced by fast neutron, gamma radiation, and chemical mutagenesis are available. These are excellent resources for delineating the function of annotated genes of agronomically important genes.

DESIGNER RICE WITH WELL-CHARACTERIZED GENES/QTLs FOR TARGET TRAITS

Advances in molecular marker technologies, genomics, and transformation have opened new ways to develop designer crops with well-defined genes and QTLs for target traits. In a broader sense, designer rice refers to the directed transfer of desirable alleles at target loci while retaining the key characteristics of recipient lines, which have unique adaptability because of either wider adaptation or region-specific quality traits. As an example, designer basmati rice can be produced by introgressing (through MAS) genes/QTLs for disease (BB and blast) resistance, insect (BPH) resistance, high grain number, sturdy stem, etc., along with retaining genes for its key basmati characteristics, such as aroma, grain length, grain elongation after cooking, alkali spreading value, amylose content, photoperiod sensitivity, etc.

The Future Outlook

Prebreeding and trait development: Prebreeding should be the priority for key agronomic traits. Good progress has been made in trait development for resistance to biotic stresses, such as BB, blast, BPH, and gall midge. Contrary to these traits, no well-defined genes/QTLs are available for abiotic stresses, such as drought, salinity, heat, or cold, along with nutrient uptake, grain nutrition, and cooking and eating quality. Milling is affected by several factors, of which chalkiness is important, resulting in heavy grain breakage and poor grain quality. A number of QTLs for chalkiness have been identified that should be used to improve milling and head rice recovery. Efforts should be made to identify diverse genes/QTLs with different mechanisms for all these traits. There is a need to develop NILs for each of these traits, map them, and make them available to breeders for stacking through MAS. So far, emphasis has been primarily on varietal development with limited emphasis on prebreeding and trait development. Equally important is trait development for newly emerging diseases, such as false smut.

Germplasm characterization: With genome sequencing becoming high-throughput and relatively inexpensive, the characterization of modern varieties and megavarieties, landraces, primitive cultivars, and wild species through whole genome sequencing is emphasized for the identification of desirable and rare alleles that have been left out of the main breeding programs.

Identification of heterotic patterns: Genome-based selection is emphasized for the identification of heterotic patterns and to enhance heterosis.

Raising yield potential: Integrating genomic tools, physiological traits, and diverse germplasm is emphasized for increasing rice productivity.

Mapping of QTLs: QTLs have been identified and mapped for only a few selected traits. With technological advances in genome sequencing, it is now easier and faster to identify and map QTLs. We suggest undertaking the mapping of QTLs for various agronomic traits and biotic and abiotic stresses on a larger scale. Breeders in collaboration with biotechnologists should seek to develop various specialized populations, such as RILs, nested association mapping (NAM), multiparent advanced generation intercross (MAGIC), and CSSLs for the mapping of QTLs. Emphasis should be given to the mapping of abiotic stress tolerance at various developmental stages, particularly the reproductive stage.

Integrating MAS in breeding programs: MAS is becoming a powerful tool for accelerating breeding not only for major genes but also for QTLs. With the availability of high-throughput genotyping platforms, MAS should become an integral part of breeding programs. Emphasis should be given to the pyramiding of genes/QTLs with different mechanisms for resistance to/ tolerance of biotic and abiotic stresses. Since several genes/QTLs have been cloned for yield component traits, priority should be given to the pyramiding of yield component QTLs for enhancing yield potential in inbreds, as well as in hybrids.

Developing haploid induction systems in indica rice: Various approaches to enhance anther culturability and haploids through chromosome elimination involving wide crosses, and the search for haploid inducer genes/stocks, should be given priority.

Nutrient-use efficiency: Molecular approaches should be used as a priority for the identification of genes for efficient nutrient uptake and transport. Emphasis should be given to understanding the role of soil microbes and their interaction with rice roots in promoting NUE.

Biological nitrogen fixation: Breeders should explore the possibilities of enhancing BNF through endophytes.

RNAi and genome editing: These two newly emerging technologies hold promise and should be explored to enhance pest resistance and improve the quality characteristics of rice cultivars.

GM technology: GM technology holds immense potential in rice improvement. Though at present some policy and regulatory constraints exist for the commercialization of GM rice, this should not impede research on and investment in transgenics.

Human resource development: Since genomics technologies are expanding at a faster rate, regular training of young breeders should be strengthened. A repeat of the 1990s model in rice biotechnology as adopted by the Rockefeller Foundation is required for training in rice genomics.

REFERENCES

Ashikari, M., Matsuoka, M., 2006. Identification, isolation and pyramiding of quantitative trait loci for rice breeding. Trends Plant Sci. 11, 344–350.

Bai, X., Wu, B., Xing, Y., 2012. Yield-related QTLs and their applications in rice genetic improvement. J. Integr. Plant Biol. 54 (5), 300–311.

Bhatia, D., Sharma, R., Vikal, Y., Mangat, G.S., Mahajan, R., Sharma, N., Lore, J.S., Singh, N., Bharaj, T.S., Singh, K., 2011. Marker-assisted development of bacterial blight resistant, dwarf and high yielding versions of two traditional basmati rice cultivars. Crop Sci. 51, 759–770.

Bhullar, N.K., Zhang, Z.Q., Wicker, T., Keller, B., 2010. Wheat gene bank accessions as a source of new alleles of the powdery mildew resistance gene *Pm3*: a large scale allele mining project. BMC Plant Biol. 10, 88.

Brar, D.S., Khush, G.S., 1997. Alien introgression in rice. Plant Mol. Biol. 35, 35–47.

Brar, D.S., Khush, G.S., 2006. Cytogenetic manipulation and germplasm enhancement of rice (*Oryza sativa* L.). In: Singh, R.J., Jauhar, P.P. (Eds.), Genetic Resources, Chromosome Engineering and Crop Improvement. CRC, Boca Raton, FL, pp. 115–158.

Brar, D.S., Khush, G.S., 2013. Biotechnological approaches for increasing productivity and sustainability of rice production. In: Bhullar, G.S., Bhullar, N.K. (Eds.), Agricultural Sustainability: Progress and Prospects in Crop Research. Academic Press, London, pp. 151–175.

Brar, D.S., Singh, K., 2011. *Oryza*. In: Kole, C. (Ed.), Wild Crop Relatives: Genomic and Breeding Resources: Cereals. Springer-Verlag, Berlin, pp. 321–365.

Brunner, S., Hurni, S., Streckeisen, P., Mayr, G., Albrecht, M., Yahiaoui, N., Keller, B., 2010. Intragenic allele pyramiding combines different specificities of wheat *Pm3* resistance alleles. Plant J. 64, 433–445.

Buerstmayr, H., Ban, T., Anderson, J.A., 2009. QTL mapping and marker-assisted selection for Fusarium head blight resistance in wheat: a review. Plant Breed. 128, 1–26.

Fan, X., Tang, Z., Tan, Y., Zhang, Y., Luo, B., Yang, M., Lian, X., Shen, Q., Millerc, A.J., Xu, G., 2016. Overexpression of a pH-sensitive nitrate transporter in rice increases crop yields. Proc. Natl. Acad. Sci. USA 113 (26), 7118–7123.

Imai, I., Kimball, J.A., Conway, B., Yeater, K.M., McCouch, S.R., McClung, A., 2013. Validation of yield-enhancing quantitative trait loci from a low-yielding wild ancestor of rice. Mol. Breed. 32, 1–20.

IRRI (International Rice Research Institute), 1989. IRRI Towards 2000 and Beyond. IRRI, Manila, Philippines.

Khush, G.S., 1995. Breaking the yield frontier of rice. GeoJournal 35, 329–332.

Khush, G.S., 2001. Green revolution: the way forward. Nat. Rev. Genet. 2, 815–822.

Khush, G.S., 2005. What it will take to feed 5.0 billion rice consumers in 2030. Plant Mol. Biol. 59, 1–6.

Khush, G.S., 2013. Strategies for increasing the yield potential of cereals: case of rice as an example. Plant Breed. 132, 433–436.

Khush, G.S., Brar, D.S., 2001. Rice genetics from Mendel to functional genomics. In: Khush, G.S., Brar, D.S., Hardy, B. (Eds.), Rice Genetics IV. Science Publishers, Inc. and Los Baños (Philippines): International Rice Research Institute, New Delhi, India, pp. 3–25.

Khush, G.S., Ling, K.C., Aquino, R.C., Aguiero, V.M., 1977. Breeding for resistance to grassy stunt in rice. Proceedings of the Third International Congress, SABRAO. Canberra, Australia.

Lorenz, A.J., Chao, S., Asoro, F.G., Heffner, E.L., Hayashi, T., Iwata, H., Smith, K.P., Sorrells, M.E., Jannink, J.L., 2011. Genomic selection in plant breeding: knowledge and prospects. Adv. Agron. 110, 77–123.

Miura, K., Ashikari, M., Matsuoka, M., 2011. The role of QTLs in the breeding of high-yielding rice. Trends Plant Sci. 16, 319–325.

Singh, K., Mangat, G.S., Kaur, R., Vikal, Y., Bhatia, D., Singh, N., Bharaj, T.S., 2014. Punjab basmati 3: a bacterial blight resistant dwarf version of basmati rice variety basmati 386. J. Res. Punjab Agric. Univ. 51, 206–207.

Uauy, C., Distelfeld, A., Fahima, T., Blechl, A., Dubcovsky, J., 2006. A NAC gene regulating senescence improves grain protein, zinc, and iron content in wheat. Science 314, 1298–1301.

Ye, X., Al-Babili, S., Kloti, A., Zhang, J., Lucca, P., Beyer, P., et al., 2000. Engineering the provitamin A (β carotene) biosynthetic pathway into (carotenoid-free) rice endosperm. Science 287, 303–305.

Yuan, L., 2001. Breeding of super hybrid rice. In: Peng, S., Hardy, B. (Eds.), Rice Research for Food Security and Poverty Alleviation. International Rice Research Institute, Los Baños, Philippines, pp. 143–149.

Chapter 7

Rice Varietal Development to Meet Future Challenges

Arvind Kumar*, Nitika Sandhu*, Shailesh Yadav*, Sharat Kumar Pradhan**,
Annamalai Anandan**, Elssa Pandit**, Anumalla Mahender**,
Tilathoo Ram[†]

*International Rice Research Institute, Los Baños, Philippines; **National Rice Research
Institute, ICAR, Cuttack, Odisha, India; [†]Indian Institute of Rice Research, ICAR, Hyderabad,
Telangana, India

INTRODUCTION

The goal of rice breeding has changed over time. The priority of rice breeding
during the Green Revolution was the generation of high yields with improved
nutrient use and lodging resistance. Since the development of the first semidwarf rice variety (IR8) in 1966, priorities have shifted toward the development
of rice varieties that are resistant to pests and diseases; most semidwarf varieties that were developed initially lacked resistance to the insects and diseases
existing in host countries. Grain quality also received attention during 1980–90;
however, limited knowledge on the traits and genes controlling grain and cooking quality has meant that very few varieties developed over the past 50 years
can match the quality of traditional varieties. During 1990–99, the focus of
breeding programs revolved around improving yield potential by developing
hybrid rice and new plant types (Peng and Khush, 2003). From the beginning
of 2000, with the realization of increasing effects of abiotic stresses, the development of climate-adapted, stress-tolerant varieties has been the main focus of
successful rice breeding programs around the world.

Ongoing climate change threatens world food security and is currently one
of the most important challenges facing global rice production. Scientists have
been pushed to develop varieties of rice that are adaptable to the changing climate to sustain sufficient yield under variable climatic conditions and increasing threats of drought, flood, salinity, and high temperature. Quantification of
climate uncertainty is an important indicator for crop yield variation in future
climate scenarios. Furthermore, concerns now exist about our ability to increase
or even sustain rice yield and quality in the face of dynamic abiotic threats

The Future Rice Strategy for India. http://dx.doi.org/10.1016/B978-0-12-805374-4.00007-5
161

that will be particularly challenging in the face of rapid global environmental change. Along with breeding and agronomic- and management-based approaches to improving food production, improvements in a crop's ability to maintain yield with a lower water supply and poor quality of water will be critical. Put simply, we need to increase the tolerance of crops of abiotic stresses, such as drought, flood, salinity, high temperature, low temperature, and low light intensity, as well as their resistance to biotic stresses, such as bacterial blight (BB), blast, brown spot, sheath blight, false smut, brown planthopper (BPH), gall midge (GM), stem borer, and so on; a hard but necessary task. These abiotic and biotic stresses are likely to show similar adverse effects in both rainfed and irrigated ecosystems. The problems of high temperature, low temperature, low light intensity, and salinity run across all ecosystems and will become more severe in the future. Water shortages have extended to irrigated areas, forcing scientists to think of innovative methods to cultivate rice with less water. Maintaining resistance to rapidly evolving pests and pathogens will also continue to be an essential mainstay of breeding programs, even more so due to the likely shift in insect–disease scenarios with climate change.

Decreasing land, water, and labor for agriculture in general, and rice in particular, will determine the way rice can be cultivated in the future. Mechanized rice cultivation practices are emerging as alternatives to the puddled transplanted systems that require a high use of both water and labor. Breeding a plant type for water-saving cultivation conditions is highly likely to become the focus of breeding programs in the coming decade. Breeding programs need to be realigned to develop rice varieties for water-deprived and mechanized systems using a new set of traits that provide better adaptability of rice under fluctuating anaerobic to aerobic conditions during crop growth stages.

With rice being the staple food of most of Asia's poor, supplementation of rice with higher concentrations of zinc, iron, protein, and vitamin A will be required in the future. Iron deficiency (anemia) has profound negative effects on human health and development, causing impaired physical growth in infants and young children and increased risk of maternal mortality and preterm delivery, retarded fetal growth, low birth weight, and increased risk of neonatal death. It is the world's most widespread nutritional disorder and affects approximately 5 billion people in developed and developing countries. Vitamin A deficiency leads to a weakening of the immune system and, in its most severe form, night blindness in 140–250 million children under the age of five. Zinc deficiency has broad adverse health consequences, especially in pregnant women. In children it is associated with poor growth, reduced motor and cognitive development, impaired immune responses, and increased susceptibility to infectious diseases. In light of this, improvements in the grain, cooking, and nutritional quality of rice are also likely to be important breeding objectives in the coming 10 or more years, until the world finds a sustainable solution to the problem of malnutrition.

Plant breeding is a continuous process. It has evolved from being more an art before the beginning of the Green Revolution to more a science as new technological innovations and DNA-based molecular markers are being used as tools to generate desired combinations of traits. The transfer of genes that confer tolerance of different biotic and abiotic stresses (drought, submergence, salinity, etc.) has led to the development of several tolerant varieties. Advances made in molecular marker technology have permitted the tracking of multiple genes/QTLs linked to various traits in an elite cultivar, making it easier and more systematic for any crop improvement program to develop appropriate rice cultivars for the future. This chapter highlights some of the looming problems and opportunities for breeding new rice varieties for sustainable rice production in changing climate scenarios under the following subheadings:

- Breeding for increased yield potential.
- Breeding for biotic stresses.
- Breeding for abiotic stresses.
- Breeding for mechanized water-labor shortage situations.
- High grain and cooking quality: essential breeding components across ecosystems.
- Breeding nutritious rice.

BREEDING FOR INCREASED YIELD POTENTIAL

The International Rice Research Institute (IRRI) released the first semidwarf rice variety (IR8) in 1966, and with this a whole new chapter in increasing yield potential in rice was written and remembered as the Green Revolution. The development of semidwarf, high-yielding varieties remained the main aim of the majority of breeding programs for at least the next 2–3 decades. The primary emphasis of rice improvement programs was later directed toward incorporating disease and insect resistance, shortening growth duration, and improving grain quality. In the late 1960s and early 1970s, the yield of IR8 and other early cultivars was reported to be 9–10 t/ha under favorable irrigated conditions in the Philippines (Chandler, 1969; De Datta et al., 1968; Yoshida and Parao, 1976). The current yield potential of high-yielding inbred rice cultivars is only 10 t/ha; therefore, rice yield potential has remained almost constant in tropical environments. The theoretical yield potential in these environments has been estimated at 15.9 t/ha based on the total amount of incident solar radiation during the growing season (Yoshida, 1981). The average yield of Asia's irrigated rice must increase from 5.0 to 8.5 t/ha by 2025. A mean yield of 8.5 t/ha is very close to the estimated climate-adjusted yield potential of current rice cultivars in the major rice-growing areas (Matthews et al., 1995). Various strategies to increase yield potential include (1) conventional hybridization and selection, (2) ideotype breeding, (3) hybrid breeding, (4) exploitation of wild species germplasm, (5) enhancement of photosynthesis, and (6) genomic approaches (Khush, 2013).

Conventional hybridization and selection is a time-tested strategy. It is still used across a number of breeding programs to develop crop varieties with a higher yield potential. Using this strategy, an approximate 1.0% increase in yield potential per year is observed in cereal crops such as wheat, barley, and rice (Peng et al., 2000). However, the future use of this strategy for the generation of recombinants with higher yield potential is limited by reduced variability and increased inbreeding through crosses between a finite number of advanced breeding lines. Achieving a large increase in yield potential using this strategy looks unattainable unless crop improvement programs introduce new diverse germplasm and use trait breeding to bring new alleles for yield into improved backgrounds.

Ideotype breeding, aimed at modifying the plant architecture, is another strategy that has been implemented to increase yield potential. Yield is a function of total biomass and harvest index (HI). Tall, traditional rice varieties could produce a biomass of 13–14 t/ha, but they tended to have very low HI at ~0.3. On the other hand, semidwarf, high-yielding varieties showed HI of 0.45–0.50, but a lower biomass. To increase the yield potential of semidwarf rice, IRRI scientists proposed a new plant type (NPT; IRRI, 1989) with the characteristics of reduced tillering (9–10 tillers for transplanted conditions), zero unproductive tillers, 200–250 grains per panicle, dark green and erect leaves, a vigorous and deep root system, growth duration of 110–130 days, multiple disease and insect resistance, and increased HI (Khush, 2013). During breeding efforts to develop the NPT in the early 1990s, tropical *japonica* (TJ) varieties (*Bulus*) from Indonesia, known for their low tillering ability, large panicles, sturdy stems, and deep root systems, were crossed with a semidwarf *japonica* breeding line (Sheng Nung 89–366) from the Shenyang Agricultural University, China. Over 2000 crosses were made, and more than 100,000 pedigree lines were evaluated. Breeding lines with desirable ideotype traits were selected and more than 500 so-called NPT-TJ lines were evaluated in yield trials. These lines had improved sink size, improved lodging resistance, and no unproductive tillers; however, the grain yield of these lines was even lower than that of elite varieties. It was observed that reduced tillering contributed to a lower biomass and lower compensation ability. Moreover, NPT-TJ lines had very poor grain filling, which was attributed to the lack of apical dominance within a panicle, the compact arrangement of spikelets, and the limited number of large vascular bundles for assimilation and transport to grains. These NPT-TJ lines were also susceptible to diseases and insects and their grain quality was not acceptable for consumers in tropical and subtropical countries. These lines were evaluated in several countries, with some showing good performance in temperate areas where *japonica* grain quality is preferred. Three of these lines (Diancho 1, 2, and 3) were released as varieties in Yunnan Province of China (Khush, 1995).

To improve the acceptability of these NPT-TJ lines for tropical conditions, and to improve their yield potential, they were crossed with elite *indica* lines and other rice varieties with disease and insect resistance and good grain quality.

More than 400 lines designated as NPT-IJ were evaluated in yield trials. Several of these lines outyielded the best *indica* varieties (i.e., IR72) by as much as 1.0–1.5 t/ha. These breeding lines have been continuously used in the breeding program and have contributed to increased genetic diversity through the introduction of *japonica* germplasm into otherwise purely *indica* breeding materials. Before the NPT project, the *japonica* gene pool was essentially excluded from breeding programs in the tropics; however, four NPT-IJ lines have now been released as varieties in the Philippines and Indonesia. Numerous NPT-IJ lines were distributed to breeding programs throughout the world through INGER nurseries. Many have been used in hybridization programs and have contributed to the widening of gene pools. For example, IR66154-521-2-2 has been used widely in hybridization programs in China and Vietnam. Stimulated by IRRI's NPT breeding program, China established a nationwide megaproject to develop super rice. Many super rice breeding lines have been developed by Chinese breeders and have been used as parents in hybrid rice breeding (Peng et al., 2008).

Exploiting hybrid vigor has long been suggested as an approach to improve rice yield. This approach has been successful in cross-pollinated species, such as maize; however, the yield increase expected through hybrid (F_1) development in rice did not surpass its target except in China. Rice hybrids were introduced in China during the mid-1970s and are now planted on 13–14 million ha or 50% of the rice area in that country. The average yield advantage of these hybrids over other varieties is 10%–15%. Rice hybrids that are adapted for tropical conditions were developed at IRRI (Virmani, 2003) and by the national programs of most of the countries in tropical Asia, but these varieties are only planted on ~2 million ha outside of China. The main reason for the lack of large-scale adoption of hybrids is their limited yield advantage over other varieties, which is caused by the use of germplasm with limited genetic diversity. The development of diverse gene pools to increase heterosis, increase seed set and production, and develop hybrids with higher grain quality could help increase yield potential, as well as the adoption of hybrids among farmers. In the super hybrid rice breeding program in China, parental lines with variable amounts of *japonica* alleles in an *indica* background were developed to obtain higher heterosis. In the United States, two-line hybrids have been produced using temperature-sensitive male sterile lines. Male sterile female lines are tropical *japonicas* and restorers are *indicas*, and generated hybrids are reported to have as much as a 25% yield advantage over other varieties. The success of seed production from such lines in tropical conditions with variable temperature differences remains to be seen. Chinese scientists under the leadership of Prof. Yuan Long Peng initiated a super hybrid rice breeding program to improve the level of heterosis. Their program was based on moderate tillering capacity; heavy (5 g per panicle) and drooping panicles at maturity; slightly taller plant height (~100 cm); flag-leaf length of ~50 and 55 cm for the second and third leaves; all three leaves remaining above panicle height and erect until maturity, leaf angles of flag, second, and third leaves of 5, 10, and 20 degrees, respectively; narrow and V-shaped leaves (2 cm when flattened); thick and dark green leaves; leaf area index of the top three leaves of ~6.0; and HI of 0.55 (Yuan, 2001). Chinese breeders have developed many hybrids of this ideotype over time. In tropical areas, most hybrid development still revolves around CMS line 25 A and *indica* as the restorer line, thus limiting diversity and heterosis. Hybrids in tropical areas need to concentrate on creating a diverse gene pool and increasing diversity between the CMS line and restorer line to achieve higher heterosis and increased yield potential. This requires long-term planning to identify diverse sources and convert these into maintainers and restorers, a strategy not yet picked up by many public and private companies.

Wild species possess many yield-attributing genes that have not been used by breeding programs. Xiao et al. (1996) reported that some backcross derivatives from a cross between an *Oryza rufipogon* accession from Malaysia and cultivated rice outyielded the recurrent parents by as much as 18%. Two QTLs that contributed to the yield increase were subsequently identified from the wild species. In several other studies, QTLs for yield have also been identified in progenies derived from wild species; however, very few programs have concentrated on the use of wild rice relatives to improve yield. A large-scale program of trait development, identification of QTLs, and validation of QTL effects across diverse genetic backgrounds is needed for the systematic use of wild relatives in breeding programs.

C_4 plants, such as maize and sorghum, are 30%–35% more efficient in photosynthesis and are more productive than C_3 plants (i.e., rice and wheat). Four enzymes in maize are known to be responsible for the C_4 photosynthetic pathway, and therefore molecular engineering strategies have been employed to systematically introduce these photosynthetic genes into rice. Overexpression of the maize PEPC gene in rice produced enzyme activity that was two- to threefold higher than that seen in maize, and the enzyme itself accounted for up to 12% of the leaf-soluble protein (Matsuoka et al., 2001). These results suggest a possible strategy for introducing the key biochemical components of the C_4 pathway of photosynthesis into C_3 rice. However, all C_4 domesticated plants have Kranz anatomy and, to achieve the goal of converting C_3 rice to the C_4 photosynthetic pathway, alteration of the leaf anatomy may be required. Rice germplasm has been screened to identify some of the components of maize leaf anatomy. For example, maize leaves have narrower vein spacing, whereas rice has wider vein spacing. Wild rice relatives *Oryza barthii* and *Oryza australiensis* have narrower vein spacing. Conversion of rice from C_3 to C_4 is a long-term, high-risk strategy; however, if successful, this would increase both the yield potential of rice and its tolerance of drought.

Completion of the rice genomic sequence has facilitated the identification and cloning of genes and QTLs for yield traits. These genes/QTLs can be pyramided in elite varieties through the use of molecular marker-assisted selection with the ultimate aim of further increasing yield potential (Xing and

TABLE 7.1 Genes for Different Yield-Enhancing Traits Pyramided at IRRI in Several Backgrounds

Traits	Genes	References
Increased grain number	*Gn1a*	Ashikari et al. (2005)
Increased panicle branching	*Spl14*	Miura et al. (2010)
Increased culm strength	*SCM2*	Ookawa et al. (2010)
Uniform heading and grain filling	*Sps*	Rao et al. (2011)
Increased grain weight	*TGW6*	Ishimaru (2003)

Yang, 2010). Three grain yield components (number of panicles per unit area, number of spikelets per panicle, and grain weight) determine sink size. Genomic approaches have allowed the identification and cloning of genes for these sink traits (Sakamoto and Matsuoka, 2008) and these include *FC1* and *htd1* for tillering; *Gn1a* and *OsSPL14* for panicle architecture; and *GS3*, *GW2*, and *tgw6* for grain weight (Ashikari and Matsuoka, 2006). At IRRI, efforts to pyramid the identified genes are in the final stages (Table 7.1) and the validation of the yield increase across several genetic backgrounds is being determined.

BREEDING FOR BIOTIC STRESSES

Insect damage to irrigated rice is estimated to cause average yield losses of 12.7% (Litsinger et al., 2005). Losses range from 5% to 71% in upland rice (Litsinger et al., 2009a) and from 2% to 88% in rainfed lowland rice (Litsinger et al., 2009b). Reducing insect-mediated yield losses has been one of the important objectives of breeding programs in the post–Green Revolution era (Zhang, 2007). The development and cultivation of rice varieties with multiple resistances to insect pests and diseases is an important aspect for minimizing yield losses.

Planthoppers and leafhoppers have long been regarded as rice pests in Asia, but their importance has increased after the introduction of semidwarf, high-yielding varieties and the associated changes in production practices (i.e., the high use of fertilizers). Various species of planthoppers and leafhoppers affect rice, including *Nephotettix cincticeps* (green rice leafhopper) and *Nephotettix virescens* (green leafhopper, GLH), *Nilaparvata lugens* (BPH), *Sogatella furcifera* (whitebacked planthopper, WBPH), and *Tagosodes orizicolus*, of which the BPH has become a serious problem causing severe yield losses in most rice-growing countries. These insect pests damage rice by removing xylem and phloem fluids, resulting in wilting, browning, and death of the plant (hopperburn). They also cause injury to the plant by transmitting serious viruses, such as rice tungro virus, *hoja blanca*, and grassy stunt virus.

Orseolia oryza (Asian GMs) and *Orseolia oryzivora* (African GMs) are rice pests that have become a significant problem in India, China, Bangladesh, Southeast Asia, Africa, Indonesia, Myanmar, and Sri Lanka. After hatching, their larvae drill into rice plants along the leaf sheath, leading to onion shoot-like galls and destruction of the plant. Larvae cause injury to the plants by feeding between the leaf sheaths while a component of the fly's saliva suppresses normal plant growth and induces the formation of galls (silver shoots), resulting in stunted and numerous nonproductive tillers.

In upland (nonflooded) rice, root-feeding arthropods frequently constrain rice production. Other important pests in this ecosystem include root aphids, termites, and white grubs. In lowland rice, flooding eliminates many potential root-feeding pests; however, the rice water weevil (*Lissorhoptrus oryzophilusis*) has adapted to live in flooded environments. Their larvae feed on or in the roots of flooded rice, resulting in reduced tillering and shoot growth in the vegetative phase of rice development and reduced panicle density and grain weight at harvest.

Grain-sucking insects, mostly bugs, feed on developing grains by sucking juices with their piercing–sucking mouthparts. They are emerging as important rice pests, as they result in empty or partially filled grains, thereby reducing grain quality and yield. Bugs in the genera *Trigonotylus* and *Leptocorisa* are common in Asian countries, whereas several species in the genus *Oebalusare* are important pests in North and South America. Larval leaffolders fold rice leaves, bind the leaves with silk, and feed on green tissue, thereby creating injury and removing photosynthetic tissues. Defoliators may be adult and larval stages of insects from different orders; various species of grasshoppers and armyworms are important defoliators. *Cnaphalocrocis medinalis* and several species of *Marasmia* are rice pests in Asia. The impact of leaf removal on yield depends on the stage at which it occurs, with leaf removal early in the season being less damaging than leaf removal during reproductive stages. Stem borers have emerged as one of the most important rice insect pests from a global perspective. Several economically important stem borers occur in rice ecosystems of Asian countries, the most important being *Chilo suppressalis* and *Scirpophaga incertulas*. In comparison, stem borers, such as *Sesamia calamistis*, *Chilo zacconius*, and *Maliarpha separatella*, are important pests in Africa.

Several diverse sources of BPH resistance have been identified and some have been used to breed resistant rice varieties. Genetic analyses revealed more than 30 major genes and several QTLs in *indica* cultivars and wild species of rice. They are designated as *Bph1* in Mudgo (Athwal et al., 1971), *bph2* in ASD 7 (Athwal et al., 1971), *Bph3* in Rathu Heenati (Lakashminarayana and Khush, 1977), *bph4* in Babawee (Lakashminarayana and Khush, 1977), *bph5* in ARC 10550 (Khush et al., 1985), *Bph6* in Swarnalata (Kabir and Khush, 1988), *bph7* in T12 (Kabir and Khush, 1988), *bph8* in Chinsaba (Nemoto et al., 1989), *Bph9* in Balamawee (Nemoto et al., 1989), *Bph10(t)* in IR65482-4-136-2-2 from *O. australiensis* (Ishii et al., 1994), *bph11* in *Oryza officinalis*

(Takita, 1996), *bph12* in *O. officinalis* (Hirabayashi et al., 1998), *Bph13* in IR54745-2-21-12-17-6 from *O. officinalis* (Renganayaki et al., 2002), *Bph14* in B5 (Huang et al., 2001), *Bph15* in B5 (Huang et al., 2001), *bph16* in *O. officinalis* (Hirabayashi et al., 1998), *Bph17* in Rathu Heenati (Sun et al., 2005), *Bph18(t)* in IR65482-7-216-1-2 from *O. australiensis* (Jena et al., 2006), *bph19* in AS 20-1 (Chen et al., 2006), *Bph20(t)* and *Bph21(t)* in IR71033-121-15 from *Oryza minuta* (Rahman et al., 2009), *Bph22(t)* in *Oryza glaberrima* (Ram et al., 2010), *Bph23(t)* in *O. minuta* (Ram et al., 2010), *bph24(t)* in *O. rufipogon* (Deen et al., 2010), *Bph25(t)* and *Bph26(t)* in ADR 52 (Myint et al., 2012), and *Bph27(t)* in GX2 183 (Huang et al., 2013). Of the 28 BPH resistance genes characterized so far, 9 are recessive (*bph2*, *bph4*, *bph5*, *bph7*, *bph8*, *bph12*, *bph16*, *bph19*, and *bph24*) and 19 are dominant in nature; 21 BPH resistance genes have been mapped with molecular markers while the remaining 7 genes (*bph5*, *Bph6*, *bph7*, *bph8*, *Bph22*, *Bph23*, and *bph24*) have yet to be mapped. In addition to major genes, a set of 63 QTLs controlling resistance was also reported (Fujita et al., 2013). QTLs conferring resistance have been detected in IR64, Kasalath, DV 85, Teqing, and *O. officinalis* (Alam and Cohen, 1998; Su et al., 2002, 2005; Xu et al., 2002). Most of the identified genes that showed a susceptible reaction against the Indian biotype (biotype 4) were tested in the Indian Institute of Rice Research, except for *Bph18*, *Bph20* + *Bph21*, *Bph22*, *Bph23*, and *Bph24*. The donors that have more than one gene showing a resistance reaction indicated that the use of multiple genes is required for developing varieties with BPH resistance.

More than 10 genes have been identified that confer rice resistance to GM, including *Gm1* in Samridhi, Ruchi, Kavya, and Asha (Chaudhary et al., 1986); *Gm2* in Phalguna (Chaudhary et al., 1986); *gm3* in RP2068-18-3-5 (Kumar et al., 1998); *Gm4* in Abhaya (Shrivastava et al., 1993); *Gm5* in ARC 5984 (Kumar et al., 1998); *Gm6* in Duokang #1 (Yang et al., 1997); *Gm7* in RP2333-156-8 (Kumar et al., 1999); *Gm8* in Jhitpiti (Kumar et al., 2000); *Gm9* in Madhuri line 9 (Shrivastava et al., 2002); *Gm10* in BG 380-2 (Kumar et al., 2005); and *Gm11* in CR 57-MR1523 (Himabindu et al., 2010). These genes have all been used in breeding programs. Of these, nine genes (*Gm1*, *Gm2*, *gm3*, *Gm4*, *Gm5*, *Gm6*, *Gm7*, *Gm8*, and *Gm11*) have been tagged and mapped. *Gm1* is mapped on chromosome 9 (Sundaram, 2007), *Gm2* (Sundaram, 2007) and *gm3* (Sama, 2011) are mapped on chromosome 4, *Gm4* is localized on chromosome 8 (Himabindu, 2009; Mohan et al., 1997; Nair et al., 1996), *Gm6* is reported on chromosome 4 (Katiyar et al., 2001), *Gm7* has been mapped on chromosome 4 (Sardesai et al., 2002), *Gm8* has been fine mapped on chromosome 8 (Jain et al., 2004; Sama et al., 2012), and *Gm11* on chromosome 12 between the SSR markers RM28574 and RM28706 (Himabindu et al., 2010). Identified and frequently used genes for different abiotic stresses are presented in Table 7.2.

It is important to note that resistance against the widespread yellow stem borer, *S. incertulas*, has been rare in rice germplasm. Moderate resistance identified from donor parent TKM6 was used in a breeding program that resulted

TABLE 7.2 Important Genes for Different Abiotic Stresses Used in Breeding Programs

Abiotic stresses	Genes/ QTLs	Chromosomes	Donors	References
BPH	*Bph1*	12	Mudgo, TKM6	Kim and Sohn (2005)
	bph2	12	IR1154-243	Murai et al. (2001)
	Bph26/ bph2	12	ADR52	Tamura et al. (2014)
	bph7	12	T12	Qiu et al. (2014)
	Bph9	12	Kaharamana	Su et al. (2006)
	Bph10(t)	12	IR65482-4-136, *O. australiensis*	Ishii et al. (1994)
	Bph18(t)	12	IR65482-7-216, *O. australiensis*	Jena et al. (2006)
	Bph21(t)	12	IR71033-121-15, *O. minuta*	Rahman et al. (2009)
	Bph12	4	B14, *O. latifolia*	Qiu et al. (2012)
	Bph15	4	B5, *O. officinalis*	Lv and Du (2014)
	Bph17	4	Rathu Heenati	Sun et al. (2005)
	Bph20(t)	4	IR71033-121-15, *O. minuta*	Rahman et al. (2009)
	Bph6	4	Swarnalata	Qiu et al. (2010)
	Bph27	4	GX2183, *O. rufipogon*	Huang et al. (2013)
	Bph27(t)	4	Balamawee	He et al. (2013)
	Bph14	3	B5, *O. officinalis*	Du et al. (2009)
	QBph3	3	IR02W101, *O. officinalis*	Hu et al. (2015)
	Bph13	3	IR54745-2-21, *O. officinalis*	Renganayaki et al. (2002)
	bph19	3	AS20-1	Chen et al. (2006)
	qBph3	3	Rathu Heenati	Kumari et al. (2010)
	Bph3	6	Rathu Heenati	Jairin et al. (2007)
	bph4	6	Babawee	Kawaguchi et al. (2001)
	Bph25(t)	6	ADR52	Myint et al. (2012)
	bph29	6	RBPH54, *O. rufipogon*	Wang et al. (2015)
	Bph6	11	IR54741-3-21-22, *O. officinalis*	Jena et al. (2003)
	Bph28(t)	11	DV85	Wu et al. (2014)

TABLE 7.2 Important Genes for Different Abiotic Stresses Used in Breeding Programs (*Cont.*)

Abiotic stresses	Genes/ QTLs	Chromosomes	Donors	References
GM	*Gm1*	9	Samridhi	Chaudhary et al. (1986)
	Gm2	9	Surekha	Chaudhary et al. (1986)
	Gm3	4	RP2068-18-3-5	Kumar et al. (1999)
	Gm4	8	Abhaya	Shrivastava et al. (1993)
	Gm5	12	ARC 5984	Kumar et al. (1999)
	Gm6	4	Duokang 1	Yang et al. (1997)
	Gm7	4	RP2333-156-8	Kumar et al. (1999)
	Gm8	8	Jhitpiti	Kumar et al. (2000)
	Gm9	8	Madhuri line 9	Shrivastava et al. (2002)
	Gm10	8	BG380-2	Kumar et al. (2005)
	Gm11		CR 57-MR1523	Himabindu et al. (2010)
Blast	*Pi-1*	11	PB-1460	Fuentes et al. (2007)
	Pi2-1	6	TY	Wang et al. (2012)
	Pi51(t)	12	TY	Wang et al. (2012)
	Pi-z(t)	6	PB-1460	Fjellstrom et al. (2006)
	Pi-ta 2	12	Tadukan	Koide et al. (2011)
				Fjellstrom et al. (2004)
				Shikari et al. (2013)
	AC134922	11	AC134922	Wang et al. (2014)
	Pi-14(t)	12	—	Kiyosawa and Ando (1990)
	Pi-k^h	11	Tetep	Sharma et al. (2005)
	Pi9	6	WHD127	Qu et al. (2006)
	Pi33	8	IR64	Berruyer et al. (2003)
	Pi25	6	Gumei2	Chen et al. (2011b)

(Continued)

TABLE 7.2 Important Genes for Different Abiotic Stresses Used in Breeding Programs (*cont.*)

Abiotic stresses	Genes/ QTLs	Chromosomes	Donors	References
BB	xa4	11	IRBB60	Ma et al. (1999)
	xa5	5	IRBB60	Perumalsamy et al. (2010)
	xa7	6	IRBB66	Porter et al. (2003)
	xa13	8	IRBB60	Chu et al. (2006)
	xa21	11	IRBB60	Song et al. (1997)
	xa23	11	IRBB23	Ona et al. (2010)
	xa25(t)	12	*Indica*	Liu et al. (2011)
	xa26(t)	11	*Indica*	Lee et al. (2003)
	xa27(t)	6	Wild spp. of *Oryza*	Gu et al. (2004)
	xa31(t)	11	*Japonica*	Wang et al. (2009)
	xa32(t)	11	Wild spp. of *Oryza*	Zheng et al. (2009)
	xa35(t)	11	Wild spp. of *Oryza*	Guo et al. (2010)
	xa36(t)	11	—	Miao et al. (2010)
	xa38	4	Wild spp. of *Oryza*	Bhasin et al. (2012)

BB, Bacterial blight; BPH, brown planthopper; GM, gall midge; TY, Tianjingyeshengdao.

in the release of rice varieties Ratna, Sasyasree, and Vikas. Some limited success has been achieved in developing varieties resistant to stem borer by using multiple donors. Resistance sources, such as Ptb 18 and Ptb 21, were used in breeding programs and tolerant varieties Suraksha, Shaktiman, and Triguna were developed. The identification of QTLs from wild species and stacking of QTLs from different sources could be effective strategies to develop resistance and overcome this emerging insect.

Many countries have reported high levels of damage due to BPH (*N. lugens*). Breeding for resistance to the pest began during the mid-1970s and many varieties have been released. For example, in India varieties such as Jyothi, Sonasali, Vajram, Chaitanya, Udaya, Manasarovar, MTU 1010, Bharani, Pant dhan-12, ASD-19, ADT-42, CORH-1, PMK-2, Jitendra, Krishnaveni, Kartika, Neela, Jyoti, Bhadra, Bharathidasan, Aruna, Bhadra, Vijetha, PHB-71, Co42, KAUM57-18-1-1, KAUM57-9-1-1, KAUM45-20-1, KAUM42-6-3, KAUM20-19-4, KAUM61-6-1-1-2, KAUM59-2-1-2, SKL-8, CSRC(S)11-5-0-2, Richa, Suruchi5401, PKV-SKL-3-11-25-30-36, Rajlaxmi, Pant sankar Dhan-3, ADT(R)-48, Onam, PTB-45, PTB-51, PTB-52, Gouri(MO20), and Reeta were released. For WBPH, Nidhi, Jitendra, Tapaswini, Dhala Heera, Sudhir, Sunil, Satyaranjan, Surya, Mahananda, Harsha, Dandi, HKR 120, and HKR 126 are recommended in the WBPH-endemic areas of India. Concerted efforts to breed for resistance to GLH (*N. virescens*) and tungro disease during the 1980s led to the release of Vikramarya and Nidhi in India.

Concerted efforts have also been made to incorporate resistance to GMs (*O. oryza*) in rice. Popular varieties Abhaya, Shyamala, Jitendra, Samaleswari, Purnendu, PHB-71, Triguna, Indur Samba, Tapaswini, ADT-43, Neeraja, Uma, Karishma, Srikakulam Sannalu, KAUM57-18-1-1, KAUM61-6-1-1-2, KAUM59-29-2-1-2, Dhanrasi, JGL-1798, PNR-546, SYE-2001, SKL-8, Shusk Samrat, PTB-49, PTB-51, PTB-52, Swetha, Gouri (MO20), Abhisek, Chandrahansini, IGKVR-1, IGKVR-2, MO 21, Mandakini, Nua Chinikamini, Phalguni, IGKVR-1244, Indira Barani dhan-1, CR Dhan500, 27P61, Arize Tej, Kanak, ADT(R)48, Kavya, Jyoti, Vikram, Surekha, Varulu, Asha, Samridhi, Ratnagiri3, Panchami, Pabitra, R636-405, Birsa Dhan-105, and Birsa Dhan-106 were recommended for GM-endemic areas in India.

Disease resistance in plants has been nature's protective measure for restricting parasites since the world began. Evolutionary selection leads to modification of the toxin formation of parasites and building up resistance in the host plants. Rice (*Oryza sativa*), consumed by 50% of the world's population, is one of the most important staple foods. The key factors limiting rice yield are the diseases caused by fungi, bacteria, viruses, and nematodes that generally result in 20%–70% yield losses (Mew et al., 1993; Ou, 1985). The most threatening rice diseases are blast, bacterial leaf blight, sheath blight, false smut, and rice yellow mosaic virus. Pesticides have traditionally been used for disease control; however, this is not a long-term reliable solution, as pathogens develop immunity and the use of pesticides is both uneconomical and environmentally unfriendly. Therefore, there is a strong need to develop new strategies that enhance the composition of plant genomes to provide long-term protection over broad geographic areas.

Advances in molecular plant breeding can be used to expand genetic diversity (through modification of gene function, resistance mechanisms, and tolerance ability), characterize genetic architecture, and increase selection efficiency. Breeding and nurturing cultivars with qualitative resistance (pathogen race–specific complete resistance; conferred by a single resistance gene) have been successfully applied against BB and blast diseases. On the other hand, efforts continue with quantitative resistance (partial resistance, conferred by QTLs/multiple genes). The incorporation of both qualitative and quantitative genes through classical plant breeding, advances in rice molecular breeding and genomics, use of the broad spectrum of resistance genes, and pyramiding genes may help to overcome the devastating global disease problems in rice.

Breakdown of blast resistance is the major cause of yield instability in several rice-growing areas and efforts are under way to develop rice varieties with durable blast resistance. Rice blast disease could be leaf blast, neck blast, or collar blast. More than 40 major genes including *Pi36* (chromosome 1; Liu et al., 2007a,b); *Pib* (chromosome 2; Wang and Leung, 1998); *Pi2, Piz, Piz-t, Piz-5, Pi8*(t), *Pi9, Pi13, Pi13*(t), *Pi25*(t), *Pi26*(t), *Pi27*(t), *Pid2, Pigm*(t), and *Pi40*(t) (chromosome 6; Jeung et al., 2007; Qu et al., 2006; Yu et al., 1991); *Pi37* (chromosome 8; Chen et al., 2005); *Pi1* (chromosome 11; Mew et al., 1993); *Pi7, Pi18, Pif, Pi34, Pi38, Pi44*(t), *PBR*, and *Pilm2* (chromosome 11); *Pi39, Pita,* and *Pita2* (chromosome 12, Liu et al., 2007a,b; Bryan et al., 2000); *Pitq6, Pi6*(t), *Pi12*(t), *Pi19*(t), *Pi20*(t), *Pi21*(t), *Pi24*(t), *Pi31*(t), *Pi32*(t), *Pi62*(t), *Pi157*(t), *IPi*, and *IPi3* (chromosome 12); as well as QTLs for blast resistance have been identified. Out of these, 22 genes (*Pib, Pita, Pik-h, Pi9, Pi2, Piz-t, Pid2, Pi36, Pi37, Pik-m, Pit, Pi5, Pid3, pi21, Pb1, Pish, Pik, Pik-p, Pia, NLS1, Pi25,* and *Pi54rh*) have been cloned (Ashikawa et al., 2008; Bryan et al., 2000; Chen et al., 2006, 2011a,b; Das et al., 2012; Fukuoka et al., 2009; Hayashi and Yoshida, 2009; Hayashi et al., 2010; Lee et al., 2009; Lin et al., 2007; Liu et al., 2007a,b; Okuyama et al., 2011; Qu et al., 2006; Sharma et al., 2005; Shang et al., 2009; Takahashi et al., 2010; Tang et al., 2011; Wang et al., 1999; Zhai et al., 2011; Zhou et al., 2006).

One of the identified blast-resistance genes, *Pi-9*, is reported to have wide-spectrum resistance to rice blast races from 43 different countries (Liu et al., 2002) while the *Pita* locus in rice has also been effectively used to manage rice blast worldwide (including India). The genes *Pita* and *Pita2* (chromosome 12, linked closely at 0.4 cM) are reported to have broad resistance (Bryan et al., 2000; Fjellstrom et al., 2004; Shikari et al., 2013). Three *indica* landrace cultivars, Tadukan from the Philippines, Tetep from Vietnam, and TeQing from China, are known as sources of *Pita* resistance worldwide (Moldenhauer et al., 1990). Blast resistant genes have also been introgressed into PRR78, Jin 23B, Luhui 17, Zhenshan 97B, IR50, G46B, CO39, MR219, Pusa Basmati 1, Pusa1602, and Pusa1603 lines through marker-assisted selection, and blast-resistant lines have been developed.

BB, caused by *Xanthomonas oryzae* pv. *oryzae* (*Xoo*), is one of the most destructive diseases active in major rice-growing countries of Asia. In these countries, rice is grown under irrigated and high-fertilizer input conditions that are conducive to disease development. In severe epidemics, yield losses in the range of 20%–40% have been reported. So far, *Xa1* (Yoshimura et al., 1998), *Xa2* (Sakaguchi, 1967), *xa3* (Sun et al., 2004), *Xa4* (Sun et al., 2004), *xa5* (Blair et al., 2003), *Xa7* (Porter et al., 2003), *Xa8* (Vikal et al., 2014), *Xa10* (Mew et al., 1982; Yoshimura et al., 1983), *Xa11* (Goto et al., 2009), *Xa12* and *xa13* (Kurata and Yamazaki, 2006; Ogawa et al., 1988), *Xa14* (Kurata and Yamazaki, 2006; Sidhu et al., 1978), *Xa15* (Nakai et al., 1988; Ogawa, 1996), *Xa16* (Kurata and Yamazaki, 2006), *Xa17* (Kurata and Yamazaki, 2006), *Xa18* (Kurata and Yamazaki, 2006), *Xa19* (Kurata and Yamazaki, 2006; Taura et al., 1992), *Xa20* (Kurata and Yamazaki, 2006; Taura et al., 1992), *Xa21* (Song et al., 1995), *Xa22*(t) (Lin et al., 1996), *Xa23* (Ona et al., 2010), *Xa27* (Gu et al., 2005), *Xa29*(*t*) (Tan et al., 2004), *Xa30*(*t*) (Cheema et al., 2008),

Xa31(t) (Wang et al., 2009), *Xa32(t)* (Ruan et al., 2008; Zheng et al., 2009), *Xa33* (Natrajkumar et al., 2012), *Xa33(t)* (Korinsak et al., 2009), *Xa34(t)* (Chen et al., 2011a), *Xa35(t)* (Guo et al., 2010), *Xa36(t)* (Miao et al., 2010), and *Xa38* (Bhasin et al., 2012) BB resistance genes have been identified. Of these, *Xa1, xa5, xa13, Xa21, Xa3/Xa26,* and *Xa27* have been cloned; another six have been physically mapped (*Xa2, Xa4, Xa7, Xa30, Xa33,* and *Xa38*) and used for marker-aided selection of resistant lines, resulting in the release of cultivars in several countries. The majority of the BB genes have been identified from cultivated rice (i.e., *O. sativa*); however, some have been introgressed from related wild species, such as *Oryza longistaminata* (*Xa21*), *O. rufipogon* (*Xa23*), *O. brachyantha* (*Xa34t*), *O. minuta* (*Xa27* and *Xa35t*), *O. officinalis* (*Xa29t*), *O. australiensis* (*Xa32t*), and *O. nivara* (*Xa33* and *Xa38*). Long-term cultivation of rice varieties carrying a single resistance gene has resulted in a significant shift in pathogen race frequency and consequent breakdown of resistance. Therefore, pyramiding of genes is a better strategy to overcome yield loss caused by BB. Loan et al. (2006) reported high resistance in combinations with three dominant resistance genes, *Xa4 + Xa7 + Xa21*, in IRBB62, followed by *Xa4 + xa5 + xa13 + Xa21* in IRBB60. Individual resistance genes as well as pyramided genes in combinations of two to four genes are available in the common variety IR24. A number of BB-resistant rice varieties have been released, including Zensho, Kogyyoku, Kono35, Norin 27, and Asakaze (Japan); IR20, IR22, IR26, IR28, IR29, IR30, NSIC Rc142, and NSIC Rc154 (Philippines); Angke (*Xa4 + xa5*) and Conde (*Xa4 + Xa7*) (Indonesia); Macassane (Mozambique); and Sambha Mahsuri, Triguna rice lines, Ratnagiri, 68-1, PR 4141, Ajaya (IET8585), and improved versions of the elite basmati variety, Pusa Basmati 1, Pusa RH10, and KMR-3R (India).

Rice brown spot (BS) is another serious disease in South and Southeast Asian countries and it affects millions of hectares of rice every growing season. BS is conventionally perceived as a secondary problem in rice crops that have experienced physiological stress (e.g., drought and poor soil fertility, especially lack of nitrogen) rather than a true infectious disease. Three QTLs (*qBS2, qBS9, qBS11*) were detected in cultivar Tadukan on chromosomes 2, 9, and 11, respectively (Sato et al., 2008), with *qBS11* considered to have a major effect.

Sheath blight caused by *Rhizoctonia solani* Kuhn is also emerging as a major disease of rice and it severely impairs both rice yield and quality (Lee and Rush, 1983). It occurs throughout temperate and tropical areas and is also known as brown-bordered leaf and sheath spot. So far, no rice variety has been found to be completely resistant to this fungus. Wild rice has been suggested to be an important source for identifying donors and genetic regions associated with sheath blight resistance, and this could be useful in reducing the yield losses attributed to this pest. The two most common smut diseases found on rice panicles are kernel smut (*Tilletia barclayana*) and panicle smut or false smut (*Ustilaginoidea virens*) (Lee and Gunnell, 1992). False smut disease is more common in areas with high relative humidity (>90%) and temperatures of

25–35°C. False smut is now widespread throughout tropical Asia, India, Egypt, Italy, Australia, and Central and South America (Atia, 2004). Early flowering and spikelet maturity stages of the rice crop are reported to be more prone to infection, leading to chalkiness of grains and reduced crop yield. False smut also causes a reduction in seed germination of up to 35%.

The intelligent design of gene/QTL pyramiding for major and minor genes is essential for long-term resistance, and this is especially pertinent with the frequent changing of insect biotypes and disease pathotypes. The availability of precision biotechnological tools is making it easier for breeders to overcome these challenges, understand biotypic/pathotypic variability, and plan gene/QTL introgression and pyramiding.

BREEDING FOR ABIOTIC STRESSES

Drought, flood, salinity, problem soils, high temperature, and low temperature are all important abiotic stresses that reduce rice yield. Drastic losses in rice yield are becoming more prevalent as weather conditions become harsher due to climate change. The complexity of genetic architecture and the unpredictable occurrence of climatic conditions pose a challenge for breeding varieties against these abiotic stresses. Abiotic stress can occur at multiple stages of plant growth and plants are frequently simultaneously affected by more than one type of stress and often in combination with biotic stresses, such as blast and BS, together with drought during the growth period.

Drought and water-limited conditions are regular problems on more than 20 million ha in Asia and more than 80% of the rice area in Africa. The Eastern Indo–Gangetic plains are one of the major drought-prone, rice-producing regions in the world (Huke and Huke, 1997; Pandey et al., 2005). Of the 20.7 million ha of rainfed rice area in India, around 13.6 million ha are prone to drought (Bansal et al., 2014; Pandey et al., 2012). It has been estimated that 36% of the total value of rice production is lost during a drought year in eastern India and the economic cost of drought has reached USD 100 million per year (Pandey and Bhandari, 2009).

Advanced molecular markers and genomics technologies have played a major role in the selection of drought-tolerance traits. Associated chromosomal regions of QTLs and genes have been identified in tolerant genotypes and significant improvement in grain yield under water-limited conditions has been successfully demonstrated. Several physiological traits have been reported to be linked to drought tolerance; however, it has been very difficult to combine these physiological mechanisms appropriately to obtain the desired yield increase under drought. Direct selection for grain yield under drought and a combination of high yield potential and good yield under drought have been suggested as efficient alternative strategies (Kumar et al., 2008, 2009a; Venuprasad et al., 2007, 2008). A marker strategy based on QTLs for secondary traits was also tried. Through MAS approaches in rice, numerous drought-tolerance traits associated with QTLs have been identified from different sources of rice germplasm. These include morphological traits, such as cell membrane stability and leaf water status; root traits, such as thickness, length, fresh weight and dry weight, root and shoot ratio and deep rooting system, and leaf rolling and drying; physiological and biochemical traits, such as stomatal regulation, stomatal closure, relative water content, photosynthetic rate, chlorophyll stability index, proline accumulation, signal transduction, abscisic acid content, osmotic adjustment, and antioxidant systems (Ananthi and Vijayaraghavan, 2012; Chutia and Borah, 2012; Szabados and Savouré, 2009); and grain yield traits, such as grain yield and 1000-grain weight (Kirigwi et al., 2007; Lafitte et al., 2004; Nezhad et al., 2012; Salem et al., 2007; Shamsudin et al., 2016; Xu et al., 2004). Among the total parameters, reduction in rice grain yield is commonly associated with an increased percentage of spikelet sterility (Fukai et al., 1999; Jongdee et al., 2002; Liu et al., 2006) and spikelet number per panicle (Boonjung and Fukai, 1996).

Targeted yield improvement under drought was not achieved using QTLs and genes for secondary traits. Therefore, the focus shifted from secondary traits to grain yield under drought through repeated experiments and screening techniques (Kumar et al., 2008, 2009a; Venuprasad et al., 2007, 2008). Grain yield under drought was found to be moderately heritable and this suggested the possibility of improving yield under water-limited conditions. This led to the identification of several large-effect QTLs for grain yield under upland and lowland reproductive-stage drought, such as $qDTY_{1.1}$ (Ghimire et al., 2012; Vikram et al., 2011), $qDTY_{2.1}$ (Venuprasad et al., 2009), $qDTY_{2.2}$ (Swamy et al., 2013; Venuprasad et al., 2007), $qDTY_{3.1}$ (Venuprasad et al., 2009), $qDTY_{4.1}$ (Swamy et al., 2013), $qDTY_{6.1}$ (Venuprasad et al., 2012), $qDTY_{9.1}$ (Swamy et al., 2013), $qDTY_{10.1}$ (Swamy et al., 2013), and $qDTY_{12.1}$ (Bernier et al., 2007). Some of these QTLs were introgressed into popular high-yielding but drought-susceptible varieties (Kumar et al., 2014). Table 7.3 lists the QTLs identified for drought along with QTLs and genes for other abiotic stresses. Several rice research institutes in India (in collaboration with IRRI-Philippines) have found a yield advantage of these introgressed lines under stress conditions. Under severe reproductive-stage drought stress, NILs of IR64 showed a grain yield advantage of 100–500 percentage points over the recurrent parent IR64 (Kumar et al., 2014). Furthermore, four DTY QTLs ($qDTY_{1.1}$, $qDTY_{2.1}$, $qDTY_{3.1}$, and $qDTY_{6.1}$) showed a positive effect in Swarna. Similarly, the introgression of qDTYs into several popular varieties, such as IR64, Sabitri, TDK1, and BR11, showed a parallel yield advantage over the parents and local popular varieties under drought (Kumar et al., 2014). On the other hand, the grouping of specific DTY QTLs helps to further improve grain yield and determine their interaction effect. Pyramiding of DTY QTLs into a single genetic background has shown a yield advantage over genotypes with a single QTL under drought. For instance, the introgression of one QTL led to a yield increase of 0.5 t/ha in Vandana and the introgression of two QTLs showed a yield advantage of >1.0 t/ha in Sambha Mahsuri and IR64. Subsequently, the addition of three QTLs in Swarna led to a further increase of >1.5 t/ha under drought (Kumar et al., 2014). IR64

TABLE 7.3 Genes and QTLs for Abiotic Stresses Identified and Regularly Used in Rice Breeding Programs

Traits	QTLs/ genes	Donors	Chromosomes	References
Drought tolerance	$qDTY_{1.1}$	N22, Dhagaddeshi	1	Vikram et al. (2011), Ghimire et al. (2012)
	$qDTY_{2.1}$	Apo	2	Venuprasad et al. (2009)
	$qDTY_{2.2}$	Aday Sel	2	Swamy et al. (2013)
	$qDTY_{3.1}$	Apo	3	Venuprasad et al. (2009)
	$qDTY_{4.1}$	Aday Sel	4	Swamy et al. (2013)
	$qDTY_{6.1}$	IR55419-04	6	Dixit et al. (2014)
	$qDTY_{12.1}$	Way Rarem	12	Bernier et al. (2007)
Anaerobic germination	$qAG_{9.1}$	IR93312-30-101-20-3-66-6	9	Angaji et al. (2010)
Submergence	$SUB1$	IR49830, IR40931	9	—
Heat tolerance	$qHTSF_{4.1}$	Giza 178, N22	4	Ye et al. (2015, 2012)
	$qHTSF_{1.1}$	N22	—	Jagadish et al. (2010), Ye et al. (2012)
Cold tolerance	$qPSST$-3	Geumobyeo	3	Suh et al. (2010)
	$qPSST$-7	Geumobyeo	7	Suh et al. (2010)
	$qPSST$-9	Geumobyeo	9	Suh et al. (2010)
	$qCTS4a$, $qCTS4b$	Geumobyeo	4	Suh et al. (2012)
Salinity tolerance	$Saltol$	Pokkali	1	Bonilla et al. (2002), Thomson et al. (2010)

introgressed with $qDTY_{2.2}$ and $qDTY_{4.1}$ was released as variety DRR dhan 42 in India, as Sukha dhan 4 in Nepal, and as Yeanelo 4, 5, 6, and 7 in Myanmar. Several drought-tolerant varieties released during the last 10 years are listed in Table 7.4. In addition to being high yielding, these new varieties provide a yield advantage of 1.0–1.5 t/ha under moderate-to-severe drought when compared with the drought-susceptible varieties that are currently grown.

Submergence and stagnant flooding are problems in rainfed lowlands during the monsoon season. The crop could face yield losses of 10%–100% depending on the duration and depth of the floodwater (Ismail et al., 2013; Mackill et al., 2012). The productivity of rice grain yield in eastern India, particularly in submergence-prone areas, is 0.5–0.8 t/ha, whereas favorable rainfed lowlands could achieve 2.0 t/ha, compared with the input-intensive irrigated system (5.0 t/ha). Therefore, these ecosystems have huge potential for more food production (Ismail et al., 2013; Langridge and Fleury, 2011). To achieve significant improvements in rice grain yield in flood-prone areas of rainfed lowlands, high-yielding, flood-tolerant rice varieties should be developed and use with appropriate management practices (Ismail et al., 2008; Singh et al., 2013; Swain et al., 2005).

In the 1950s flood- and submergence-tolerant landraces FR13A and FR43B were reported in Odisha, India. Other similar genotypes with resilience to complete submergence were also reported in Sri Lanka (Mackill et al., 1996; Vergara and Mazaredo, 1975). Of these landraces, FR13A was found to be the most tolerant and the tolerance trait was subsequently introduced into breeding line IR49830-7-1-2-2 at IRRI (Mackill et al., 1993; Mishra et al., 1996). In the 1990s a major QTL, *SUBMERGENCE 1* (*SUB1*), responsible for tolerance was identified on chromosome 9, contributing up to 70% of the phenotypic variation (PV; Xu and Mackill, 1996). Furthermore, cloning and sequencing of the SUB1 region revealed that the *SUB1A* gene encodes a putative ethylene response factor (ERF) DNA-binding protein with a single ERF/APETELA2 domain (Xu et al., 2006). Two additional ERFs, designated *SUB1B* and *SUB1C*, were identified in *japonica* intolerant cultivars, whereas tolerant *indica* and aus accessions showed the presence of the *SUB1A* gene. Two forms of *SUB1* (*SUB1A-1* and *SUB1A-2*) were observed in tolerant and intolerant *indica* and aus accessions. These encode identical proteins, with the exception of Ser_{186} in the tolerant allele and Pro_{186} in the intolerant allele, which contributes to varied levels of gene expression. The *SUB1* QTL was introgressed into megavarieties, such as Swarna, Sambha Mahsuri, IR64, Thadokkam 1 (TDK1), CR1009, and BR11, by marker-assisted backcrossing (MABC) at IRRI (Septiningsih et al., 2009). Multiple evaluation trials of these Sub1 lines in several locations have confirmed the complete tolerance of submergence for 14–17 days when compared with their original parents (Sarkar et al., 2009). It has been estimated that Swarna is cultivated in 30%–40% of the rainfed lowlands of India. Several submergence-tolerant varieties have been released in South and Southeast Asia over the last 10 years (Table 7.4) and these varieties provide a yield advantage of 1.0–2.0 t/ha under submergence for 14–17 days vis-à-vis the submergence-susceptible varieties.

TABLE 7.4 Varieties for Abiotic Stresses Released in Different Countries of South and Southeast Asia and Africa

Study number	Abiotic stresses	Names	Designations	Environment, countries
1	Drought	Sahod Ulan 1	IR74371-54-1-1	Rainfed lowland, Philippines
2	Drought	Sahbhagi dhan	IR74371-70-1-1	Rainfed lowland, India
3	Drought	BRRI dhan56	IR74371-70-1-1	Rainfed lowland, Bangladesh
4	Drought	Sukha dhan 1	IR74371-46-1-1	Rainfed lowland, Nepal
5	Drought	Sukha dhan 2	IR74371-54-1-1	Rainfed lowland, Nepal
6	Drought	Sukha dhan 3	IR74371-70-1-1	Rainfed lowland, Nepal
7	Drought	Katihan 1	IR79913-B-176-B-4	Upland, Philippines
8	Drought	Sahod Ulan 3	IR81412-B-B-82-1	Rainfed lowland, Philippines
9	Drought	Sahod Ulan 5	IR81023-B-116-1-2	Rainfed lowland, Philippines
10	Drought	Sahod Ulan 6	IR72667-16-1-B-B-3	Rainfed lowland, Philippines
11	Drought	Inpago Lipi GO1	IR79971-B-191-B-B	Upland, Indonesia
12	Drought	Inpago Lipi GO2	IR79971-B-227-B-B	Upland, Indonesia
13	Drought	M'ziva	IR77080-B-34-3	Rainfed lowland, Mozambique
14	Drought	Upia 3	IR74371-54-1-1	Rainfed lowland, Nigeria
15	Drought	Sukha dhan 4	IR64 + $qDTY_{2.2}$, $qDTY4.1$	Rainfed lowland, Nepal

TABLE 7.4 Varieties for Abiotic Stresses Released in Different Countries of South and Southeast Asia and Africa (*cont.*)

Study number	Abiotic stresses	Names	Designations	Environment, countries
16	Drought	Sukha dhan 5	IR83388-B-B-108-3	Rainfed lowland, Nepal
17	Drought	Sukha dhan 6	IR83383-B-B-129-4	Rainfed lowland, Nepal
18	Drought	DRR dhan 42	IR64 + $qDTY_{2.2}$, $qDTY_{4.1}$	Rainfed lowland, India
19	Drought	DRR dhan 44	IR93376-B-B-130	Lowland, India
20	Drought	Kaithan 2	IR82635-B-B-47-2	Upland, Philippines
21	Drought	Kaithan 3	IR86857-101-2-1-3	Upland, Philippines
22	Drought	Inpago Lipi GO4	IR79971-B-102-B-B	Upland, Indonesia
23	Drought	BRRI dhan66	IR82635-B-B-75-2	Rainfed lowland, Bangladesh
24	Drought	Yeanelo 4	IR87707-44-B-B-B	Rainfed lowland, Myanmar
25	Drought	BRRI dhan71	IR82589-B-B-84-3	Rainfed lowland, Bangladesh
26	Drought	Swarna Shreya	IR84899-B-179-16-1-1-1-1	Rainfed lowland, India
27	Drought	Sahod Ulan 15	IR83383-B-B-129-4	Rainfed lowland, Philippines
28	Drought	Sahod Ulan 20	IR86781-3-3-1-1	Rainfed lowland, Philippines
29	Submergence	Improved Swarna	Swarna + Sub1	India
30	Submergence	INPARA-5	Swarna + Sub1	Indonesia
31	Submergence	BRRI dhan51	Swarna + Sub1	Bangladesh

(*Continued*)

TABLE 7.4 Varieties for Abiotic Stresses Released in Different Countries of South and Southeast Asia and Africa (*cont.*)

Study number	Abiotic stresses	Names	Designations	Environment, countries
32	Submergence	Swarna-Sub1	Swarna + Sub1	Nepal
33	Submergence	Yemyoke Khan	Swarna + Sub1	Myanmar
34	Submergence	NSIC Rc194	IR64 + Sub1	Philippines
35	Submergence	INPARA-4	IR64 + Sub1	Indonesia
36	Submergence	BRRI dhan52	BR 11 + Sub1	Bangladesh
37	Submergence	INPARA-5	Ciherang + Sub1	Indonesia
38	Submergence	BINA Dhan 11	Ciherang + Sub1	Bangladesh
39	Submergence	Ciherang-Sub1	Ciherang + Sub1	Nepal
40	Submergence	S. Mahsuri-Sub1	Sambha Mahsuri + Sub1	India
41	Submergence	S. Mahsuri-Sub1	Sambha Mahsuri + Sub1	Nepal
42	Submergence	BINA dhan12	Sambha Mahsuri + Sub1	Bangladesh
43	Submergence	CR1009-Sub1	CR1009 + Sub1	India
44	Salinity	CSR 36	CSR 36	India
45	Salinity	BRRI dhan47	IR63307-4B-4-3	Bangladesh
46	Salinity	NDRK 5088	—	India
47	Salinity	BINA Dhan 8	IR66946-3R-149-1-1	Bangladesh
48	Salinity	BRRI dhan53	BR5778-156-1-3-HR14	Bangladesh
49	Salinity	BRRI dhan54	BR5999-82-3-2-HR1	Bangladesh
50	Salinity	CSR 43	CSR-89IR-8	India
51	Salinity	BRRI dhan55	IR73678-6-9-B (AS996)	Bangladesh

TABLE 7.4 Varieties for Abiotic Stresses Released in Different Countries of South and Southeast Asia and Africa (*cont.*)

Study number	Abiotic stresses	Names	Designations	Environment, countries
52	Salinity	BINA Dhan 10	IR64197-3B-14-2	Bangladesh
53	Salinity	CR Dhan 405 (Luna Sankhi)	CR dhan 405	India
54	Salinity	CR Dhan 406 (Luna Barial)	CR2092-158-3	India
55	Salinity	BRRI dhan61	BR7105-4R-2	Bangladesh
56	Salinity	BRRI dhan65	OM1490	Bangladesh
57	Salinity	BRRI dhan69	BR7100-R-6-6	Bangladesh
58	Salinity	Gosaba (Gozaba) 5	IR55179-3B-11-3	India

Stagnant partial flooding is another major abiotic problem facing farmers in the rainfed ecosystem. It is well established that floodwater depth has a strong effect on internode elongation pattern, leaf blade, and sheath area. Elongation pattern is an escape strategy adopted by rice to resume aerobic metabolism and to improve carbon fixation (Anandan et al., 2012). However, the escape strategy varies among genotypes. Therefore, breeders should consider these mechanisms when breeding rice for different flood-prone environments. Prolonged stagnant flooding significantly affects the number of ear-bearing tillers, spikelet fertility percentage, and basal culm girth (Anandan et al., 2015). The whole-plant demand of photosynthates is reliant on the plant portion above the water level; therefore, photosynthesis, transpiration rate, and intercellular CO_2 of the aerial portion of the plant have a strong influence on spikelet fertility percentage and culm strength (Anandan et al., 2015). To escape deep water, a contrasting mechanism to *SUB1* is needed. The genes that promote elongation of internodes were identified from deepwater rice and named *SNORKEL1* (*SK1*) and *SNORKEL2* (*SK2*) (Hattori et al., 2009). *SK1/SK2* are two ERFs that work in contrast to *SUB1A* by regulating ethylene-mediated GA responsiveness (Gao et al., 2011). In India, many stagnant flood–tolerant varieties have been released that have the ability to yield 3.5–4.0 t/ha. Popular varieties, such as Varshadhan and CR Dhan 500, were developed for semideepwater conditions and Jalamani and Jayanti dhan were developed for deepwater conditions. Combining *SUB1* and *SK* into a single genetic background would be an enlightened and improved option for both flash flood and stagnant flooding. However, the modes of mechanism are in contrast to each other.

High temperatures, particularly during the flowering stage, drastically affect the seed setting rate and floret sterility of rice, resulting in high grain-yield losses (Morita et al., 2005; Peng et al., 2004). Frequent changes in the occurrence of heat waves have been observed in many parts of the world and particularly in tropical regions. Peng et al. (2004) reported that annual mean maximum and minimum temperatures increased by 0.35 and 1.13°C, respectively, during 1979–2003, and for each 1°C increase in minimum temperature, grain yield declined by 10%. The adverse effect of temperature on rice yield is seen across Pakistan, India, Bangladesh, China, Thailand, Japan, Australia, and the United States. The optimal temperature range for growing rice is 22–28°C; temperatures above 35°C in the booting and flowering stages reduce pollen viability, lower seed setting rate, and increase spikelet sterility, which in turn leads to significant grain yield losses, low grain quality, and a low HI (Jagadish et al., 2007). During the reproductive phase, the milky stage (6–16 days after heading) is the most sensitive to high temperature during grain filling (Zhu et al., 2005) and high temperatures here can result in serious damage to grain quality (Kondo et al., 2012; Lanning and Siebenmorgen, 2011) because of the increase in the proportion of chalky and unfilled grains. The occurrence of chalky grains under high temperatures is generally attributed to the inhibition of starch accumulation (Morita and Nakano, 2011). Ceccarelli et al. (2010) estimated that ~41% of all yield losses by the end of the 21st century will be due to high temperature. Therefore, immediate action has to be taken to identify and develop tolerance of high temperature in rice. To date, very few donors with tolerance of high temperature have been reported and few QTLs have been identified (Table 7.3).

Tolerance of high temperature is a complex quantitative trait and is influenced by many environmental factors, such as humidity (Weerakoon et al., 2008), wind speed (Ishimaru et al., 2012), CO_2 (Madan et al., 2012), and water stress (Rang et al., 2011). Therefore, improving the accuracy of phenotyping and associating the traits with genomic data are essential for crop improvement programs. Recent trends in the development of molecular marker technology have played a major role in dissecting complex traits in rice. Using a RIL population from Bala (tolerant) × Azucena (susceptible) lines, Jagadish et al. (2010) identified 10 QTLs for traits related to spikelet fertility under heat stress at anthesis. Of these 10 QTLs, the most significant heat-responsive QTL ($qtl_{1.1}$) was contributed by Bala and explained up to 18% of the PV. Using the mapping population of IR64 × N22, Ye et al. (2012) reported two major QTLs ($qHTSF_{1.1}$ and $qHTSF_{4.1}$) on chromosomes 1 and 4. These QTLs accounted for 12.6% and 17.6% of the PV for spikelet fertility under temperatures of 38 and 24°C for 14 days at the flowering stage (Table 7.3). Zhao et al. (2016) exploited the chromosome segment substitution lines of a Sasanishiki (*japonica* subspecies, heat susceptible) and Habataki (*indica* subspecies, heat tolerant) cross to identify 11 QTLs for heat tolerance. Among these, two QTLs for spikelet fertility were identified on chromosomes 2 and 4; four QTLs for daily flowering time were detected on chromosomes 3, 8, 10, and 11; and another five QTLs for pollen shedding level were reported on chromosomes 1, 4, 5, 7, and 10.

Rice is extremely susceptible to low temperature. Rice is principally a crop of tropical and semitropical regions, but it is also grown in many cold regions of the world in late spring to summer. Efforts to genetically dissect cold tolerance in rice using DNA markers have resulted in the discovery and mapping of many QTLs. Rice cultivars for improved cold tolerance are mainly evaluated during the seedling and reproductive stages. Cold tolerance in plants is known to be a complex trait associated with a number of biochemical and physiological events. QTLs for cold tolerance at seed germination and vegetative and booting stages have also been identified in different rice mapping populations (Table 7.3). Many researchers have reported QTLs controlling responses to an array of cold stress conditions at the seedling stage (Table 7.3). Breeding cold-tolerant rice varieties has been a tough challenge for plant breeders. Despite the availability of major-effect QTLs for seedling- and reproductive-stage cold tolerance, their use in marker-assisted selection was not successful because of their inconsistency across varying environments, the gap between greenhouse research and field application, the lack of strong donors in the *indica* germplasm pool, and various inferior traits incorporated through *japonica* cultivars.

Problem and poor soils. Salinity is one of the most serious problems in arid zones and it ranks second after moisture stress in damaging rice yield. Soil problems related to salt can vary from coastal to inland areas. Salinity (also known as sodicity or alkalinity) is found to be widespread in inland areas because of the deposition of existing salt or the use of irrigation water rich in sodium. Siphoning of excess water from deep bore wells followed by intrusion of seawater may also account for the rapid buildup of salt in the soil. The intrusion of tidal water and mixing with fresh water, deposition of salt carried by wind, and the rise in sea level due to climate change are expected to aggravate the problem of salinity in coastal regions (Gregorio et al., 2002). Excess salt affects almost all metabolic processes of the plant and thereby reduces growth and vigor. Uptake of excess salt disrupts homeostasis and ion sharing, leading to an arrest in plant growth, molecular damage, death of tissues, and finally death of the whole plant (Javid et al., 2011). The soil's high ionic concentration disrupts osmosis and interferes with nutrient and water uptake. Therefore, to overcome this problem, developing suitable varieties that have a direct effect in increasing productivity is essential.

When plants are initially exposed to salt, they experience water stress, which in turn reduces leaf expansion; with long-term exposure to salt, plants experience ionic stress, which leads to premature senescence of mature leaves. The accumulation of Na^+ in the tissue also alters the uptake of other nutrients, such as potassium, which is essential for preserving membrane potential, enzyme activities, and cell turgor. The adaptability of plants under salinity was categorized into three types: osmotic stress tolerance, Na^+ or Cl^- exclusion, and tolerance of accumulated Na^+ or Cl^- (Munns and Tester, 2008). Of the two subspecies of *O. sativa*, *indica* varieties were found to be more tolerant than *japonica* varieties, thereby excluding Na^+ more efficiently, maintaining a low Na^+/K^+ ratio, and absorbing high K^+. The tolerant rice genotypes were found to have a higher rate of germination, shoot length, root length, and vigor index.

Salinity tolerance is a complex trait controlled by many QTLs and genes (Shannon et al., 1985; Yeo and Flowers, 1986) and dissecting these complex quantitative genetic traits has been done by molecular marker technology (Thomson et al., 2010). Employing conventional breeding methods for selection of salinity tolerance is a difficult and time-consuming process because of environmental interaction and the low heritability of salt tolerance.

The seedling, booting, and pollination stages are the most salt-sensitive growth stages in rice. At the booting stage, salinity drastically reduces pollen viability, leading to a reduction in the percentage of filled grains (poor fertilization). This significantly reduces plant yield and marks salinity-susceptible genotypes at the flowering stage (Khatun and Flowers, 1995). Landraces, such as Kalarata, Cheriviruppu, Pokkali, and Bhirpala (Das et al., 2015), as well as improved variants, such as IR72046-B-R-7-3-1-2, IR4630-22-2-5-1-3, and CN499-160-13-6, have been reported to be tolerant during the flowering stage because of improved pollen viability and higher grain yield (up to 49%).

Several attempts have been made to develop salt-tolerant rice varieties through conventional breeding techniques. Combining together tolerance traits and high yield through conventional breeding techniques is a challenging task, but several successful varieties have been released (Table 7.4). Five rice varieties suitable for reclaimed saline soils and 30 varieties adaptable for coastal saline soils have been released in India. Popular tolerant varieties CSR 10, CSR 13, CSR 27, Narendra usar 2, and Narendra usar 3 were released for reclaimed saline soils and CSR 2, CSR 3, CSR 13, CSR 22, CSR 23, CSR 26, CSR 27, Basmati CSR 30, Panvel 1, Panvel 2, Panvel 3, Vytilla 1, and Vytilla 2 were identified as suitable for coastal saline soils (Sankar et al., 2011; Shahbaz and Ashraf, 2013).

The dissection of salt-tolerance traits has led to the identification of major and minor QTLs located on different rice chromosomes. Using RILs derived from a cross between Pokkali and IR29, Gregorio (1997) identified a major QTL for salt tolerance on chromosome 1. The QTL, controlled by the Na^+–K^+ absorption ratio, accounted for 64.3%–80.2% of the PV for salt tolerance with a LOD >14.5. Bonilla et al. (2002) saturated the identified regions with markers using RFLP and SSR and reported three major QTLs for Na^+, K^+, and the Na^+–K^+ absorption ratio accounting for 39.2%, 43.9%, and 43.2% of PV, respectively. Based on these studies, the major QTL on chromosome 1 was identified and designated as *Saltol*. The same segment of chromosome 1 was further fine mapped by Niones (2004) using NILs of the backcross population derived from Pokkali and IR29. A breakthrough in K^+ ion homeostasis came from the mapping of *SKC1* on chromosome 1 in the salinity-tolerant cultivar

(Nona Bokra) under saline conditions (Ren et al., 2005). Thomson et al. (2010) also identified a few promising markers in the *Saltol* QTL region (AP3206, RM8094, and RM3412) and flanking markers from the telomeric to the *Saltol* region (RM1287 and RM10694) and centromeric to the *Saltol* (RM493 and RM10793) region. These markers were reported to be useful and powerful for developing MABC lines by introgressing *Saltol* to acquire tolerance. Among the markers reported within the *Saltol* region, RM8094 was found to be the best at discriminating tolerant and intolerant genotypes (Islam et al., 2012).

Poor soil with low phosphorus is a problem for resource-poor farmers. In subtropical India, Pakistan, Latin America, and Turkey, imbalances of P and Zn have resulted in crop yield losses of >40% (Alloway, 2009; Khorgamy and Farnia, 2009). China and India are the largest consumers of global phosphate fertilizer, using 34% and 19%, respectively (Tirado and Allsopp, 2012). Around 50% of all rice soils were estimated to be P-deficient, accounting for 5.7 billion ha of arable land (Batjes, 1997); more recently, Krishnamurthy et al. (2014) reported that as much as 85% of the soils may be deficient in phosphorus and this is a major hurdle in rice production. The effects of low P vary from seedling to maturity stage. In the early stage to seedling stage (21–35 days), slow emergence of the seed through the water, an erect narrow leaf with brittle and grayish green younger leaves, and rolled and spear-like older leaves with chlorosis may be observed (Dobermann and Fairhurst, 2000). During active tillering to maturity, an accumulation of anthocyanin leaves the plant with stunted growth, reduced number of tillers, reduced root development, high sterility with poor grain quality, and delayed maturity (5–10 days). The rice plant adopts several mechanisms to increase the uptake and use efficiency of P, such as enhancing root hair density, Pi transporters, enzyme activities, secretion of organic acids, phosphatases, ribonucleases, and phosphodiesterases in the rhizosphere, along with alteration of cellular P metabolism (Sánchez-Calderón et al., 2005; Ticconi et al., 2004). A recent step forward in the field of low P was the identification of the major QTL *Phosphorus uptake 1* (*Pup1*) from the low phosphorus–tolerant traditional aus type Kasalath (Li et al., 2009; Shimizu et al., 2004, 2008; Van Kauwenbergh, 2010; Wissuwa et al., 1998, 2002). This QTL controlling P uptake was mapped on chromosome 12 between 15.1 and 15.6 Mb and explained approximately 80% of the PV (Wissuwa et al., 2002). This was therefore considered as one promising QTL for developing tolerance of phosphorus deficiency in rice breeding programs. Fine mapping of *Pup1* identified the *P starvation tolerance 1* (*PSTOL1*) gene encoding a PSI protein kinase. This protein kinase conferred P starvation tolerance by enhancing early root elongation, thus increasing grain yield in P-deficient soils (Gamuyao et al., 2012). However, the specific in vivo protein substrate of *PSTOL1* remains to be identified. Indirect evidence suggests that the phosphorylation of transcription factors by *PSTOL1* promoted the induction of several genes that enhance early root growth. Therefore, *PSTOL1* has become a candidate gene for improving crop tolerance of low-P conditions (Gamuyao et al., 2012).

Surveying a set of germplasm confirmed that the *Pup1* QTL is confined within upland drought-tolerant lines and is largely absent from irrigated rice varieties (Chin et al., 2011; Pradhan et al., 2016). This shows that varieties developed for upland conditions were indirectly selected for phosphorus-use efficiency. Furthermore, several institutes started working on *PSTOL1* and are trying to introgress *PUP1* into elite varieties by marker-assisted selection.

BREEDING FOR MECHANIZED WATER-LABOR SHORTAGE SITUATIONS

The global population is set to expand by 80 million people every year. In line with this, the demand for fresh water is set to increase by ~64 billion m^3/year (WWAP, 2009). Global water availability is already seeing a downward trajectory from 12,900 m^3 per capita per year in 1970 to 9,000 m^3 in 1990, and ~7,000 m^3 in 2000 [United Nations Environment Programme (UNEP), 2008]. Groundwater levels have been dangerously exploited over the last 50 years, and freshwater withdrawals for energy production are set to increase by 20% in 2035 (IEA, 2012). Globally, agriculture needs to produce 60% more food by 2050, with this level needing to increase to 100% in some developing countries (Alexandratos and Bruinsma, 2012), all using less land and less water. If the trend of increasing population with declining resources and increasing consumption continues, the water requirement will go beyond its availability by 56% (World Water Organization, 2010) and 1.8 billion people (UN News Centre, 2009) will suffer water scarcity by 2025. An increase in agricultural productivity through a combination of improved water control, proper land management, improved agronomic practices, use of water-saving technologies, mechanization, development of new data sources, superior models, powerful data analysis tools and techniques, and the design of adaptive management approaches can help in responding effectively to changing and uncertain conditions.

The agricultural labor force is declining at an annual rate of 0.1%–0.4% with an average of 0.2% per year in Asia, 0.25%–0.40% in Bangladesh, Malaysia, and Thailand, and 0.18% in the Philippines and Cambodia. From 1961 to 2008 the labor force decreased from 45% to 25% in Bangladesh and from 22% to 6% in Malaysia. Likewise, from 1960 to 2011 the labor force decreased from 40% to 30% in Thailand and from 40% to 35% in Vietnam (Kumar and Ladha, 2011). Both this and rising labor wages are equally important constraints for puddled transplanted rice (PTR). The rising labor costs of establishing nurseries, puddling fields, and transplanting rice seedlings have increased the costs of transplanting and these costs are expected to increase further in the future. All of this makes the PTR cultivation system uneconomical in the rice bowls of South Asian and Southeast Asian countries. Reports say that PTR needs 114 person-days per hectare vis-à-vis the 68 person-days per hectare for direct-seeded rice (DSR; Yamano et al., 2013). Based on the season, location, and cultivation method of DSR, total labor requirements can be reduced from 11% to 66% compared to PTR (Kumar et al., 2009b; Rashid et al., 2009) with mechanization of seeding, weeding, herbicides, and fertilizer applications.

The traditional water-labor energy-intensive method of rice cultivation involves transplanting seedlings in puddled soil (PTR). The rise in scarcity of water and labor, increased cost of irrigation, expense of labor, change in rainfall distribution, reduced nutrient availability under dry conditions, increased greenhouse gas emissions, and the drudgery associated with manual transplanting (a job largely done by female farmers) are major factors that compel farmers to move toward water-saving technologies. Several water-saving technologies, such as alternate wetting and drying (AWD; Li, 2001; Tabbal et al., 2002), groundcover systems (Shan et al., 2002), the system of rice intensification (SRI; Stoop et al., 2002), aerobic rice (Bouman et al., 2002), raised beds (Singh et al., 2002), and DSR (Kumar and Ladha, 2011), have all been proposed as sustainable alternatives to PTR. Rice varieties for these water-saving technologies require a different set of traits than those present in semidwarf rice varieties suited to PTR.

DSR is emerging as an efficient, resourceful, mechanized, and economically viable alternative to PTR with the potential to ensure sustainability of rice cultivation. It benefits irrigated areas suffering from water and labor scarcity, it benefits the succeeding crop by allowing timely planting and improved soil conditions, and it results in increased agricultural productivity (Kumar and Ladha, 2011). The improved opportunities for mechanization and resource management (including weed control) offered by DSR mean that this cultivation practice is gaining in popularity in South Asia (India, Nepal, and Bangladesh), in Southeast Asia (Philippines, Cambodia, Vietnam, Laos, and Myanmar), and to some extent in West Africa.

The lack of suitable varieties and lack of understanding of suitable traits are major constraints to achieving adaptability and maximum yield potential under DSR. The traits likely to be important for direct-seeded conditions are anaerobic germination; early uniform seedling emergence from deeper soil depth; early vegetative vigor; high mesocotyl length; increased numbers of nodal roots; increased root length density; lateral root length and branching with more root hairs for proper nutrient uptake under dry conditions; root plasticity (i.e., the ability of roots to behave as per the system); erect leaves with low specific leaf area; high chlorophyll content; medium plant height; early flowering; strong, thick, and sturdy culm (lodging resistance); uniform maturity; long-branched panicles with more fertile spikelets with good grain quality; and high grain yield.

Recent studies at IRRI have identified genotypes with increased ability to germinate from deeper soil depth and with longer coleoptile length; QTLs and donors are for traits related to early uniform emergence ($qEMM_{1.1}$; Dixit et al., 2015), lodging resistance ($qLDG_{3.1}$, $qLDG_{4.1}$; Dixit et al., 2015), nodal root and root hair density for higher nutrient uptake under DSR ($qNR_{4.1}$, $qNR_{5.1}$, $qRHD_{1.1}$, $RHD_{5.1}$, $qN_{5.1}$, $qP_{5.2}$, $qFe_{5.2}$; Sandhu et al., 2015), and higher grain yield ($qGY_{1.1}$, $qGY_{8.1}$, $qGY_{10.1}$; Sandhu et al., 2015). QTLs for higher nutrient uptake (Al, Fe, and P), higher nutrient concentration (Al, Fe, N, and P), and root hair density were colocalized in the same region (Sandhu et al., 2015) (Table 7.5 and Fig. 7.1). Efforts continue with the marker-assisted breeding program involving traits (Table 7.5) leading to better adaptation to dry direct-seeded conditions to develop better rice varieties with high yield potential. IRRI has also been actively involved in conventional breeding programs involving multiple donors (having DSR adaptability traits) to open the window for increases in productivity under direct-seeded conditions. Under DSR conditions, a grain yield advantage of lines from complex mapping populations over lines from biparental crosses has been observed. Successful introgression of these traits

TABLE 7.5 Traits and Genetic Regions Associated With Adaptability of Direct-Seeded Rice (DSR)

Study number	Traits	Donors	QTLs/genes	References
1	Anaerobic germination	IR93312-30-101-20-3-66-6	$qAG_{9.1}$, $qAG_{9.2}$	Angaji et al. (2010), Kretzschmar et al. (2015)
2	Early uniform emergence	IR91648-B-32-B	$qEMM_{1.1}$	Dixit et al. (2015)
3	Nodal root	IR94225-B-82-B, IR94226-B-177-B	$qNR_{4.1}$, $qNR_{5.1}$	Sandhu et al. (2015)
4	Early vigor	IR94226-B-177-B	$qEVV_{9.1}$	Sandhu et al. (2015)
5	Root hair density	IR94225-B-82-B, IR94226-B-177-B	$qRHD_{1.1}$, $RHD_{5.1}$, $qRHD_{8.1}$	Sandhu et al. (2015)
6	Nutrient uptake	IR94225-B-82-B	$qN_{5.1}$, $qP_{5.2}$, $qFe_{5.2}$	Sandhu et al. (2015)
7	Grain yield DS	IR94225-B-82-B, IR94226-B-177-B	$qGY_{1.1}$, $qGY_{6.1}$, $qGY_{10.1}$	Sandhu et al. (2015)
8	Lodging resistance	IR91648-B-289-B	$qLDG_{4.1}$	Dixit et al. (2015)

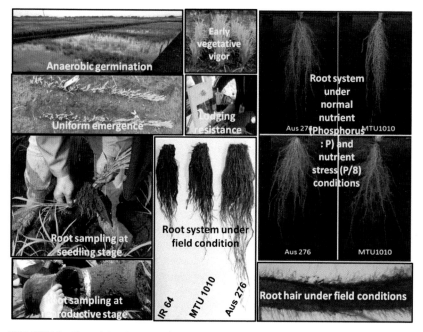

FIGURE 7.1   Essential traits related to direct-seeded rice adaptation.

can provide a unique opportunity for breeders/molecular biologists/physiologists to combine and study the interaction of these traits/QTLs/genes efficiently in research aimed at increasing the yield and adaptability of rainfed rice to direct-seeded conditions. Such trait complementarities may result in improved performance beyond the sum of that conferred by the individual traits. The development of a complete package of water-saving, labor-saving, energy-saving, resource-efficient, mechanized, sustainable, new-generation climate-smart rice varieties is urgently needed in the near future. This could be of significant benefit to farmers in regions with unstable rainfall and those struggling with a lack of water, labor, and resources. This could also help to reduce global warming.

AWD technology tackles water scarcity in irrigated rice cultivation and has the potential to contribute to more sustainable and effective water and energy use. Instead of keeping paddies in a permanently flooded state, this practice allows the field to periodically dry and become wet again (when the water table goes 15 cm below the soil surface; safe AWD) throughout the growing season. AWD uses a single device designed to observe the water level in the rice field to dictate the time of irrigation. The number of days of nonflooded soil between irrigations can vary from 1 to > 10 depending on factors such as soil type, weather, and crop growth stage. For AWD, it is important (more so than for flooded or aerobic cultivation) that the field be leveled, as differences in water depth will be experienced during every cycle. AWD can lower water use for irrigated rice by 15%–30% (Zhang et al., 2009); increase rice yield by ~10% relative to permanent flooding (Yang et al., 2009; Zhang et al., 2009); increase nutritional status; and decrease toxic elements, such as cadmium (by ~20%), that can be problematic in rice (Yang et al., 2009). One disadvantage of AWD is that the water requirement is still relatively high, as initial land preparation consists of soaking followed by wet plowing or puddling of saturated soil; however, the water requirement still remains lower than in PTR. A well-developed root system can help reduce the water requirement to a larger extent, and for AWD a root system that is more developed from 15 to 25 cm below the soil surface is needed to capture water and nutrient under fluctuating aerobic and anaerobic conditions.

Using conventional breeding approaches, several varieties for water-saving AWD aerobic conditions have been developed and released in South Asia (Table 7.6). The rice varieties from the complex mapping populations, using conventional or marker-assisted forward breeding combining different traits, provide better adaptation and nutrient availability to rice under DSR. Breeding for AWD has not been used yet as a separate breeding program by any institute. Root traits below 15 cm become very important for AWD and plant type, or more specifically root traits required in rice plants under AWD will help develop better rice varieties.

HIGH GRAIN QUALITY: AN ESSENTIAL BREEDING COMPONENT ACROSS ECOSYSTEMS

Rice grain quality creates demand for a particular type of grain in a region. It is quite evident that grain quality, or more specifically the cooking quality, of traditional varieties is far superior than most of the semidwarf varieties, with the exception of a few varieties, such as IR64 and IR36. Grain quality is an important goal for any rice breeding program of modern high-yielding varieties. Several of the varieties showing high yield potential could not find a place in farmers' fields because of their poor grain quality. The choice of rice grain type for consumption varies from region to region, and from country to country, with slender types preferred in India, China, the United States, and most Asian countries (compared to short and round types preferred in northern China, Korea, and Japan). The quality of rice is an important characteristic that determines the economic value in the export market and consumer acceptance (Pingali et al., 1997). Grain quality can be categorized based on the (1) grain appearance and (2) cooking and eating quality (Juliano and Villarreal, 1993). Grain appearance is determined by length, breadth, length–breadth ratio, and endosperm translucency; cooking and eating quality is related to aroma, volume expansion, cooked kernel elongation, firmness or stickiness [related to amylose content (AC)], gelatinization temperature (GT, measured as alkali spreading value), and mouth feel. The most important traits for consumer preference are grain appearance and cooking and eating quality; alongside milling and head rice recovery, these traits can affect the market value of rice grain (Nelson et al., 2011).

TABLE 7.6 Rice Varieties for Water-Saving Dry DSR and Alternate Wetting and Drying (AWD) Conditions

Study number	Names	Designations	Maturity (days)	Plant height (cm)	Grain types	Conditions; countries
1	Tarahara 1	IR80411-B-49-1	125	100	Medium slender	AWD; Nepal
2	ARB 6	Not known	115	110	Medium bold	DSR, aerobic; Karnataka, India
3	MAS 26	UASMAS26	125	90	Medium slender	DSR, aerobic; Karnataka, India
4	MAS 946-1	UASMAS945	130	90	Medium slender	DSR, aerobic; Karnataka, India
5	CR Dhan 200	IR55423-01	105	115	Medium bold	DSR, aerobic; Odisha, India
6	CR Dhan 201	IR83380-B-B-124-1	115	100	Long slender	DSR, aerobic; Chhattisgarh, Bihar, India
7	CR Dhan 202	IR84899-B-154	115	100	Short bold	DSR, aerobic; Jharkhand, Madhya Pradesh, Bihar, Odisha, India
8	CR Dhan 204	IR83927-B-B-279	115	100	Medium slender	DSR, aerobic; Chhattisgarh, Madhya Pradesh, India
9	CR Dhan 205	IR86931-B-578	106	110	Short bold	DSR, aerobic; Tamil Nadu, Gujarat, Odisha, Madhya Pradesh, Punjab, India
10	Sahod Ulan 12	IR81047-B-106-2-4	105	119	Medium slender	DSR, rainfed lowland dry season; Philippines
11	CAR 14	IR80463-B-39-3	110	115	Medium slender	DSR, rainfed; Cambodia

Over the past 2 decades, many QTLs controlling grain dimensions have been mapped using several populations. QTLs were reported on chromosome 7 (Zhang et al., 2013), chromosome 5 (Lin et al., 1995), chromosome 2 (Zhang et al., 2013), chromosome 1 (Amarawathi et al., 2008; Zheng et al., 2007), chromosome 4 (Zhang et al., 2014), chromosome 10 (Wan et al., 2008), and chromosome 6 (Zhang et al., 2014), with significant PV for physical grain appearance traits in rice. The validation of QTLs and their use in improving grain quality should be the next step.

Grain chalkiness is a key factor that determines grain quality and price. Chalam and Venkateswarlu (1965) classified chalkiness into four groups based on the opaqueness in the milled rice grains and associated this with the high degree of damage to the kernel during milling, which leads to a high percentage of broken grains and thus a substantial reduction in head rice recovery. In rice genome QTL databases, more than 82 QTLs govern chalkiness-related traits. These include 30 QTLs for percentage grain with white core (PGWC), 26 QTLs for the degree of endosperm chalkiness, 12 QTLs for the area of endosperm chalkiness, 11 QTLs for whitebacked kernel, and 3 QTLs for basal white (Kobayashi et al., 2007). Chandusingh et al. (2013) reported two significant QTLs for PGWC (*qPGWC6*) and chalkiness score (CS; *qCS6*) that were consistently found during all three seasons. Peng et al. (2014) identified 79 QTLs associated with 6 chalkiness traits (white core rate, white core area, white belly rate, white belly area, chalkiness rate, and chalkiness area) on chromosome 12. Validation of these QTLs, the identification of prominent QTLs, and their use in breeding programs should be the next step.

Grain milling quality is very important from a commercial point of view. Milling percentage affects nutritional value and is associated with grain size, shape, and chalkiness. During the past 15 years, molecular markers associated with milling quality traits and rice grain chalkiness QTLs have been identified (Nelson et al., 2011; Zheng et al., 2012) and more than 80 QTLs have been detected using several mapping populations (http://archive.gramene.org/qtl/). Grain milling quality QTLs have also been mapped in intraspecific (*O. sativa* subsp. *indica* × *O. sativa* subsp. *japonica*) crosses (Dong et al., 2004; Zheng et al., 2007) and interspecific crosses between cultivated rice and wild rice relatives such as *O. glaberrima* and *O. rufipogon* (Li et al., 2004; Septiningsih et al., 2003). Dong et al. (2004) reported two QTLs (*qBRP-9* and *qBRP-*10) for brown rice, two QTLs (*qMRP-*11 and *qMRP-*12) for milled rice percentage, and three QTLs (*qMHP-1*, *qMHP-3*, and *qMHP-5*) for milled head-rice percentage.

Improving the cooking and eating quality traits of rice has been the main concern of plant breeders. Cooking and eating quality is mainly related to the physical and chemical properties of starch in the endosperm, namely, AC, gel consistency (GC), and GT or alkali spreading value. Other traits that influence the quality of cooked rice are water uptake, the volume expansion ratio, kernel length after cooking, and the elongation ratio (Juliano, 1985; Swamy et al., 2011; Tan et al., 1999). Cooking quality properties are of major importance to plant breeders as they dictate the market value of the commodity and play an important role in the adoption of new varieties (Fitzgerald et al., 2009). AC is considered as an important property of the cooking and processing quality of rice grain and is classified into five groups: waxy (0%–5%), very low (6%–11%), low (12%–19%), intermediate (20%–25%), and high (>25%). A high amount of AC leads to a decline in viscosity, softness, shine, and palatability; however, AC does not absolutely determine the texture of cooked rice because the palatability of cooked rice with similar AC may vary greatly. An intermediate amylose range is preferred as a breeding target, but higher AC can be combined with soft GC and low GT. GT is another important quality predictor in the cooking quality of rice for determining the time and water required for cooking (Sabouri, 2009). The relationship between cooking time and final texture of the cooked rice depends on water uptake, volume expansion, and linear kernel elongation when GT is in the 55–85°C range (Pandey et al., 2012). The existence of high GT leads to an enormous amount of extra energy during the process of cooking. It has been estimated that reducing the cooking time of rice by 1 min each day would globally represent 2500 years of cooking time saved per day (Fitzgerald et al., 2009). Thus, lowering gelatinization time would lead to significant savings in fuel and a measureable reduction in the carbon footprint of rice. Molecular analysis of GT in rice controlled by the *starch synthase IIa* (*SSIIa*) gene located on chromosome 6 categorized the germplasm into two groups (high and low GT) based on the allelic variation in *SSIIa* (Bao et al., 2006). GC is a good index for assessing the texture and quality of cooked rice for intermediate- and high-amylose rice, which determines good mouth feel (Cagampang et al., 1973). Most rice consumers prefer a soft gel consistency.

Several reports suggested that AC is controlled by one major gene with modifiers, stipulating that high AC is completely dominant over low AC (Chauhan and Nanda, 1983). On the other hand, Stansel (1966) reported the involvement of two complementary genes, the major gene being an allele of *Wx* that accounted for 91.9% of the total variation. Later, two major QTLs and five minor QTLs controlling AC were observed on chromosomes 3 and 6 by Liu et al. (2000). Among the total QTLs, one major QTL located on chromosome 6 was found near the *Wx* locus. Using a cross between KDML105 and CT9993, Lanceras et al. (2000) detected four QTLs for AC on chromosomes 3, 4, 6, and 7 and three QTLs for GC on chromosomes 2 and 6. Aluko et al. (2004) reported a major QTL accounting for 73% of the PV for AC on chromosome 6 with an additive effect of 2.6. However, AC of rice is determined by the *Wx* gene. This gene is located on chromosome 6, consists of 13 exons and 12 introns, and encodes the granule-bound starch synthase (GBSSI) enzyme (Biselli et al., 2014). GBSSI in the endosperm is directly associated with the amount of AC in rice grain (Mikami et al., 2008) and null mutants of GBSSI produce glutinous rice with no amylose (Dobo et al., 2010). The *Wx* gene is the sole major gene affecting both AC and GC; as a minor gene, it also affects GT.

Several studies proved that QTLs for AC, GC, and GT are found in the vicinity of the *waxy* locus and this suggests that all three traits might be controlled by the Wx locus. The Wx locus has two alleles, Wx^a and Wx^b, differentiating *indica* and *japonica* cultivars, respectively, and these alleles are associated with differences in AC (Hori et al., 2007). Chen et al. (2008b) examined a combination of two SNP markers in the Wx gene that explained 86.7% of the variation in apparent AC in 171 international germplasm accessions. Therefore, this Indel might be used as a molecular marker to characterize the waxy grain trait in rice. These novel QTL markers of cooking and taste traits could be used as an indirect selection tool in breeding programs according to the preference of consumers.

Fragranced and soft-textured rice usually achieve higher prices in national and international markets (Yi et al., 2009). The active key volatile compound 2-acetyl-1-pyrroline (2AP) determines aromatic or scented rice (Wongpornchai et al., 2004). Different methods were adopted to detect and quantify the 2AP compound: tasting, smelling, or chewing individual grains (Reinke et al., 1991); gas chromatography (Mahatheeranont et al., 2001); stable isotope dilution (Yoshihashi et al., 2002); and histochemical localization (Nadaf et al., 2006). The limitations of these methods are the difficult and time-consuming processes, the requirement of large samples, and the unreliability of fragrance. Hence, the development and use of PCR-based molecular markers in breeding programs would have many advantages over sensory or chemical detection methods.

The genetic control of aroma in rice has been reported to be monogenically inherited and controlled by a recessive nuclear gene (Dong et al., 2000). Many PCR-based markers have been developed to detect fragrance and assist breeders in their selection of this trait through MAS breeding programs. Ahn et al. (1992) reported that RFLP marker RG28 was linked to grain fragrance in aromatic rice cultivar Lemont and controlled by a recessive gene (*fgr*) located on chromosome 8 at a distance of 4.5 cM. Lorieux et al. (1996) subsequently confirmed the linkage between RG28 and *fgr* (5.8 cM) and identified two more QTLs on chromosomes 4 and 12 from the aromatic rice of jasmine and basmati. Tomar and Prassad (1997) reported that fragrance was associated with a dominant aroma gene located on chromosome 11 in cultivar Baspatri, an Indian rice landrace. These reports suggest that varying numbers of dominant or recessive genes are involved in the aroma of rice.

Studies have also identified *badh2* (recessive allele, chromosome 8) as a candidate gene for aroma (Amarawathi et al., 2008), which codes for the enzyme betaine aldehyde dehydrogenase (BADH). The candidate gene was reported to be an analog of the functional *badh2* gene, as it has an 8-bp deletion (5′-GATTATGG-3′) and three SNPs in exon 7 (Kovach et al., 2009; Sakthivel et al., 2009). Kumari et al. (2012) developed three markers (BAD2, BADEX7-5, and SCUSSR1) located on chromosome 8 from the RIL population of CSR 10 and Taraori Basmati. Among these markers, BAD2 amplified aroma-specific alleles with 256 bp, BADEX7-5 with 95 bp, and SCUSSR1 with 129 bp, and had a strong correlation between aroma traits on chromosome 8. The potential of these functional markers derived from fragrance is sufficient for plant breeding using MAS approaches.

BREEDING NUTRITIOUS RICE

Rice grain consists of ~80% starch and its quality is dependent on the combination of several micronutrients. Rice grain is relatively low in some essential micronutrients, such as Fe, Zn, and Ca, compared with other food crops, such as wheat, maize, legumes, and tubers (Adeyeye et al., 2000). Rice bran is also an important source of protein, vitamins, minerals, antioxidants, and phytosterols (Renuka and Arumughan, 2007). Rice bran protein has great potential in the food industry owing to its unique nutraceutical properties (Saunders, 1990). Improving these constituents of the food grain could be useful for reducing malnutrition in India.

Grain protein content (PC) is very low (8.5%) in rice compared with that in other cereals, such as wheat (12.3%), barley (12.8%), and millet (13.4%), and is also low in milled rice (7%–8%). The total seed PC of rice is composed of 60%–80% glutelin and 20%–30% prolamin, controlled by 15 and 34 genes, respectively (Xu and Messing, 2009). Rice supplies ~40% of dietary protein to humans in developing countries and the quality of PC in rice is high because it is rich in lysine (3.8%) (Shobha Rani et al., 2006). Therefore, improvement of PC in rice grain should be a major target for plant breeders and biotechnologists. Very limited success has been achieved in improving PC due to the involvement of many QTLs and the influence of environment on the trait. Germplasm lines ARC 10063 and ARC 10075 were reported to contain 16.41% and 15.27% crude protein in brown rice (Mohanty et al., 2011). Rice variety CR Dhan 310 has been developed with a PC of 11% and it was rich in threonine and lysine using ARC 10075 as a donor. Genes for PC in rice have been mapped (Yun et al., 2014; Zhong et al., 2011).

In brown rice, the micronutrient concentrations of Fe and Zn range from 6.3–24.4 to 13.5–28.4 µg/g, respectively (Gregorio et al., 2002). The recommended dietary intake of Fe and Zn is 10–15 and 12–15 mg/kg, respectively. Polished rice contains only 2 mg/kg of Fe and 12 mg/kg of Zn (FAO, 2001). A germplasm survey for Fe concentration in grain showed a range of 1.20–1.72 mg/100 g (Shabbir et al., 2008), whereas Zn concentration varied from 11.30 to 67.34 ppm in rice germplasm (Vishnu et al., 2014). Enhancing the concentrations of these micronutrients needs to be achieved in modern varieties by taking advantage of the presence of large variability for these traits in the rice germplasm. Fe and Zn inheritance is controlled by many QTLs and it is affected by environmental factors. DRR Dhan 45 and Chhattisgarh Zinc Rice-1 are two rice varieties released in India with enriched Zn content. Several QTLs related to nutritional quality traits have been reported in rice from different genetic backgrounds of intraspecific and interspecific crosses (Cheng et al., 2014; Lee et al., 2014; Peng et al., 2014; Yun et al., 2014). Enhancement of Fe in the starchy endosperm of rice was obtained by transferring the *ferritin* gene of soybean (Vasconcelos et al., 2003) and common bean (Lucca et al., 2001) using transgenic approaches, resulting in a doubling of Fe concentration.

Phenolic and flavonoid compounds in rice grain help to decrease toxic compounds, reduce the risk of developing chronic diseases including cardiovascular disease and type-2 diabetes, reduce oxidative stress, and help prevent some cancers (Shao et al., 2011). Red rice has phenolic compounds in the range of 165.8–731.8 mg gallic acid equivalent per 100 g (Shen et al., 2009), and black/purple rice exhibits higher amounts of Fe, Zn, Ca, Cu, and Mg than red rice (Meng et al., 2005). Pigmented rice cultivars are reported to have a higher amount of antioxidative activity (Hiemori et al., 2009). Phenolic compounds are mainly associated with pericarp color: the darker the pericarp, the higher the amount of polyphenols (Yawadio et al., 2007). Shen et al. (2009) characterized the color of rice grain (white, red, and black rice) in a wide collection of rice germplasm and found that it was significantly associated with total phenolic, flavonoid, and antioxidant capacity. Two loci, *Rc* (Furukawa et al., 2007) and *Rd* (Furukawa et al., 2007), were found to be responsible for the formation of pericarp color in rice. *Rc* produces brown pericarp and seed coat, *Rc* along with *Rd* develops red pericarp and seed coat, whereas *Rd* alone has no color phenotype.

Glycemic index (GI) is an indicator of blood sugar level and is measured from the amount of carbohydrate consumption (after ingestion), which can be measured by rapidly available glucose. Rice contains 80% starch and increased consumption leads to an increased risk of type-2 diabetes (Courage, 2010), a disease that is predicted to affect almost 330 million people by 2030 (Mishra et al., 2010). Food grains have been categorized as low (<55), medium (56–69), or high (>70) GI foods (Brand-Miller et al., 2009), with GI values of 54–121 found among rice germplasm accessions (Manay and Shadaksharaswamy, 2001). The amount of AC, *Waxy* haplotype, and digestibility of rice are significantly correlated (Fitzgerald et al., 2011) and it has been observed that AC along with GT plays a key role in the rate of starch digestion and GI (Kharabian-Masouleh et al., 2012). The breakdown of the starch molecule and GI is controlled by the *Waxy* gene (Chen et al., 2008a; Wang et al., 1995), and it has different allelic changes in the respective exons and introns of the gene sequence. Therefore, the incorporation of the low glycemic trait in the popular rice varieties of low-earning countries, such as India, Bangladesh, Indonesia, Malaysia, and Sri Lanka, could offer an inexpensive method to manage the disease (Fitzgerald et al., 2011). In India, three rice varieties (Sampda, Sambha Mahsuri, and Lalat) have been reported to possess low GI (data unpublished).

Different breeding strategies based on the existing variability are required to improve the nutritional content of rice. For iron, conventional breeding could be an appropriate strategy. First, iron occurs naturally in rice grains, and second, the high variability in grain iron content allows the selection of high-iron parents for crossbreeding. Breeding for zinc is similar to that for iron in that conventional breeding techniques can be used. Substantial genetic variation exists in germplasm banks and research has shown that high zinc density and high iron density are positively correlated. However, the stability of Fe and Zn content across different soil types has been an issue. This requires the identification of donors with high and stable Fe and Zn content for use in conventional or marker-assisted breeding. Conventional breeding techniques cannot be used for vitamin A–enriched rice because vitamin A does not occur naturally in rice grains. Biotechnology can make a unique contribution to incorporate vitamin A genes into high-yielding rice varieties.

FUTURE BREEDING PRIORITIES

Plant breeding has moved from being an art to a science, from breeding by chance recombination to breeding with accurate prediction of recombinations, and from using conventional breeding strategies to using marker-assisted breeding and genomic breeding strategies. More accurate breeding strategies will continue to evolve to enable plant breeders to tackle the increased food demand and threats to sustainability from climate change-related challenges. Many examples have been cited in this chapter regarding the identification of genes, QTLs, and tightly linked markers. Many examples have also been given regarding the improvement in popular rice varieties for individual biotic stresses (BB, blast, GM, and BPH) and abiotic stresses (submergence and drought), as well as increasing yield potential following marker-assisted backcross breeding or marker-assisted selection. Moreover, the positive effect of many of the identified genes for individual stresses has been validated. With the availability of tightly linked markers for different genes, breeding needs to move forward to combine a larger number of genes for a set of different traits required to develop stable genotypes for any specific ecosystem or region so as to keep rice production sustainable under decreasing land, water, labor, and energy; deteriorating soil nutritional status; and increasing climate-related variability. This will require more accurate strategies to understand the interaction between large numbers of genes and the identification of a set of genes with a high positive effect against diverse genetic backgrounds. The sequenced rice genome has provided new opportunities to identify superior alleles for different traits using functional genomics, transcriptomics, and proteomics knowledge. Molecular tools, such as whole-genome SNP array, genomewide association mapping, genomic-based genotyping platforms and resequencing, genome-guided RNA-seq, map-based cloning approaches, transcriptome profiling, sequencing-by-synthesis, next-generation sequencing technologies, and massive parallel signature sequencing, could be strategically exploited to understand molecular mechanisms and their relationship between genotypic and phenotypic traits. Sequencing information will also help breeders to select diverse genotypes during breeding and reduce inbreeding, which has been widely practiced since the start of the Green Revolution. Keeping genomic breeding strategies at the forefront, increasing yield potential and reducing yield losses from abiotic and biotic stresses will continue to be important future breeding objectives. Developing better rice varieties for mechanized water-saving conditions will need immediate attention from breeding programs. This will need systematic study to identify and develop traits, QTLs, genes, and plant types for increased adaptability of rice to mechanized water-saving AWD and DSR conditions. The strategy should also involve the identification of risks associated with large-scale shifting of rice cultivation from PTR to DSR [i.e., the emergence of new diseases (rice root-knot nematodes), nutrient deficiency, and problems of weed control], as well as devising smaller implements better suited to the small field sizes prevalent in most of Asia.

Significant achievements have been made in genetic studies of grain appearance, cooking traits, and nutritional value, but more research is needed for processing and curative properties. Generating knowledge on grain quality and cooking quality and incorporating the knowledge systematically into breeding programs should be an important strategy to develop better rice genotypes. Some progress has been made in genetic studies of grain protein and amino acid content, vitamin and mineral content, GI value, phenolic and flavonoid compounds, and zinc and iron content, along with the QTLs linked to these traits. The development of genomic technologies may augment the improvement of grain and cooking quality in rice when this goes hand in hand with breeding programs. The development of nutrition-dense rice will be a worthy step in providing a solution to malnutrition in rice-consuming populations. The current gain in knowledge and further research on gene-based markers linked to grain and nutritional quality will help in developing genotypes with desired grain and cooking quality traits.

Long-term large-scale genetics and breeding programs are needed to facilitate the jump in increasing rice yield, under conditions more adverse than those that existed 60 years ago. This will require committed funding from donors and government agencies. To reduce the gap between yield at the research station and in farmers' fields, many donors have shifted their attention to funding seed production and disseminating technologies. Scientific communities need to reach donors and convince them to maintain a better balance between funding research and dissemination so that science can continue to develop new knowledge and adequately serve the world community.

REFERENCES

Adeyeye, E.I., Arogundade, L.A., Akintayo, E.T., 2000. Calcium, zinc and phytate interrelationship in some foods of major consumption in Nigeria. Food Chem. 71 (4), 435–441.

Ahn, S.N., Bollich, C.N., Tanksley, S.D., 1992. RFLP tagging of a gene for aroma in rice. Theor. Appl. Genet. 84, 825–828.

Alam, S.N., Cohen, M.B., 1998. Detection and analysis of QTLs for resistance to the brown planthopper, *Nilaparvata lugens*, in a doubled haploid rice population. Theor. Appl. Genet. 97, 1370–1379.

Alexandratos, N., Bruinsma, J., 2012. World agriculture towards 2030/2050: the 2012 revision. ESA Working Paper No. 12-03. Food and Agriculture Organization of the United Nations (FAO), Rome.

Alloway, B.J., 2009. Soil factors associated with zinc deficiency in crops and humans. Environ. Geochem. Health 31, 537–548.

Aluko, G., Martinez, C., Tohme, J., Castaño, C., Bergman, C., Oard, J.H., 2004. QTL mapping of grain quality traits from the interspecific cross *Oryza sativa* × *O. glaberrima*. Theor. Appl. Genet. 109, 630–639.

Amarawathi, Y., Singh, R., Singh, A.K., Singh, V.P., Mohapatra, T., Sharma, T.R., 2008. Mapping of quantitative trait loci for basmati quality traits in rice (*Oryza sativa* L.). Mol. Breeding 21, 49–65.

Anandan, A., Pradhan, S.K., Das, S.K., Behera, L., Sangeetha, G., 2015. Differential responses of rice genotypes and physiological mechanism under prolonged deepwater flooding. Field Crops Res. 172, 153–163.

Anandan, A., Rajiv, G., Ramarao, A., Prakash, M., 2012. Internode elongation pattern and differential response of rice genotypes to varying levels of flood water. Funct. Plant Biol. 39, 137–145.

Ananthi, K., Vijayaraghavan, H., 2012. Soluble protein, nitrate reductase activity and yield responses in cotton genotypes under water stress. Biochem. Insights 2, 1–4.

Angaji, S.A., Septiningsih, E.M., Mackill, D.J., Ismail, A.M., 2010. Identification of QTLs associated with tolerance of anaerobic conditions during germination in rice (*Oryza sativa* L.). Euphytica 172, 159–168.

Ashikari, M., Matsuoka, M., 2006. Identification, isolation and pyramiding of quantitative trait loci for rice breeding. Trends Plant Sci. 11, 344–350.

Ashikari, M., Sakakibara, H., Lin, S., Yamamoto, T., Takashi, T., Nishimura, A., Angeles, E.R., Qian, Q., Kitano, H., Matsuoka, M., 2005. Cytokinin oxidase regulates rice grain production. Science 309 (5735), 741–745.

Ashikawa, I., Hayashi, N., Yamane, H., Kanamori, H., Wu, J., Matsumoto, T., Ono, K., Yano, M., 2008. Two adjacent nucleotide-binding site-leucine-rich repeat class genes are required to confer *Pikm*-specific rice blast resistance. Genetics 180, 2267–2276.

Athwal, D.S., Pathak, M.D., Bacalangco, E.H., Pura, C.D., 1971. Genetics of resistance to brown planthopper and green leafhoppers in *Oryza sativa* L. Crop Sci. 11, 747–750.

Atia, M.M.M., 2004. Rice false smut (*Ustilaginoidea virens*) in Egypt. J. Plant Dis. Prot. 111 (1), 71–82.

Bansal, S., Harrington, C.A., Gould, P.J., St Clair, J.B., 2012. Climate-related genetic variation in drought-resistance of Douglas-fir (*Pseudotsuga menziesii*). Glob. Change Biol. 21 (2), 947–958.

Bao, J.S., Corke, H., Sun, M., 2006. Microsatellites, single nucleotide polymorphisms and a sequence tagged site in starch-synthesizing genes in relation to starch physicochemical properties in nonwaxy rice (*Oryza sativa* L.). Theor. Appl. Genet. 113, 1185–1196.

Batjes, N.H., 1997. A world data set of derived soil properties by FAO-UNESCO soil unit for global modelling. Soil Use Manage. 13, 9–16.

Bernier, J., Kumar, A., Ramaiah, V., Spaner, D., Atlin, G., 2007. A large-effect QTL for grain yield under reproductive-stage drought stress in upland rice. Crop Sci. 7, 507–518.

Berruyer, R., Adreit, H., Milazzo, J., Gaillard, S., Berger, A., Dioh, W., Lebrun, M.H., Tharreau, D., 2003. Identification and fine mapping of *Pi33* the rice resistance gene corresponding to the *Magnaporthe grisea* avirulence gene *ACE1*. Theor. Appl. Genet. 107, 1139–1147.

Bhasin, H., Bhatia, D., Raghuvanshi, S., Lore, J.S., Sahi, G.K., Kaur, B., 2012. New PCR-based sequence-tagged site marker for bacterial blight resistance gene *Xa38* of rice. Mol. Breeding 30, 607–611.

Biselli, C., Cavalluzzo, D., Perrini, R., Gianinetti, A., Bagnaresi, P., Urso, S., 2014. Improvement of marker-based predictability of apparent amylose content in *japonica* rice through GBSSI allele mining. Rice 7, 1.

Blair, M.W., Garris, A.J., Iyer, A.S., Chapman, B., Kresovich, S., McCouch, S.R., 2003. High resolution genetic mapping and candidate gene identification at the *xa5* locus for bacterial blight resistance in rice (*Oryza sativa* L.). Theor. Appl. Genet. 107, 62–73.

Bonilla, P., Dvorak, J., Mackill, D., Deal, K., Gregorio, G., 2002. RFLP and SSLP mapping of salinity tolerance genes in chromosome 1 of rice (*Oryza sativa* L.) using recombinant inbred lines. Philipp. Agric. Sci. 65 (1), 68–76.

Boonjung, H., Fukai, S., 1996. Effects of soil water deficit at different growth stages on rice growth and yield under upland conditions. II. Phenology, biomass production and yield. Field Crops Res. 48, 47–55.

Bouman, B.A.M., Wang, H., Yang, X., Zhao, J.F., Wang, C.G., 2002. Aerobic rice (Han Dao): a new way of growing rice in water-short areas. In: Proceedings of the Twelfth International Soil Conservation Organization Conference. 26–31 May, 2002, Beijing, China, pp. 175–181.

Brand-Miller, J., McMillan-Price, J., Steinbeck, K., Caterson, I., 2009. Dietary glycemic index: health implications. J. Am. Coll. Nutr. 28, 446S–449S.

Bryan, G.T., Wu, K.S., Farrall, L., Jia, Y.L., Hershey, H.P., McAdams, S.A., Faulk, K.N., Donaldson, G.K., Tarchini, R., Valent, B., 2000. A single amino acid difference distinguishes resistant and susceptible alleles of the rice blast resistance gene *Pi-ta*. Plant Cell 12, 2033–2045.

Cagampang, G.B., Perez, C.M., Juliano, B.O., 1973. A gel consistency test for eating quality of rice. J. Sci. Food Agric. 24, 1589–1594.

Ceccarelli, S., Grando, S., Maatougui, M., Michael, M., Slash, M., Haghparast, R., Mimoun, H., Nachit, M., 2010. Plant breeding and climate changes. Cambridge J. Agric. Sci. 148, 627–637.

Chalam, G.V., Venkateswarlu, J., 1965. Introduction to agricultural botany in India. Asia Publishing House, New Delhi, (460 p.).

Chandler, R.F., 1969. Improving the rice plant and its culture. Nature 221, 1007–1010.

Chandusingh, P.R., Singh, N.K., Prabhu, K.V., Vinod, K.K., Singh, A.K., 2013. Molecular mapping of quantitative trait loci for grain chalkiness in rice. Indian J. Genet. 73 (3), 244–251.

Chaudhary, B.P., Shrivastava, P.S., Shrivastava, M.N., Khush, G.S., 1986. Inheritance of resistance to gall midge in some cultivars of rice. Rice Genet. 1, 523–528.

Chauhan, J.S., Nanda, J.S., 1983. Inheritance of amylose content and its association with grain yield and yield contributing characters in rice. Oryza 20, 81–85.

Cheema, K., Grewal, N., Vikal, Y., Sharma, R., Lore, J.S., Das, A., Bhatia, D., Mahajan, R., Gupta, V., Bharaj, T.S., Singh, K., 2008. A novel bacterial blight resistance gene from *Oryza nivara* mapped to 38 kb region on chromosome 4L and transferred to *Oryza sativa* L. Genet Res. 90, 397–407.

Chen, M., Bergman, C., Pinson, S., Fjellstrom, R., 2008a. Waxy gene haplotypes: associations with apparent amylose content and the effect by the environment in an international rice germplasm collection. J. Cereal Sci. 47, 536–545.

Chen, M.H., Bergman, C.J., Pinson, S.R.M., Fjellstrom, R.G., 2008b. Waxy gene haplotypes: associations with pasting properties in an international rice germplasm collection. J. Cereal Sci. 48, 781–788.

Chen, S., Liu, X., Zeng, L., Ouyang, D., Yang, J., Zhu, X., 2011a. Genetic analysis and molecular mapping of a novel recessive gene *xa34(t)* for resistance against *Xanthomonas oryzae* pv. *oryzae*. Theor. Appl. Genet. 122, 1331–1338.

Chen, J., Shi, Y., Liu, W., Chai, R., Fu, Y., Zhuang, J., Wu, J., 2011b. A *Pid3* allele from rice cultivar Gumei-2 confers resistance to *Magnaporthe oryzae*. J. Genet. Genomics 38, 209–216.

Cheng, J., Wang, L., Du, W., 2014. Dynamic quantitative trait locus analysis of seed dormancy at three development stages in rice. Mol. Breeding 34 (2), 501–510.

Chen, J.W., Wang, L., Pang, F., 2006. Genetic analysis and fine mapping of a rice brown planthopper (*Nilaparvata lugens* Stål) resistance gene *bph19(t)*. Mol. Genet. Genomics 275, 321–329.

Chen, S., Wang, L., Que, Z.Q., Pan, R.Q., Pan, Q.H., 2005. Genetic and physical mapping of *Pi37(t)*, a new gene conferring resistance to rice blast in the famous cultivar St. No. 1. Theor. Appl. Genet. 111, 1563–1570.

Chin, J.H., Gamuyao, R., Dalid, C., Bustamam, M., Prasetiyono, J., Moeljopawiro, S., Wissuwa, M., Heuer, S., 2011. Developing rice with high yield under phosphorus deficiency: *Pup1* sequence to application. Plant Physiol. 156, 1202–1216.

Chu, Z., Fu, B., Yang, H., Xu, C., Li, Z., Sanchez, A., Park, Y.J., Bennetzen, J.L., Zhang, Q., Wang, S., 2006. Targeting *xa13*, a recessive gene for bacterial blight resistance in rice. Theor. Appl. Genet. 112, 455–461.

Chutia, J., Borah, S.P., 2012. Water stress effects on leaf growth and chlorophyll content but not the grain yield in traditional rice (*Oryza sativa* L.) genotypes of Assam, India. II. Protein and proline status in seedlings under PEG induced water stress. Am. J. Plant Sci. 3, 971–980.

Courage, H.K., 2010. White rice raises risk of type 2 diabetes. Observations Scientific American, Blog Network (June 14).

Das, P., Nutan, K.K., Singla-Pareek, S.L., Pareek, A., 2015. Understanding salinity responses and adopting "omics-based" approaches to generate salinity tolerant cultivars of rice. Front. Plant Sci. 6, 712.

Das, A., Soubam, D., Singh, P.K., Thakur, S., Singh, N.K., Sharma, T.R., 2012. A novel blast resistance gene, *Pi54rh* cloned from wild species of rice, *Oryza rhizomatis* confers broad spectrum resistance to *Magnaporthe oryzae*. Funct. Integr. Genomics 12, 215–228.

De Datta, S.K., Tauro, A.C., Balaoing, S.N., 1968. Effect of plant type and nitrogen level on growth characteristics and grain yield of *indica* rice in the tropics. Agron. J. 60, 643–647.

Deen, R.K., Ramesh Gautam, K.S., Viraktamath, B.C., Brar, D.S., Ram, T., 2010. Identification of new gene for BPH resistance introgressed from *O. rufipogon*. Rice Genet. Newsl. 25, 70–72.

Dixit, S., Grondin, A., Lee, C.R., Henry, A., Olds, T.M., Kumar, A., 2015. Understanding rice adaptation to varying agro-ecosystems: trait interactions and quantitative trait loci. BMC Genomics 16 (1), 86.

Dixit, S., Singh, A., Sta Cruz, M.T., Maturan, P.T., Amante, M., Kumar, A., 2014. Multiple major QTLs lead to stable yield performance of rice cultivars across variable drought intensities. BMC Genet. 15, 16.

Dobermann, A., Fairhurst, T., 2000. Rice: Nutrient Disorders and Nutrient Management. PPI/PPIC; IRRI, Singapore; Philippines.

Dobo, M., Ayres, N., Walker, G., Park, W.D., 2010. Polymorphism in the *GBSS* gene affects amylose content in US and European rice germplasm. J. Cereal Sci. 52, 450–456.

Dong, Y., Tsuzuki, E., Lin, D., Kamiunten, H., Terao, H., Matsuo, M., Cheng, S., 2004. Molecular genetic mapping of quantitative trait loci for milling quality in rice (*Oryza sativa* L.). J. Cereal Sci. 40, 1185–1196.

Dong, J.Y., Tsuzuki, E., Terao, H., 2000. Inheritance of aroma in four rice cultivars (*Oryza sativa* L.). IRRI. International Rice Research Notes 25, 2.

Du, B., Zhang, W.L., Liu, B.F., et al., 2009. Identification and characterization of *Bph14*, a gene conferring resistance to brown planthopper in rice. Proc. Natl. Acad. Sci. USA 106, 22163–22168.

FAO, 2001. Human Vitamin & Mineral Requirements. Report of a Joint FAO/WHO Expert Consultation, Bangkok, Thailand.

Fitzgerald, M.A., McCouch, S.R., Hall, R.D., 2009. Not just a grain of rice: the quest for quality. Trends Plant Sci. 14, 133–139.

Fitzgerald, M.A., Rahman, S., Resurreccion, A.P., Morrel, M.K., Bird, A.R., 2011. Identification of a major genetic determinant of glycemic index in rice. Rice 4, 66.

Fjellstrom, R., Concetta, A., Conaway-Bormans, A., McClung, M., Marchetti, A. M., Shank, R.A., William, D.P., 2004. Development of DNA markers suitable for marker assisted selection of three *Pi* genes conferring resistance to multiple *Pyricularia grisea* pathotypes. Crop Sci. 44, 1790–1798.

Fjellstrom, R., McClung, A.M., Shank, A.R., 2006. SSR markers closely linked to the *Pi-z* locus are useful for selection of blast resistance in a broad array of rice germplasm. Mol. Breeding 17, 149–157.

Fuentes, J.L., Correa-Victoria, F.J., Escobar, F., Prado, G., Aricapa, G., Duque, M.C., Tohme, J., 2007. Identification of microsatellite markers linked to the blast resistance gene *Pi-1(t)* in rice. Euphytica 160, 295–304.

Fujita, D., Kohli, A., Horgan, F.G., 2013. Rice resistance to planthoppers and leafhoppers. CRC Crit. Rev. Plant Sci. 32, 162–191.

Fukai, S., Pantuwan, G., Jongdee, C., Cooper, M., 1999. Screening for drought resistance in rainfed lowland rice. Field Crops Res. 64, 61–74.

Fukuoka, S., Saka, N., Koga, H., Ono, K., Shimizu, T., Ebana, K., Hayashi, N., Takahashi, A., Hirochika, H., Okuno, K., Yano, M., 2009. Loss of function of a proline-containing protein confers durable disease resistance in rice. Science 325, 998–1001.

Furukawa, T., Maekawa, M., Oki, T., Suda, I., Iida, S., et al., 2007. The *Rc* and *Rd* genes are involved in proanthocyanidin synthesis in rice pericarp. Plant J. 49, 91–102.

Gamuyao, R., Chin, J.H., Pariasca-Tanaka, J., Pesaresi, P., Dalid, C., Slamet-Loedin, I., Tecson-Mendoza, E.M., Wissuwa, M., Heuer, S., 2012. The protein kinase OsPSTOL1 from traditional rice confers tolerance of phosphorus deficiency. Nature 488, 535–539.

Gao, X.H., Xiao, S.L., Yao, Q.F., Wang, Y.J., Fu, X.D., 2011. An updated GA signaling "relief of repression" regulatory model. Mol. Plant 4, 601–606.

Ghimire, K.H., Quiatchon, L.A., Vikram, P., Swamy, B.P.M., Dixit, S., Ahmed, H., Hernandez, J.E., Borromeo, T.H., Kumar, A., 2012. Identification and mapping of a QTL (*qDTY_{1.1}*) with a consistent effect on grain yield under drought. Field Crops Res. 131, 88–96.

Goto, T., Matsumoto, T., Furuya, N., Tsuchiya, K., Yoshimura, A., 2009. Mapping of bacterial blight resistance gene *Xa11* on rice chromosome 3. Jpn. Agric. Res. Q. 43 (3), 221–225.

Gregorio, G.B., 1997. Tagging Salinity Tolerant Genes in Rice Using Amplified Fragment Length Polymorphism (AFLP). Ph.D. Dissertation. University of the Philippines Los Baños, College, Laguna, Philippines (118 p.).

Gregorio, G.B., Senadhira, D., Mendoza, R.D., Manigbas, N.L., Roxas, J.P., Guerta, C.Q., 2002. Progress in breeding for salinity tolerance and associated abiotic stresses in rice. Field Crops Res. 76, 91–101.

Gu, K., Tian, D., Yang, F.W.U.L., Sreekala, C., Wang, D., 2004. High-resolution genetic mapping of *Xa27(t)*, a new bacterial blight resistance gene in rice, *Oryza sativa* L. Theor. Appl. Genet. 108, 800–807.

Gu, K., Yang, B., Tian, D., Wu, L., Wang, D., Sreekala, C., Yang, F., Chu, Z., Wang, G.L., White, F.F., Yin, Z., 2005. R gene expression induced by a type III effector triggers disease resistance in rice. Nature 435, 1122–1125.

Guo, S., Zhang, D., Lin, X., 2010. Identification and mapping of a novel bacterial blight resistance gene *Xa35(t)* originated from *Oryza minuta*. Sci. Agric. Sin. 43, 2611–2618.

Hattori, Y., Nagai, K., Furukawa, S., Song, X.J., Kawano, R., Sakakibara, H., Wu, J., Matsumoto, T., Yoshimura, A., Kitano, H., Matsuoka, M., Ashikari, M., 2009. The ethylene response factors SNORKEL1 and SNORKEL2 allow rice to adapt to deep water. Nature 2460, 1026–1030.

Hayashi, K., Yoshida, H., 2009. Refunctionalization of the ancient rice blast disease resistance gene *Pit* by the recruitment of a retrotransposon as a promoter. Plant J. 57, 413–425.

Hayashi, N., Inoue, H., Kato, T., Funao, T., Shirota, M., Shimizu, T., Kanamori, H., Yamane, H., Hayano-Saito, Y., Matsumoto, T., Yano, M., Takatsuji, H., 2010. Durable panicle blast resistance gene *Pb1* encodes an atypical CC-NBS-LRR protein and was generated by acquiring a promoter through local genome duplication. Plant J. 64, 498–510.

He, J., Liu, Y.Q., Liu, Y.L., et al., 2013. High-resolution mapping of brown planthopper (BPH) resistance gene Bph27(t) in rice (Oryza sativa L.). Mol. Breeding 31, 549–557.

Hiemori, M., Koh, E., Mitchell, A.E., 2009. Influence of cooking on anthocyanins in black rice (Oryza sativa L. japonica var. SBR). J. Agric. Food Chem. 57, 1908–1914.

Himabindu, K., 2009. Identification, Tagging and Mapping of Rice Gall Midge Resistance Genes Using Microsatellite Markers. Ph.D. Thesis. Department of Biotechnology, Acharya Nagarjuna University, Guntur, India (58 p.).

Himabindu, K., Suneetha, K., Sama, V.S.A.K., Bentur, J.S., 2010. A new rice gall midge resistance gene in the breeding line CR57-MR1523, mapping with flanking markers and development of NILs. Euphytica 174, 179–187.

Hirabayashi, H., Angeles, E.R., Kaji, R., Ogawa, T., Brar, D.S., Khush, G.S., 1998. Identification of brown planthopper resistance gene derived from O. officinalis using molecular markers in rice. Breed. Sci. 48 (Suppl. 1), 82–86.

Hori, K., Sato, K., Takeda, K., 2007. Detection of seed dormancy QTL in multiple mapping populations derived from crosses involving novel barley germplasm. Theor. Appl. Genet. 115, 869–876.

Hu, J., Xiao, C., Cheng, M.X., et al., 2015. Fine mapping and pyramiding of brown planthopper resistance genes QBph3 and QBph4 in an introgression line from wild rice O. officinalis. Mol. Breeding 35, 3.

Huang, N., He, G., Shu, L., Li, X., Zhang, Q., 2001. Identification and mapping of two brown planthopper genes in rice. Theor. Appl. Genet. 102, 929–934.

Huang, D., Qiu, Y., Zhang, Y., et al., 2013. Fine mapping and characterization of BPH27, a brown planthopper resistance gene from wild rice (Oryza rufipogon Griff.). Theor. Appl. Genet. 126, 219–229.

Huke, R.E., Huke, E.H., 1997. Rice Area by Type of Culture: South, Southeast, and East Asia. International Rice Research Institute, Los Baños, Philippines.

International Energy Agency (IEA), 2012. The IEA Model of Short-Term Energy Security (MOSES), Primary Energy Sources, and Secondary Fuels. Working Paper. OECD/I, Paris.

International Rice Research Institute (IRRI), 1989. IRRI Towards 2000 and Beyond. IRRI, Manila, Philippines, (pp. 36–37).

Ishii, T., Brar, D.S., Multani, D.S., Khush, G.S., 1994. Molecular tagging of genes for brown planthopper resistance and earliness introgressed from Oryza australiensis into cultivated rice, O. sativa. Genome 37, 217–221.

Ishimaru, K., 2003. Identification of a locus increasing rice yield and physiological analysis of its function. Plant Physiol. 133 (3), 1083–1090.

Ishimaru, Y., Takahashi, R., Bashir, K., Shimo, H., Senoura, T., Sugimoto, K., Ono, K., Yano, M., Ishikawa, S., Arao, T., Nakanishi, H., Nishizawa, N.K., 2012. Characterizing the role of rice NRAMP5 in manganese, iron and cadmium transport. Sci. Rep. 2, 286.

Islam, M.M., Begum, S.N., Emon, R.M., Halder, J., Manidas, A.C., 2012. Carbon isotope discrimination in rice under salt affected conditions in Bangladesh. International Atomic Energy Agency; Joint FAO/IAEA Division of Nuclear Techniques in Food and Agriculture; Soil and Water Management and Crop Nutrition Section, Vienna, Austria, (pp. 7–23).

Ismail, A.M., Singh, U.S., Singh, S., Dar, M.H., Mackill, D.J., 2013. The contribution of submergence-tolerant (Sub1) rice varieties to food security in flood-prone rainfed lowland areas in Asia. Field Crops Res. 152, 83–93.

Ismail, A.M., Thomson, M.J., Singh, R.K., Gregorio, G.B., Mackill, D.J., 2008. Designing rice varieties adapted to coastal area of South and Southeast Asia. J. Indian Soc. Coastal Agric. Res. 26, 69–73.

Jagadish, S.V.K., Craufurd, P.Q., Wheeler, T.R., 2007. High temperature stress and spikelet fertility in rice. J. Exp. Bot. 58, 1627–1635.

Jagadish, S.V.K., Muthurajan, R., Oane, R., Wheeler, T.R., Heuer, S., Bennett, J., Craufurd, P.Q., 2010. Physiological and proteomic approaches to address heat tolerance during anthesis in rice (*Oryza sativa* L.). J. Exp. Bot. 61, 143–156.

Jain, A., Ariyadasa, R., Kumar, A., Srivastava, M.N., Mohan, M., Nair, S., 2004. Tagging and mapping of a rice gall midge resistance gene, *Gm8*, and development of SCARs for use in marker-aided selection and gene pyramiding. Theor. Appl. Genet. 109, 1377–1384.

Jairin, J., Teangdeerith, S., Leelagud, P., et al., 2007. Detection of brown planthopper resistance genes from different rice mapping populations in the same genomic location. Sci. Asia 33, 347–352.

Javid, M.G., Sorooshzadeh, A., Moradi, F., Sanavy Seyed, A.M.M., Allahdadi, I., 2011. The role of phytohormones in alleviating salt stress in crop plants. Aust. J. Crop Sci. 5 (6), 726–734.

Jena, K.K., Jeung, J.U., Lee, J.H., et al., 2006. High-resolution mapping of a new brown planthopper (BPH) resistance gene, *Bph18(t)*, and marker-assisted selection for BPH resistance in rice (*Oryza sativa* L.). Theor. Appl. Genet. 112, 288–297.

Jena, K.K., Pasalu, I.C., Rao, Y.K., et al., 2003. Molecular tagging of a gene for resistance to brown planthopper in rice (*Oryza sativa* L.). Euphytica 129, 81–88.

Jeung, J.U., Kim, B.R., Cho, Y.C., Han, S.S., Moon, H.P., Lee, Y.T., Jena, K.K., 2007. A novel gene, *Pi40(t)*, linked to the DNA markers derived from NBS-LRR motifs confers broad spectrum of blast resistance in rice. Theor. Appl. Genet. 115, 1163–1177.

Jongdee, B., Fukai, S., Cooper, M., 2002. Leaf water potential and osmotic adjustment as physiological traits to improve drought tolerance in rice. Field Crops Res. 76, 153–163.

Juliano, B.O., 1985. Rice Chemistry and Technology, second ed. American Association of Cereal Chemists, Inc., St. Paul, MN.

Juliano, B.O., Villarreal, C.P., 1993. Grain Quality Evaluation of World Rice. International Rice Research Institute, Los Baños, Philippines, (148 p.).

Kabir, M.A., Khush, G.S., 1988. Genetic analysis of resistance to brown planthopper in rice, *Oryza sativa* L. Plant Breed. 100, 54–58.

Katiyar, S., Verulkar, S., Chandel, G., Zhang, Y., Huang, B.C., Bennett, J., 2001. Genetic analysis and pyramiding of two gall midge resistance genes (*Gm-2* and *Gm-6t*) in rice (*Oryza sativa* L.). Euphytica 122, 327–334.

Kawaguchi, M., Murata, K., Ishii, T., et al., 2001. Assignment of a brown planthopper (*Nilaparvata lugens* Stål) resistance gene bph4 to the rice chromosome 6. Breed. Sci. 51, 13–18.

Kharabian-Masouleh, A., Waters, D.L., Reinke, R.F., Ward, R., Henry, R.J., 2012. SNP in starch biosynthesis genes associated with nutritional and functional properties of rice. Sci. Rep. 2, 557.

Khatun, S., Flowers, T.J., 1995. Effects of salinity on seed set in rice. Plant Cell Environ. 18, 61–67.

Khorgamy, A., Farnia, A., 2009. Effect of phosphorus and zinc fertilisation on yield and yield components of chick pea cultivars. Afr. Crop Sci. Conf. Proc. 9, 205–208.

Khush, G.S., 1995. Breaking the yield frontier of rice. GeoJournal 35, 329–332.

Khush, G.S., 2013. Strategies for increasing the yield potential of cereals: case of rice as an example. Plant Breed. 132, 433–436.

Khush, G.S., Karim, A.R., Angeles, E.R., 1985. Genetics of resistance of rice cultivar ARC 10550 to Bangladesh brown planthopper biotype. J. Genet. 64, 121–125.

Kim, S.M., Sohn, J.K., 2005. Identification of rice gene (*Bph1*) conferring resistance to brown plant hopper (*Nilaparvata lugens* Stål) using STS markers. Mol. Cells 20, 30–34.

Kirigwi, F.M., Van, G.M., Brown, G.G., Gill, B.S., Paulsen, G.M., Fritz, A.K., 2007. Markers associated with a QTL for grain yield in wheat under drought. Mol. Breeding 20, 401–413.

Kiyosawa, S., Ando, I., 1990. Blast resistance. Sci. Rice Plant 3, 361–385.

Kobayashi, A., Bao, G., Ye, S., Tomita, K., 2007. Detection of quantitative trait loci for white-back and basal-white kernels under high temperature stress in *japonica* rice varieties. Breed. Sci. 2007 (7), 107–116.

Koide, Y., Ebron, L.A., Kato, H., Tsunematsu, H., Yanoria, M.J.T., Kobayashi, N., Yokoo, M., Maruyama, K., Imbe, T., Fukuta, Y., 2011. A set of near-isogenic lines for blast resistance genes with an *indica*-type rainfed lowland elite rice (*Oryza sativa* L.) genetic background. Field Crops Res. 123 (1), 19–27.

Kondo, M., Iwasawa, N., Yoshida, H., Nakagawa, H., Ohno, H., Nakazono, K., Usui, Y., Tokida, T., Hasegawa, T., Kuwagata, T., Morita, S., Nagata, K., 2012. Factors influencing the appearance quality in rice under high temperature in 2010. Jpn. J. Crop Sci. 81 (Extra issue 1), 120–121.

Korinsak, S., Sriprakhon, S., Sirithanya, P., Jairin, J., Korinsak, S., Vanavichit, A., 2009. Identification of microsatellite markers (SSR) linked to a new bacterial blight resistance gene *xa33(t)* in rice cultivar "Ba7". Maejo Int. J. Sci. Tech. 3, 235–247.

Kovach, M.J., Calingacion, M.N., Fitzgerald, M.A., McCouch, S.R., 2009. The origin and evolution of fragrance in rice. Proc. Natl. Acad. Sci. USA 106, 14444–14449.

Kretzschmar, T., Pelayo, M.A., Trijatmiko, K.R., Gabunada, L.F., Alam, R., Jimenez, R., Mendioro, M.S., Slamet-Loedin, I.H., Sreenivasulu, N., Bailey-Serres, J., Ismail, A.M., 2015. A trehalose-6-phosphate phosphatase enhances anaerobic germination tolerance in rice. Nat. Plants 1, 15124.

Krishnamurthy, P., Panigrahy, M., Rao, D.N., Yugandhar, P., Raju, N.S., Voleti, S.R., et al., 2014. Hydroponic experiment for identification of tolerance traits developed by rice Nagina 22 mutants to low-phosphorus in field condition. Arch. Agron. Soil Sci. 60, 565–576.

Kumar, V., Ladha, J.K., 2011. Direct seeding of rice: recent developments and future research needs. Adv. Agron. 111, 297–413.

Kumar, A., Bhandarkar, S., Pophlay, D.J., Shrivastava, M.N., 2000. A new gene for gall midge resistance in rice accession Jhitpiti. Rice Genet. Newsl. 17, 83–84.

Kumar, A., Bernier, J., Verulkar, S., Lafitte, H.R., Atlin, G.N., 2008. Breeding for drought tolerance: direct selection for yield, response to selection, and use of drought-tolerant donors in upland and lowland adapted populations. Field Crops Res. 107, 221–231.

Kumar, A., Dixit, S., Ram, T., Yadaw, R.B., Mishra, K.K., Mandal, N.P., 2014. Breeding high-yielding drought-tolerant rice: genetic variations and conventional and molecular approaches. J. Exp. Bot. 65, 6265–6278.

Kumar, A., Jain, A., Sahu, R.K., Shrivastava, M.N., Nair, S., Mohan, M., 2005. Genetic analysis of resistance genes for the rice gall midge in two rice genotypes. Crop Sci. 45 (4), 1631–1635.

Kumar, V., Ladha, J.K., Gathala, M.K., 2009. Direct drill-seeded rice: a need of the day. In: Annual Meeting of Agronomy Society of America. 1–5 November, 2009, Pittsburgh. Available from: http://a-c-s.confex.com/crops/2009am/webprogram/Paper53386.html

Kumar, A., Shrivastava, M.N., Sahu, R.K., 1998. Genetic analysis of ARC 5984 for gall midge resistance: a reconsideration. Rice Genet. Newsl. 15, 142–143.

Kumar, A., Shrivastava, M.N., Shukla, B.C., 1999. A new gene for resistance to gall midge in rice cultivar RP2333-156-8. Rice Genet. Newsl. 16, 85–87.

Kumar, A., Verulkar, S.B., Dixit, S., Chauhan, B., Bernier, J., Venuprasad, R., Zhao, D., Shrivastava, M.N., 2009b. Yield and yield-attributing traits of rice (*Oryza sativa* L.) under lowland drought and suitability of early vigor as a selection criterion. Field Crops Res. 114, 99–107.

Kumari, P., Ahuja, U., Jain, S., Jai, R.K., 2012. Fragrance analysis among recombinant inbred lines of rice. Asian J. Plant Sci. 11 (4), 190–194.

Kumari, S., Sheba, J.M., Marappan, M., Ponnuswamy, S., Seetharaman, S., Pothi, N., Subbarayalu, M., Muthurajan, R., Natesan, S., 2010. Screening of IR50 × Rathu Heenati F_7 RILs and identification of SSR markers linked to brown planthopper (*Nilaparvata lugens* Stål) resistance in rice (*Oryza sativa* L.). Mol. Biotechnol. 46, 63–71.

Kurata, N., Yamazaki, Y., 2006. Oryzabase: an integrated biological and genome information database for rice. Plant Physiol. 140, 12–17.

Lafitte, H.R., Price, A.H., Courtois, B., 2004. Yield response to water deficit in an upland rice mapping population: associations among traits and genetic markers. Theor. Appl. Genet. 109, 1237–1246.

Lakashminarayana, A., Khush, G.S., 1977. New genes for resistance to the brown planthopper in rice. Crop Sci. 17, 96–100.

Lanceras, J.C., Huang, Z.L., Naivikul, O., Vanavichit, A., Ruanjaichon, V., Tragoonrung, S., 2000. Mapping of genes for cooking and eating qualities in Thai jasmine rice (KDML105). DNA Res. 7, 93–101.

Langridge, P., Fleury, D., 2011. Making the most of "omics" for crop breeding. Trends Biotechnol. 29, 33–40.

Lanning, S.B., Siebenmorgen, T.J., 2011. Comparison of milling characteristics of hybrid and pure-line rice cultivars. Appl. Eng. Agric. 27 (5), 787–795.

Lee, F.N., Gunnell, P.S., 1992. False smut. In: Webster, R.K., Gunnell, P.S. (Eds.), Compendium of Rice Diseases. APS Press, St. Paul, MN, pp. 5–28.

Lee, F.N., Rush, M.C., 1983. Rice sheath blight: a major rice disease. Plant Dis. 67 (7), 829–833.

Lee, G.H., Kang, I., Kim, K.M., 2016. Mapping of novel QTL regulating grain shattering using doubled haploid population in rice (*Oryza sativa* L.). Int. J. Genomics 2016, (2128010).

Lee, K.S., Choi, W.Y., Ko, J.C., Kim, T.S., Gregorio, G.B., 2003. Salinity tolerance of *japonica* and *indica* rice (*Oryza sativa* L.) at the seedling stage. Planta 216, 1043–1046.

Lee, S.K., Song, M.Y., Seo, Y.S., Kim, H.K., Ko, S., Cao, P.J., Suh, J.P., Jeon, J.S., 2009. Rice *Pi5*-mediated resistance to *Magnaporthe oryzae* requires the presence of two coiled nucleotide-binding-leucine-rich repeat genes. Genetics 181, 1627–1638.

Li, Y.H., 2001. Research and practice of water-saving irrigation for rice in China. In: Barker, R., Li, Y., Tuong, T.P. (Eds.), Water-Saving Irrigation for Rice, Proceedings of an International Workshop. 23–25 March, 2001, Wuhan, China. International Water Management Institute, Colombo, Sri Lanka, pp. 135–144.

Li, J., Xie, Y., Dai, A., Liu, L., Li, Z., 2009. Root and shoot traits responses to phosphorus deficiency and QTL analysis at seedling stage using introgression lines of rice. J. Genet. Genomics 36, 173–183.

Li, J.M., Xiao, J.H., Grandillo, S., Jiang, L.Y., Wan, Y.Z., Deng, Q.Y., Yuan, L.P., McCouch, S.R., 2004. QTL detection for rice grain quality traits using an interspecific backcross population derived from cultivated Asian (*O. sativa* L.) and African (*O. glaberrima* L.) rice. Genome 47, 697–704.

Lin, F., Chen, S., Que, Z., Wang, L., Liu, X., Pan, Q., 2007. The blast resistance gene *Pi37* encodes a nucleotide binding site leucine-rich repeat protein and is a member of a resistance gene cluster on rice chromosome 1. Genetics 177, 1871–1880.

Lin, H.X., Min-Shao, K., Xiong, Z.M., Qian, H.R., Zhuang, J.Y., Lu, J., Huang, N., Zheng, K., Lin, H.X., Min, S.K., Xiong, Z.M., Qian, H.R., Zhuang, J.Y., Lu, J., Huang, N., Zheng, K.L., 1995. RFLP mapping of QTLs for grain shape traits in *indica* rice (*Oryza sativa* L. subsp. *indica*). Sci. Agric. 28, 1–7.

Lin, X.H., Zhang, D.P., Xie, Y.F., Gao, H.P., Zhang, Q., 1996. Identifying and mapping a new gene for bacterial blight resistance in rice based on RFLP markers. Phytopathology 86, 1156–1159.

Litsinger, J.A., Bandong, J.P., Canapi, B.L., Dela Cruz, C.G., Pantua, P.C., Alviola, III, A.L., Batayan, E., 2005. Evaluation of action thresholds against chronic insect pests of rice in the Philippines. I. Less frequently occurring pests and overall assessment. Int. J. Pest Manage. 51, 45–61.

Litsinger, J.A., Canapi, B.L., Bandong, J.P., Lumaban, M.D., Raymundo, F.D., Barrion, A.T., 2009a. Insect pests of rainfed wetland rice in the Philippines: population densities, yield loss, and insecticide management. Int. J. Pest Manage. 55, 221–242.

Litsinger, J.A., Libetario, E.M., Canapi, B.L., 2009. Eliciting farmer knowledge, attitudes, and practices in the development of integrated pest management programs for rice in Asia. In: Peshin, R., Dhawan, A.K. (Eds.). Integrated Pest Management: Dissemination and Impact, vol. 2. Springer Science + Media B.V., Berlin, pp. 119–273.

Liu, J.X., Liao, D.Q., Oane, R., Estenor, L., Yang, X.E., Li, Z.C., Bennett, J., 2006. Genetic variation in the sensitivity of anther dehiscence to drought stress in rice. Field Crops Res. 97, 87–100.

Liu, X., Lin, F., Wang, L., Pan, Q., 2007a. The in silico map-based cloning of *Pi36* a rice coiled-coil nucleotide-binding site leucine-rich repeat gene that confers race-specific resistance to the blast fungus. Genetics 176, 2541–2549.

Liu, G., Lu, G., Zeng, L., Wang, G.L., 2002. Two broad-spectrum blast resistance genes, *Pi9(t)* and *Pi2(t)*, are physically linked on rice chromosome 6. Mol. Genet. Genomics 267 (4), 472–480.

Liu, G., Xiong, Y.L., Butterfield, D.A., 2000. Chemical, physical, and gel-forming properties of oxidized myofibrils and whey- and soy-protein isolates. J. Food Sci. 65, 811–818.

Liu, X., Yang, Q., Lin, F., Hua, L., Wang, C., Wang, L., Pan, Q., 2007b. Identification and fine mapping of *Pi39(t)*, a major gene conferring the broad-spectrum resistance to *Magnaporthe oryzae*. Mol Genet. Genomics 278, 403–410.

Liu, Q., Yuan, M., Zhou, Y., Li, X.G., Xiao, J., Wang, S., 2011. A paralog of the MtN3/sativa family recessively confers race-specific resistance to *Xanthomonas oryzae* in rice. Plant Cell Environ. 34, 1658–1669.

Loan, L.C., Ngan, V.T., Van, P.D., 2006. Preliminary evaluation on resistance genes against rice bacterial leaf blight in Can Tho province—Vietnam. Omon Rice 14, 44–47.

Lorieux, M., Petrov, N., Huang, N., Guiderdoni, E., Ghesquier, A., 1996. Aroma in rice: genetic analysis of a quantitative trait. Theor. Appl. Genet. 93, 1145–1151.

Lucca, P., Hurrell, R., Potrykus, I., 2001. Genetic engineering approaches to improve the bioavailability and the level of iron in rice grains. Theor. Appl. Genet. 102, 392–397.

Lv, W.T., Du, B., Shang, Guan, et al., 2014. BAC and RNA sequencing reveal the brown planthopper resistance gene *BPH15* in a recombination cold spot that mediates a unique defense mechanism. BMC Genet. 15, 674–689.

Ma, B.J., Wang, W.M., Zhao, B., Zhou, Y.L., Zhu, L.H., Zhai, W.X., 1999. Study on the PCR marker for the rice bacterial blight resistance gene *Xa4*. Hereditas 21 (3), 9–12.

Mackill, D.J., Amante, M.M., Vergara, B.S., Sarkarung, S., 1993. Improved semidwarf rice lines with tolerance to submergence of seedlings. Crop Sci. 33, 749–753.

Mackill, D.J., Coffman, W.R., Garrity, D.P., 1996. Rainfed Lowland Rice Improvement. International Rice Research Institute, Los Baños, Philippines.

Mackill, D.J., Ismail, A.M., Singh, U.S., Labios, R.V., Paris, T.R., 2012. Development and rapid adoption of submergence-tolerant (Sub1) rice varieties. Adv. Agron. 115, 299–352.

Madan, P., Jagadish, S.V.K., Craufurd, P.Q., Fitzgerald, M., Lafarge, T., Wheeler, T.R., 2012. Effect of elevated CO_2 and high temperature on seed-set and grain quality of rice. J. Exp. Bot. 63, 3843–3852.

Mahatheeranont, S., Keawsa-ard, S., Dumri, K., 2001. Quantification of the rice aroma compound, 2-acetyl-1-pyrroline, in uncooked Khao Dawk Mali105 brown rice. J. Agric. Food Chem. 49, 773–779.

Manay, N.S., Shadaksharaswamy, M., 2001. Food Facts and Principles, second ed. New Age International Pvt. Ltd. Publishers, New Delhi, (pp. 232–240).

Matsuoka, M., Furbank, R.T., Fukayama, H., Miyao, M., 2001. Molecular engineering of C4 photosynthesis. Annu. Rev. Physiol. Plant Mol. Biol. 52, 297–314.

Matthews, R.B., Horie, T., Kropff, M.J., Bachelet, D., Centeno, H.G., Shin, J.G., Mohandass, S., Singh, S., Defeng, Z., Moon, H.L., 1995. A regional evaluation of the effect of future climate change on rice production in Asia. In: Matthews, R.B., et al. (Eds.), Modeling the Impact of Climate Change on Rice Production in Asia. IRRI and CAB International, Wallingford, pp. 95–139.

Meng, F., Wei, Y., Yang, X., 2005. Iron content and bioavailability in rice. J. Trace Elem. Med. Biol. 18 (4), 333–338.

Mew, T.W., Alvarez, A.M., Leach, J.E., Swings, J., 1993. Focus on bacterial blight of rice. Plant Dis. 77, 5–12.

Mew, T.W., Vera Cruz, C.M., Reyes, R.C., 1982. Interaction of *Xanthomonas campestris* pv. *oryzae* and a resistant rice cultivar. Phytopathology 72, 786–789.

Miao, L., Wang, C., Zheng, C., Che, J., Gao, Y., Wen, Y., 2010. Molecular mapping of a new gene for resistance to rice bacterial blight. Sci. Agric. Sin. 43, 3051–3058.

Mikami, I., Uwatoko, N., Ikeda, Y., Yamaguchi, J., Hirano, H., Suzuki, Y., Sano, Y., 2008. Allelic diversification at the *wx* locus in landraces of Asian rice. Theor. Appl. Genet. 116, 979–989.

Mishra, P.K., Givvimani, S., Metreveli, N., Tyagi, S.C., 2010. Attenuation of beta2-adrenergic receptors and homocysteine metabolic enzymes cause diabetic cardiomyopathy. Biochem. Biophys. Res. Commun. 401, 175–181.

Mishra, S.B., Senadhira, D., Manigbas, N.L., 1996. Genetics of submergence tolerance in rice (*Oryza sativa* L.). Field Crops Res. 46, 177–181.

Miura, K., Ikeda, M., Matsubara, A., Song, X.J., Ito, M., Asano, K., Matsuoka, M., Kitano, H., Ashikari, M., 2010. OsSPL14 promotes panicle branching and higher grain productivity in rice. Nat. Genet. 42 (6), 545–549.

Mohan, M., Sathyanarayanan, P.V., Kumar, A., Srivastava, M.N., Nair, S., 1997. Molecular mapping of a resistance-specific PCR-based marker linked to a gall midge resistance gene (*Gm4t*) in rice. Theor. Appl. Genet. 95, 777–782.

Mohanty, R., Tripathi, B.K., Panda, T., 2011. Status, distribution and conservation of some threatened indigenous rice varieties cultivated in Odisha, India. Int. J. Conserv. Sci. 2, 269–274.

Moldenhauer, K.A.K., Lee, F.N., Norman, R.J., Helms, R.S., Well, R.H., Dilday, R.H., Rohnian, P.C., Marchetti, M.A., 1990. Registration of "Katy" rice. Crop Sci. 30, 747–748.

Morita, S., Nakano, H., 2011. Nonstructural carbohydrate content in the stem at full heading contributes to high performance of ripening in heat-tolerant rice cultivar Nikomaru. Crop Sci. 51, 818–828.

Morita, S.Y., Jun, I., Jun, T., 2005. Growth and endosperm cell size under high night temperatures in rice (*Oryza sativa* L.). Ann. Bot. 95, 695–701.

Munns, R., Tester, M., 2008. Mechanisms of salinity tolerance. Annu. Rev. Plant Biol. 59, 651–668.

Murai, H., Hashimoto, Z., Sharma, P., et al., 2001. Construction of a high resolution linkage map of a rice brown planthopper (*Nilaparvata lugens* Stål) resistance gene *bph2*. Theor. Appl. Genet. 103, 526–532.

Myint, K., Fujita, D., Matsumura, M., et al., 2012. Mapping and pyramiding of two major genes for resistance to the brown planthopper (*Nilaparvata lugens* [Stål]) in the rice cultivar ADR52. Theor. Appl. Genet. 124, 495–504.

Nadaf, A.B., Krishnan, S., Wakte, K.V., 2006. Histochemical and biochemical analysis of major aroma compound (2-acetyl-1-pyrroline) in basmati and other scented rice (*Oryza sativa* L.). Curr. Sci. 91 (11), 1533–1536.

Nakai, H., Nakamura, K., Kuwahara, S., Saito, M., 1988. Genetic studies of an induced rice mutant resistant to multiple races of bacterial leaf blight. Rice Genet. Newsl. 5, 101–103.

Nair, S., Kumar, A., Srivastava, M.N., Mohan, M., 1996. PCR based DNA markers linked to a gall midge resistance gene, Gm4t, has potential for marker aided selection in rice. Theor. Appl. Genet. 92, 660–665.

Natrajkumar, P., Sujatha, K., Laha, G.S., Srinivasarao, K., Mishra, B., Viraktamath, B.C., 2012. Identification and fine-mapping of Xa33, a novel gene for resistance to Xanthomonas oryzae pv. oryzae. Phytopathology 102, 222–228.

Nelson, J.C., McClung, A.M., Fjellstrom, R.G., Moldenhauer, K.A., Boza, E., Jodari, F., Oard, J.H., Linscombe, S., Scheffler, B., Yeater, K.M., 2011. Mapping QTL main and interaction influences on milling quality in elite US rice germplasm. Theor. Appl. Genet. 122, 291–309.

Nemoto, H., Ikeda, R., Kaneda, C., 1989. New genes for resistance to brown planthopper Nilaparvata lugens Stål in rice. Jpn. J. Breed. 39, 23–28.

Nezhad, K.Z., Weber, W., Röder, M., Sharma, S., Lohwasser, U., Meyer, R., Saal, B., Börner, A., 2012. QTL analysis for thousand-grain weight under terminal drought stress in bread wheat (Triticum aestivum L.). Euphytica 186, 127–138.

Niones, J.M., 2004. Fine mapping of the salinity tolerance gene on chromosome 1 of rice (Oryza sativa L.) using near isogenic lines. MS dissertation. University of the Philippines Los Baños, College, Laguna, Philippines.

Ogawa, T., 1996. Monitoring race distribution and identification of genes for resistance to bacterial leaf blight. In: Rice genetics III, Proceedings of the Third International Rice Genetics Symposium, pp. 456–459.

Ogawa, T., Yamamoto, T., Khush, G.S., Mew, T.W., Kaku, H., 1988. Near-isogenic lines as international differentials for resistance to bacterial blight of rice. Rice Genet. Newsl. 5, 106–109.

Okuyama, Y., Kanzaki, H., Abe, A., Yoshida, K., Tamiru, M., Saitoh, H., Fujibe, T., Matsumura, H., Shenton, M., Galam, D.C., Undan, J., Ito, A., Sone, T., Terauchi, R., 2011. A multifaceted genomics approach allows the isolation of rice Pia blast resistance gene consisting of two adjacent NBS-LRR protein genes. Plant J. 66, 467–479.

Ona, I., Casal, C., dela Paz, M., Zhang, Q., Xie, F., Vera Cruz, C.M., 2010. Bacterial blight-resistant lines with Xa23 are comparable with IRBB pyramid lines. Plant Breeding in International Rice Research Notes.(0117-4185).

Ookawa, T., Hobo, T., Yano, M., Murata, K., Ando, T., Miura, H., Asano, K., Ochiai, Y., Ikeda, M., Nishitani, R., Ebitani, T., 2010. New approach for rice improvement using a pleiotropic QTL gene for lodging resistance and yield. Nat. Commun. 30 (1), 132.

Ou, S.H., 1985. Rice Diseases, second ed. Commonwealth Mycological Institute; CAB International, Kew, Surrey, England; Farnham Royal, Slough.

Pandey, S., Bhandari, H., 2009. Drought, coping mechanisms, and poverty: insights from rainfed rice farming in Asia. Seventh Discussion Papers, Asia and the Pacific Division, IFAD.

Pandey, S., Behura, D.D., Villano, R., Naik, D., 2005. Economic costs of drought and farmers' coping mechanisms: a study of rainfed rice systems in Eastern India. IRRI Discussion Paper Series 39. International Rice Research Institute, Los Baños, Philippines (35 p.).

Pandey, A., Misra, P., Chandrashekar, K., Trivedi, P.K., 2012. Development of AtMYB12-expressing transgenic tobacco callus culture for production of rutin with biopesticidal potential. Plant Cell Rep. 31, 1867–1876.

Peng, S., Khush, G.S., 2003. Four decades of breeding for varietal improvement of irrigated lowland rice in the International Rice Research Institute. Plant Prod. Sci. 6, 157–164.

Peng, S., Huang, J., Sheehy, J.E., Laza, R.C., Visperas, R.M., Zhong, X., Khush, G.S., Cassman, K.G., 2004. Rice yield decline with higher night temperature from global warming. In: Redoña, E.D., Castro, A.P., Llanto, G.P. (Eds.), Rice Integrated Crop Management: Towards a Rice Check System in the Philippines. PhilRice, Nueva Ecija, Philippines, pp. 46–56.

Peng, S., Khush, G.S., Virk, P., Tang, Q., Zou, Y., 2008. Progress in ideotype breeding to increase rice yield potential. Field Crops Res. 108, 32–38.

Peng, S., Laza, R.C., Cassman, K.G., Khush, G.S., 2000. Grain yield of rice cultivars and lines developed in the Philippines since 1966. Crop Sci. 40, 307–314.

Peng, C., Wang, Y.H., Liu, F., Ren, Y.L., Zhou, K.N., Lv, J., Zheng, M., Zhao, S.L., Zhang, L., Wang, C.M., Jiang, L., Zhang, X., Guo, X.P., Bao, Y.Q., Wan, J.M., 2014. FLOURY ENDO-SPERM6 encodes a CBM48 domain-containing protein involved in compound granule formation and starch synthesis in rice endosperm. Plant J. 77, 917–930.

Perumalsamy, S., Bharani, M., Sudha, M., Nagarajan, P., Arul, L., Saraswathi, R., Balasubramanian, P., Ramalingam, J., 2010. Functional marker-assisted selection for bacterial leaf blight resistance genes in rice (*Oryza sativa* L.). Plant Breed. 129 (4), 400–406.

Pingali, P.L., Hossain, M., Gerpacio, R.V., 1997. Asian Rice Bowls: The Returning Crisis? CAB International, Wallingford, United Kingdom.

Porter, B.W., Chittoor, J.M., Yano, M., Sasaki, T., White, F.F., 2003. Development and mapping of markers linked to the rice bacterial blight resistance gene *Xa7*. J. Crop Sci. 43, 1484–1492.

Pradhan, A., Idol, T., Roul, P.K., 2016. Conservation agriculture practices in rainfed uplands of India improve maize-based system productivity and profitability. Front. Plant Sci. 7, 1008.

Qiu, Y., Guo, J., Jing, S., et al., 2010. High-resolution mapping of the brown planthopper resistance gene *Bph6* in rice and characterizing its resistance in the 93-11 and Nipponbare near isogenic backgrounds. Theor. Appl. Genet. 121, 1601–1611.

Qiu, Y., Guo, J., Jing, S., et al., 2012. Development and characterization of *japonica* rice lines carrying the brown planthopper-resistance genes *Bph12* and *Bph6*. Theor. Appl. Genet. 124, 485–494.

Qiu, Y.F., Guo, J.P., Jing, S.L., et al., 2014. Fine mapping of the rice brown planthopper resistance gene *Bph7* and characterization of its resistance in the 93-11 background. Euphytica 198, 369–379.

Qu, S., Liu, G., Zhou, B., Bellizzi, M., Zeng, L., Dai, L., Han, B., Wang, G.L., 2006. The broad spectrum blast resistance gene *Pi9* encodes a nucleotide-binding site leucine-rich repeat protein and is a member of a multigene family in rice. Genetics 172, 1901–1914.

Rahman, M.L., Jiang, W., Chu, S.H., et al., 2009. High-resolution mapping of two rice brown planthopper resistance genes, *Bph20(t)* and *Bph21(t)*, originating from *Oryza minuta*. Theor. Appl. Genet. 119, 1237–1246.

Ram, T., Deen, R., Gautam, S.K., Ramesh, K., Rao, Y.K., Brar, D.S., 2010. Identification of new genes for brown planthopper resistance in rice introgressed from *O. glaberrima* and *O. minuta*. Rice Genet. Newsl. 25, 67–69.

Rang, Z.W., Jagadish, S.V.K., Zhou, Q.M., Craufurd, P.Q., Heuer, S., 2011. Effect of high temperature and water stress on pollen germination and spikelet fertility in rice. Environ. Exp. Bot. 70, 58–65.

Rao, I.S., Srikanth, B., Kishore, V.H., Suresh, P.B., Chaitanya, U., Sundaram, R.M., Madhav, M.S., 2011. Indel polymorphism in sugar translocation and transport genes associated with grain filling of rice (*Oryza sativa* L.). Mol. Breeding 28 (4), 683–691.

Rashid, M.H., Alam, M.M., Khan, M.A.H., Ladha, J.K., 2009. Productivity and resource use of direct-(drum)-seeded and transplanted rice in puddled soils in rice-rice and rice-wheat ecosystem. Field Crops Res. 113, 274–281.

Reinke, R.F., Welsh, L.A., Reece, J.E., Lewin, L.G., Blakeney, A.B., 1991. Procedures for the quality selection of aromatic rice varieties. Int. Rice Res. Newsl. 16 (5), 10–11.

Ren, Z.H., et al., 2005. A rice quantitative trait locus for salt tolerance encodes a sodium transporter. Nat. Genet. 37, 1141–1146.

Renganayaki, K., Fritz, A.K., Sadasivam, S., et al., 2002. Mapping and progress toward map-based cloning of brown planthopper biotype-4 resistance gene introgressed from *Oryza officinalis* into cultivated rice, *O. sativa*. Crop Sci. 42, 2112–2117.

Renuka, D.R., Arumughan, C., 2007. Antiradical efficacy of phytochemical extracts from defatted rice bran. Food Chem. Toxicol. 45 (10), 2014–2021.

Ruan, H.H., Yan, C.Q., An, D.R., Liu, R.H., Chen, J.P., 2008. Identifying and mapping new gene *xa32(t)* for resistance to bacterial blight (*Xanthomonas oryzae* pv. *oryzae*, *Xoo*) from *Oryzae meyeriana* L. Acta Agric. Borealioccidentalis Sin. 17, 170–174.

Sabouri, H., 2009. QTL detection of rice grain quality traits by microsatellite markers using an *indica* rice (*Oryza sativa* L.) combination. Indian Acad. Sci. 88, 81–85.

Sakaguchi, S., 1967. Bull. Natl. Inst. Agric. Sci. Ser. D16:1–18.(In Japanese with English summary).

Sakamoto, T., Matsuoka, M., 2008. Identifying and exploiting grain yield genes in rice. Curr. Opin. Plant Biol. 11, 209–223.

Sakthivel, K., Sundaram, R.M., Rani, N.S., Balachandran, S.M., Neereja, C.N., 2009. Genetic and molecular basis of fragrance in rice. Biotechnol. Adv. 27, 468–473.

Salem, M.A., Kakani, V.G., Koti, S., Reddy, K.R., 2007. Pollen-based screening of soybean germplasm for high temperatures. Crop Sci. 47, 219–231.

Sama, V.S.A.K., 2011. Identification, tagging and mapping of new resistance gene(s) against the Asian rice gall midge *Orseolia oryzae* in rice varieties. Ph.D. Thesis. Department of Genetics, Osmania University, Hyderabad (107 p.).

Sama, V.S.A.K., Himabindu, K., Bhaskar Naik, S., Sundaram, R.M., Viraktamath, B.C., Bentur, J.S., 2012. Mapping and marker-assisted breeding of a gene allelic to the major Asian rice gall midge resistance gene *Gm8*. Euphytica 187, 393–400.

Sánchez-Calderón, L., López-Bucio, J., Chacón-López, A., Cruz-Ramírez, A., Nieto-Jacobo, F., Dubrovsky, J.G., Herrera-Estrella, L., 2005. Phosphate starvation induces a determinate developmental program in the roots of *Arabidopsis thaliana*. Plant Cell Physiol. 46, 174–184.

Sandhu, N., Torres, R.O., Cruz, M.T., Maturan, P.C., Jain, R., Kumar, A., Henry, A., 2015. Traits and QTLs for development of dry direct-seeded rainfed rice varieties. J. Exp. Bot. 21, e413.

Sankar, P.D., Saleh, M.A., Selvaraj, C.I., 2011. Rice breeding for salt tolerance. Res. Biotechnol. 2, 1–10.

Sardesai, N., Kumar, A., Rajyashri, K.R., Nair, S., Mohan, M., 2002. Identification and mapping of an AFLP marker linked to *Gm7*, a gall midge resistance gene and its conversion to a SCAR marker for its utility in marker-aided selection in rice. Theor. Appl. Genet. 105, 691–698.

Sarkar, R.K., Panda, D., Reddy, J.N., Patnaik, S.S.C., Mackill, D.J., Ismail, A.M., 2009. Performance of submergence tolerant rice genotypes carrying the *Sub1* QTL under stressed and non-stressed natural field conditions. Indian J. Agric. Sci. 79, 876–883.

Sato, H., Ando, I., Hirabayashi, H., Takeuchi, Y., Arase, S., Kihara, J., 2008. QTL analysis of brown spot resistance in rice (*Oryza sativa* L.). Breed. Sci. 58, 93–96.

Saunders, R., 1990. Rice bran and rice bran oil. Lipid Technol. 2 (3), 72–76.

Septiningsih, E.M., Pamplona, A.M., Sanchez, D.L., Neeraja, C.N., Vergara, G.V., Heuer, S., Ismail, A.M., Mackill, D.J., 2009. Development of submergence tolerant rice cultivars: the *Sub1* locus and beyond. Ann. Bot. 103, 151–160.

Septiningsih, E.M., Trijatmiko, K.R., Moeljopawiro, S., McCouch, S.R., 2003. Identification of quantitative trait loci for grain quality in an advanced backcross population derived from the *Oryza sativa* variety IR64 and the wild relative *O. rufipogon*. Theor. Appl. Genet. 107, 1433–1441.

Shabbir, M.A., Anjum, F.M., Zahoor, T., Nawaz, H., 2008. Mineral and pasting characterization of *indica* rice varieties with different milling fractions. Int. J. Agric. Biol. 10 (5), 556–560.

Shahbaz, M., Ashraf, M., 2013. Improving salinity tolerance in cereals. Crit. Rev. Plant Sci. 32, 237–249.

Shamsudin, N.A.A., Swamy, B.P.M., Ratnam, W., et al., 2016. Pyramiding of drought yield QTLs into a high quality Malaysian rice cultivar MRQ74 improves yield under reproductive stage drought. Rice 9, 21.

Shan, L., Dittert, K., Tao, H., Kreye, C., Xu, Y., Shen, Q., Fan, X., Sattelmacher, B., 2002. The ground cover rice production system (GCRPS): a successful new approach to save water and increase nitrogen fertilizer efficiency? In: Bouman, B.A.M., Hengsdijk, H., Hardy, B., Bindraban, P.S., Tuong, T.P., Ladha, J.K. (Eds.), Water-Wise Rice Production. International Rice Research Institute, Los Baños, Philippines, pp. 187–196.

Shang, J., Tao, Y., Chen, X., Zou, Y., Lei, C., Wang, J., Li, X., Zhao, X., Zhang, M., Lu, Z., Xu, J., Cheng, Z., Wan, J., Zhu, L., 2009. Identification of a new rice blast resistance gene *Pid3* by genome-wide comparison of paired nucleotide binding site-leucine-rich repeat genes and their pseudogene alleles between the two sequenced rice genomes. Genetics 182, 1303–1311.

Shannon, M.C., McCreight, J.D., Draper, J.H., 1985. Screening tests for salt tolerance in lettuce. J. Am. Soc. Hortic. Sci. 108, 225–230.

Shao, Y., Jin, L., Zhang, G., Bao, J., 2011. Association mapping of grain color, phenolic content, flavonoid content and antioxidant capacity in dehulled rice. Theor. Appl. Genet. 122, 1005–1016.

Sharma, T., Madhav, M., Jana, T., Singh, A., Gaikwad, K., Upreti, H., 2005. High-resolution mapping, cloning and molecular characterization of the *Pi-kh* gene of rice, which confers resistance to *Magnaporthe grisea*. Mol. Genet. Genomics 274, 569–578.

Shen, Y., Jin, L., Xiao, P., Lu, Y., Bao, J.S., 2009. Total phenolics, flavonoids, antioxidant capacity in rice grain and their relations to grain color, size and weight. J. Cereal Sci. 49, 106–111.

Shikari, A.B., Khanna, A., Krishnan, S.G., Singh, U.D., Rathour, R., Tonapi, V., Sharma, T.R., Nagarajan, M., Prabhu, K.V., Singh, A.K., 2013. Molecular analysis and phenotypic validation of blast resistance genes *Pita* and *Pita2* in landraces of rice (*Oryza sativa* L.). Indian J. Genet. 73 (2), 131–141.

Shimizu, A., Kato, K., Komatsu, A., Motomura, K., Ikehashi, H., 2008. Genetic analysis of root elongation induced by phosphorus deficiency in rice (*Oryza sativa* L.): fine QTL mapping and multivariate analysis of related traits. Theor. Appl. Genet. 117, 987–996.

Shimizu, A., Yanagihara, S., Kawasaki, S., Ikehashi, H., 2004. Phosphorus deficiency-induced root elongation and its QTL in rice (*Oryza sativa* L.). Theor. Appl. Genet. 109, 1361–1368.

Shobha Rani, N., Subba Rao, L.V., Viraktamath, B.C., 2006. National Guidelines for the Conduct of Tests for Distinctness, Uniformity and Stability: Rice (*Oryza sativa* L.). Directorate of Rice Research, Rajendranagar, Hyderabad, India.

Shrivastava, M.N., Kumar, A., Bhandarkar, S., Shukla, B.C., Agrawal, K.C., 2002. A new gene for resistance to Asian rice gall midge (*Orseolia oryzae* Wood Mason) biotype 1 population at Raipur. Euphytica 130 (1), 143–145.

Shrivastava, M.N., Kumar, A., Shrivastava, S.K., Sahu, R.K., 1993. A new gene for resistance to gall midge in rice variety Abhaya. Rice Genet. Newsl. 10, 79–80.

Sidhu, G.S., Khush, G.S., Mew, T.W., 1978. Genetic analysis of bacterial blight resistance in seventy-four cultivars of rice, *Oryza sativa* L. Theor. Appl. Genet. 53 (3), 105–111.

Singh, A.K., Choudury, B.U., Bouman, B.A.M., 2002. The effect of rice establishment techniques and water management on crop-water relations. In: Bouman, B.A.M., Hengsdijk, H., Hardy, B., Bindraban, P.S., Tuong, T.P., Ladha, J.K. (Eds.), Water-Wise Rice Production. International Rice Research Institute, Los Baños, Philippines, pp. 237–244.

Singh, U.S., Dar, M.H., Singh, S., Zaidi, N.W., Bari, M.A., Mackill, D.J., Collard, B.C.Y., Singh, V.N., Singh, J.P., Reddy, J.N., Singh, R.K., Ismail, A.M., 2013. Field performance, dissemination, impact and tracking of submergence tolerant (Sub1) rice varieties in South Asia. SABRAO J. Breed. Genet. 45, 112–131.

Song, W.Y., Pi, L.Y., Wang, G.L., Gardner, J., Hoisten, T., Ronald, P.C., 1997. Evolution of the rice *Xa21* disease resistance gene family. Plant Cell 9, 1279–1287.

Song, W.Y., Wang, G.L., Chen, L.L., Kim, H.S., Pi, L.Y., Holsten, T., Gardner, J., Wang, B., Zhai, W.X., Zhu, L.H., Ronald, P.C., 1995. A receptor kinase-like protein encoded by the rice disease resistance gene, *Xa21*. Science 270, 1804–1806.

Stansel, J.W., 1966. The influence of heredity and environment on endosperm characteristics of rice. Diss. Abstr. 27, 48B.

Stoop, W., Uphoff, N., Kassam, A., 2002. A review of agricultural research issues raised by the system of rice intensification (SRI) from Madagascar: opportunities for improving farming systems for resource-poor farmers. Agric. Syst. 71, 249–274.

Su, C.C., Cheng, X.N., Zhai, H.Q., Wan, J.M., 2002. Detection and analysis of QTL for resistance to brown planthopper, *Nilaparvata lugens* (Stål), in rice (*Oryza sativa* L.) using backcross inbred lines. Acta Genet. Sin. 29, 332–338.

Su, C.C., Wan, J., Zhai, H.Q., Wang, C.M., Sun, L.H., Yasui, H., Yoshimura, A., 2005. A new locus for resistance to brown planthopper identified in *indica* rice variety DV85. Plant Breed. 124, 93–95.

Su, C.C., Zhai, H.Q., Wang, C.M., et al., 2006. SSR mapping of brown planthopper resistance gene *Bph9* in Kaharamana, an *indica* rice (*Oryza sativa* L.). Acta Genet. Sin. 33, 262–268.

Suh, J., Jeung, J., Lee, J., Choi, Y., Yea, J., Virk, P., Mackill, D., Jena, K., 2010. Identification and analysis of QTLs controlling cold tolerance at the reproductive stage and validation of effective QTLs in cold-tolerant genotypes of rice (*Oryza sativa* L.). Theor. Appl. Genet. 120 (5), 985–995.

Suh, J., Lee, C., Lee, J., Kim, J., Kim, S., Cho, Y., Park, S., Shin, J., Kim, Y., Jena, K., 2012. Identification of quantitative trait loci for seedling cold tolerance using RILs derived from a cross between *japonica* and tropical *japonica* rice cultivars. Euphytica 184 (1), 101–108.

Sun, X., Cao, Y., Yang, Z., Xu, C., Lie, X., Wang, S., Zhang, Q., 2004. *Xa26*, a gene conferring resistance to *Xanthomonas oryzae* pv. *oryzae* in rice, encodes an LRR receptor kinase-like protein. Plant J. 37, 517–527.

Sun, L., Su, C., Wang, C., et al., 2005. Mapping of a major resistance gene to brown planthopper in the rice cultivar Rathu Heenati. Breed Sci. 55, 391–396.

Sundaram, R.M., 2007. Fine mapping of rice gall midge resistance genes *Gm1* and *Gm2* and validation of the linked markers. Ph.D. Thesis. Department of Plant Sciences, University of Hyderabad, Hyderabad, India (123 p.).

Swain, D.K., Herath, S., Pathirana, A., Mittra, B.N., 2005. Rainfed lowland and flood-prone rice: a critical review on ecology and management technology for improving the productivity in Asian rainfed lowland and flood-prone rice. Available from: www.mekongnet.org/images/e/e9/Dillip.pdf

Swamy, B.P.M., Kaladhar, K., Anuradha, K., Batchu, A.K., Longvah, T., Viraktamath, B.C., Sarla, N., 2011. Enhancing iron and zinc concentration in rice grains using wild species. ADNAT Convention and International Symposium on Genomics and Biodiversity. 23–25 February, 2011, CCMB, Hyderabad, India, p. 71.

Swamy, B.P.M., Ahmed, H.U., Henry, A., Dixit, S., Vikram, P., Ram, T., Verulkar, S.V., Perraju, P., Mandal, N.P., Variar, M., Mishra, K.K., Lata, T.A., Karmakar, B., Mauleon, R., Satoh, K., Moumeni, A., Kikuchi, S., Leung, H., Kumar, A., 2013. Genetic, physiological, and gene expression analyses reveal multiple QTLs that enhance yield of rice mega-variety IR64 under drought. PLoS One 8, e62795.

Szabados, L., Savouré, A., 2009. Proline: a multifunctional amino acid. Trends Plant Sci. 15, 89–97.

Tabbal, D.F., Bouman, B.A.M., Bhuiyan, S.I., Sibayan, E.B., Sattar, M.A., 2002. On-farm strategies for reducing water input in irrigated rice: case studies in the Philippines. Agric. Water Manage. 56 (2), 93–112.

Takahashi, A., Hayashi, N., Miyao, A., Hirochika, H., 2010. Unique features of the rice blast resistance *Pish* locus revealed by large scale retrotransposon-tagging. BMC Plant Biol. 10, 175.

Takita, T., 1996. A new dominant gene for brown planthopper resistance found in an improved Japanese rice strain. Breed Sci. 46, 207–211.

Tamura, Y., Hattori, M., Yoshioka, H., et al., 2014. Map-based cloning and characterization of a brown planthopper resistance gene *Bph26* from *Oryza sativa* L. ssp. *indica* cultivar ADR52. Sci. Rep. 4, 58–72.

Tan, Y.F., Li, J.X., Yu, S.B., Xing, Y.Z., Xu, C.G., Zhang, Q.F., 1999. The three important traits for cooking and eating quality of rice grains are controlled by a single locus in an elite rice hybrid Shanyou 63. Theor. Appl. Genet. 99, 642–648.

Tan, G.X., Ren, X., Weng, Q.M., Shi, Z.Y., Zhu, L.L., He, G.C., 2004. Mapping of a new resistance gene to bacterial blight in rice line introgressed from *Oryza officinalis*. Yi Chuan Xue Bao 31, 724–729, (in Chinese).

Tang, J., Zhu, X., Wang, Y., Liu, L., Xu, B., Li, F., Fang, J., Chu, C., 2011. Semi-dominant mutations in the CC-NB-LRR-type R gene *Nls1* lead to constitutive activation of defense responses in rice. Plant J. 66, 996–1007.

Taura, S., Ogawa, T., Yoshimura, A., Ikeda, R., Iwata, N., 1992. Identification of a recessive resistance gene to rice bacterial blight of mutant line XM6, *Oryza sativa* L. Jpn. J. Breed. 42, 7–13.

Thomson, M.J., de Ocampo, M., Egdane, J., Rahman, M.A., Sajise, A.G., Adorada, D.L., Tumimbang-Raiz, E., Blumwald, E., Seraj, Z.I., Singh, R.K., Gregorio, G.B., 2010. Characterizing the saltol quantitative trait locus for salinity tolerance in rice. Rice 3 (2–3), 148–160.

Ticconi, C.A., Delatorre, C.A., Lahner, B., Salt, D.E., Abel, S., 2004. *Arabidopsis pdr2* reveals a phosphate-sensitive checkpoint in root development. Plant J. 37, 801–814.

Tirado, R., Allsopp, M., 2012. Phosphorus in Agriculture: Problems and Solutions. Greenpeace Research Laboratories; Greenpeace International, Amsterdam, The Netherlands.

Tomar, J.B., Prasad, S.C., 1997. Genetic analysis of aroma in rice. Oryza 34, 3191–3195.

UN News Centre, UN News Service, 2009. Majority of world population faces water shortages unless action taken, warns Migiro. Available from: www.un.org/apps/news/story.asp?News ID= 29796&Cr=water&Cr1=agriculture

United Nations Environment Programme (UNEP), 2008. An overview of the state of the world's fresh and marine waters. Annual Report, United Nations Environment Program 2008, second ed. Available from: www.unep.org/dewa/vitalwater/article186.htm; www.unep.org

Van Kauwenbergh, S.J., 2010. World phosphate rock reserves and resources. International Fertilizer Development Center. Available from: www.ifdc.org

Vasconcelos, M., Datta, K., Oliva, N., Khalekuzzaman, M., Torrizo, L., Krishnan, S., Oliveira, M., Goto, F., Datta, S.K., 2003. Enhanced iron and zinc accumulation in transgenic rice with the ferritin gene. Plant Sci. 6, 371–378.

Venuprasad, R., Bool, M.E., Quiatchon, L., Sta Cruz, M.T., Amante, M., Atlin, G.N., 2012. A large-effect QTL for rice grain yield under upland drought stress on chromosome 1. Mol. Breeding 30, 535–547.

Venuprasad, R., Dalid, C.O., Marilyn, D.V., Bool, M.E., Zhao, D., Espiritu, M., Sta Cruz, M.T., Amante, M., Kumar, A., Atlin, G.N., 2009. Identification and characterization of large-effect quantitative trait loci (QTL) for grain yield under lowland drought stress and aerobic conditions in rice using bulk-segregant analysis (BSA). Theor. Appl. Genet. 120, 177–190.

Venuprasad, R., Lafitte, H.R., Atlin, G.N., 2007. Response to direct selection for grain yield under drought stress in rice. Crop Sci. 47, 285–293.

Venuprasad, R., Sta-Cruz, M.T., Amante, M., Magbanua, R., Kumar, A., Atlin, G.N., 2008. Response to two cycles of divergent selection for grain yield under drought stress in four rice breeding populations. Field Crops Res. 107, 232–244.

Vergara, B.S., Mazaredo, A.M., 1975. Screening for resistance to submergence under greenhouse conditions. In: Proceedings of the International Seminar on Deepwater Rice. Bangladesh Rice Research Institute, Dhaka, Bangladesh, pp. 67–70.

Vikal, Y., Chawla, H., Sharma, R., Lore, J.S., Singh, K., 2014. Mapping of bacterial blight resistance gene *xa8* in rice (*Oryza sativa* L.). Indian J. Genet. Plant Breed. 74 (4s), 589–595.

Vikram, P., Swamy, B., Dixit, S., Ahmed, H., Cruz, M.T.S., Singh, A., Kumar, A., 2011. *qDTY1.1*, a major QTL for rice grain yield under reproductive-stage drought stress with a consistent effect in multiple elite genetic backgrounds. BMC Genet. 12, 89.

Virmani, S.S., 2003. Advances in hybrid rice research and development in the tropics. In: Virmani, S.S., Mao, C.X., Hardy, B. (Eds.), Hybrid Rice for Food Security, Poverty Alleviation, and Environmental Protection. International Rice Research Institute, Manila, Philippines, pp. 7–20.

Vishnu, V.N., Robin, S., Sudhakar, D., Rajeswari, S., Raveendran, M., Subramanian, K.S., Shalini, A.P., 2014. Genotypic variation for micronutrient content in traditional and improved rice lines and its role in biofortification programme. Indian J. Sci. Technol. 7 (9), 1414–1425.

Wan, X.Y., Weng, J.F., Zhai, H.Q., Wang, J.K., Lei, C.L., Liu, X.L., Guo, T., Jiang, L., Su, N., Wan, J.M., 2008. Quantitative trait loci (QTL) analysis for rice grain width and fine mapping of an identified QTL allele *gw-5* in a recombination hotspot region on chromosome 5. Genetics 179, 2239–2252.

Wang, G.L., Leung, H., 1998. Molecular biology of host-pathogen interactions in rice diseases. In: Shimamoto, K. (Ed.), Molecular Biology of Rice. Springer, Tokyo, pp. 201–232.

Wang, Y., Cao, L.M., Zhang, Y.X., et al., 2015. Map-based cloning and characterization of *Bph29*, a B3 domain-containing recessive gene conferring brown planthopper resistance in rice. J. Exp. Bot. 66, 6035–6045.

Wang, D., Guo, C., Huang, J., Yang, S., Tian, D., Zhang, X., 2014. Allele-mining of rice blast resistance genes at AC134922 locus. Biochem. Biophys. Res. Commun. 446, 1085–1090.

Wang, Y., Wang, D., Deng, X., et al., 2012. Molecular mapping of the blast resistance genes *Pi2-1* and *Pi51(t)* in the durably resistant rice Tianjingyeshengdao. Phytopathology 102, 779–786.

Wang, C.T., Wen, G.S., Lin, X.H., Liu, X.Q., Zhang, D.P., 2009. Identification and fine mapping of the new bacterial blight resistance gene, *Xa31*(t), in rice. Eur. J. Plant Pathol. 123, 235–240.

Wang, Z.X., Yano, M., Yamanouchi, U., Iwamoto, M., Monna, L., Hayasaka, H., Katayose, Y., Sasaki, T., 1999. The *Pib* gene for rice blast resistance belongs to the nucleotide-binding and leucine-rich repeat class of plant disease resistance genes. Plant J. 19, 55–64.

Wang, Z.Y., Zheng, F.Q., Shen, G.Z., Zhang, J.L., Hong, M.M., 1995. The amylose content in rice endosperm is related to the post-transcriptional regulation of the waxy gene. Plant J. 7 (4), 613–622.

Weerakoon, W.M., Maruyama, A., Ohba, K., 2008. Impact of humidity on temperature-induced grain sterility in rice (*Oryza sativa* L.). J. Agron. Crop Sci. 194, 135–140.

Wissuwa, M., Yano, M., Ae, N., 1998. Mapping of QTLs for phosphorus-deficiency tolerance in rice (*Oryza sativa* L.). Theor. Appl. Genet. 97, 777–783.

Wissuwa, M., Wegner, J., Ae, N., Yano, M., 2002. Substitution mapping of *Pup1*: a major QTL increasing phosphorus uptake of rice from a phosphorus-deficient soil. Theor. Appl. Genet. 105, 890–897.

Wongpornchai, S., Dumri, K., Jongkaewwattana, S., Siri, B., 2004. Effects of drying methods and storage time on the aroma and milling quality of rice (*Oryza sativa* L.) cv. Khao Dawk Mali 105. Food Chem. 87, 407–414.

World Water Organization, 2010. Water facts and water stories from across the globe. Available from: www.theworldwater.org/water_facts.php

Wu, H., Liu, Y., He, J., Liu, Y., Jiang, L., Liu, L., Wang, C., Cheng, X., Wan, J., 2014. Fine mapping of brown planthopper (*Nilaparvata lugens* Stål) resistance gene *Bph28(t)* in rice (*Oryza sativa* L.). Mol. Breeding 33, 909–918.

World Water Assessment Programme (WWAP), 2009. Chapter 2: Water in a changing world. UN Water Development Report 3. Available from: www.unesco.org/water/wwap/wwdr/wwdr3/pdf/WWDR3_Facts_and_Figures.pdf4

Xiao, J., Grandillo, S., Ahn, S.N., McCouch, S.R., Tanksley, S.D., Li, Z., Yuan, L., 1996. Genes from wild rice improve yield. Nature 384, 1223–1224.

Xing, Y., Yang, Q., 2010. Genetic and molecular bases of rice yield. Annu. Rev. Plant Biol. 61, 421–442.

Xu, K., Mackill, D.J., 1996. A major locus for submergence tolerance mapped on rice chromosome 9. Mol. Breeding 2, 219–224.

Xu, J.H., Messing, J., 2009. Amplification of prolamin storage protein genes in different subfamilies of the Poaceae. Theor. Appl. Genet. 119, 1397–1412.

Xu, Z.J., Chen, W.F., Ma, D.R., Lu, Y.N., Zhou, S.Q., Liu, L.X., 2004. Correlations between rice grain shapes and main qualitative characteristics. Acta Agron. Sin. 30, 894–900.

Xu, X.F., Mei, H.W., Luo, L.J., Cheng, X.N., Li, Z.K., 2002. RFLP-facilitated investigation of the quantitative resistance of rice to brown planthopper (*Nilaparvata lugens*). Theor. Appl. Genet. 104, 248–253.

Xu, K., Xu, X., Fukao, T., Canlas, P., Maghirang-Rodriguez, R., Heuer, S., Ismail, A.M., Bailey-Serres, J., Ronald, P.C., Mackill, D.J., 2006. *Sub1A* is an ethylene-response factor-like gene that confers submergence tolerance to rice. Nature 442, 705–708.

Yamano, T., Baruah, S., Sharma, R., Kumar, A., 2013. Factors affecting the adoption of direct-seeded rice in the northeastern Indo-Gangetic Plain. Report. Cereal Systems Initiative for South Asia.

Yang, J., Huang, D., Duan, H., Tan, G., Zhang, J., 2009. Alternate wetting and moderate soil drying increases grain yield and reduces cadmium accumulation in rice grains. J. Sci. Food Agric. 89, 1728–1736.

Yang, D.C., Parco, A., Nandi, S., Zhu, Y.G., Wang, G.L., Huang, L., 1997. Construction of a bacterial artificial chromosome (BAC) library and identification of overlapping BAC clones with chromosome 4-specific RFLP markers in rice. Theor. Appl. Genet. 98, 1147–1154.

Yawadio, R., Tanimori, S., Morita, N., 2007. Identification of phenolic compounds isolated from pigmented rices and their aldose reductase inhibitory activities. Food Chem. 101, 1644–1653.

Ye, C., Argayoso, M.A., Redoña, E.D., Sierra, S.N., Laza, M.A., Dilla, C.J., 2012. Mapping QTL for heat tolerance at flowering stage in rice using SNP markers. Plant Breed. 131 (1), 33–41.

Ye, C., Tenorio, F.A., Argayoso, M.A., Laza, M.A., Koh, H., Redoña, E.D., Jagadish, K.S.V., Gregorio, G., 2015. Identifying and confirming quantitative trait loci associated with heat tolerance at flowering stage in different rice populations. BMC Genet. 16, 41.

Yeo, A.R., Flowers, T.J., 1986. Salinity resistance in rice (*Oryza sativa* L.) and a pyramiding approach to breeding varieties for saline soils. Aust. J. Plant Physiol. 13 (1), 161–173.

Yi, M., Than, K., Vanavichit, A., Chai-arree, W., Toojinda, T., 2009. Marker assisted backcross breeding to improve cooking quality traits in Myanmar rice cultivar Manawthukha. Field Crops Res. 113, 178–186.

Yoshida, S., 1981. Fundamentals of Rice Crop Science. International Rice Research Institute; CAB International, Los Baños, Philippines; Wallingford, United Kingdom, (269 p.).

Yoshida, S., Parao, F.T., 1976. Climatic influence on yield and yield components of lowland rice in the tropics. International Rice Research Institute, Los Baños, Philippines, (pp. 471–494).

Yoshihashi, T., Kabaki, N., Nguyen, T.T.H., Inatomi, H., 2002. Formation of flavor compound in aromatic rice and its fluctuations with drought stress. Res. Highlights JIRCAS, 32–33.

Yoshimura, A., Mew, T.W., Khush, G.S., Moura, T., 1983. Inheritance of resistance to bacterial blight in rice cultivar Cas 209. Phytopathology 73, 1409–1412.

Yoshimura, S., Yamanouchi, U., Katayose, Y., Toki, S., Wang, Z., Kono, I., Kurata, N., Yano, M., Iwata, N., Sasaki, T., 1998. Expression of *Xa1*, a bacterial blight resistance gene in rice, is induced by bacterial inoculation. Proc. Natl. Acad. Sci. USA 95, 1663–1668.

Yu, Z.H., Mackill, D.J., Bonman, J.M., Tanksley, S.D., 1991. Tagging genes for blast resistance in rice via linkage to RFLP markers. Theor. Appl. Genet. 81, 471–476.

Yuan, L., 2001. Breeding of super hybrid rice. In: Peng, S., Hardy, B. (Eds.), Rice Research for Food Security and Poverty Alleviation. International Rice Research Institute, Los Baños, Philippines, pp. 143–149.

Yun, B., Kim, M., Handoyo, T., Kim, K., 2014. Analysis of rice grain quality-associated quantitative trait loci by using genetic mapping. Am. J. Plant Sci. 5, 1125–1132.

Zhai, C., Lin, F., Dong, Z., He, X., Yuan, B., et al., 2011. The isolation and characterization of *Pik*, a rice blast resistance gene which emerged after rice domestication. New Phytol. 189, 321–334.

Zhang, Q., 2007. Strategies for developing green super rice. Proc. Natl. Acad. Sci. USA 104, 16402–16409.

Zhang, G.H., Li, S.Y., Wang, L., et al., 2014. LSCHL4 from *japonica* cultivar, which is allelic to NAL1, increases yield of *indica* super rice. Mol. Plant. 7 (8), 1350–1364.

Zhang, H., Xue, Y., Wang, Z., Yang, J., Zhang, J., 2009. An alternate and moderate soil drying regime improves root and shoot growth in rice. Crop Sci. 49, 2246–2260.

Zhang, Y.D., Zhang, Y.H., Zhen, Z., et al., 2013. QTL mapping for grain size traits based on extra-large grain rice line TD70. Rice Sci. 20 (6), 400–406.

Zhao, L., Zhao, C.F., Zhou, L.H., Matsui, T., Wang, C.L., 2016. QTL mapping of dehiscence length at the basal part of thecae related to heat tolerance of rice (*Oryza sativa* L.). Euphytica 209, 715.

Zheng, Z., Li, M., Fang, Y., Liu, F., Lu, Y., Meng, Q., et al., 2012. Diversification of the *Waxy* gene is closely related to variations in rice eating and cooking quality. Plant Mol. Biol. Rep. 30, 462–469.

Zheng, C.K., Wang, C.L., Yu, Y.J., Liang, Y.T., Zhao, K.J., 2009. Identification and molecular mapping of *Xa32*(t), a novel resistance gene for bacterial blight (*Xanthomonas oryzae* pv. *oryzae*) in rice. Acta Agron. Sin. 35, 1173–1180.

Zheng, T.Q., Xu, J.L., Li, Z.K., Zhai, H.Q., Wan, J.M., 2007. Genomic regions associated with milling quality and grain shape identified in a set of random introgression lines of rice (*Oryza sativa* L.). Plant Breed. 126, 158–163.

Zhong, R., Lee, C., McCarthy, R.L., Reeves, C.K., Jones, E.G., Ye, Z.H., 2011. Transcriptional activation of secondary wall biosynthesis by rice and maize NAC and MYB transcription factors. Plant Cell Physiol. 52, 1856–1871.

Zhou, B., Qu, S., Liu, G., Dolan, M., Sakai, H., Lu, G., Bellizzi, M., Wang, G.L., 2006. The eight amino-acid differences within three leucine-rich repeats between Pi2 and Piz-t resistance proteins determine the resistance specificity to *Magnaporthe grisea*. Mol. Plant-Microbe Interact. 19, 1216–1228.

Zhu, X., Gong, H., Chen, G., Wang, S., Zhang, C., 2005. Different solute levels in two spring wheat cultivars induced by progressive field water stress at different developmental stages. J. Arid Environ. 62, 1–14.

FURTHER READING

Fuller, D.Q., Qin, L., Zheng, Y., Zhao, Z., Chen, X., Hosoya, L.A., Sun, G.P., 2009. The domestication process and domestication rate in rice: spikelet bases from the Lower Yangtze. Science 323, 1607–1610.

Hasegawa, T., Kuwagata, T., Nishimori, M., Ishigooka, Y., Murakami, M., Yoshimoto, M., Kondo, M., Ishimaru, T., Sawano, S., Masaki, Y., Matsuzaki, H., 2009. In: Hasegawa, T., Sakai, H. (Eds.). Recent warming trends and rice growth and yield in Japan. Proceeding of the MARCO Symposium. National Institute for Agro-Environmental Sciences, Tsukuba, Japan (pp. 44–51).

Liu, B., Zhang, S.H., Zhu, X.Y., Yang, Q.Y., Wu, S.Z., Mei, M.T., Mauleon, R., Leach, J., Mew, T., Leung, H., 2004. Candidate defense genes as predictors of quantitative blast resistance in rice. Mol. Plant-Microbe Interact. 17, 1146–1152.

Yu, J., Pressoir, G., Briggs, H.W., Yamasaki, M., Doebley, J.F., McMullen, M.D., Gaut, B.S., Nielsen, D.M., Holland, J.B., Kresovich, S., Bucker, E.S., 2006. A unified mixed-model method for association mapping that accounts for multiple levels of relatedness. Nat. Genet. 2, 203–208.

Chapter 8

Growing Rice in Eastern India: New Paradigms of Risk Reduction and Improving Productivity

Sudhir-Yadav*, Virender Kumar*, Sudhanshu Singh*, Rapolu M. Kumar**,
Sheetal Sharma*, Rahul Tripathi†, Amaresh K. Nayak†, J.K. Ladha*
*International Rice Research Institute, Los Baños, Philippines; **Indian Institute of Rice
Research, ICAR, Hyderabad, Telangana, India; †National Rice Research Institute, ICAR, Cuttack,
Odisha, India

BACKGROUND

Rice plays a critical role in the food security and economic growth of India. With 44 million hectares of land and 107 million tons of production (2013–14) (GOI, 2014), India has the largest area under rice worldwide and is the second-largest producer after China. Rice has a central role in culture, diet, and employment, and is considered as an instrumental crop that changed India's status from food deficient to major grain exporter.

Manual transplanting of rice seedlings is the most common method of rice establishment in India (Pandey and Velasco, 2002). Before transplanting, the soil in the main field is puddled (churning of soil under saturated conditions) to minimize water loss by percolation, destroy weeds, and soften the soil for transplanting (Adachi, 1992; Singh et al., 1995). After transplanting, fields are normally kept flooded until shortly before harvest. Other traditional methods of direct seeding are also prevalent in most of the eastern Indian states, for instance, "*beushening*" is performed in Odisha and broadcasting in Bihar. *Beushening* is practiced in rainfed environments (lowland and medium land) where farmers broadcast a high amount of seeds (up to 150 kg/ha) before the onset of the monsoon season. The rice seed germinates in the presence of rain and after about 1 month, when 15–20 cm of rainwater are held in rice fields, crossplowing of the standing rice crop is performed, followed by laddering and seedling redistribution. This practice loosens the soil, suppresses weeds, reduces insect pests

The Future Rice Strategy for India. http://dx.doi.org/10.1016/B978-0-12-805374-4.00008-7

(Gautam et al., 2013), improves water-use efficiency, improves nutrient uptake, and allows for a better distribution of seedlings. Broadcasting of the seed prior to the monsoon facilitates the efficient use of early rain for crop establishment, while crossplowing the field ensures optimum control of weeds, insects, and rodents. The widespread availability of irrigation facilities for rice fields has encouraged many farmers to shift from traditional methods to puddled transplanted rice (PTR) even though it requires a large amount of water, labor, and energy. These resources, however, are becoming scarce and expensive, therefore threatening the sustainability and profitability of rice systems.

The production of rice and wheat in the northwest Indo-Gangetic plains of India is critical for the nation's food security yet the sustainability of the traditional production system is threatened by the rapid degradation of natural resources including the depletion of groundwater (Rodell et al., 2009), soil health concerns (Shukla et al., 2005), and environmental problems (Aggarwal et al., 2004). The productivity of rice in northwest India has remained stagnant (Ladha et al., 2003; Pathak et al., 2003) for the past decade, which has compelled the country to throw extra efforts into increasing production and productivity in eastern India. Flagship schemes such as the National Food Security Mission, Rashtriya Krishi Vikas Yojna, and Bringing Green Revolution to Eastern India (BGREI) have been rolled out in the last two five-year plans. These schemes aimed to improve rice productivity in eastern India with the hope of successfully meeting the challenges posed to food security by the fast-growing population, and they also aimed to reduce the production pressure on northwest India. These efforts have led to an average rice yield increase of more than 40%, with seven eastern Indian states (eastern Uttar Pradesh, Bihar, Odisha, Jharkhand, Chhattisgarh, West Bengal, and Assam) producing an additional 16 million tons of rice between 2004–05 and 2012–13 (Mohanty, 2015).

The sustainability of the rice sector growth rate is a prime concern for the country as it is relied upon to maintain food security and to support economic growth. At the current growth rate, the nation's rice exports could decrease by 2% in 2024 (International Baseline Data 2015). This projected loss could be compensated for by enhancing productivity to maintain the same growth rate in exports, by reducing production costs to maintain the same profit levels, or by using a combination of both of them. Thus, gains in productivity have to be achieved with reduced resources including land, labor, water, energy, and chemicals, and a lower environmental footprint. These emerging issues warrant a paradigm shift in the traditional rice-growing system being practiced in eastern India, giving way to more resource-efficient, capital-efficient, and profitable systems. These systems may include improved germplasm integration coupled with transformative agronomic management to enable sustainability of the system (Ladha et al., 2016).

In this chapter we describe (1) some of the key drivers that are triggering change in the way rice is cultivated and (2) various technological opportunities that have the potential to bring about the needed changes. The key technological opportunities include holistic agronomic interventions and mechanization solutions integrated with high-yielding rice varieties that are climate smart against various biotic and abiotic stresses. The major emphasis will be on exploring alternatives to the inefficient tillage and crop establishment practices currently used by farmers. We also discuss decision tools (fertilizer, weed, and water management) to guide large-scale dissemination. A futuristic outlook of rice management to achieve social, environmental, and economic sustainability with a geographic focus on northeastern India is also presented in this chapter.

MAJOR DRIVERS OF CHANGE IN RICE PRODUCTION TECHNOLOGIES

Rising Labor Scarcity and Labor Wages

Rising labor scarcity and associated increased wages are the basic and flagrant issues that call for a shift from conventional PTR to alternate crop establishment methods. Rapid economic growth in India has fueled the demand for labor in nonagricultural sectors, resulting in decreased availability of agricultural laborers. In the changing socioeconomic environment, youth have also indicated a preference for nonagricultural work over farming. Furthermore, the availability of migrant labor from eastern India has decreased considerably in northwest India after the introduction of policies to discourage en masse labor migration, such as "The Mahatma Gandhi National Rural Employment Guarantee Act (MNREGA)." This flagship scheme was introduced in 2005 and guarantees 100 days of paid unskilled work in a person's home village. All these factors have led to increasing labor scarcity in agriculture and rising labor wages in the past 2 decades. For example, minimum labor wages under MNREGA increased from INR 55 to 100/day in 2005–06 to INR 160 to 250/day in 2015–16 in the major rice-growing states of India, a 108%–224% increase in 10 years (Table 8.1). Factors such as the high labor demand at transplanting time and rising scarcity and rising labor wages are forcing farmers to shift from the labor-intensive method of crop establishment (i.e., PTR) to more labor-efficient methods (Kumar and Ladha, 2011).

Rising Water Scarcity

The second important factor driving the shift toward alternate rice crop establishment methods is rising water stress. Water is becoming an increasingly scarce resource in India and its share for agriculture has been declining steadily as competition from other nonagricultural sectors, including domestic and industrial use, increases. Additional evidence to support the observation of increasing water scarcity in India is the fast-depleting groundwater resources

TABLE 8.1 Minimum Labor Wages Under the MNREGA Scheme in the Major Rice-Producing States of India From 2005–06 to 2015–16

State	2005– 06	2011– 12	2012– 13	2013– 14	2014– 15	2015– 16	Increase in 10 years (%)
	Wage rate (INR/day)						
Haryana	95	179	191	214	236	251	164
Punjab	101	153	166	184	200	210	108
Bihar	68	120	122	138	158	162	138
Uttar Pradesh	58	120	125	142	156	161	178
Odisha	55	125	126	143	164	174	216
West Bengal	67	130	136	151	169	174	160
Karnataka	63	125	155	174	191	204	224
Andhra Pradesh	80	121	137	149	169	180	125
Tamil Nadu	80	119	132	148	167	183	129

(Humphreys et al., 2010; Kumar and Ladha, 2011). The groundwater table is falling in all major rice-growing states of India because of its heavy use for rice production, this is especially evident in Punjab, Haryana, Tamil Nadu, and Karnataka. In a study conducted by NASA, Rodell et al. (2009) reported a groundwater table decline rate of 0.33 m/year in northwestern India. Over a period of 6 years (2002–08), these authors estimated a net loss of 109 km^3 of groundwater; double the capacity of India's largest surface reservoir. Indiscriminate and unrestricted use of groundwater for irrigation, as evidenced from a fast increase in the number of tube wells over a short period of time, has created this alarming situation.

The puddling of soil requires a large amount of water (Fig. 8.1). Farmers keep the field flooded for the majority of the time and thus a large proportion of irrigation water is lost in the form of evaporation, seepage, and percolation. This is a nonproductive loss from the field. In India, seasonal water input in puddled transplanted systems varies depending on monsoon rain and soil type, ranging from 1566 mm in clay loam soil to 2262 mm in sandy loam soil, and from 1144 mm in Bihar to 1500–2800 mm in Haryana (Gathala et al., 2013; Gupta et al., 2002). This necessitates exploring new alternative establishment options that inherently require less water and are more efficient for long-term sustainability of irrigated rice-based systems.

FIGURE 8.1 **Drivers of change from puddled transplanted rice to improved crop establishment methods.** PTR, Puddled transplanted rice.

Economics

High input cost is another major reason why farmers are compelled to shift away from PTR. PTR is highly labor-, water-, and tillage-intensive, all of which are becoming scarce and expensive, thus making this method less profitable (Kumar and Ladha, 2011). For example, the labor cost for rice transplanting has increased by INR 1500–2000/ha in the past 6 years (Fig. 8.2).

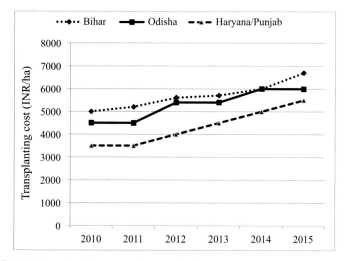

FIGURE 8.2 **Transplanting cost per hectare in Bihar, Odisha, and Haryana/Punjab from 2010 to 2015.**

PATHWAYS FOR IMPROVED SUSTAINABILITY OF RICE-BASED SYSTEMS

During the 1960s to the 1980s management work mainly focused on improving the productivity of irrigated systems through rice megavarieties, controlling the use of fertilizers, and reducing pest destruction. The long decades of production during the Green Revolution in and around the irrigated northwestern region of the country (Haryana, Punjab, and western Uttar Pradesh) had gradually started to show signs of a slump. The unsustainable use of natural resources (water) and indiscriminate use of external inputs (fertilizers and pesticides) were adversely affecting the total factor productivity and sustainability of wheat and rice in the region. A consensus among policymakers gradually started building up by the late 1990s on sustainability and the limited capacity of irrigated environments to meet the food demand of the growing population. This prompted policymakers to look toward the unfavorable regions of eastern India as the future food basket for the country. This realization finally culminated in the rolling out of flagship schemes (i.e., BGREI) by the Government of India in 2011. Focus was placed on rice growing in the eastern regions, which are water abundant, mostly rainfed, and not reliant on agricultural inputs. Although the Green Revolution in northwestern Indian states achieved self-sufficiency in food grain production, efforts in eastern India focused on sustainability by using resources wisely and attempting to overcome the constraints in rainfed rice production.

Until the early 20th century, crop and nutrient management recommendations for unfavorable regions of eastern India were generic and based on broader characterization of the environment (rainfed upland, rainfed lowland, irrigated, etc.). For instance, blanket crop and nutrient management recommendations for improved varieties were available in rainfed lowland but they were produced without any consideration for the different classes of prevailing stress-prone subsystems (drought, submergence, and salinity, etc.). More importantly, farmers in these regions continued to use traditional varieties that suited their local needs. The first change, which happened in unfavorable regions, was to understand the different stresses present in rice fields (drought, submergence, semideep water, or deepwater) and the importance of the appropriate variety and matching crop and nutrient management recommendations for each domain. The first breakthrough of higher yield in unfavorable regions came through the development of stress-tolerant rice varieties (STRVs) for each category of field, which has been further improved through the development and adaptation of variety (type)-specific best management practices (Fig. 8.3).

Quantifying and semiquantifying rice environments in unfavorable regions will become easier as the use of geographic information systems, better weather forecasts, and hydrology and crop models allow us to better understand the predictability of the environment (i.e., understanding of stress domains). Until the predictability of rice environments such as drought, flood, salinity, pests, and diseases improves at the farm scale, STRVs will remain the main option to

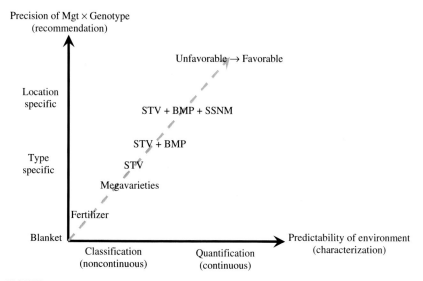

FIGURE 8.3 Target setting and optimizing system management in unfavorable environments. *BMP*, Best management practices; *CA*, conservation agriculture; *SSNM*, site-specific nutrient management; *STV*, stress-tolerant varieties; *x*-axis, the degree of understanding of the stress domain; *y*-axis, the degree of precision in management-variety combination.

buffer against hazardous weather. Reducing risk by using STRVs has a positive effect on the adoption of more labor-intensive establishment methods, more appropriate fertilizer usage, conservation agriculture practices, and mechanization (Emerick et al., 2016). A large share of the expected yield gains in unfavorable environments would come from combining STRVs with investments in management (Dar et al., 2013). Thus, a transition from blanket to location-specific recommendations using a suitable tolerant variety in a properly characterized stress-prone field could ultimately provide yield in unfavorable environments close to that of favorable environments. Obtaining sustainability in rice production in unfavorable environments should rely on this principle.

Stress-Tolerant Varieties and Conforming Management Practices

Considerable progress has been made in developing rice varieties with increased tolerance of submergence, drought, and salinity stress (Table 8.2). For flood-affected areas, the submergence-tolerance gene *SUB1* was identified, fine-mapped, and transferred into several Indian rice megavarieties such as Swarna (MTU7029), Samba Mahsuri (BPT5204), IR64, and CR1009 using marker-assisted backcross (MABC) approaches to enhance conventional plant breeding (Septiningsih et al., 2009; Xu et al., 2006). *SUB1*-containing modern varieties are identical to the original varieties in nearly all traits. The survival of Sub1 lines is substantially higher than that of non-Sub1 varieties

TABLE 8.2 Important STRVs Released in India for Cultivation

Submergence tolerant		Drought tolerant		Salinity tolerant	
Narendra Dhan 8002	2005	Sushk Samrat	2006	DRR 39	2009
Varsha Dhan	2005	CR Dhan 40	2008	CSR36	2005
Narendra Mayank	2009	Sahbhagi Dhan	2010	CR Dhan 405 (Luna Sankhi)	2012
Narendra Jalpushp	2009	DRR 42 (IR64-Drt1)	2014	CR Dhan 405 (Luna Sankhi)	2012
Narendra Naraini	2009	DRR 43	2014	CSR43	2013
Swarna-Sub1	2009	DRR 44	2014	Gosaba (Gozaba) 5	2014
Samba-Sub1	2014	CR Dhan 203	2014		
CR1009-Sub1	2014				
Bina Dhan 11	2015				

STRVs, Stress-tolerant rice varieties.
Source: Available from: http://strasa.irri.org/resources/varietal-releases

following submergence, resulting in a yield advantage of 1–3 t/ha depending on the stage at which submergence occurred and the severity. These varieties were quickly adopted by farmers within a few years of their release in rainfed stress-prone areas of South Asia (Mackill et al., 2012; Singh et al., 2009a; Singh et al., 2011a, 2014a).

India has seen the release of a few very promising drought-tolerant rice varieties such as Sahbhagi Dhan, DRR 42 (IR64-Drt1), Sushk Samrat, DRR 44, and CR Dhan 40. Under normal conditions they produce yield equal to that of any popular rice variety, but under severe drought these lines still manage to produce 1–1.5 t/ha when the yield of other popular varieties fails completely (Dar et al., 2014). Salt-tolerant rice varieties such as CSR23, CSR30, CSR36, CSR43, Narendra Usar dhan1, Narendra Usar dhan2, and Narendra Usar dhan3 for inland salinity and CR Dhan 405, CR Dhan 406, and Canning7 for coastal salinity have been released for cultivation (Singh et al., 2014b) and farmers are taking full advantage of these tolerant materials in their salt-affected fields. Many areas in rainfed lowlands experience both floods and droughts at different times within the same cropping season. Multiple-stress-tolerant genotypes that can withstand both submergence and drought have been developed at the IRRI and are under evaluation at different sites in India. Efforts in developing varieties able to tolerate flooding during germination (referred to as anaerobic germination genotypes) are now at a very advanced stage of field evaluation.

This may help to overcome the risk of poor crop establishment in direct-seeded rice systems associated with field inundation caused by continuous rains immediately after the dry seeding (Ismail et al., 2009, 2013).

Stress-prone areas in India are estimated to have around 30 million small holder (average <1 ha) farmers who are affected by different stresses. Out of these farmers, more than 3 million have already been provided with the seeds of stress-tolerant rice, which equates to a region of 1.1 million ha (Singh et al., 2013a,b; Dar et al., 2014). The impact of these varieties in the field is visible. Scientists from the University of California conducted randomized control trials in 128 villages of Odisha. They reported that crop submerged for 10 days led to a yield gain of 45% for Swarna-Sub1 when compared to its parent Swarna. In these impact assessment studies on Swarna-Sub1, Dar et al. (2013) argued that flood-tolerant Sub1 rice could deliver efficiency gains through reduced yield variability, higher yield, and equity gains by disproportionately benefiting marginal groups of farmers, as low-lying areas prone to flooding are more heavily occupied by people belonging to scheduled caste social groups. Farmers with access to the new risk-reducing technologies (e.g., Sub1) use more inputs and agricultural credit and shift away from cheap and less effective technologies. These investments translate into increases in productivity even during years when flooding is absent.

Variety-driven changes will help to achieve higher rice production and sustainability of rainfed lowlands; however, blanket recommendations are insufficient for the needs of these STRVs. Improved rice varieties can only provide stable yields when coupled with improved agronomic practices (crop establishment methods, nursery management practices, and fertilizer management) specifically developed for rainfed drought, submergence, or salinity-prone environments. Proper nursery management using appropriate seed density and a combination of organic and inorganic fertilizers can produce healthier seedlings. Such seedlings show lower mortality under stress, recover faster from transplanting shock, and show better poststress growth (Bhowmick et al., 2014; Singh et al., 2016). Main field and poststress nutrient management also helps in ensuring good plant establishment, proper growth, and higher yields for STRVs. When combined with improved management practices, these varieties produce significantly higher yields in all environments compared with STRVs grown using a farmers own management systems (Sarangi et al., 2015, 2016; Singh et al., 2016), therefore proving appropriate crop and nutrient management practices is essential for harnessing the full potential of tolerant varieties under stress-prone conditions. Thus, a combination of research on germplasm × environment × management aspects is necessary to develop the best management practices for a given rainfed environment. This approach can transform rice-based systems in rainfed lowlands by making them more productive and increasing and stabilizing farmers' income (Haefele et al., 2016).

Farm Mechanization

Mechanization at the system level is one of the most important aspects required to reform Indian agriculture. India has made significant progress in field preparation by using tractors with matching implements for plowing and puddling. This was primarily driven by the reduction in draft animal power from 0.133 kW/ha in 1971–72 to 0.094 kW/ha in 2012–13. During this period the respective shares of tractors, power tillers, diesel engines, and electric motors increased from 0.020 to 0.844, 0.001 to 0.015, 0.053 to 0.300, and 0.041 to 0.494 kW/ha. However, the spread of mechanization is limited to land preparation and to an extent harvesting of the crop. Mechanization has seen some developments in other activities of rice production.

Component Technologies of Mechanization

Laser Leveling

Proper land leveling is considered as an entry point for mechanization. It not only facilitates good and uniform germination but also increases cultivation area because of fewer bunds.

Flooding is the most common irrigation method in irrigated environments, but it consumes a large amount of water. In addition, deep drainage and evaporation during application because of poor management (inadequate flow rates) and uneven fields (Kahlown et al., 2002) results in a significant proportion (10%–25%) of the water being lost. Land leveling is a prerequisite for achieving high efficiency of flood irrigation. Rice fields are puddled and leveled every year prior to transplanting; however, such fields are still uneven, and field slopes vary from 1 to 5 degrees across the region (Jat et al., 2006). Laser leveling can reduce irrigation input by 100–200 mm for rice and by 50–100 mm for wheat (Humphreys et al., 2010) because of the reduction in shallower drainage; however, the magnitude of the reduction will depend upon site conditions, water management, and irrigation systems (flow rates relative to field size). In general, laser leveling has a small effect on rice yield but a much larger effect on increasing wheat yield in the rice–wheat system (Jat et al., 2009; Kahlown et al., 2006). The savings in irrigation water also help reduce costs by decreasing energy consumption. Based on an assumption of average irrigation water savings of 150 mm, laser leveling of the entire rice-growing area of 2.8 million ha in Punjab could reduce irrigation input by 0.42 million ha m (0.42×10^9 m^3) and save 247 GWh of electricity (based on computations defined by Sidhu and Vatta, 2010).

One of the key constraints to production in rainfed environments is poor crop stand, especially in regions where farmers broadcast rice before the onset of the monsoon season. The fields in rainfed environments are highly undulated, which results in patchy crop growth. The undulating topography not only affects crop establishment but also results in nonuniform distribution of fertilizer and high and patchy weed infestation. Therefore, leveling of the field also helps improve nutrient-use efficiency and weed control.

Challenges and Opportunities

Although laser leveling has been a mature technology in India, its adoption has remained limited in eastern India. This might be because the direct benefits are less visibility in this region; limited availability of machines, high cost, and poor awareness of the benefits of the technology mean they may not be thought of as appropriate. Lack of adoption may also be related to small landholdings. In northwest India, where the field size is relatively larger than in eastern India, the applicability and benefits of a four-wheel, tractor-driven, laser-leveling machine are better. The reduced availability of four-wheel tractors in the region also limits technology adoption. The two-wheel, tractor-driven, laser leveler has been widely tested on a research scale but hasn't been commercialized yet.

Along with machinery, it is important to explore the adoption environment of these technologies in rainfed environments. In northwest India, the spread and adoption of this technology occurred mainly because of the service provider model. A similar approach needs to be tested and strengthened in eastern India to improve the adoption rate.

Crop Establishment Methods

Different methods can be used to establish the rice crop, as shown in Fig. 8.4.

Direct-Seeded Rice

Significant developments have been made in mechanized options for direct seeding of rice. Direct seeding can be done under wet [wet-seeded rice (WSR)] or dry [dry-seeded rice (DSR)] conditions. Although WSR is more

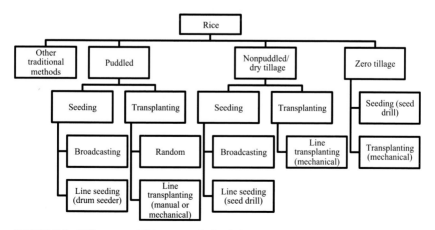

FIGURE 8.4   Different establishment methods of rice.

FIGURE 8.5   Wet seeding of rice with a drum seeder.

suited for irrigated environments, DSR is equally good for irrigated and rain-fed environments.

Wet-Seeded Rice

In WSR, sprouted rice seeds are broadcast or line seeded on puddled soil just af-ter drainage. The broadcasting is generally done manually whereas line seeding can be done with a drum seeder. A drum seeder is a simple, manually-operated implement for the seeding of rice on puddled soil (Fig. 8.5). It generally consists of six drums (25 cm long and 55 cm in diameter) connected one after the other to an iron rod having two wheels at either end (Khan et al., 2009). Some motor-ized blowers are available in the market for broadcasting seeds but they have had very limited adoption.

Challenges and Opportunities

Drainage and leveling are important to achieve reliable crop establishment with WSR. Too much standing water creates anaerobic conditions and inhibits the germination of seeds because of the reduction in oxygen. Discovering rice gen-otypes with better germination ability under anaerobic conditions will be the key to success and expansion of WSR.

Wet seeding of rice using a drum seeder is effective and could enhance yield and net returns for farmers in favorable rainfed and irrigated environments. In-tegrating the use of stress-tolerant varieties with higher yield and better grain quality with drum seeding and nutrient management could increase the produc-tivity of boro rice in the coastal areas (Sarangi et al., 2014).

Dry-Seeded Rice (DSR)

DSR, in which dry seeds are either broadcast or line seeded with a country plow or seed drill, is considered as one of the resource conservation tech-nologies. In contrast to the widely practiced transplanting, dry seeding allows early establishment, which in turn enables better use of early rains, reduc-es labor time for crop establishment, and reduces the risk from late-season drought due to early harvesting. The labor requirement can be decreased to

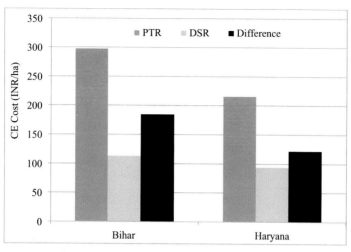

FIGURE 8.6   Crop establishment cost (including seed costs) of PTR and DSR in Bihar and Haryana/Punjab.

3–5 person-days/ha with DSR compared with the 25–50 person-days/ha required for crop establishment of PTR (Kumar and Ladha, 2011). Crop establishment cost (land preparation + nursery raising including seed + transplanting/sowing) was INR 7,900–12,000/ha higher in PTR than DSR in the Indian states of Bihar and Haryana/Punjab, making DSR economically attractive to farmers (Fig. 8.6). Average savings in production cost ranged from USD 26 to 31/ha for WSR to USD 48 to 125/ha for DSR when compared to PTR (Kumar and Ladha, 2011).

Although DSR is merely an alternative crop establishment method, with the correct water management practices it has the potential to save irrigation water (primarily because of avoiding the water requirement for puddling) in irrigated environments. Many research studies in India have demonstrated a 10%–50% saving in irrigation water with DSR vis-à-vis PTR when irrigation was scheduled based on the appearance of hairline cracks or based on soil matric potential (−20 kPa at 15–17 cm depth) (Gathala et al., 2013; Sudhir-Yadav et al., 2011a,b). DSR can also reduce total labor requirements by 11%–66% depending on the season, location, and type of DSR compared with PTR (Isvilanonda 2002; Kumar et al., 2009). There has been a large increase in the area under DSR in northwest and eastern India. Along with labor scarcity, advancements in the development and refinement of seed drills and weed management has also played a significant role in the spread of this technology. A multicrop planter fitted with an inclined-plate seed-metering system and its ability to drill both seed and fertilizer has helped improve crop establishment (Fig. 8.7). DSR can also be sown with a conventional seed-cum-fertilizer drill with a fluted-roller seed-metering mechanism, but the seeds are not spaced evenly in this system, and the system requires a higher seed rate to compensate for seed breakage. The inclined-plate

FIGURE 8.7 (A–B) Dry-seeded rice with a two-wheel and four-wheel tractor and (C) with inclined-plate and (D) fluted-roller seed-metering system.

seed-metering mechanism also provides an opportunity to use primed seed as the problem of seed breakage is negated with this technology.

Challenges and Opportunities

In countries such as India, population pressure and food demand mean that land productivity (yield) is a prime consideration in determining the adoption of alternative crop production technologies. To date, most research has indicated an average yield penalty of around 10% with DSR compared with PTR, but losses can be as much as 33% (Table 8.3). The higher yield penalties in DSR are primarily due to high weed infestation (Yadav et al., 2008), micronutrient deficiency (Choudhury et al., 2007; Kreye et al., 2009), and nematode infestation (Choudhury et al., 2007; Singh et al., 2002). Weed management is the most important factor for better performance of DSR. Challenges and opportunities of weed management in DSR are covered in the following section. Refinement in agronomy will continue to be crucial but the development of cultivars adapted to DSR conditions is also equally important for attaining optimal grain yield in DSR.

In the submergence-prone rainfed lowland environment, farmers are using direct seeding through broadcasting as the cultivation technique, and although it produces less yield than transplanting, it also requires less labor. In areas where dry seeding is practiced, farmers may encounter flooding or waterlogging immediately after seeding. This leads to a severe reduction in yield or even failure of crop establishment because of the high sensitivity of rice to the anaerobic

TABLE 8.3 Grain Yield Response of DSR Relative to PTR in Different Soils at Different Locations in the IGP

Soil type	Grain yield of PTR (t/ha)	Yield response of DSR (% change)	Location	References
Silty loam	4.2	−33.3	Faizabad, India	Yadav et al. (2008)
Sandy loam	5.5	−23.6	New Delhi, India	Choudhury et al. (2007)
Loam	5.4	−22.2	New Delhi, India	Choudhury et al. (2007)
Silty clay loam	5.7	−9.3	Pantnagar, India	Singh et al. (2004)
Silty clay loam	5.7	−8.6	Pantnagar, India	Tripathi et al. (2005)
Silty loam	6.1	−8.2	Pantnagar, India	Hobbs et al. (2002)
Clay loam	7.2	−8.1	Kaithal, India	Saharawat et al. (2010)
Silty clay loam	5.4	−7.0	Pantnagar, India	Sharma et al. (2005)
Silty clay loam	6.9	−5.8	Pantnagar, India	Singh et al. (2008)
Sandy loam	5.6	−5.4	Pantnagar, India	Hobbs et al. (2002)
—	5.5	−5.0	Karnal, India	Goel and Verma (2000)
Silty loam	7.3	−4.1	Modipurum, India	Bhushan et al. (2007)
Sandy loam to loam	4.9	−4.1	Meerut, Ghaziabad, and Bhulandshar, India	Saharawat et al. (2009)
Sandy loam to clay loam	6.3	−1.6	Karnal, India	Saharawat et al. (2009)
Silty clay loam	5.3	1.9	Bhairahawa, Nepal	Hobbs et al. (2002)
Sandy loam	6.6	3.8	Ludhiana, India	Gill (2008)
24 Villages	3.9	12.4	Ballia, India	Singh et al. (2009b)
Sandy loam	7.1	−10.0	Karnal, India	Kumar et al. (2015)

Source: Adapted from Sudhir-Yadav 2011. Effect of crop establishment method and irrigation schedule on productivity and water use of rice. School of Agriculture, Food and Wine. The University of Adelaide, Adelaide, Australia.

conditions caused by flooding during germination. Varieties that can germinate in flooded soils could be beneficial in flood-prone areas and they could even be used in intensive irrigated systems where flooding after direct seeding can be used as a tool to effectively suppress most weeds. Loci that determine rice's ability to germinate and establish under flooding were found recently and good progress is being made in developing high-yielding varieties capable of germinating under anaerobic conditions (Ismail et al., 2013). These new rice cultivars need to be widely evaluated in terms of their tolerance of flooding during germination in submergence-prone regions of South Asia, and management packages should be developed for their effective use.

Mechanical Transplanting of Rice

The decreased availability and increased cost of labor have been major factors for the switch to an alternative crop establishment method. In addition, poor aeration in randomly transplanted crops results in complex problems including insect pests and diseases, and ultimately results in lower yields. In the past, machine transplanting has been successful in East Asian countries (i.e., Korea and Japan) but at the time little interest was generated in India, primarily because of the availability of relatively cheap labor. However, the past decade saw significant progress in the development and refinement of the mechanical transplanter in India. Alongside labor savings, mechanical transplanting brings additional benefits and opportunities (Malik et al., 2011) including:

- Synchronization of transplanting time in a region.
- Transplanting of seedlings at the optimal age (14–18 days).
- Uniform spacing and optimum plant density (26–28 hills/m^2 with 2–3 seedlings per hill).
- Reduced transplanting shock, early seedling vigor, and uniform crop stand.
- Reduced stress, drudgery, and health risks for farm laborers, especially women.
- Better employment and business opportunities for rural youth through the development of custom service business.

Most of the efforts in mechanized transplanting were made in puddled conditions, which has limitations due to the problem of soil settling (loose soil), resulting in missing seedlings. Attempts have been made in mechanized transplanting of rice under nonpuddled conditions (Kamboj et al., 2013), resulting in further savings in resources (water and energy).

Key Components of Technology

Mat Nursery In the mat nursery, seedlings are raised on a thin layer of soil and farmyard manure or compost mixture is placed on a polythene sheet (Fig. 8.8). The polythene sheet prevents the seedling roots from penetrating the underlying soil, thus creating a dense mat. This type of nursery is a prerequisite for machine transplanting. The mat can be cut into desired shapes and sizes to fit into the

FIGURE 8.8 (A) Dry and (B) wet bed mat nursery.

trays of the transplanter. Seedlings are ready for planting within 14–18 days after sowing (DAS).

Mechanical Transplanter Transplanting is done with a self-propelled mechanical rice transplanter. The transplanter commonly available in India is either a walk behind or riding type (Fig. 8.9). The technical specifications of this machine vary with different models; however, common features include adjustment options in plant-to-plant spacing, number of seedlings/hill, and planting depth.

Key Challenges and Opportunities Crop establishment with machine transplanting depends on the depth of standing-water at the time of transplanting. It is recommended to drain the water before transplanting for good anchoring of the seedlings. Crop establishment is generally poor if standing-water depth surpasses 1–2 cm, therefore this poses a problem in rainfed lowland areas. The optimum seedling height for transplanting is generally 15–18 cm. Lowland areas, where water remains stagnant at the time of crop establishment, are not suitable for this technology. In these environments, farmers instead practice manual transplanting of 30- to 40-day-old seedlings. The high capital cost

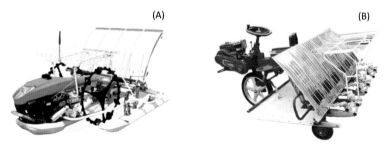

FIGURE 8.9 The mechanical transplanter: (A) walk-behind and (B) riding type.

of machines has restricted the spread of mechanical transplanting in India. However, some case studies have shown successful adoption of the technology with a "one-to-many" service provision model. This business model has been used for the mat nursery, machine transplanting, or package of both operations (most common).

Precision Crop Management

Fertilizer Management

Over the past 4 decades crop management in Asia has been driven by the increasing use of external inputs and blanket recommendations for fertilizer use over wide areas. In addition, about half of the rice-growing areas in South Asia, particularly in India, are rainfed and prone to abiotic stresses such as flooding, drought, and excess salt in soils. Across South Asia, fertilizer guidelines are generally blanket recommendations and do not reflect differences in indigenous soil fertility, prevailing crop management practices, yield responses, or attainable yield potential across sites, years, stresses, or varieties. To optimize the potential of any improved variety, appropriate management practices are important. This is especially true for stress-prone areas. It is the package of the variety along with appropriate nutrient and crop management practices that can have a real impact on productivity in farmers' fields. An adequate nutrient supply is essential to optimize the potential of improved high-yielding varieties and recently introduced stress-tolerant varieties. Soil testing is one approach advocated for improving nutrient-use efficiency although the approach has not been recognized as the most effective because of difficulties in understanding soil test reports and the long delay in obtaining reports from the soil testing laboratories (Mohapatra and Kameswari, 2014; Wollenhaupt et al., 1997). On-farm research has demonstrated that large variability exists between rice farms or single rice fields in terms of soil nutrient supply, nutrient-use efficiency, and crop response to nutrients (Adhikari et al., 1999; Cassman et al., 1996). Managing this variability has become a principal challenge for further increasing the productivity of these intensive rice areas. Future gains in productivity and input-use efficiency will require soil and crop management technologies that are more knowledge intensive and tailored to the specific characteristics of individual farms and fields (Dobermann and White, 1999). Site-specific nutrient management (SSNM) enables farmers to dynamically adjust fertilizer use by supplying optimum amounts of nutrients at critical time points in the crop's growth to produce high yields.

The original concept for SSNM was developed in 1996 (Dobermann and White, 1999; Dobermann et al., 1996) and was tested on 205 irrigated rice farms across China, India, Indonesia, the Philippines, Thailand, and Vietnam. In the approach described here, SSNM was a general concept for optimizing the supply and demand of nutrients according to their variation in time and space.

An important part of field testing was to continuously collect data that could be used to improve the approach. Making use of the availability of this expanding database, SSNM was refined in recent years by integrating different approaches for determining phosphorus (P) and potassium (K) requirements. For fields without a certain yield gain, fertilizer P and K needs can be determined by a partial maintenance approach (i.e., fertilizer input < output in nutrient balance), which considers the nutrient supply mediated through soil processes and balances trade-offs between financial loss with full maintenance rates and risk of excessive nutrient depletion without nutrient application. When yield gains for an added nutrient are certain, partial maintenance plus yield gain can be used to determine fertilizer requirements. The SSNM-based approach and algorithms enable the rapid development of field-specific K and P management recommendations (Buresh et al., 2010).

The Development of Decision Tools

The dissemination of SSNM practices on a wider scale requires transformation of the principles of SSNM into locally adapted tools that enable extension workers, crop advisers, and farmers to rapidly access, adopt, and implement best management practices for specific fields and growing conditions. Web- and mobile phone-based decision support tools are options for addressing this futuristic cause.

IRRI used the principles of SSNM and developed the concept and framework for a web-based decision tool called *Crop Manager*. Various versions of the tool for favorable and unfavorable environments have been developed and evaluated in some states (Odisha, Bihar, eastern Uttar Pradesh, and Tamil Nadu) in collaboration with state agricultural universities, ICAR centers, international NGOs, and other international research organizations (CIMMYT). The tool has been developed for rice as well as rice-based systems (i.e., rice–wheat and rice–maize). The tool is a web-based application with a simple, user-friendly interface providing personalized fertilizer and crop management guidance for small-scale farmers and extension workers. The farmers have to provide information about their fields by responding to a set of 12–15 brief questions regarding field location, planting method, seed variety, typical yield, choice of fertilizer, method of harvesting, and other factors. The tool has been designed for use by extension workers, crop advisers, input providers, and service providers.

Based on 129 initial on-farm evaluation trials, *Rice–Wheat Crop Manager* (RWCM) increased rice yield in eastern Uttar Pradesh and Bihar by 0.5 t/ha compared with the farmers' current practices, with a reduction in nitrogen (N) and phosphorous pentoxide (P_2O_5) by 9 and 35 kg/ha, respectively, and an increase in potassium oxide (K_2O) by 12 kg/ha. Similarly, *Rice Crop Manager* in Odisha increased rice yield by 0.7 t/ha with a decrease in P_2O_5 and K_2O by 16 and 14 kg/ha, respectively, and a 30 kg/ha increase in N (Tables 8.4 and 8.5).

TABLE 8.4 Performance of *Rice–Wheat Crop Manager* Compared With Current Farmers' Practice and State Fertilizer Recommendation in Eastern Uttar Pradesh and Bihar (*N* = 129) (Unpublished data[a])

Treatment	Yield (kg/ha)		N (kg/ha)		P_2O_5 (kg/ha)		K_2O (kg/ha)		Added net income (INR/ha)
FP	5100	(0)[b]	135	(0)	61	(0)	31	(0)	0
SFR	5300	(+200)	120	(−15)	49	(−12)	37	(−6)	5550
RWCM	5600	(+500)	126	(−9)	26	(−35)	43	(−12)	8100

FP, Farmers' practice; SFR, state fertilizer recommendation; RWCM, based on *Rice–Wheat Crop Manager*.
[a]*The unpublished data is from the evaluation trials conducted by IRRI in farmers' fields across the several districts of respective states. The three treatments: (1) farmers' practice (which was defined by the farmer and captured using RCM); (2) state fertilizer recommendation (which is the blanket fertilizer recommendation given by state agriculture universities of each state common for all farmers in each state); and (3) RCM recommendations (generated through the RCM tool through an interview of the farmer specific to each farmer's field) were tested in farmers' fields.*
[b]*Values in parentheses represent the change from farmers' practice.*

TABLE 8.5 Performance of *Rice Crop Manager* Compared With Current Farmers' Practice and State Fertilizer Recommendation in Odisha (*N* = 84) (Unpublished data)

Treatment	Yield (kg/ha)		N (kg/ha)		P_2O_5 (kg/ha)		K_2O (kg/ha)		Added net income (INR/ha)
FP	4726	(0)[a]	71	(0)	46	(0)	46	(0)	0
SFR	5203	(+477)	80	(+9)	40	(−6)	40	(−6)	4226
RCM	5442	(+716)	101	(+30)	30	(−16)	32	(−14)	10107

FP, Farmers' practice; SFR, state fertilizer recommendation; RCM, based on *Rice Crop Manager*.
[a]*Values in parentheses represent the change from farmers' practice.*

The Way Forward

It is exemplary how scientific collaboration can be effectively used to provide farmers with field-specific fertilizer and crop management at the right time using recent advances in ICT. These tools have the flexibility to incorporate various advances in knowledge and bring about a paradigm shift in how knowledge is disseminated. Government of India (GOI) schemes such as the "Soil health program" provide an opportunity to make the best use of data generated from soil tests and to develop a web-based tool as an alternative method for providing field-specific nutrient and crop management. A recent initiative of GOI called "Digital India" paves the way for these tools to reach a large number of farmers with fertilizer and crop management information at the time needed.

The positive results received during the pilot initiatives pave the way for the future success of web- and mobile phone-based decision tools for precise nutrient and crop management to improve the livelihoods of small and marginal farmers. With the increase in mobile services and extension of Internet services to rural areas, the scope is large for the success of such tools for knowledge dissemination to the right people at the right time.

Weed Management

Despite multiple benefits of alternative crop establishment methods, weeds remain the major biological constraint to their wider adoption in India. Here, we discuss the major weed management challenges as we shift from PTR to DSR and how we can overcome these challenges.

Minimizing Yield Losses due to Weeds and Weed Control Costs

Weeds in DSR are more diverse and difficult to control than in PTR, which results in higher yield losses and higher costs (Kumar et al., 2013; Ladha et al., 2009; Rao et al., 2007). Weeds are more problematic in DSR because (1) simultaneously emerging DSR seedlings are less competitive with weeds than 25- to 30-day-old seedlings in transplanted crops, and (2) early flooding, which is used to control the initial flush of weeds in PTR, is not possible in DSR (Kumar and Ladha, 2011). Singh et al. (2011b) in their medium-term study reported 85%–98% yield losses when using DSR in unweeded controls compared with just 12% in PTR. Weed control costs in DSR have been reported to be USD 20–50/ha higher than in PTR (Ladha et al., 2009). Therefore, one of the biggest weed management challenges in DSR is reducing yield losses caused by weeds while also minimizing the cost of weed control.

Predicting Shifts in Weed Flora

Shifting from PTR to DSR often result in changes in weed composition and diversity because of changes in tillage, the crop establishment method, and water and weed management. Our inability to predict and manage weeds for those species that will dominate with this shift poses a major threat to the sustainability of DSR production systems. Adoption of DSR may result in shifts in weed flora toward more difficult-to-control and competitive grasses and sedges. Weedy rice has emerged as a major threat for DSR in countries where it has been widely adopted (Kumar and Ladha, 2011). In India, in addition to *Echinochloa* species, other difficult-to-control grasses such as *Leptochloa chinensis*, *Eragrostis japonica*, *Dactyloctenium aegyptium*, and weedy and volunteer rice have started showing dominance in DSR.

Preventing/Delaying Herbicide Resistance

Herbicide resistance has not yet appeared in PTR production systems mainly because of the integrated approaches followed in this method, including puddling (wet tillage), transplanting (the age difference gives an advantage to the crop), continuous flooding, need-based herbicides (pre or post), and hand weeding. DSR depends on herbicides for weed control as it has to compensate for the loss of cultural methods of weed suppression such as flooding, transplanting, and wet tillage. Therefore, an increase in DSR adoption in the region increases the risk of evolution of herbicide resistance in weeds. In Asian countries where DSR is widely adopted, cases of herbicide resistance have evolved and increased with its introduction (Kumar and Ladha, 2011). For example, in Malaysia, Korea, Thailand, the Philippines, and Sri Lanka, the number of herbicide resistance cases increased from zero (prior to DSR introduction) to 10, 10, 5, 2, and 3, respectively, after DSR introduction (Kumar and Ladha, 2011). These examples demonstrate that preventative measures should be taken to stop or delay the risk of herbicide resistance in weeds with the introduction of DSR.

Integrated Weed Management

In DSR, no single method can provide an effective and sustainable weed management solution and therefore an integrated weed management (IWM) strategy is needed. This strategy should build upon the foundation of good knowledge of weed biology and ecology to comprise preventive, cultural/agronomic, and physical methods in conjunction with the judicious use of modern herbicides and different modes of action. Effective weed control can be achieved in DSR by integrating the following components of IWM: (1) promoting fatal germination of rice weeds using stale seedbed techniques or cover crop rotation, (2) using certified seeds free from weed seed contamination, (3) using fallow and bund management to prevent weed seed production, (4) using crop mulch and soil or dust mulch for weed suppression, (5) using weed-competitive cultivars, (6) using water and nutrient management to improve the crop's competitive advantage relative to weeds, (7) using crop management practices that enhance weed seed predation and decay (e.g., zero tillage with residue retention), (8) rotating other rice establishment methods that are more weed suppressive (e.g., PTR mechanically or manually after a couple of cycles of DSR), (9) manipulating seed rate and crop geometry to shift the competitive advantage in favor of the crop, (10) using mechanical tools and need-based hand weeding to remove escaped weeds and prevent seed production, and (11) using appropriate herbicides/tank mixes effective under DSR conditions with appropriate rotation with different modes of action (Kumar et al., 2013).

Studies suggest that practices that promote germination of rice weeds and killing them prior to rice planting may be extremely beneficial for reducing weed populations, particularly for problematic species that are difficult to control with herbicides such as weedy rice (*Oryza sativa* L.), *D. aegyptium, Leptochloa chinensis*, etc. For example, inclusion of mungbean during the fallow period between wheat and rice resulted in an 84% and 40% reduction in the population of *D. aegyptium* in the subsequent rice crop under ZT and CT systems, respectively (Kumar et al., 2015, unpublished). Similarly, a stale seedbed using one presowing irrigation reduced the weed density and biomass in the rice crop by 44%–68% and 60%, respectively, compared with farmers' practice without a stale seedbed. When the stale seedbed was practiced with two presowing irrigations, weed density further decreased to 77%–85% and weed biomass decreased >85%.

Initial investigations also suggests that creating dust or soil mulch is effective for suppressing weeds and conserving soil moisture in DSR. It has been observed that DSR fields established after presowing irrigation under conventional tillage (dust mulching) have less weed infestation compared with when they are established in dry conditions followed by immediate postsowing irrigation. This is attributed to the dust mulch effect created in the former (Malik et al., 2015). Irrigation immediately after sowing creates conditions favorable for weed emergence and growth. In DSR with dust mulch, rice is seeded after presowing irrigation followed by tillage, the first postsowing irrigation is delayed for almost 2–3 weeks. Broken capillaries (as a result of tillage and no irrigation) minimize the continuum of moisture loss from the lower soil layer to the atmosphere and create dust mulch; hence, conditions are less favorable for weeds to emerge. Dust mulching was effective in reducing weed density by 62%–65% and weed biomass by 43%–60% at 15–60 days after sowing (Kumar and Malik et al., 2015, unpublished). Similarly, crop residue mulch can also provide weed suppression in DSR. In a field study conducted in India, 5 t/ha wheat residue as mulch reduced the emergence of grass, broadleaf, and sedge species by >70%, 65%, and 22%–70%, respectively, compared with bare soil (no residue mulch), ultimately resulting in a 70% reduction in weed biomass (Kumar et al., 2013).

Several herbicide combinations have shown promise at successfully controlling weeds in DSR. Based on studies in Bihar, eastern Uttar Pradesh, Odisha, and Haryana, preemergence (pendimethalin, oxadiargyl, or pretilachlor with safener) followed by postemergence (bispyribac or bispyribac-based tank mixture including bispyribac + pyrazosulfuron/azimsulfuron/2,4-D, or fenoxaprop with safener or fenoxaprop-based tank mixture including fenoxaprop + ethoxysulfuron), or penoxsulam or penoxsulam-based tank mixture herbicide application are found to be effective in controlling weeds in DSR (Kumar and Ladha, 2011, Mishra et al., unpublished). In the absence of preemergence application, the bispyribac or bispyribac-based tank mixture aforementioned followed by one hand weeding or mechanical weeding can effectively control weeds. Nutsedge (*Cyperus rotundus*)-dominated weed flora in rice, which are commonly found in eastern India, were effectively controlled with a tank mix combination of bispyribac with pyrazosulfuron, applied 20–25 days after sowing (DAS). Fenoxaprop with safener or metamifop is effective in controlling emerging problematic weeds in DSR, such as *L. chinensis*, *D. aegyptium*, *Eragrostis* sp., etc. (Kumar, personal communication).

In the rainfed environment of eastern India, weed management in DSR is even more challenging and complex because (1) weeds are more intensified and diversified, (2) the time of weed management is uncertain because of doubts about rains and the lack of assured irrigation facilities, and (3) the lack of access to herbicides because of poor market development and lack of knowledge on safe handling and application methodologies. Current experiences in these environments clearly reveal that an IWM approach using suitable pre- and postemergence herbicides supplemented by need-based manual or mechanical weeding (Cono-weeder or Power-weeder) and other cultural practices, would provide more effective weed management than any of these options in isolation. In these environments, effective land preparation is critical to achieving a weed-free seedbed. Farmers generally keep their fields unplowed in summer and start land preparation just prior to rice establishment before the onset of the rainy season. At this point plowing or tillage is relatively less effective at killing existing weeds because of the presence of sufficient soil moisture, which allows weeds to regrow. Therefore, summer plowing could be more effective at killing existing weeds, as the dry period creates congenial conditions for the desiccation of uprooted weeds. In conditions where tillage/land preparation has to be delayed until the rain starts and weed pressure is high, glyphosate application (1–2 days prior to tillage) followed by tillage has been found to be effective at killing existing weeds more so than by tillage alone (Ashok Yadav, personal communication).

Development of a Decision Tool for Weed Control

Weed management in DSR is knowledge intensive. Farmers generally have limited knowledge on proper herbicide handling, selection of herbicide molecules and their time of application, application technologies for better efficacy, and environmental and human health risks associated with their incorrect use. Farmers usually follow blanket recommendations for herbicide application as per the advice of input dealers or agrovets or general state recommendations, and this is less effective in DSR because of the diverse weed flora. Therefore, it is important to recommend the right herbicides and the appropriate time for their application based on the type of weed flora present in farmers' fields, as this will increase the efficacy of herbicides and reduce their use. An ICT-based decision tool or Android-based application for weed identification and weed management is needed. This could be used by extension agents, service providers, and private-sector partners, including private pesticide companies and their dealers and distributors, and help all stakeholders to identify weed problems and make accurate weed management recommendations based on the problem. IRRI is therefore working with national and international partners to develop mobile phone-based weed ID and management tools for providing site-specific weed management guidelines and recommendations based on the type of weed flora. The aim is to integrate this decision tool with *Crop Manager*, another ICT-based tool for site-specific crop management and nutrient recommendations.

Water Management

Rice is traditionally grown with continuous flooding in irrigated environments, which requires large amounts of irrigation water. High seepage and percolation losses account for the much higher water usage in flooded rice than in other crops. A reduction in hydrostatic pressure is an important means for reducing such losses (Bouman et al., 1994) and the driving principle behind alternate wetting and drying (AWD). In AWD, the field is not continuously flooded; instead, the soil is allowed to dry out for one to several days after the disappearance of ponded water before it is flooded again. Development of this technology began in India in the 1970s and 1980s as "intermittent irrigation," for which it was recommended to irrigate 2–3 days after the disappearance of ponded water (Sandhu et al., 1980) (Fig. 8.10). However, this practice does not take into account soil type and crop water requirement, which is determined by growth rate and evaporative demand. The other common method for irrigation scheduling is visual observation of the appearance of hairline cracks. The appearance of hairline cracks mainly depends on soil type and this approach also doesn't take into account crop water requirement.

In 2002 IRRI, through the Water-Savings Workgroup of the Irrigated Rice Research Consortium, developed a set of simple guidelines coupled with an easy-to-use and practical tool (the field water tube) to allow farmers to reduce irrigation water input while maintaining yield. More recently, Punjab Agricultural University has developed a farmer-friendly tensiometer to schedule irrigation. The tensiometer tube is calibrated to a color-coded scale; as the soil tension increases to 15 kPa, the internal water level drops from the green zone to the yellow zone and to the red zone if the soil further dries beyond 15 kPa (Kukal et al., 2007).

It has been well established that irrigation input can be decreased by using safe AWD (i.e., AWD managed to avoid yield loss). The reduction is in the range of 10%–40% of the amount of water applied to a continuously flooded field, depending on soil hydraulic conductivity and depth to the water table.

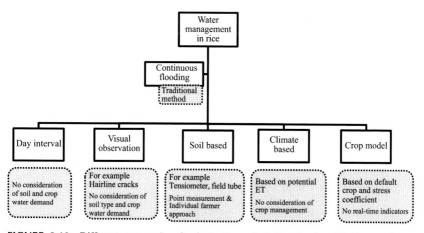

FIGURE 8.10 **Different approaches for irrigation scheduling in rice.** ET, Evaporation + transpiration.

However, these approaches (Panipipe, tensiometer) are based on "point" measurements, which don't reflect the soil water status for the entire field. New tools for irrigation scheduling are required that can consider the variability of the entire field and can indicate both current and projected water demand.

In rainfed environment, where rice cultivation largely depends on rainfall, crop management options to improve water-use efficiency play a significant role in reducing risk and improving productivity. Bundling of knowledge on improved crop management (e.g., cultivar, field design, establishment method, and so on) along with understanding of weather patterns can be effective for improving rainwater-use efficiency and avoiding abiotic stresses. In favorable rainfed environments, best use of supplementary irrigation plays a critical role in improving productivity and net profit.

Irrigation Method

The conventional method of irrigation, flood irrigation, is generally highly inefficient when flow rates are too slow to complete the irrigation quickly (a couple of hours). This inefficiency is due to deep drainage below the root zone. Flood irrigation also causes temporary waterlogging, with adverse effects on nonrice crops such as wheat, maize, and legumes. Waterlogging is more prolonged and more severe on heavy-textured soils and on soils used for rice culture because of the well-developed shallow hard pan (slowly permeable) as a result of puddling. This leads to aeration stress in upland crops, especially in wheat (Kukal and Aggarwal, 2003). Modern pressurized irrigation systems (center pivot and lateral-move sprinklers, microsprinklers, surface drips, and subsurface drips) have the potential to increase irrigation water-use efficiency by matching water provision to crop requirements, reducing runoff and deep drainage, and generally keeping the root zone moist. Drier soil also means less waterlogging, reduced soil evaporation, and increased capacity to capture rainfall, thus further reducing runoff and deep drainage. Sharda et al. (2016) reported a 40% saving in irrigation water when using drip irrigation while maintaining rice yield similar to that of flood irrigation in South Asia.

As water is almost a free commodity in Asia, farmers don't usually think about transforming irrigation methods, especially for cereal crops. The high efficiency and savings with these modern technologies also comes with a significant cost that is hard for small and marginal farmers to afford. There have been some developments in low-cost drip irrigation systems but their adoption is limited to vegetables and other high-value crops.

Understanding of Water Savings and Trade-off

Water entering the field in the form of irrigation or rain may be lost from the root zone in a number of ways including through deep drainage beyond the root zone, surface runoff, evaporation, and transpiration. All losses are nonproductive losses, with the exception of transpiration as this is directly related to biomass production and yield. Deep drainage and runoff losses are not considered as real losses from the soil–groundwater system if they can be reused. This is often the case in groundwater-irrigated areas such as the rice–wheat areas of northwest India, where deep drainage flows back into the groundwater. Here, reducing deep drainage through practices such as AWD in rice reduces the water loss at the farmer field level, but, from a regional groundwater resource perspective, there is no net water saving (just an energy saving in terms of less pumping) or effect on groundwater depletion (Humphreys et al., 2010). However, in the regions where groundwater is of poor quality, the water lost as drainage can't be used again. Surface runoff from one field is often captured by downstream fields or other users, and reducing runoff at the field level does not necessarily mean that more water is available for other users (Hafeez et al., 2007). However, reducing ET (evaporation + transpiration) is always a real water saving. Technologies are available that can reduce irrigation input at the farmer field level by reducing drainage and runoff, but these are unlikely to have a significant effect on real water savings. Thus, from the perspective of regional water resource management in the rice–wheat systems of northwest India and the rice–rice system in eastern India, where the amount of available water is declining fast because of overextraction, increasing water productivity based on ET losses needs to be given priority.

System-Based Approach

Besides large fallow lands in the dry season, the mismatch between wet-season (kharif/aman) and dry-season (rabi/boro) crops and their respective varieties is one major reason for low productivity in rainfed areas. Rice planting, being dependent on rainfall, requires a careful varietal selection because this not only determines productivity in the wet season but also affects the productivity of the following dry-season crops. Appropriate planting techniques such as DSR and minimum tillage paired with appropriate-duration varieties can lead to early crop maturity, thus bringing about opportunities to maximize the use of residual water for postrice crops. Improved varieties of dry-season crops, notably pulses, allow for a range of options in rainfed areas. The use of varieties suited to late-sown conditions (e.g., late-sown wheat varieties in the rice–wheat system) or other crops, notably pulses and oilseeds, to complement cropping systems based on late-maturing submergence-tolerant rice varieties needs to be explored in submergence-prone environments.

Surface seeding (in excess soil moisture conditions) and zero tillage are other resource-conserving technologies for the sustainable intensification of cropping systems as they allow the advancement of planting dates and avoid fallow in the dry season after late harvesting of the rice crop. For example, surface seeding of linseed was performed under excess moisture conditions after the harvest of long-duration rice cultivar Swarna-Sub1 on the experimental farm of Banaras Hindu University in Eastern Uttar Pradesh. Under these conditions, linseed produced the highest rice equivalent system yield (12.9 t/ha), followed by wheat (10.7 t/ha), gram (5.5 t/ha), and lentil (5.4 t/ha), thus demonstrating that the surface seeding of linseed or wheat before the harvesting of Sub1 rice varieties could be a profitable option for flood-prone lowlands that mostly remain underused because of excess soil moisture at the time of establishing dry-season crops (Singh et al., 2014c). In contrast to conventional tillage in the rice–wheat system, ZT in wheat facilitates early wheat planting (by reducing the turnaround period between rice harvesting and planting of wheat) and minimizes the risks of terminal heat stress. This results in a robust yield gain of 0.5 t/ha (19%) and economic gain of INR 7334/ha (Keil et al., 2015). In the rice–rice system, early planting of rice in the wet season coupled with timely planting of boro rice reduces the risk of terminal heat and improves system productivity.

Choosing an appropriate dry-season crop and establishment method may help in overcoming the rice-fallow problem in drought-prone environments of eastern India. Short-duration rice varieties (Sahbhagi Dhan and IR64-Drought1, etc.) and climate-resilient ones with establishment methods (DSR) can provide options for diversification of postrice crops (Haefele et al., 2016). The early maturity of drought-tolerant rice varieties could allow farmers to use a rice–pulse system because, at the time of harvest of these varieties, the residual moisture is appropriate for pulse germination and growth (Dar and Singh, 2013). Farmers can even think about growing three crops (rice and early peas followed by late-sown wheat varieties, or shorter-duration rice followed by mustard followed by mungbean) if supplementary irrigation is available, thus increasing their annual production and income. In areas with assured irrigation, some farmers prefer potato (because it fetches a good market price in March) followed by cowpea, chillies, or mungbean (Dar et al., 2012).

In the sodic soils of the Indo-Gangetic plains rice is the most favored crop during the wet (kharif) season because it can tolerate standing water and medium ESP. In the rabi season, only shallow-rooted crops such as wheat, barley, berseem, and mustard could be grown in the initial years of reclamation/use, since only surface soil (0–15 cm depth) is considered for the reclamation of sodic soils. Crop rotation of rice in the wet season followed by wheat in the dry season and a *sesbania* green manure crop in summer (because of its tolerance of sodicity) has been found most suitable and is being adopted on a large scale for the reclamation of sodic soils. Varietal pairing of salt-tolerant rice (CSR36 and CSR43) and wheat (KRL19 and KRL213) has also been found to be good for salt-affected areas and highly accepted by the communities in the demonstration areas (IRRI, 2013). This tolerant rice–wheat rotation should be continued for at least the first 3 years and the field should not be left fallow to ensure continuity of the reclamation process and to avoid a reversion to sodic conditions. After substantial improvement in soil conditions, other less tolerant but more remunerative crops such as mustard could be included in the cropping system. Recently released short-duration, salt-tolerant varieties, such as CSR43 may enable farmers to grow a short-duration (70 days) crop of toria (*Brassica compestris*) or vegetables (such as spinach or lady's finger) in between the traditional rice–wheat cropping system, thus providing an additional income of about USD 125/ha in the sodic soils of Eastern Uttar Pradesh and Bihar (Singh and Mishra, 2013; Singh et al., 2016).

Reducing the Environmental Footprint

The country is supporting a paradigm shift in the traditional rice-growing system to support food security and sustainability. The elements of a transformative agronomic management system are closely linked with environmental impact or global warming. The introduction of high-yielding varieties alongside the intensive use of agrochemicals (mainly fertilizer) resulted in several environmental problems linked with the groundwater table, soil fertility changes, and greenhouse gas emissions.

The conventional practice of growing rice (transplanting in puddled conditions) is not only resource inefficient but it also emits a significant amount of greenhouse gases (GHGs), especially methane (Grace et al., 2003). The major factors that affect greenhouse gas emissions are tillage (Drury et al., 2006), moisture and aeration (Jiao et al., 2007), N fertilization (McSwiney and Robertson, 2005), and C supply (Rochette et al., 2006). All of these factors interact and the level of interaction guides the amount of GHG emissions.

The flooded PTR system contributes 10%–20% (50–100 Tg/year) of total global annual methane emissions (Houghton et al., 1996; Wassmann and Aulakh, 2000), thus making rice an important target crop for mitigating GHG emissions, especially methane. Methane emissions were lower with DSR (both dry and wet) than with conventional PTR; however, the reduction was predicted to be greater in dry-DSR (24%–79%) than in wet-DSR (8%–22%), compared with PTR under continuous flooding water management (Kumar and Ladha, 2011). Under intermittent irrigation, the reduction in methane emissions increased further by 43%–75% in dry-DSR vis-à-vis conventional PTR. When DSR (dry or wet) is combined with midseason drainage or intermittent irrigation, the reduction in methane increased further compared with flooded PTR. In a study conducted in Haryana, Padre et al. (2016) reported that substituting PTR with ZT-DSR reduced methane emissions by 56%–73%. In the Philippines, Corton et al. (2000) observed that the reduction in methane emissions increased from 16% to 22% under continuous flooding to 82%–92% under midseason drainage or intermittent irrigation compared with flooded PTR. Similarly, AWD is a water management practice that saves 30% of irrigation water and also reduces methane emissions by 48% without affecting yield (Richard and Sander, 2014).

PTR is considered to be a less important source of atmospheric nitrous oxide (N_2O) emissions but there are some trade-offs with increased emissions of N_2O in alternative crop establishment methods (e.g., DSR) and water-saving technologies (e.g., AWD). Although DSR and AWD help reduce methane emissions through enhanced aerobic conditions, the risk of N_2O emissions increases as aerobic conditions of the soil increase. N_2O emissions increase at soil redox potential above 250 mV (Hou et al., 2000). In a study conducted in Modipuram, N_2O emissions increased from 0.31 to 0.39 kg N/ha in PTR to 0.90 to 1.10 kg N/ha and 1.30 to 2.20 kg N/ha under conventional-till dry-DSR and ZT-DSR, respectively (Kumar and Ladha, 2011). Similarly Padre et al. (2016) observed higher N_2O emissions under ZT-DSR and PTR with midseason drainage when compared with PTR alone. The N_2O emissions were 1.66–1.92 kg N_2O/ha in PTR, but 3.70–4.04 kg N_2O/ha in PTR with midseason drainage and 2.67–3.35 kg N_2O/ha in ZT-DSR (Padre et al., 2016). These results suggest developing and deploying effective management strategies, especially for water and nutrients, to reduce N_2O emissions from dry-DSR and other water-efficient technologies and minimize negative impacts on the environment. For example, it is important to develop water management practices that keep the soil redox potential in an intermediate range (−100 to +200 mV) to keep emissions of both methane and N_2O in check (Hou et al., 2000). The critical soil redox potential identified for N_2O and methane emissions is +250 mV and −150 mV, respectively. Other strategies that improve nitrogen-use efficiency also help to mitigate N_2O emissions, including matching N application (rate and time) with crop needs, using nitrification inhibitors and slow-release N fertilizers (Akiyama et al., 2010; Linquist et al., 2012), and improving N-fertilizer placement (Liu et al., 2006). For example, it has been estimated that applying N on a per crop basis (using a leaf color chart) could reduce N_2O emissions and global warming potential (GWP) by 11%–14% (Bhatia et al., 2012). Linquist et al. (2012) reported that dicyandiamide, a nitrification inhibitor, reduced N_2O emissions in rice by 29%.

The overall effect of any technology or practice on GWP depends on the total emissions of all three GHGs (methane, N_2O, and CO_2). In a study conducted in Haryana, Padre et al. (2016) reported that transitioning from conventional PTR to DSR practices could reduce the total GWP of rice by 23% or by 1.28 Tg CO_2 eq/year. Another study conducted in Punjab, estimated the total GWP of PTR was in the 2.0–4.6 t CO_2 eq/ha range and in the 1.3–2.9 t CO_2 eq/ha range for DSR (Pathak et al., 2013). To take full advantage of DSR and water-saving technologies in the environment, the development of optimal water and nutrient management practices is critical and requires more research to buffer the risks of higher N_2O emissions associated with these resource-efficient technologies/practices.

Unintended consequences in water and chemical use and soil degradation have had serious environmental impacts (Burney et al., 2010). These environmental costs are widely recognized as a potential threat to long-term sustainability. Subsidized inputs—especially fertilizer, pesticide, and irrigation water—blurred the gains at the farm level for adopting practices that would enhance efficiency in input use and thereby contribute to sustaining the agricultural resource base (Pingali, 2012). However, there are examples of changed behavior in adopting sustainable practices where the country (e.g., Indonesia) has changed the policy on input subsidy. Although agricultural growth is improving in eastern India, it needs to center around closing yield gaps and a reduced environmental footprint.

SUMMARY AND CONCLUSIONS

Rice and rice-based systems will continue to play a significant role in the food security and economy of India. To meet future food demand, it is estimated that many developing countries, including those in South Asian, will have to double their food production from 2010 to 2050 (Alexandratos and Bruinsma, 2012). This additional future food production will be limited by the availability of resources (land, water, and energy), therefore, decoupling future agricultural growth from the unsustainable use of these resources for increasing food production, and this has become one of the cornerstones for a new sustainable development agenda (Dobermann et al., 2013). Agricultural research should now focus on reducing risk and improving productivity, using eco-efficient approaches for long-term sustainability, and efficiently using resources while making nutrition-inclusive and safe food for all. This can be best achieved by combining agroecological intensification approaches with best management practices and modern genetic improvement (Ladha et al., 2016).

The favorable rainfed lowlands in India can benefit from irrigated rice technologies already available throughout the system; therefore, futuristic rice research should focus on stress-prone subsystems (drought, submergence, and salinity, etc.) of rainfed lowlands. Emphasis should be on the identification, adaptation, and validation of technology options that address key constraints for better and more stable rice production systems in these stress-prone environments. With stress-tolerant varieties becoming widely available to farmers, opportunities exist to raise productivity through complementary crop management practices (establishment and crop nutrition and weed management), and in more diversified and intensified production systems to reduce risks and improve the market value of their produce. These tolerant varieties not only reduce downside risk by providing good tolerance but they also have positive effects on the adoption of a more labor-, water-, and energy-efficient planting method, area cultivated, fertilizer usage, and credit demand. A large share of the expected gains from new technology comes from joining with other investments. Therefore, improved technologies that reduce risk by protecting production in bad years have the potential to increase agricultural productivity in normal years in stress-prone subsystems (Emerick et al., 2016) and should become a part of any futuristic strategy on rice management in unfavorable areas.

With mechanized labor-saving land management and crop establishment on center stage, ICT-based decision support tools will play a crucial role in transforming conventional agriculture toward precision agriculture at scale in South Asia and other regions. In addition, the integration of noncereal crops such as a legume in the system would strike a better nutritional balance and could improve soil and plant health and system productivity.

ACKNOWLEDGMENTS

We would like to acknowledge Yoichiro Kato, Ashok Kumar, and R.K. Malik for their suggestions and for providing information on different sections of this chapter.

REFERENCES

Adachi, K., 1992. Effect of puddling on rice-soil physics: softness of puddled soil and percolation. In: Murty, V.V.N., Koga, K. (Eds.), Soil and Water Engineering for Paddy Field Management. Asian Institute of Technology, Bangkok, Thailand, pp. 220–231.

Adhikari, C., Bronson, K., Panuallah, G., Regmi, A., Saha, P., Dobermann, A., Olk, D., Hobbs, P., Pasuquin, E., 1999. On-farm soil N supply and N nutrition in the rice-wheat system of Nepal and Bangladesh. Field Crops Res. 64, 273–286.

Aggarwal, P.K., Joshi, P.K., Ingram, J.S.I., Gupta, R.K., 2004. Adapting food systems of the Indo-Gangetic plains to global environmental change: key information needs to improve policy formulation. Environ. Sci. Policy 7, 487–498.

Akiyama, H., Yan, X., Yagi, K., 2010. Evaluation of effectiveness of enhanced efficiency fertilizers as mitigation options for N_2O and NO emissions from agricultural soils: meta-analysis. Glob. Change Biol. 16, 1837–1846.

Alexandratos, N., Bruinsma, J., 2012. World agriculture towards 2030/2050: the 2012 revision. ESA Working Paper No. 12-03. Rome, Italy: FAO. 1-147.

Bhatia, A., Aggarwal, P.K., Jain, N., Pathak, H., 2012. Greenhouse gas emission from rice and wheat-growing areas in India: spatial analysis and upscaling. Greenhouse Gas Sci. Technol. 2, 115–125.

Bhowmick, M.K., Dhara, M.C., Singh, S., Dar, M.H., Singh, U.S., 2014. Improved management options for submergence-tolerant (Sub1) rice genotypes in flood-prone rainfed lowlands of West Bengal. Am. J. Plant Sci. 5, 14–23.

Bhushan, L., Ladha, J.K., Gupta, R.K., Singh, S., Tirol-Padre, A., Saharawat, Y.S., Gathala, M., Pathak, H., 2007. Saving of water and labor in a rice–wheat system with no-tillage and direct seeding technologies. Agron. J. 99, 1288–1296.

Bouman, B., Wopereis, M., Kropff, M., Tenberge, H., Tuong, T., 1994. Water-use efficiency of flooded rice fields. II. Percolation and seepage losses. Agr. Water Manage. 26, 291–304.

Buresh, R., Pampolino, M., Witt, C., 2010. Field-specific potassium and phosphorus balances and fertilizer requirements for irrigated rice-based cropping systems. Plant Soil 335, 35–64.

Burney, J.A., Davis, S.J., Lobell, D.B., 2010. Greenhouse gas mitigation by agricultural intensification. Proc. Natl. Acad. Sci. USA 107, 12052–12057.

Cassman, K., Gines, G., Dizon, M., Samson, M., Alcantara, J., 1996. Nitrogen-use efficiency in tropical lowland rice systems: contributions from indigenous and applied nitrogen. Field Crops Res. 47, 1–12.

Choudhury, B.U., Bouman, B.A.M., Singh, A.K., 2007. Yield and water productivity of rice-wheat on raised beds at New Delhi, India. Field Crops Res. 100, 229–239.

Corton, T.M., Bajita, J., Grospe, F., Pamplona, R., Wassmann, R., Lantin, R.S., 2000. Methane emission from irrigated and intensively managed rice fields in Central Luzon (Philippines). Nutr. Cycl. Agroecosys. 58, 37–53.

Dar, M.H., Singh, U.S., 2013. Cluster demonstration of stress-tolerant rice varieties in stress-prone parts of India. Annual Report 2012–13. National Food Security Mission, Ministry of Agriculture, Govt. of India, International Rice Research Institute, Pusa, New Delhi, India, 152 p.

Dar, M.H., Singh, S., Zaidi, N.W., Shukla, S., 2012. Sahbhagi Dhan: science's answer to drought problems. STRASA News 5, 1–3.

Dar, M., de Janvry, A., Emerick, K., Raitzer, D., Sadoulet, E., 2013. Flood-tolerant rice reduces yield variability and raises expected yield, differentially benefiting socially disadvantaged groups. Sci. Rep. 3, 3315.

Dar, M.H., Singh, S., Singh, U.S., Zaidi, N.W., Ismail, A.M., 2014. Stress-tolerant rice varieties making headway in India. SATSA Mukhapatra, Annual Technical Issue. Seed: Hope to Harvest 18, 1–14.

Dobermann, A., White, P., 1999. Strategies for nutrient management in irrigated and rainfed lowland rice systems. Nutr. Cycl. Agroecosys. 53, 1–18.

Dobermann, A., Cassman, K.G., Peng, S., Tan, P.S., Phung, C.V., Sta. Cruz, P.C., Bajita, J.B., Adviento, M.A.A., Olk, D.C., 1996. Precision nutrient management in intensive irrigated rice systems. In: Proceedings of the International Symposium on Maximizing Sustainable Rice Yields Through Improved Soil and Environmental Management, 11–17 November 1996, Khon Kaen, Thailand. Department of Agriculture, Soil, and Fertilizer Society of Thailand, Department of Land Development, ISSS, Bangkok, Thailand, pp. 133–154.

Dobermann, A, Nelson, R, Beever, D., 2013. Solutions for sustainable agriculture and food systems. Technical Report for the Post-2015 Development Agenda. Sustainable Development Solutions Network, New York (pp. 1–99). Available from: www.unsdsn.org

Drury, C.F., Reynolds, W.D., Tan, C.S., Welacky, T.W., Calder, W., McLaughlin, N.B., 2006. Emissions of nitrous oxide and carbon dioxide: influence of tillage type and nitrogen placement depth. Soil Sci. Soc. Am. J. 70, 570–581.

Emerick, K., Janvry, A., Sadoulet, E., Dar, M.H., 2016. Technological innovations, downside risk, and the modernization of agriculture. Am. Econ. Rev. 106, 1537–1599.

Gathala, M.K., Kumar, V., Sharma, P.C., Saharawat, Y.S., Jat, H.S., Singh, M., Kumar, A., Jat, M.L., Humphreys, E., Sharma, D.K., Sharma, S., Ladha, J.K., 2013. Optimizing intensive cereal-based cropping systems addressing current and future drivers of agricultural change in the northwestern Indo-Gangetic plains of India. Agric. Ecosyst. Environ. 177, 85–97.

Gautam, P., Lal, B., Katara, J.L., Joshi, E., 2013. Beushening: a traditional method of rice crop establishment in Eastern India. Popular Kheti 1, 1–4.

Gill, M.S., 2008. Productivity of direct-seeded rice (Oryza sativa) under varying seed rates, weed control and irrigation levels. Indian J. Agr. Sci. 78, 766–770.

Goel, A.C., Verma, K.S., 2000. Comparative study of direct seeding and transplanting of rice. Indian J. Agr. Res. 34, 194–196.

GOI, Government of India, 2014. Agricultural Statistics at a Glance. Ministry of Agriculture, Government of India. Oxford University Press, New Delhi.

Grace, P.R., Jain, M.C., Harrington, L., Philip, G., 2003. Long-term sustainability of the tropical and subtropical rice–wheat system: an environmental perspective. In: Ladha, J.K., Hill, J., Gupta, R.K., Duxbury, J., Buresh, R.J. (Eds.), Improving the Productivity and Sustainability of Rice–Wheat Systems: Issues and Impact. ASA Special Publication 65. American Society of Agronomy, Madison, WI, pp. 27–43.

Gupta, R.K., Naresh, R.K., Hobbs, P.R., Ladha, J.K., 2002. Adopting conservation agriculture in rice–wheat systems of the Indo-Gangetic Plains—New opportunities for saving on water. In: Bouman, B.A.M., Hengsdijk, H., Hardy, B., Bindraban, B., Toung, T.P., Ladha, J.K. (Eds.), Proceedings of the International Workshop on Water-Wise Rice Production. International Rice Research Institute, Los Baños, Philippines, pp. 207–222.

Haefele, S.M., Kato, Y., Singh, S., 2016. Climate ready rice: augmenting drought tolerance with best management practices. Field Crops Res. 190, 60–69.

Hafeez, M.M., Bouman, B.A.M., Van de Giesen, N., Vlek, P., 2007. Scale effects on water use and water productivity in a rice-based irrigation system (UPRIIS) in the Philippines. Agr. Water Manage. 92, 81–89.

Hobbs, P.R., Singh, Y., Giri, G.S., Lauren, J.G., Duxbury, J.M., 2002. Direct-seeding and reduced-tillage options in the rice–wheat systems of the Indo-Gangetic plains of South Asia. In: Pandey, S., Mortimer, M., Wade, L., Tuong, T.P., Lopez, K., Hardy, B. (Eds.), International Workshop on Direct Seeding in Asian Rice Systems: Strategies, Research Issues, and Opportunities. International Rice Research Institute, Los Baños, Philippines, pp. 201–215.

Hou, A.X., Chen, G.X., Wang, Z.P., Van Cleemput, O., Patrick, Jr., W.H., 2000. Methane and nitrous oxide emissions form a rice field in relation to soil redox and microbiological processes. Soil Sci. Soc. Am. J. 64, 2180–2186.

Houghton, J.T., Meira Filho, L.G., Callander, B.A., Harris, N., Katterberg, A., Maskell. K. 1996. IPCC report on climate change: the science of climate change. WG1 Contribution to the IPCC Second Assessment Report on Methane Emission from Rice Cultivation. Cambridge University Press, Cambridge, UK.

Humphreys, E., Kukal, S.S., Christen, E.W., Hira, G.S., Balwinder, S., Sudhir, Y., Sharma, R.K., 2010. Halting the groundwater decline in north-west India—Which crop technologies will be winners? Adv. Agron. 109, 155–217.

IRRI (International Rice Research Institute), 2013. Cluster demonstrations of stress-tolerant rice varieties in stress-prone parts of India. Annual Report (2012–13), NFSM, GOI. IRRI, New Delhi.

Ismail, A., Ella, E., Vergara, G., Mackill, D., 2009. Mechanisms associated with tolerance to flooding during germination and early seedling growth in rice (*Oryza sativa*). Ann. Bot. 103, 197–209.

Ismail, A., Singh, U., Singh, S., Dar, M., Mackill, D., 2013. The contribution of submergence-tolerant (Sub1) rice varieties to food security in flood-prone rainfed lowland areas in Asia. Field Crops Res. 152, 83–93.

Isvilanonda, S. 2002. Development trends and farmers' benefits in the adoption of wet seeded rice in Thailand. In: Pandey, S., Mortimer, M., Wade, L., Tuong, T.P., Lopez, K., Hardy, B. (Eds.). Direct Seeding: Research Strategies and Opportunities. Proceedings of the International Workshop on Direct Seeding in Asian Rice Systems: Strategic Research Issues and Opportunities, 25–28 January 2000, Bangkok, Thailand. International Rice Research Institute, Los Baños, Philippines. pp. 115–124.

Jat, M.L., Chandana, P., Sharma, S.K., Gill, M.A., Gupta, R.K., 2006. Laser land leveling: a precursor technology for resource conservation. Rice–Wheat Consortium Technical Bulletin Series 7. Rice–Wheat Consortium for the Indo-Gangetic Plains, New Delhi, India, 48 p.

Jat, M.L., Gathala, M.K., Ladha, J.K., Saharawat, Y.S., Jat, A.S., Kumar, V., Sharma, S.K., Kumar, V., Gupta, R., 2009. Evaluation of precision land leveling and double zero-till systems in the rice–wheat rotation: water use, productivity, profitability, and soil physical properties. Soil Till. Res. 105, 112–121.

Jiao, Z., Hou, A., Shi, Y., Huang, G., Wang, Y., Chen, X., 2007. Water management influencing methane and nitrous oxide emissions from rice field in relation to soil redox and microbial community. Commun. Soil Sci. Plant Anal. 37, 1889–1903.

Kahlown, M.A., Gill, M.A., Ashraf, M., 2002. Evaluation of Resource Conservation Technologies in Rice–Wheat System of Pakistan. Pakistan Council of Research in Water Resources (PCRWR), Research Report-I, Islamabad, Pakistan.

Kahlown, M.A., Azam, M., Kemper, W.D., 2006. Soil management strategies for rice–wheat rotations in Pakistan's Punjab. J. Soil Water Conserv. 61, 40–44.

Kamboj, B.R., Yadav, D.B., Yadav, A., Goel, N.K., Gill, G., Malik, R.K., Chauhan, B.S., 2013. Mechanized transplanting of rice (*Oryza sativa* L.) in nonpuddled and no-till conditions in the rice–wheat cropping system in Haryana, India. Am. J. Plant Sci. 4, 2409–2413.

Keil, A., D'Souza, A., McDonald, A., 2015. Zero-tillage as a pathway for sustainable wheat intensification in the Eastern Indo-Gangetic plains: does it work in farmers' fields? Food Secur. 7, 983–1001.

Khan, M.A.H., Alam, M.M., Hossain, M.I., Rashid, M.H., Mollah, M.I.U., Quddus, M.A., Miah, M.I.B., Sikder, M.A.A., Ladha, J.K., 2009. Validation and delivery of improved technologies in the rice–wheat ecosystem in Bangladesh. In: Ladha, J.K., Singh, Y., Erenstein, O., Hardy, B. (Eds.), Integrated Crop and Resource Management in the Rice–Wheat System of South Asia. International Rice Research Institute, Los Baños, Philippines, pp. 197–220.

Kreye, C., Bouman, B.A.M., Castaneda, A.R., Lampayan, R.M., Faronilo, J.E., Lactaoen, A.T., Fernandez, L., 2009. Possible causes of yield failure in tropical aerobic rice. Field Crops Res. 111, 197–206.

Kukal, S., Aggarwal, G., 2003. Puddling depth and intensity effects in rice-wheat system on a sandy loam soil. II. Water use and crop performance. Soil Till. Res. 74, 37–45.

Kukal, S.S., Sidhu, A.S., Hira, G.S., 2007. Tensiometer-Based Irrigation Scheduling to Rice. Extension Bulletin 2007/02. Department of Soils, PAU, Ludhiana.

Kumar, V., Ladha, J.K., 2011. Direct seeding of rice. Adv. Agron. 111, 297–413.

Kumar, V., Ladha, J.K., Gathala, M.K., 2009. Direct drill-seeded rice: a need of the day. In: Annual Meeting of Agronomy Society of America, 1–5 November, 2009, Pittsburgh. Available from http://a-c-s.confex.com/crops/2009am/webprogram/Paper53386.html

Kumar, V., Singh, S., Chhokar, R., Malik, R., Brainard, D., Ladha, J.K., 2013. Weed management strategies to reduce herbicide use in zero-till rice–wheat cropping systems of the Indo-Gangetic plains. Weed Technol. 27, 241–254.

Kumar, A., Kumar, S., Dahiya, K., Kumar, S., Kumar, M., 2015. Productivity and economics of direct seeded rice (*Oryza sativa* L.). J. Appl. Nat. Sci. 7, 410–416.

Ladha, J.K., Dawe, D., Pathak, H., Padre, A., Yadav, R., Singh, B., Singh, Y., Singh, P., Kundu, A., Sakal, R., Ram, N., Regmi, A., Gami, S., Bhandari, A., Amin, R., Yadav, C., Bhattarai, E., Das, S., Aggarwal, H., Gupta, R., Hobbs, P., 2003. How extensive are yield declines in long-term rice–wheat experiments in Asia? Field Crops Res. 81, 159–180.

Ladha, J.K., Kumar, V., Alam, M.M., Sharma, S., Gathala, M., Chandna, P., Saharawat, Y.S., Balasubramanian, V., 2009. Integrating crop and resource management technologies for enhanced productivity, profitability, and sustainability of the rice–wheat system in South Asia. In: Ladha, J.K., Singh, Y., Erenstein, O., Hardy, B. (Eds.), Integrated Crop and Resource Management in the Rice–Wheat system of South Asia. International Rice Research Institute, Los Baños, Philippines, pp. 69–108.

Ladha, J.K., Rao, A.N., Raman, A.K., Padre, A.T., Dobermann, A., Gathala, M., Kumar, V., Saharawat, Y., Sharma, S., Piepho, H.P., Alam, M.M., Liak, R., Rajendran, R., Reddy, C.K., Parsad, R., Sharma, P.C., Singh, S.S., Saha, A., Noor, S., 2016. Agronomic improvements can make future cereal systems in South Asia far more productive and result in a lower environmental footprint. Glob. Chang. Biol. 22, 1054–1074.

Linquist, B., Adviento-Borbe, M., Pittelkow, C., van Kessel, C., van Groenigen, K., 2012. Fertilizer management practices and greenhouse gas emissions from rice systems: a quantitative review and analysis. Field Crops Res. 135, 10–21.

Liu, X.J., Mosier, A.R., Halvorson, A.D., Zhang, F.S., 2006. The impact of nitrogen placement and tillage on NO, N_2O, CH_4, and CO_2 fluxes from a clay loam soil. Plant Soil 280, 177–188.

Mackill, D.J., Ismail, A.M., Singh, U.S., Labios, R.V., Paris, T.R., 2012. Development and rapid adoption of submergence-tolerant (Sub1) rice varieties. Adv. Agron. 115, 299–352.

Malik, R., Kamboj, B., Jat, M., Sidhu, H., Bana, A., Singh, V., Sharawat, Y., Pundir, A., Sahnawaz, R., Anuradha, T., Kumaran, N., Gupta, R., 2011. No-Till and Unpuddled Mechanical Transplanting of Rice. Cereal Systems Initiative for South Asia, New Delhi, India, Available from: http://repository.cimmyt.org/xmlui/handle/10883/1296

Malik, R.K., Kumar, V., Yadav, A., McDonald, A., 2015. Weed science and sustainable intensification of cropping systems in South Asia. 25th Asian Pacific Weed Science Society Conference, Hyderabad, India.

McSwiney, C., Robertson, G., 2005. Nonlinear response of N_2O flux to incremental fertilizer addition in a continuous maize (*Zea mays* L.) cropping system. Glob. Chang. Biol. 11, 1712–1719.

Mohanty, S., 2015. India reaches the pinnacle in rice exports. Rice Today 14 (2), 43–45.

Mohapatra, L., Kameswari, V.L.V., 2014. Knowledge level of soil management practices and their adoption by farmers of Odisha. Int. J. Farm Sci. 4, 240–246.

Padre, A., Rai, M., Kumar, V., Gathala, M., Sharma, P.C., Sharma, S., Nagar, R.K., Deshwal, S., Singh, L.K., Jat, H.S., Sharma, D.K., Wassmann, R., Ladha, J.K., 2016. Quantifying changes to the global warming potential of rice-wheat systems with adoption of conservation agriculture in northwestern India. Agric. Ecosyst. Environ. 219, 125–137.

Pandey, S., Velasco, L., 2002. Economics of direct seeding in Asia: patterns of adoption and research priorities. In: Pandey, S., Mortimer, M., Wade, L., Tuong, T.P., Lopez, K., Hardy, B. (Eds.), Direct seeding: Research Issues and Opportunities. Proceedings of the International Workshop on Direct Seeding in Asian Rice Systems: Strategic Research Issues and Opportunities. International Rice Research Institute, Los Baños, Philippines, pp. 3–14.

Pathak, H., Ladha, J., Aggarwal, P., Peng, S., Das, S., Singh, Y., Singh, B., Kamra, S., Mishra, B., Sastri, A., Aggarwal, H., Das, D., Gupta, R., 2003. Trends of climatic potential and on-farm yields of rice and wheat in the Indo-Gangetic plains. Field Crops Res. 80, 223–234.

Pathak, H., Sankhyan, S., Dubey, D.S., Bhatia, A., Jain, N., 2013. Dry direct-seeding of rice for mitigating greenhouse gas emission: field experimentation and simulation. Paddy Water Environ. 11, 593–601.

Pingali, P.L., 2012. Green Revolution: impacts, limits, and the path ahead. Proc. Natl. Acad. Sci. USA 109 (31), 12302–12308.

Rao, A.N., Johnson, D.E., Sivaprasad, B., Ladha, J.K., Mortimer, A.M., 2007. Weed management in direct-seeded rice. Adv. Agron. 93, 153–255.

Richard, M., Sander, B.O., 2014. Alternate wetting and drying in irrigated rice: implementation guidance for policymakers and investors. Practice Brief: Climate-Smart Agriculture. April 2014. Available from: https://cgspace.cgiar.org/rest/bitstreams/34363/retrieve

Rochette, P., Angers, D.A., Chantigny, M.H., Gagnon, B., Bertrand, N., 2006. N_2O fluxes in soils of contrasting textures fertilized with liquid and soil dairy cattle manures. Can. J. Soil Sci. 88, 175–187.

Rodell, M., Velicogna, I., Famiglietti, J.S., 2009. Satellite-based estimates of groundwater depletion in India. Nature 460, 999–1002.

Saharawat, Y.S., Gathala, M., Ladha, J.K., Malik, R.K., Singh, S., Jat, M.L., Gupta, R.K., Pathak, H., Singh, K., 2009. Evaluation and promotion of integrated crop and resource management in the rice–wheat system in northwest India. In: Ladha, J.K., Yadvinder-Singh, Erenstein, O., Hardy, B. (Eds.), Integrated Crop and Resource Management in the Rice–Wheat System of South Asia. International Rice Research Institute, Los Baños, Philippines, pp. 151–176.

Saharawat, Y.S., Singh, B., Malik, R.K., Ladha, J.K., Gathala, M., Jat, M.L., Kumar, V., 2010. Evaluation of alternative tillage and crop establishment methods in a rice–wheat rotation in NorthWestern IGP. Field Crops Res. 116, 260–267.

Sandhu, B.S., Khera, K.L., Prihar, S.S., Singh, B., 1980. Irrigation needs and yield of rice on a sandy-loam soil as affected by continuous and intermittent submergence. Indian J. Agric. Sci. 50, 492–496.

Sarangi, S., Maji, B., Singh, S., Sharma, D., Burman, D., Mandal, S., Ismail, A., Haefele, S., 2014. Crop establishment and nutrient management for dry season (*boro*) rice in coastal areas. Agron. J. 106, 2013–2023.

Sarangi, S.K., Maji, B., Singh, S., Burman, D., Mandal, S., Sharma, D.K., Singh, U.S., Ismail, A.M., Haefele, S.M., 2015. Improved nursery management further enhances the productivity of stress-tolerant rice varieties in coastal rainfed lowlands. Field Crops Res. 174, 61–70.

Sarangi, S.K., Maji, B., Singh, S., Sharma, D.K., Burman, D., Mandal, S., Singh, U.S., Ismail, A.M., Haefele, S.M., 2016. Using improved variety and management enhances rice productivity in stagnant flood-affected tropical coastal zones. Field Crops Res. 190, 70–81.

Septiningsih, E., Pamplona, A., Sanchez, D., Neeraja, C., Vergara, G., Heuer, S., Ismail, A., Mackill, D., 2009. Development of submergence-tolerant rice cultivars: the Sub1 locus and beyond. Ann. Bot. 103, 151–160.

Sharda, R., Mahajan, G., Siag, M., Chauhan, B.S., 2016. Performance of drip-irrigated dry-seeded rice (*Oryza sativa* L.) in South Asia. Paddy Water Environ. doi: 10.1007/s10333-016-0531-5.

Sharma, P., Tripathi, R.P., Singh, S., 2005. Tillage effects on soil physical properties and performance of rice–wheat-cropping system under shallow water table conditions of Tarai, Northern India. Eur. J. Agron. 23, 327–335.

Shukla, A.K., Sharma, S.K., Tiwari, R., Tiwari, K.N., 2005. Nutrient depletion in the rice–wheat cropping system of the Indo-Gangetic plains. Better Crops 89, 28–31.

Sidhu, R.S., Vatta, K., 2010. Economic valuation of innovative agriculture conservation technologies and practices in Indian Punjab. In: 12th Annual BIOECON Conference: From the Wealth of Nations to the Wealth of Nature: Rethinking Economic Growth. Centro Culturale Don Orione Artigianelli, Venice, Italy, September 27–28, 2010. Available from: www.ucl.ac.uk/bioecon/12th_2010/

Singh, Y.P., Mishra, V.K., 2013. Effect of short duration salt tolerant variety of rice on cropping intensity under different sodic environments. Annual Report 2013. ICAR-Central Soil Salinity Research Institute, Karnal, India (pp. 86–87).

Singh, R., Gajri, P.R., Gill, K.S., Khera, R., 1995. Puddling intensity and nitrogen-use efficiency of rice (*Oryza sativa*) on a sandy-loam soil of Punjab. Indian J. Agric. Sci. 65, 749–751.

Singh, A.K., Choudhury, B.U., Bouman, B.A.M., 2002. Effects of rice establishment methods on crop performance, water use, and mineral nitrogen. In: Bouman, B.A.M., Hengsdijk, H., Hardy, B., Bindraban, B., Tuong, T.P., Ladha, J.K. (Eds.), In: Proceedings of the International Workshop on Waterwise Rice Production. International Rice Research Institute, Los Baños, Philippines, pp. 237–246.

Singh, S., Tripathi, R.P., Sharma, P., Kumar, R., 2004. Effect of tillage on root growth, crop performance and economics of rice (*Oryza sativa*)–wheat (*Triticum aestivum*) system. Indian J. Agric. Sci. 74, 300–304.

Singh, V.P., Singh, G., Singh, S.P., Kumar, A., Singh, Y., Johnson, D.E., 2008. Direct seeding and weed management in the irrigated rice–wheat production system. In: Singh, Y., Singh, V.P., Chauhan, B., Orr, A., Mortimer, A.M., Johnson, D.E., Hardy, B. (Eds.), Direct seeding and weed management in the irrigated rice–wheat cropping system of the Indo-Gangetic Plains. International Rice Research Institute and Pantnagar (India): Directorate of Experiment Station, G.B. Pant University of Agriculture and Technology, Los Baños, Philippines, pp. 131–138.

Singh, S., Mackill, D.J., Ismail, A.M., 2009a. Responses of SUB1 rice introgression lines to submergence in the field: yield and grain quality. Field Crops Res. 113, 12–23.

Singh, U.P., Singh, Y., Virender-Kumar, Ladha, J.K., 2009b. Evaluation and promotion of resource-conserving tillage and crop establishment techniques in the rice–wheat system of eastern India. In: Ladha, J.K., Yadvinder-Singh, Erenstein, O., Hardy, B. (Eds.), Integrated Crop and Resource Management in the Rice–Wheat System of South Asia. International Rice Research Institute, Los Baños, Philippines, pp. 151–176.

Singh, S., Mackill, D.J., Ismail, A.M., 2011a. Tolerance of longer-term partial stagnant flooding is independent of the SUB1 locus in rice. Field Crops Res. 121, 311–323.

Singh, Y., Singh, V.P., Singh, G., Yadav, D.S., Sinha, R.K.P., Johnson, D.E., Mortimer, A.M., 2011b. The implications of land preparation, crop establishment method, and weed management on rice yield variation in the rice–wheat system in the Indo-Gangetic plains. Field Crops Res. 121, 64–74.

Singh, U.S., Dar, M.H., Singh, S., Ismail, A.M., 2013a. Transforming rice production in flood-affected areas: development of the Swarna-Sub1 variety using marker-assisted backcrossing and its deployment in India. In: Ghosh, K. et al., (Eds.), Case Studies of Use of Agricultural Biotechnologies in Developing Countries. FAO, Rome, pp. 63–70.

Singh, U.S., Dar, M.H., Singh, S., Zaidi, N.W., Bari, M.A., Mackill, D.J., Collard, B.C.Y., Singh, V.N., Singh, J.P., Reddy, J.N., Singh, R.K., Ismail, A.M., 2013b. Field performance, dissemination, impact, and tracking of submergence tolerant (Sub1) rice varieties in South Asia. SABRAO J. Breed. Genet. 45, 112–131.

Singh, S., Mackill, D.J., Ismail, A.M., 2014a. Physiological basis of tolerance to complete submergence in rice involves genetic factors in addition to the *SUB1* gene. Aob Plants 6, plu060.

Singh, Y.P., Nayak, A., Sharma, D., Gautam, R., Singh, R., Singh, R., Mishra, V., Paris, T., Ismail, A., 2014b. Farmers' participatory varietal selection: a sustainable crop improvement approach for the 21st century. Agroecol. Sust. Food. 38, 427–444.

Singh, U.P., Singh, S., Padmavathi, J., Sutaliya, J.M., Ravi K., H.S., Singh, H., Singh, U.S., Haefele, S.M., 2014c. Crop establishment and cropping system options for submergence-tolerant (Sub1) rice varieties in flood-prone areas. In: Fourth International Rice Congress. Oct 27,– Nov 1, 2014, Bangkok, Thailand. Abstract No. IRC 14-0323. p. 614.

Singh, Y.P., Mishra, V.K., Singh, S., Sharma, D.K., Singh, D., Singh, U.S., Singh, R.K., Haefele, S.M., Ismail, A.M., 2016. Productivity of sodic soils can be enhanced through the use of salt tolerant rice varieties and proper agronomic practices. Field Crops Res. 190, 82–90.

Sudhir-Yadav, Gill, G., Humphreys, E., Kukal, S.S., Walia, U.S., 2011a. Effect of water management on dry seeded and puddled transplanted rice. Part 1: crop performance. Field Crops Res. 120, 112–122.

Sudhir-Yadav, Humphreys, E., Kukal, S.S., Gill, G., Rangarajan, R., 2011b. Effect of water management on dry seeded and puddled transplanted rice: Part 2. Water balance and water productivity. Field Crops Res. 120, 123–132.

Tripathi, R.P., Sharma, P., Singh, S., 2005. Tilth index: an approach to optimize tillage in rice–wheat system. Soil Till. Res. 80, 125–137.

Wassmann, R., Aulakh, M., 2000. The role of rice plants in regulating mechanisms of methane missions. Biol. Fertil. Soils 31, 20–29.

Wollenhaupt, N.C., Mulla, D.J., Gotway, C.A., 1997. Soil sampling and interpolation techniques for mapping spatial variability of soil properties. In: Pierce, F.J., Sadler, E.J. (Eds.), The State of Site Specific Management for Agriculture. J. Am. Soc. Agron, Madison, WI, pp. 19–53.

Xu, K., Xu, X., Fukao, T., Canlas, P., Maghirang-Rodriguez, R., Heuer, S., Ismail, A., Bailey-Serres, J., Ronald, P., Mackill, D., 2006. Sub1A is an ethylene-response-factor-like gene that confers submergence tolerance to rice. Nature 442, 705–708.

Yadav, D.S., Sushant, Mortimer, A.M., Johnson, D.E., 2008. Studies on direct seeding of rice, weed control and tillage practices in the rice–wheat cropping system in eastern Uttar Pradesh. In: Singh, Y., Singh, V.P., Chauhan, B., Orr, A., Mortimer, A.M., Johnson, D.E., Hardy, B. (Eds.), Direct Seeding and Weed Management in the Irrigated Rice–Wheat Cropping System of the Indo-Gangetic plains. International Rice Research Institute and Pantnagar (India): Directorate of Experiment Station, G.B. Pant University of Agriculture and Technology, Los Baños, Philippines, pp. 139–150.

Chapter 9

Technological Innovations, Investments, and Impact of Rice R&D in India

Alka Singh*, Suresh Pal**, Anbukkani Perumal*
*ICAR—Indian Agricultural Research Institute, New Delhi, India; **ICAR—National Institute of Agricultural Economics and Policy Research, New Delhi, India

INTRODUCTION

Rice is grown in more than 100 countries around the world and is a major staple food for the majority of the population in a developing country. This is particularly true for people in South and Southeast Asia where the crop supplies dietary energy needs and livelihood security to most of the rural poor (FAO, 2004). The spectacular global advances in rice productivity started after the introduction of semidwarf, high-yielding varieties during the mid-1960s. This enabled many rice-growing economies, including India, to attain self-sufficiency in rice production. Since then, rapid production advances have been made through vertical expansion of technology by replacing traditional low-yielding varieties with modern rice varieties with higher and more stable yields. The new technology mainly relied on the strategy of reducing plant height to decrease proneness to lodging, enhancing fertilizer responsiveness, and producing early-maturing and photo-insensitive plants to enable a double cropping of rice. The rapid expansion in rice area and production was also made possible by enhanced public investment in irrigation and infrastructure, fertilizers, and price policy support. High-yielding varieties of rice were initially adopted in regions well endowed with resources. Later, the thrust of rice-breeding programs moved toward incorporating resistance to biotic and abiotic stresses and improving adaptability to relatively unfavorable production environments. These initiatives further provided impetus to the horizontal expansion of technology toward diverse production environments and problem soils. This phase of technological breakthroughs was further extended by enhancing the potential yield of the semidwarf varieties through exploitation of hybrid vigor and other varietal traits, such as improved grain types, aroma, nutritional content, cooking quality, etc. These

The Future Rice Strategy for India. http://dx.doi.org/10.1016/B978-0-12-805374-4.00009-9

technological, institutional, and political initiatives enabled India to not only attain an all-time record for rice production (106.64 million tons, 2013–14), but also to become a global leader in rice exportation (Government of India, 2015).

Concerted efforts by researchers coupled with favorable policy responses toward higher rice production have created an atmosphere of optimism; however, the journey of transition is still full of concern and apprehension. Studies show that depleting soil fertility, reduced groundwater levels, and increased incidence of pests and diseases, mainly in the intensive rice–wheat areas of irrigated regions, have started to pose serious challenges to the majority of rice farmers (Timsina and Connor, 2001). Erratic monsoon patterns, extreme weather conditions, and the inability of resource-poor farmers to use advanced technologies have also added woes in rainfed rice production systems. Rice production systems are likely to undergo major changes to bridge these production constraints, cope with additional pressure from the intense competition for land and water, absorb the higher prices of energy and fertilizer, and meet the greater demand for a reduced environmental footprint. Furthermore, it is anticipated that the future demand for rice as a staple food will increase against the backdrop of a rising population and per capita income and the changing global trade scenario; therefore, there is an urgent need to reorient the research portfolio of rice-based production systems to ensure productivity enhancement in a sustainable way (ADB, 2013).

In light of these changing realities and the future challenges ahead, this chapter aims to document the dynamics of India's agricultural R&D industry, emerging trends, and the orientation of the public and private research portfolio, with an emphasis on the investments made in reference to rice research. It also aims to highlight the contributions and success stories from the public and private sectors toward rice research, as well as identify pathways to strengthen rice research capacity and the demand-driven technology needs to achieve a sustainable rise in rice production.

RICE PRODUCTIVITY TRENDS AND PRODUCTION CONSTRAINTS IN MAJOR ECOSYSTEMS

India possesses the world's largest area under rice (44 million ha) and is the second largest producer after China. Rice is the most important cereal crop, occupying 24% of the gross cropped area and contributing 42% of the total food grain production. Rice production in India has more than doubled, from 39.27 million tons in TE 1970 to 105.72 million tons in TE 2014, by registering a growth of 2.31% per annum during this period. Although the country also witnessed a doubling of yield (from 1.06 to 2.4 t/ha) during this time (Table 9.1), levels were still lower than the global average. Rice is currently being grown in diverse production environments, mainly grouped into irrigated and rainfed systems, with the former being considered most favorable. The rainfed system is further subdivided into subsystems, such as shallow-, mid-, and deepwater rainfed lowlands and rainfed uplands. Productivity in these systems varies widely on account of the different socioeconomic and resource endowments

TABLE 9.1 Trend in Rice Area, Production, and Productivity in India (1980–2014)

	TE 1970	TE 1980	TE 1990	TE 2000	TE 2014
Production (million tons)	39	50	67	86	106
Yield (kg/ha)	1060	1235	1633	1935	2423
Area (million ha)	37	40	41	45	44

Source: Data taken from Government of India. 2015. Directorate of Economics and Statistics, Department of Agriculture, Cooperation and Farmers Welfare, Ministry of Agriculture and Farmers Welfare. http://eands.dacnet.nic.in.

of the regions. The area share and production share were reported to be highest in the irrigated system, followed by rainfed lowland, rainfed upland, and deepwater systems. The irrigated ecosystem contributes maximum production (63.5%) on account of its high area and productivity (3 t/ha). The states of Punjab, Haryana, Western Uttar Pradesh, Jammu and Kashmir, Andhra Pradesh, Karnataka, Himachal Pradesh, and Gujarat fall into this system. Parts of Uttar Pradesh, Bihar, West Bengal, Odisha, Jharkhand, Chhattisgarh, and Assam represent predominantly rainfed lowland systems that register an average productivity of 2 t/ha and account for one-fourth of the total rice area in the country. Rainfed upland, which forms the least productive system, covers about 16% of the total rice area and has the lowest productivity (1.1 t/ha). Deepwater and semideepwater rice ecosystems constitute only 3.5% of the total rice area with a productivity of 1.4 t/ha (Fig. 9.1). Rainfed lowlands and rainfed deepwater systems generally experience high rainfalls and severe floods, as well as droughts,

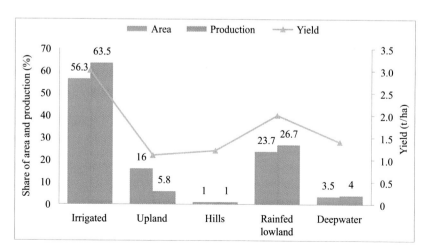

FIGURE 9.1   Rice area, production, and productivity in the major rice ecosystems of India, 2013. *(Data modified from National Rice Research Institute, 2011. Vision 2030. Cuttack, Odisha.)*

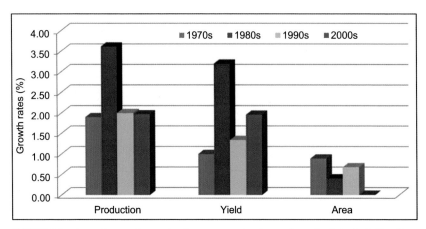

FIGURE 9.2 **Annual growth rates in rice area, production, and productivity in India.** *(Data taken from Government of India. 2015. Directorate of Economics and Statistics, Department of Agriculture, Cooperation and Farmers Welfare, Ministry of Agriculture and Farmers Welfare. http://eands.dacnet.nic.in.)*

almost every other year and, as a result, crop losses are considerably high. In upland areas, the crop faces setbacks mostly because of moisture stress. The yield gap varies in the range of 20%–30% in the irrigated ecosystem, 30%–50% in favorable shallow lowlands, and more than 50% in uplands and other unfavorable environments (National Rice Research Institute, 2011).

The decadal growth pattern (Fig. 9.2) shows that rice production grew at a maximum (3.62%) during the 1980s, and most of this was achieved through yield growth (3.20%). However, yield growth momentum decelerated during the 1990s and dropped to almost half (1.34%), before rebounding slightly (1.97%) during the 2000s. The contribution of area growth remained minimal during all periods. Among the major rice-growing states of the country, the largest share of rice area was reported in Uttar Pradesh (5.6 million ha), followed by West Bengal (5.3 million ha), Odisha (4.2 million ha), Andhra Pradesh, Punjab, and Tamil Nadu in 2013–14. However, West Bengal attained the highest rice production (15.4 million tons), followed by Uttar Pradesh (14.6 million tons), Punjab (11.3 million tons), and Haryana (4.0 million tons) during the same period (Government of India, 2015). At the all-India level, productivity increased from 0.68% pre-2004 to 1.99% post-2004, whereas rice area continued to show negative growth rates in both periods (Table 9.2). The figures show that production and yield growth declined substantially in the major irrigated rice-growing regions, with West Bengal, Punjab, and Haryana indicating a recent slowdown. This raises the important issue that declining yield growth may be due to technology fatigue, resource degradation, and production inefficiency in intensively cultivated rice areas. In contrast to this, the results revealed that relatively low-productivity states, such as Uttar Pradesh, Assam, Bihar, and Odisha, registered an upward increase in yield and production growth during 2005–06 to 2013–14 compared with 1993–94 to 2004–05 (Table 9.2).

TABLE 9.2 State-Specific Trends in Growth and Instability in Area, Production, and Yield of Rice

State	Period	Area (% growth)	Production (% growth)	Yield (% growth)
Andhra Pradesh	1994–95 to 2004–05	−2.38	−0.32	2.11
	2005–06 to 2013–14	0.36	0.31	−0.05
Assam	1994–95 to 2004–05	0.07	1.46	1.39
	2005–06 to 2013–14	1.07	6.51	5.38
Bihar	1994–95 to 2004–05	−5.03	−4.48	0.58
	2005–06 to 2013–14	−0.93	5.88	6.87
Haryana	1994–95 to 2004–05	2.35	3.16	0.78
	2005–06 to 2013–14	2.34	2.54	0.19
Odisha	1994–95 to 2004–05	−0.13	−0.35	−0.22
	2005–06 to 2013–14	−1.40	0.16	1.58
Punjab	1994–95 to 2004–05	2.02	3.59	−3.52
	2005–06 to 2013–14	1.24	1.24	0.00
Tamil Nadu	1994–95 to 2004–05	−3.19	−2.06	1.16
	2005–06 to 2013–14	−2.23	−0.79	1.47
Uttar Pradesh	1994–95 to 2004–05	0.01	0.03	0.02
	2005–06 to 2013–14	0.44	3.62	3.16
West Bengal	1994–95 to 2004–05	−0.06	2.23	2.29
	2005–06 to 2013–14	−1.01	0.21	1.24
All India	1994–95 to 2004–05	−0.21	0.47	0.68
	2005–06 to 2013–14	−0.14	1.85	1.99

Source: Data taken from Government of India. 2015. Directorate of Economics and Statistics, Department of Agriculture, Cooperation and Farmers Welfare, Ministry of Agriculture and Farmers Welfare. http://eands.dacnet.nic.in.

Modern varieties of rice have replaced traditional low-yielding varieties in almost all states. Although the rice productivity gains through varietal improvement are most important, the results show that in spite of several new releases, varietal turnover in the major rice-growing states is still low. The average age of rice varieties in the seed chain of most rice-producing states is more than 20 years, which goes against the generally accepted norm of 6–10 years (Appendix I). Furthermore, varietal concentration is quite high, as reflected by the share of the top 10 rice varieties in total seed availability. Hence, it can be concluded that the pace of replacement of older generation varieties has been slow. A similar trend was also observed in other rice-growing countries, such as Bangladesh (Baffes and Gautam, 2001). Thus, although significant investments have been made in

APPENDIX I Share of Public-Sector Availability of Certified Paddy Seeds, 2015–16

State	Name of top five varieties (kharif season)	Total number of varieties in seed supply chain	Average age of top five rice varieties grown (years)	Total availability of certified seed (1000 qtls)	Share of public sector (%)	Share of top 10 varieties in total seed availability (%)
Andhra Pradesh	Samba Mahsuri (BPT-5204)	38	20	1076	37	61
	MTU-1010					
	MTU-7029					
	Vijetha (MTU-1001)					
	Nellore Mahsuri (NLR-34449)					
Punjab	Pusa Basmati 1121	12	18	173	11	98
	PUSA-44					
	PR-118					
	HKR-47					
	PR-111					
Haryana	Pusa Basmati 1121	13	17	173	7	97
	Pusa Basmati 1 (IET-10364)					
	CSR-36					
	PR-114					
	PUSA-44					

State	Variety					
Odisha	MTU-7029	93	16	1146	100	77
	Vijetha (MTU-1001)					
	Pooja (IET-12241)					
	Rani Dhan (IET-19148)					
	Swarna-Sub1 (CR 2539-1) IET-20266					
Tamil Nadu	Savitri (IET-5897) (CR 1009)	52	23	701	34	24
	ADT-43 (IET-14878)					
	ADT-39					
	Samba Mahsuri (BPT-5204)					
	ADT (R) 45 (IET-15924)					
Uttar Pradesh	Birupa (IET-8620)	76	27	897	13	47
	Sarjoo-52					
	Samba Mahsuri (BPT-5204)					
	MTU-7029					
	Narendra Dhan-359 (NDR-359)					
West Bengal	MTU-7029	55	14	1337	36	43
	Pratikshya (ORS 201-5) (IET-15191)					
	Gontra Bidhan-1 (IET 17430)					
	Swarna-Sub1 (CR 2539-1) IET-20266					
	Shatabdi (IET-4786)					

Source: Data taken from Government of India, 2016. Farmers' Portal, Department of Agriculture, Cooperation and Farmers Welfare, Ministry of Agriculture and Farmers Welfare. Available from: http://farmer.gov.in/

developing newer-generation varieties, their adoption by farmers is slow. This raises the question of weak dissemination of new technology, farmers' limited access to new technology, as well as the low appropriateness of new technologies to farmers' needs. A demand-driven rice research agenda along with fostering of a strong public–private sector partnership for the promotion of proven technologies needs to be aggressively pushed for speedy technology adoption.

INDIA'S NATIONAL AGRICULTURAL RESEARCH SYSTEM AND R&D INVESTMENTS

Agricultural research holds great potential for raising agricultural productivity and growth, which in turn is a pivotal determinant of long-term growth in food supply and prices. Extensive empirical evidence demonstrates that agricultural R&D greatly contributes to economic growth and poverty reduction. Through its network of institutions and state agricultural universities (SAUs), the Indian Council of Agricultural Research (ICAR) of the Ministry of Agriculture represents the major public-sector agricultural R&D organizations in the country. Besides these institutions, other public organizations, such as the Department of Science and Technology, the Council of Scientific and Industrial Research organizations, and other general universities undertake agricultural research activities. CGIAR centers have also developed strong links with the Indian agricultural research system. Almost all funds from the central government are routed through ICAR while state governments provide annual block grants to their respective SAUs. Combining actual expenditure from both constituents of the national agricultural research system, the total real expenditure (2011–12 prices) on agricultural research and education (Ag R&E) is estimated to have risen almost fourfold in the last 2 decades. The all-India real expenditure rose from INR 24 billion in TE 1995 to INR 94 billion in TE 2014 by witnessing a 7.4% annual growth (Fig. 9.3). Expenditure expressed in per hectare of gross cropped area and per

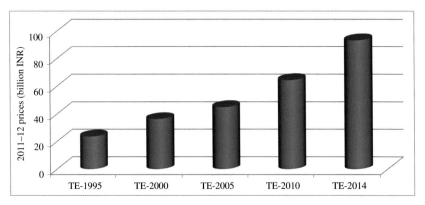

FIGURE 9.3 Trends in real government expenditure on agricultural research and education, all India (2011–12 prices). *(Modified from Singh, A., Pal, S., 2015. Emerging trends in the public and private investments in agricultural research in India. Agric. Res. 4 (2), 121–131.)*

agricultural worker has also improved considerably since the 1990s. Although the trend clearly shows the country's improved agricultural research efforts in terms of monetary resources in recent years, the sector is still underinvested as the share of research investment in agricultural gross domestic product continues to hover around 0.6%, which is far below that of other developing countries, such as Brazil (1.8%) and South Korea (2.3%) (Singh and Pal, 2015).

The resource allocation pattern of Ag R&E varied across different agroclimatic regions of the country, with increased R&D attention toward hills and the northeastern region (having research intensity close to 1%) in 2010–11. This region needs more research resources because of the difficult terrain and the need for location-specific research. The semiarid region, which has the highest share of the country's cropped area and agricultural workers, also had the largest share of research resources (45%), but the arid region received only 11% of the total resources. The demand for agricultural research continues to be strong in these regions, as their agroecological conditions are more complex and risk prone, and new breakthroughs are needed in water-saving technologies in rice cultivation, enhancement of land productivity, natural resource management, and climate-resilient agriculture. The irrigated agroecoregion, which received the highest resource priority during the Green Revolution era and covers one-fourth of the gross cropped area, now receives 0.29% of AgGDP (Singh and Pal, 2015). This should be viewed in the knowledge that the recent growth revival in this region has been weak, with yields of rice and wheat almost reaching a plateau. Therefore, achieving growth by accelerating productivity without infusing new technology, which demands higher research investments, seems difficult. The wet semiarid region comprising rainfed lowlands is marred by twin problems: drought and flood. The region is mainly agrarian (it constitutes one-fourth of the total agricultural workforce) and encompasses poverty-stricken states, such as Assam, Bihar, West Bengal, and Odisha. These states tend to have poor infrastructure and a higher concentration of subsistence producers, which makes the private sector hesitant to invest. This indicates that current resource allocation for the region is not in congruence with its needs; hence, enhanced public support is justified and it could yield high research payoffs.

Research capacity is defined in terms of the number of qualified scientists engaged in research, and the Indian national agricultural research system is one of the largest in terms of qualified personnel, with public R&D institutions employing the bulk of these scientists. The number of full-time equivalent (FTE) researchers showed a declining trend from 12,805 in 1996 to 11,286 in 2009 (Pal et al., 2012). SAUs alone employed 55% of these research scientists (FTE), whereas ICAR institutes shared one-third of the total FTE researchers. The rest were engaged in other public agricultural research institutions. Crop improvement research, the traditional strength of public R&D institutions, still receives the highest research attention by ICAR and the SAUs. The commodity focus of researchers, when separated by institute, further reveals that the research portfolio of ICAR institutes is more diversified than that of the SAUs. Livestock, natural resource management, and fisheries research received relatively greater attention in the ICAR institutes than in the SAUs (Singh and Pal, 2015).

NATIONAL RICE RESEARCH SYSTEMS: FUNDING, HUMAN CAPITAL, AND INNOVATIONS

Rice research in India has a long history and the rice R&D program has been largely in the public domain, receiving prime importance in the agricultural research portfolio. Rice R&D is being carried out by various public sector–led institutes under the aegis of ICAR, SAUs, general universities, other public and private organizations, and multinational and international research organizations. The Indian Institute of Rice Research (IIRR), Hyderabad, and National Rice Research Institute, Cuttack, are two ICAR institutes exclusively engaging in rice research for irrigated and rainfed environments, respectively. The IIRR also coordinates rice research across the country under the All India Coordinated Research Improvement Project on Rice (AICRIP). This network has several exclusively funded research stations affiliated to SAUs and departments of agriculture, with more than 350 scientific personnel. These efforts were further supplemented by the intensive rice improvement programs of SAUs aimed at enhancing the yield frontier, improving tolerance of abiotic and biotic stresses, and improving grain and nutritional quality to meet the future demand for rice in domestic and international markets. The very recent entry of the private sector into rice research and seed markets, especially in hybrids and basmati rice, is an additional opportunity to expand partnership with the public sector. The size of funds, scientific strength, efficiency of research organizations, networks, and collaborations are major factors that influence the success of research programs. It is difficult to determine the magnitude of funds spent exclusively on rice research in the multidisciplinary nature of the research portfolio; however, estimates show annual rice research expenditure[a] was INR 4.879 billion in 2012–13. This estimated rice expenditure included 22% of the crop science research expenditure of ICAR and 10% of the all India Ag R&E expenditure for that year. This amounted to INR 111 per hectare of rice area or INR 5,857 per 100,000 rural dwellers. In addition to the monetary resources, substantial human capital is also available for rice research. Approximately 780 scientists are working directly or indirectly (Janaiah and Hossain, 2004) on different aspects of rice research, which amounts to 17 scientists per million hectares of rice area. In contrast to this, countries such as the Philippines, Lao PDR, and Sri Lanka have a much higher number of scientists (25–53 scientists per million ha of rice) working on rice research (Takeshima, 2014). The majority of scientific efforts are devoted to varietal development, but other areas, such as plant protection, natural resource management, mechanization, and transfer of technology, are also supported.

---

a. Rice and wheat have been principal food crops; hence, the majority of the agricultural and food policy initiatives in terms of resource allocation over the period largely centered on rice and wheat. Pal and Byerlee (2006) estimated that the major proportion of SAU funds is being used for carrying out education, administrative, and extension activities and only 45% of the funds go to research compared to 73% in ICAR institutes. Furthermore, it was assumed that, out of total resources for research, 50% is being allocated for rice research in the case of SAUs and 30% in the case of IARI.

TABLE 9.3 Trends in Rice Varieties Developed in India by Different Rice Ecosystems (2000–12)

Period	Total number of notified rice varieties	Share of rice varieties developed by major rice ecosystems (%)				
		Irrigated	Rainfed upland	Rainfed lowland	Hills	Coastal/ waterlogged
2000–05	135	65	14	10	3	8
2006–12	225	78	8	5	3	6

Source: Data taken from Government of India, 2015. Directorate of Rice Development, Patna, Bihar, Department of Agriculture, Cooperation and Farmers Welfare, Ministry of Agriculture and Farmers Welfare.

Over the past 35 years, research efforts have led rice breeders and other scientists to develop and report nearly 680 varieties and hybrids of rice[b] for cultivation in different agroecological regions of India. Trends showed that the number of rice varieties released in the 2000s was significantly higher than in the previous decades. Of these varieties, the majority were recommended for use in the irrigated ecosystem, followed by rainfed uplands. This clearly suggests that, even in recent years, the irrigated region has received more rice research attention from the public R&D system compared to the rainfed lowlands and waterlogged rice ecosystem, which comprise 30% of the total rice area in the country (Table 9.3). The latter regions exhibited not only low rice yields, but also highly variable yields because of a range of biotic and abiotic stresses, the occurrence of flash floods, and stagnant water for extended periods of time. In terms of crop traits, the major research focus until the 1970s was largely improvement in yield, whereas subsequent breeding efforts also focused on improving resistance to pests and diseases (Table 9.4). Recent research efforts have concentrated on developing varieties and hybrids with decreased crop duration, resistance to pests and diseases, and fine grain type. A higher number of varieties with medium and long slender grain type and short-to-medium duration were released during the 2000s (Fig. 9.4). These changing priorities seem to be in line with market and consumer preferences.

However, technology adoption decisions are contingent upon farmers' perceptions of the performance of a new technology relative to that of a technology currently being practiced. Farmers may assess an improved variety not only in terms of higher yield, but also by a range of attributes, such as grain quality and taste, straw yield, input requirements (Kelley et al., 1996; Traxler and Byerlee, 1993), marketability, and compatibility with the existing cropping system. Hence, ex ante socioeconomic analysis of farmers' perceptions of varietal attributes should find an important place in crop improvement

b. Compiled from the Indian Institute of Rice Research.

TABLE 9.4 Trends in Rice Varieties Developed in India by Different Traits and Rice Ecosystems (1981–2014)

	1981–90	1991–2000	2001–14
Number of varieties developed	171	212	338
Varieties with fine grain (%)	33	37	45
Varieties resistant to diseases (%)	62	69	73
Varieties tolerant of insect pests (%)	45	58	69
Varieties of short to medium duration (%)	64	68	80

Source: Data taken from Government of India, 2015. Directorate of Rice Development, Patna, Bihar, Department of Agriculture, Cooperation and Farmers Welfare, Ministry of Agriculture and Farmers Welfare.

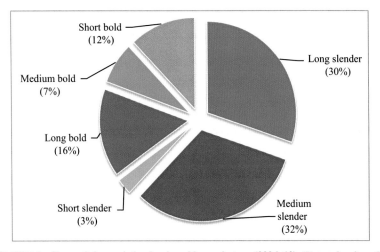

FIGURE 9.4   Share of rice varieties developed by grain type (2006–12). *(Data taken from Government of India, 2015. Directorate of Rice Development, Patna, Bihar, Department of Agriculture, Cooperation and Farmers Welfare, Ministry of Agriculture and Farmers Welfare.)*

programs. Information on the traits desired by farmers and their knowledge of the production system could be invaluable in setting the goals of a breeding program, delineating the target environment, identifying the parents for breeding, and defining the management treatment for breeding work (Eyzaguirre and Iwanaga, 1996). The recent introduction of submergence tolerance into the most popular rice varieties of the rainfed lowland rice ecosystem (i.e., Swarna and Samba Mahsuri) is considered a major breakthrough for the region. The potential adoption of recently released flood-tolerant varieties, such as Swarna-Sub1 and Samba Mahsuri-Sub1, is expected to deliver efficiency gains through reduced yield variability and higher expected yield, and equity gains by benefiting the majority of small and marginal farmers of the region (Dar et al., 2013). Studies amply demonstrated that rice varietal improvement research over the past 2 decades has generated high payoffs and accounted for 14%–24% of the total value of rice production, in addition to lifting a large number of people above the poverty line (Fan et al., 2005).

EMERGENCE OF THE PRIVATE SECTOR IN RICE R&D

Although the corporate sector has been playing a dominant role in the delivery of myriad farm inputs (seed, fertilizer, pesticide, animal feed, etc.), of late its presence has expanded to agricultural research and technology dissemination, particularly in the seed and agribiotechnology sector. Open access to public material, import of seed and planting material, and liberalization policies in the early 1990s spurred the interest of the private sector in agricultural R&D. The evidence of private R&D investment trends shows that private-sector investment in agricultural research is mainly concentrated in the areas of plant breeding; production of seed and planting materials, agrochemicals, and fertilizer; biotechnological applications; improvements in animal and livestock health; and agricultural machinery (Pray and Nagarajan, 2012). The entire market for transgenics in the country is catered to by the private sector, which grew rapidly after the commercialization of *Bt* cotton in the country in 2002. On average, the agribiotechnology industry was estimated to spend 12% of its total sales on research and development in 2011 (Singh and Pal, 2015).

Within seed and planting material, private R&D tends to focus on a limited number of commodities with particular traits. In Asia, for instance, private R&D concentrates on cash crops, such as oil palm, rubber, tea, and vegetables; hybrids of rice, sorghum, millet, and maize; and livestock and poultry (Gerpacio, 2003; Morris, 1998; Pray and Fuglie, 2001). In India, the early efforts of the private sector in seed came primarily with hybrid maize. The economic viability of hybrid maize established a stepping stone for a budding seed industry in the country. Now, hybrid seeds of cotton, vegetables, and cereals, especially hybrid maize and rice, dominate the sector. The share of proprietary hybrids developed by private R&D in total hybrids has increased significantly (more than 80%) for pearl millet, sorghum, cotton, rice, maize, sunflower, and vegetables. The area under proprietary rice hybrids, especially in eastern India (Eastern Uttar Pradesh, Bihar, and Odisha), has seen an impressive expansion (covering 6% of the total rice area). Industry data show that almost 95% of the hybrid rice market was controlled by the private sector in 2009 compared with less than 30% in 1995. In India, research on hybrid rice started in the early 1990s and 61 rice hybrids were released for commercial cultivation during the 2000s. Of these, 39 were developed by the private sector (Table 9.5), and 20 are now being aggressively marketed by the private sector, mostly in eastern India. Annual hybrid rice R&D investments by private firms were estimated at USD 9 million in 2009 (Spielman

TABLE 9.5 Institutional Share in the Release of Hybrid Rice in India (1994–2015)

S. no.	Type of institutions	Hybrid rice released during the period		
		1994–2000	2001–15	Overall
1	State agricultural universities	11	16	27
2	ICAR institutes	1	6	7
3	Indian private firms	1	29	30
4	Multinationals	1	10	11
5	All	14	61	75

Source: Data taken from Government of India, 2015. Directorate of Rice Development, Patna, Bihar, Department of Agriculture, Cooperation and Farmers Welfare, Ministry of Agriculture and Farmers Welfare.

et al., 2013) and the incremental rice output was estimated to be 989 million kg, having a value of USD 266 million in 2008–09 (Pray, 2011).

Private sector participation in seed research was further encouraged after legal IPR protection for their innovative outputs was granted under the Protection of Plant Varieties and Farmers' Rights (PPV & FR) Act. A total of 2068 applications were received for the protection of new plant varieties up to 2015 and nearly 80% of these were from the private sector, mostly for the crops in which hybrids are most common. Plant variety protection has provided confidence to private plant breeders regarding the protection of their material, hence providing a useful indicator of the level and growth of private R&D in rice. Table 9.6 shows that 946 new rice varieties were issued plant variety protection certificates during 2009–15, which was much higher than for other cereals, such as maize and wheat. Out of the total plant variety protection certificates issued, farmers claimed about three-fourths, followed by ICAR institutes and private firms. Private firms, including multinational companies (MNCs), claimed the

TABLE 9.6 Institutional Share of Plant Variety Protection Certificates Issued for New Varieties in India, Feb. 2009–Dec. 2015

Crop	Total	Indian firms	MNCs	SAUs	ICAR institutes	Farmers	NGOs
Rice	946	78	19	42	109	697	1
Wheat	117	5	—	12	97	3	—
Maize	165	46	30	2	81	5	—

Source: Data taken from Government of India, 2016. Ministry of Agriculture, "List of Certificates." Protection of Plant Varieties and Farmers' Rights Authority, Department of Agriculture, Cooperation and Farmers Welfare, Ministry of Agriculture and Farmers Welfare. Available from: www.plantauthority.gov.in/List_of_Certificates.htm

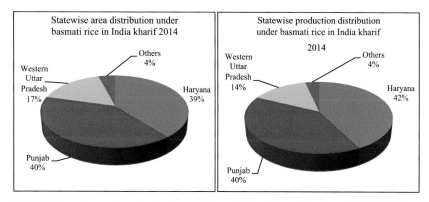

FIGURE 9.5   Area and production share of basmati rice. *(Data taken from APEDA, 2014. Basmati Acreage and Production Final Compiled Report (Kharif 2014), Report Vol. 6, Basmati Export Development Foundation (APEDA), New Delhi. Available from: http://apeda.gov.in/apedawebsite/trade_promotion/BSK-2014/Report-Volume-VI.pdf)*

highest number of certificates for rice (more than maize and wheat combined) during the period.

Having realized the scope and potential for the export of quality rice in niche markets, the rice research focus during the 2000s aggressively shifted toward genetic enhancement of superfine-quality aromatic rice. In this arena, public-sector organizations played a leadership role, which led to the release of 30 export-quality basmati and aromatic rice varieties. Many of these varieties possess tolerance of, or resistance to, major insect pests and diseases. Superfine best-quality basmati rice has traditionally been grown in the northern and northwestern part of the Indian subcontinent for many centuries. Traditional tall basmati rice varieties, such as HBC 370, Pakistani basmati, T-3, Taraori, etc., were extensively cultivated in the region before they were replaced by nontraditional newly evolved modern basmati rice varieties, such as P B-1, Pusa Sugandha, Pusa Basmati 1121, Pusa Basmati 1509, etc. With yield approximately fourfold that of traditionally grown tall basmati rice cultivars, newly evolved basmati varieties now occupy 2.13 million ha of land and contributed 8.77 million tons of basmati production in 2014. These varieties yield more than 4 t/ha in farmers' fields and are considered a major breakthrough in basmati rice research. Pusa Basmati 1121 is a major success story for basmati rice, claiming the highest area among modern basmati varieties. Since its release in 2003, the area under Pusa 1121 has consistently showed an upward trend, claiming more than 85% of basmati rice production in Punjab, Haryana, Uttaranchal, and Western Uttar Pradesh (Fig. 9.5).

A FUTURISTIC RICE R&D AGENDA

On the one hand, India's rice research program aims to enhance potential yield and input-use efficiency and overcome yield losses caused by biotic and abiotic stresses, while on the other hand it strives to improve nutrition and grain quality through the harnessing of modern scientific tools. However, several potential technological options need urgent research attention for increasing the productivity of relatively unfavorable rice-based systems. To translate these challenges into scientific innovations and specific technologies and accelerate them, a threefold strategy is suggested. First is the increased investment in public R&D. This should come from international and national funding, be aimed at upstream research, and address the major biotic and abiotic challenges with advanced scientific tools and technology systems. Second is the derivation of a set of policy initiatives that incentivize higher private investments. The private sector has already shown its willingness to invest in those technologies that have strong appropriateness; hence, further incentives will have a catalytic effect in promoting rice R&D. Public policy could also be directed into strengthening public–private partnerships, thus further augmenting research resources in the sector. Stronger IPR policies and enforcement could not only encourage complementary private investment but also accelerate the dissemination and commercialization of research. A third set of policy innovations should relate to public intervention in input and output markets, especially related to strengthening seed and procurement policies. This is needed to effectively translate innovations made in seed technology into realizable output through higher technology adoption.

Most rice production systems have witnessed increasing yield risk because of erratic weather and rainfall patterns. The frequent occurrence of abiotic stresses (i.e., drought and submergence) has been identified as the most important reason for the low productivity of rainfed ecosystems. This calls for a higher allocation of research resources to develop varieties and management technologies to better cope with extreme and erratic weather conditions. Apart from this, emphasis on crop and nutrient management practices, such as enhancing nitrogen-use efficiency in rainfed lowlands and deepwater systems, and imparting biological nitrogen fixation, could also help in improving rice productivity to a greater extent. Studies have shown that biotic constraints to rice production, such as insects and diseases of major economic importance, cause major yield losses, and chemical control continues to be the most common alternative for their management. Urgent research attention should be given to not only deploying host–plant resistance in the crop, but also strengthening IPM technologies for storage pests and pesticide residues, especially in export-quality rice. Alongside this, reducing the arsenic content in grain through crop management technologies should also be focused on, especially in arsenic-contaminated areas of irrigated ecosystems.

Furthermore, the future composition of rice demand is likely to transform because of changing consumer preferences and the demand for high-quality rice and diversified rice products on account of urbanization and income growth. Hence, it is not just a matter of producing more rice, but also meeting the rising demand for specialty rice (such as diabetic rice with a low glycemic index, medicinal rice, colored rice, etc.), fine-grain rice, and high-quality residue-free rice. Research on biofortification (i.e., the genetic enhancement of rice with nutrient enrichment, such as vitamin A) can be effective in decreasing malnutrition as part of a multipronged strategy that includes dietary diversification and fortification. Other important R&D areas could be precision farming, mechanization, and improvements in postharvest value chains for efficiency gains and loss reductions. Strengthening of rice value chains should be given priority by diversifying rice and rice-based product options for the welfare of rice farmers, consumers, and other stakeholders in the rice production and supply chain.

REFERENCES

Asian Development Bank (ADB), 2013. Food security in Asia and the Pacific. Available from: www.adb.org/sites/default/files/publication/30349/food-security-asia-pacific.pdf

Baffes, J., Gautam, M., 2001. Assessing the sustainability of rice production growth in Bangladesh. Food Policy 26, 515–542.

Dar, M.H., de Janvry, A., Emerick, K., Raitzer, D., Sadoulet, E., 2013. Flood-tolerant rice reduces yield variability and raises expected yield, differentially benefitting socially disadvantaged groups. Sci. Rep. 3, 3315.

Eyzaguirre, P., Iwanaga, M. (Eds.), 1996. Participatory plant breeding. Proceedings of a Workshop on Participatory Plant Breeding. 26–29 July, 1995, Wageningen, Netherlands; International Plant Genetic Resources Institute, Rome, Italy.

Fan, S., Chan-Kang, C., Qian, K., Krishnaiah, K., 2005. National and international agricultural research and rural poverty: the case of rice research in India and China. Agric. Econ. 33 (3), 369–379.

FAO, 2004. The state of food insecurity in the world. Monitoring Progress Towards the World Food Summit and Millennium Development Goals.

Gerpacio, R.V., 2003. The roles of public sector versus private sector in R&D and technology generation: the case of maize in Asia. Agric. Econ. 29 (3), 319–330.

Government of India. 2015. Directorate of Economics and Statistics, Department of Agriculture, Cooperation and Farmers Welfare, Ministry of Agriculture and Farmers Welfare. http://eands.dacnet.nic.in.

Janaiah, A., Hossain, M., 2004. Partnership in the public sector agricultural R&D: evidence from India. Econ. Polit. Wkly. 39 (50), 5327–5334.

Kelley, T., Parthasarathy Rao, P., Weltzien, E., Purohit, M.L., 1996. Adoption of improved cultivars of pearl millet in an arid environment: straw yield and quality considerations in western Rajasthan. Exp. Agric. 32, 161–171.

Morris, M.L. (Ed.), 1998. Maize Seed Industries in Developing Countries. Lynne Riener Publishers and CIMMYT, Boulder, CO.

National Rice Research Institute, 2011. Vision 2030. Cuttack, Odisha.

Pal, S., Byerlee, D., 2006. The funding in organizations of agricultural research in India: evolution and emerging policy issues. In: Alston, J.M., Piggott, R.R., Pardey, P.G. (Eds.), Agricultural R&D Policy in the Developing World. International Food Policy Research Institute, Washington, DC.

Pal, S., Rahija, M., Beintema, N.M., 2012. India: Recent Developments in Agricultural Research. ASTI Country Note, Agricultural Science and Technology Indicators. International Food Policy Research Institute, Washington, DC.

Pray, C.E., 2011. Impact of private R&D returns and productivity of hybrid rice in India. Presented at the ICABR Meeting. 27–29 June, 2011, Rome, Italy.

Pray, C.E., Fuglie, K., 2001. Private investments in agricultural research and international technology transfer in Asia. Agricultural Economics Report. No. 805. Economic Research Service, USDA. Available from: www.ers.usda.gov/media/463983/aer805_1_.pdf

Pray, C.E., Nagarajan, L., 2012. Innovation and research by private agribusiness in India. IFPRI Discussion Paper 01181. International Food Policy Research Institute, Washington, DC.

Singh, A., Pal, S., 2015. Emerging trends in the public and private investments in agricultural research in India. Agric. Res. 4 (2), 121–131.

Spielman, D.J., Kolady, D.E., Ward, P.S., 2013. The prospects for hybrid rice in India. Food Sec. 5 (5), 651–665.

Takeshima, H., 2014. Importance of rice research and development in rice seed policies: insights from Nigeria. IFPRI Discussion Paper 01343. International Food Policy Research Institute, Washington, DC.

Timsina, J., Connor, D.J., 2001. The productivity and sustainability of rice–wheat cropping systems: issues and challenges. Field Crops Res. 69, 93–132.

Traxler, G., Byerlee, D., 1993. A joint-product analysis of the adoption of modern cereal varieties in developing countries. Am. J. Agric. Econ. 75, 98–99.

Chapter 10

Intellectual Property Rights, Innovation, and Rice Strategy for India

Sachin Chaturvedi, Krishna Ravi Srinivas
Research and Information System for Developing Countries (RIS), New Delhi, India

INTRODUCTION

The important role of intellectual property rights (IPR) in agricultural innovation dictates the need to integrate IPR policy into India's rice strategy. However, application of IPR to plant varieties and seeds is new in India and is a first for cereals such as rice. There is a need to strike a balance between the needs and demands of different stakeholders in rice so that, while innovation is facilitated and IPR is protected, access for farmers and plant breeders is ensured and quality seeds are available at affordable prices. This should be the guiding principle for rice strategy in India. This chapter situates the intellectual property (IP) issues for rice in the broader context of IP in plant varieties and seeds and takes into account the role played by geographical indications in agriculture. Furthermore, it provides an overview of the IP regime for plants and plant varieties in India and examines the performance of the regime with respect to rice. It suggests some policy measures to facilitate innovation in rice in India, taking into account the role of IPR and the new technological options. We need to bear in mind that, although legal regimes are important, technological developments open up new possibilities for IP protection and only a deeper understanding of the legal regimes, including the IP regime and technological developments, will enable crafting a policy that balances demands from multiple stakeholders and promotes innovation without adversely affecting public interest and welfare.[a]

---

a. For an overview of IP and agriculture, see Srinivas (2015).

The Future Rice Strategy for India. http://dx.doi.org/10.1016/B978-0-12-805374-4.00010-5

IP PROTECTION FOR PLANTS AND PLANT VARIETIES

Until the 1930s, IP protection was not available for plants and plant varieties in the United States as they were considered as products of nature and hence not eligible for IP protection.[b] However, demands from the private seed sector, the advent of hybridization in plant breeding, and industrialization of agriculture meant that IP protection for plant varieties slowly but steadily became a norm.

The Plant Patent Act (PPA) of 1930 enabled patent protection to be granted to breeders of novel varieties of asexually reproducing plants. Later, the Plant Variety Protection Act (PVPA) of 1970 conferred patent-like protection to novel, sexually reproduced plants.[c] However, as this mode of protection provided for a farmers' exemption and breeders' exemption, it was perceived to be a weak mode of protection. Later, patent protection was extended to plant varieties, whereby genetically modified (GM) plants, plants, and plant varieties could be protected under patents and plant variety protection. Patenting of genes, microorganisms, biological processes, and components of plants ensured that, in addition to patenting of plant varieties, related components and processes could also be patented.[d]

Patents, plant variety protection, geographical indication, and trade secrets are the main modes of IP protection for plants and plant varieties. In some jurisdictions, such as the United States and Europe, more than one mode of protection is available for plants and plant varieties. For example, a new plant variety can be protected under patents, plant variety protection, and trade secrets in the United States. Patent protection is available for inventions and not all countries offer such a mode of protection for plants and plant varieties. Plant variety protection is now available in most countries on account of the implementation of Trade Related Intellectual Property Rights (TRIPS) by members of the World Trade Organization (WTO).[e]

India is a party to WTO Agreements, including the TRIPS Agreement, and hence is bound to adhere to TRIPS norms for IP protection. Although the move to grant IP protection for plants and plant varieties began in the 1930s, IP protection for plants and plant varieties was not available in all countries until TRIPS was negotiated and signed. Rapid changes in the IP regime in the United States, the formation of the Union for Protection of New Varieties (UPOV) in 1961, and commercial opportunities opening up for new technologies were some of the factors that led to IPR for plants and plant varieties being negotiated in the Uruguay Round (September 1986–April 1994). In the international debates over access to germplasm and sharing of germplasm in the 1980s, developing nations pointed out that although germplasm was considered as a common heritage of

---

b. See Aoki (2008) and Kloppenburg (2004).
c. See Kloppenburg (2004).
d. See Blakeney (2011), Dunwell (2010), and Yadav et al. (2014) for details and examples.
e. Kloppenburg (2004) gives an excellent analysis of developments in the United States while Llewelyn and Adcock (2006) describe the regimes in Europe.

mankind, IP protection was available for plant varieties and seeds developed from that germplasm. In the Uruguay Round, developing nations resisted the move by developed nations to expand the scope and reach of IPR and limit the powers available to governments in deciding on the grant of, enforcement of, and use of IPR. IP protection for plants and plant varieties was a contentious issue.[f] The TRIPS Agreement was finally concluded and as a compromise it provided options under Article 27.3(b). Article 27.3(b) indicates the options the parties to TRIPS have in offering IP protection to plant varieties and what can be excluded in that.

Under Article 27.3(b), "Members may exclude from patentability plants and animals other than microorganisms, and essentially biological processes for the production of plants and animals other than nonbiological and microbiological processes. However, members shall provide for the protection of plant varieties either by patents or by an effective sui generis system or by any combination thereof."

TRIPS does not define what an effective sui generis system is. The options available to members under this section include:

1. Providing patent protection for plants and plant varieties. This has been the practice in many countries that have joined UPOV or adhered to the UPOV 1991 Convention.
2. Implementing a sui generis system that can be independent of the regime for patents and through this providing IP protection for plant varieties.
3. Providing for dual protection, that is, patent protection and protection through a sui generis system for plants and plant varieties.[g]

Even as TRIPS was being negotiated, the UPOV was becoming the leading international organization in plant variety protection. Established in 1961, UPOV set the standards for plant variety protection and countries joining UPOV were obligated to give effect to the UPOV Conventions.[h] The UPOV Convention of 1978 was revised in 1991 and members joining UPOV now have to give effect to this updated convention. TRIPS did not mandate that countries that are parties to TRIPS join UPOV, nor did they recognize it as a binding convention or a model regime for plant variety protection. However, many countries that are parties to TRIPS are also parties to UPOV and vice versa. India has not become a member of UPOV. Countries that are parties to UPOV have to abide by the 1991 UPOV Convention, and therefore do not have the option to enact a sui generis system of IP protection for plant variety protection. Instead, their laws on plant variety protection have to be compatible with TRIPS and UPOV. Although the 1991 UPOV Convention provides stronger protection than what is envisaged under TRIPS Article 27.3(b), adhering to UPOV would not be a violation of TRIPS as TRIPS provides only the minimum standards for IP protection and

---

f. See Aoki (2008) and Fowler and Mooney (1990).
g. See Narasimhan (2008) and Santilli (2012) on how countries have addressed these options.
h. See Dutfield (2008) on the UPOV.

countries are free to go beyond this. These are called TRIPS-plus norms and they have been incorporated in many bilateral and regional trade agreements, including the recently concluded Trans-Pacific Partnership or Trans-Pacific Partnership Agreement. Although India is not a party to any of the agreements that mandate TRIPS-plus norms for plant variety protection, in future negotiations involving TRIPS-plus norms, this could be demanded from India.[i]

As India became a party to the WTO Agreements, changing its laws to reflect the commitments under TRIPS was inevitable. India became a party to the Convention on Biological Diversity (CBD) of 1992. This convention overruled the concept of common heritage of mankind that enabled unbridled access to germplasm and other biological resources and instead introduced access and benefit sharing (ABS) with prior informed consent. The Bonn Guidelines and the Nagoya Protocol (2010) under the CBD elaborated these further and introduced rules and regulations on ABS. The CBD mandated that genetic resources be subject to national sovereignty and sovereign states have the right to regulate access to and use of genetic resources within their borders. The CBD thus established a framework to regulate access to and use of genetic resources. Plant genetic resources are covered under the definition of genetic resources under the CBD. India became a party to the International Treaty on Plant Genetic Resources for Food and Agriculture (ITPGRFA), which was concluded in 2001. ITPGRFA envisaged a multilateral system (MLS) for accessing and sharing of plant genetic resources with provisions for ABS and it recognized farmers' rights.[j]

Therefore, India had to evolve a new regime for IP protection for plant varieties that had to be compatible with TRIPS, the CBD, and ITPGRFA. Instead of amending its Patent Act to include patent protection for plant varieties, India chose to create a sui generis system for plant variety protection by enacting a new law. It also provided for relevant provisions in the Patent Act so that it used the flexibilities in the TRIPS Agreement and options under Article 27.3(b). To give effect to its commitments under the CBD, India enacted the Biodiversity Act of 2002 and, through that act, the National Biodiversity Authority (NBA) was established in 2003. The NBA was to oversee ABS and regulate access to genetic resources. A gazette notification dated December 17, 2014, informed that prior approval from the NBA was not necessary for accessing the germplasm of crops covered under ITPGRFA for use in research, breeding, and training for food and agriculture. The Indian focal point for the implementation of this treaty is the Department of Agriculture and Cooperation, Ministry of Agriculture. According to an office memorandum dated February 16, 2015:

> *Ministry of Environment & Forest and Climate Change vide its notification No. S.O. 3232 dated December 17, 2014, has, in consultation with the National Biodiversity Authority, declared that the Department of Agriculture & Cooperation*

---

i. See Drexl et al. (2014) for a discussion of TRIPS-plus.

j. See Frison et al. (2011) on ITPGRFA and Oguamanam (2014) on farmers' rights. See Halewood et al. (2013) for a discussion on crop genetic resources as commons.

may, from time to time, specify such crops as it considers necessary from amongst the crops listed in the Annex-I of the International Treaty on Plant Genetic Resources for Food and Agriculture (ITPGRFA), being food crops and forages covered under the multilateral system thereof; and accordingly exempt them from sections 3 & 4 of the said act, for the purpose of utilization and conservation for research, breeding, and training for food and agriculture. Accordingly, the Department of Agriculture & Cooperation, Ministry of Agriculture, declares all the crops listed in the Annex-I of the ITPGRFA to be exempted from Sections 3 & 4 of the Biological Diversity Act, 2002 (18 of 2003), for the purpose of utilization and conservation for research, breeding, and training for food and agriculture. The exempted crops will be governed by the guidelines to facilitate the exchange of Plant Genetic Resources under the Multilateral System of ITPGRFA issued by this Department OM of event no. dated July 30, 2014. A copy of the guidelines along with the above-mentioned notification is enclosed.[k]

Rice is listed in Annex-I of the ITPGRFA, and therefore in light of its exemption from Sections 3 and 4 of the Biodiversity Act of 2002 it can be presumed that the NBA has a very limited role in accessing rice germplasm from India.

To give effect to its commitments under TRIPS and under the ITPGRFA, a new law for Protection of Plant Varieties and Farmers' Rights Act (PPVFRA) was enacted in 2001. This act led to the creation of "The Protection of Plant Varieties and Farmers' Rights Authority" in 2005. The authority, based in New Delhi, is the go-to authority for matters related to plant variety protection in India. PPVFRA provides plant variety protection for varieties registered with the authority in the form of plant breeders' rights (PBR). It uses distinctness, uniformity, and stability (DUS) criteria to check whether a variety is eligible for protection, and testing for DUS is mandatory for the granting of PBR for new varieties. The same rules are applicable for farmers, institutions, the private sector, or for any other applicant.

The guidelines to test for rice DUS were issued by the Directorate of Rice Research, now known as the Indian Institute of Rice Research. According to the guidelines:

A new variety shall be deemed to be:

(a)*Novel, if, at the date of filing of the application for registration for protection, the propagating or harvested material of such variety has not been sold in India, earlier than one year, or outside India, earlier than four years, before the date of filing such application.*

(b)*Distinct, if it is clearly distinguishable by at least one essential characteristic from any other variety whose existence is a matter of common knowledge in any other country at the time of filing of the application. An essential characteristic is a heritable trait which is determined by one or more genes or other heritable*

---

k. www.indiaenvironmentportal.org.in/files/file/Guidelines%20for%20the%20Implementation%20of%20International%20Treaty%20for%20Plant%20Genetic%20Resources.pdf

determinants that contribute to the principal features, performance, or value of the plant variety.

(c) *Uniform, if subject to the variation that may be expected from the particular features of its propagation, it is sufficiently uniform in its essential characteristics.*

(d) *Stable, if its characteristics remain unchanged after repeated propagation or, in the case of a particular cycle of propagation, at the end of each such cycle.*

(pp. 3–4, Directorate of Rice Research, 2006).

Although the Directorate has been renamed, we have retained the old name here as the document is listed as a publication of the Directorate of Rice Research.

In addition to this, the act recognizes extant varieties and farmer varieties. While the DUS criteria are followed by the UPOV Convention, in most countries, PBR provide protection only for novel varieties whereas varieties that fulfill the norms for inventions can be patented. In the PVP regime in India, such a rigid approach is not followed. To encourage innovation by farmers and to provide for registration for varieties developed and in use, the act provides PBR for farmers' varieties and extant varieties. As will be shown later, this has resulted in the registration of many varieties by farmers and public-sector institutions. In addition to PBR, the act provides for farmers' rights over seeds and for the use of varieties to develop new varieties by plant breeders. Such a use by plant breeders is known as the plant breeders' exemption and was provided by the UPOV Convention 1978.

Under the PVPFRA, (1) farmer means any person who cultivates crops by cultivating the land himself; or (2) cultivates crops by directly supervising the cultivation of land through any other person; or (3) conserves and preserves any wild species or traditional varieties, or adds value to such wild species or traditional varieties through selection and identification of their useful properties.

Farmers' variety means a variety that has been traditionally cultivated and evolved or developed by the farmer or farming community in their fields; or it is a wild relative, landrace, or variety about which the farmers possess common knowledge and special features compared to others.

PBR give exclusive rights to produce, market, sell, distribute, and export or import seeds of the protected variety. Researchers' rights enable a researcher (often a plant breeder) to use a protected variety for experimental purposes, to develop a new variety, or to use that variety as an initial source for developing another variety. Farmers' rights take into account the multiple roles played by farmers as users of seeds, savers of seeds, innovators, and conservers of plant genetic resources including traditional varieties and landraces. Hence, the rights available to farmers under the act include the right to save, reuse, and exchange seeds.

In the last 2 decades, such an exemption or right for breeders has been eliminated or its scope has been limited in many countries. Farmers' rights over seeds have also been limited or eliminated for all practical purposes in many countries, and under the 1991 UPOV Convention it is left to the discretion of the member country to provide for such rights to farmers. In the European Union, small-scale farmers do not pay royalties for the reuse of seeds of protected varieties while large-scale farmers have to pay these royalties. In the United States, exchange of seeds between farmers is disallowed, but farmers can reuse seeds for their own purposes for varieties protected under plant variety protection law. However, in reality, as many countries allow both patent protection and plant variety protection, the scope for such rights for farmers is practically nil as patent law does not permit such exemptions or use by farmers. The introduction of essentially derived varieties (EDVs) in UPOV 1991 has further eroded the scope of the breeders' exemption.

The extent to which patent law provides exemptions for research purposes varies between different countries. For example, in J.E.M. Ag Supply, Inc. v. Pioneer Hi-Bred International, Inc. (534 U.S. 124), the U.S. Supreme Court confirmed that both modes of protection, that is, patent protection and plant variety protection, were permissible for plant varieties. Pioneer had patented hybrid corn seeds. J.E.M. Ag Supply, conducting business under the name of Farm Advantage, bought these patented hybrid corn seeds from Pioneer in bags. These were resold without authorization from Pioneer to do so. Pioneer contended that such sales infringed on the patents it owned. The seeds were sold under a license that included the following conditions:

1. The license was granted solely to produce grain and/or forage;
2. The license did not extend to the use of seed from such crop or the progeny thereof for seed multiplication or propagation; and
3. Use of such seed or the progeny thereof for propagation was prohibited. Similarly, use for production or development of a hybrid or another variety of seed was prohibited.

J.E.M. Ag Supply argued that, as sexually reproducing plants were not patentable subject matter within the scope of 35 U.S.C. § 101, the PVPA was the applicable law and it was the PVPA that provided the legal means for the protection of plants. In a 6:2 judgment, the Supreme Court rejected this contention and said: "…, we hold that newly developed plant breeds fall within the terms of § 101, and that neither the PPA nor the PVPA limits the scope of § 101's coverage. As in *Chakrabarty*, we decline to narrow the reach of § 101 where Congress has given us no indication that it intends this result." This effectively meant that patent protection was available for plants, even if they were protected under the PPA or PVPA.

In 2013, in Bowman v. Monsanto Co., the Supreme Court held that the patent exhaustion doctrine did not apply to patented seeds and hence farmers did not have the right to reproduce patented seeds through planting and harvesting. The three major laws on plant variety protection in the United States offer different modes of protection for different categories of plant varieties. The PVPA covers plant varieties that undergo sexual reproduction, including tuber-propagated plants and first-generation (F_1) hybrids. Protection for inventions that meet the requirements of utility, novelty, and inventive step is available under patent protection. Plant varieties can be protected under this and this mode of protection is available for GM plants. The PPA covers plant varieties that are vegetative and are reproduced through asexual propagation, excluding edible tubers. Ornamentals, fruit, and forest trees can be protected under this act. Edible tubers, such as potatoes, were perhaps excluded from the PPA to prevent private firms from holding monopoly rights over important sources of food.

In India the scope of protection under PBR is limited when compared with Europe and the United States and, in the absence of patent protection, plant variety protection through PBR is much weaker than what is available under UPOV. Trade secrets is another mode of protection available for plant varieties and germplasm; however, in India, there is no specific law to regulate trade secrets and hence this mode of protection is not available for plant varieties per se.

According to Section 3(b) of the Indian Patents Act, inventions contrary to public order/morality or prejudicial to human, animal, plant life, health, or environment are not patentable. Furthermore, Section 3(j) mandates that no patents should be granted for plants in whole or in part (other than microorganisms), including seeds, varieties, species, and their production and propagation by essentially biological processes. Although the Patents Act does not indicate any specific rule regarding GM plants, inventions and processes relating to GM plant technology are patentable as long as they meet the criteria for patentability. In fact, patents have been granted in India on various technologies and processes related to *Bt* cotton.[1] Although transgenic plants are not patentable, transgenic plant varieties, if sought to be registered as EDVs under the PPV-FRA, may be granted IP protection. Since no such application has been made so far, this should be taken as a hypothesis. The number of EDVs registered so far is negligible.

However, as the rights conferred under the PVPFRA are not as strong as rights under the Patents Act, an innovator will prefer to obtain patents on gene sequences, processes, and GM microorganisms that are essential for the development of a transgenic plant or event and try to enforce them. Monsanto adopted this strategy and used patents on *Bt* cotton technology to license the use of the gene conferring the trait(s) and entered into contract with seed companies rather than taking the EDV route. However, to succeed in this, the company should have patents on many related technologies so that its patent portfolio covers all the technologies related to genetic modification, development of a transgenic plant, and conferring the trait through genetic modification in a plant. Monsanto had this advantage and the strategy to develop a strong patent portfolio. When companies have multiple patents and develop patent thickets around a technology, inventing around these becomes difficult, if not impossible.

1. See Sastry et al. (2011) for details.

In other words, although the Indian Patents Act explicitly prohibits patenting of plants, seeds, and varieties, and essential biological processes related to their production and propagation, it does not prohibit patenting of GM technology related to plants and plant varieties per se. Thus, there is no contradiction between the Patents Act and PVPFRA. If GM rice is permitted in India in the future, it is likely that the strategy adopted by Monsanto for *Bt* cotton will be adopted by others, particularly those in the private sector. This approach will also be useful for the public sector as patents can be used as a defensive mechanism. The public sector, with a strong patent portfolio, need not opt for hybrid varieties or exorbitant license fees and instead can promote open-pollinated varieties (OPVs). Table 10.1 provides a comparative analysis of UPOV 1991, TRIPS, Patents Act, and the PVPFRA.

To sum up, the coming into effect of the TRIPS Agreement radically transformed IP protection in plants and plant varieties. It is no longer an option but a mandate for most countries that are party to TRIPS.

PLANT VARIETY PROTECTION IN INDIA WITH A FOCUS ON RICE

The impacts of the PVPFRA cannot be studied in isolation of innovation and investment in the seed sector and the use of the act and its provisions by different stakeholders, such as farmers, public-sector institutions (including ICAR and state agricultural universities, SAUs), and the private sector. The PPVFRA was enacted as a response to India's commitments under TRIPS with many objectives. When the PPVFRA and the authority came into effect, the seed industry in India and globally was undergoing a transformation.[m] Studies have noted that the private sector's position in the seed sector has been further strengthened over the past 20 years and its market share in seeds for cash crops has been on the rise. Although the protection under PBR in the act is relatively weak, it still provided a useful option for the private sector. Moreover, registration of varieties was important for protection because, in the absence of registration availability, no protection will be available and misappropriation and planting of illegal/ unauthorized varieties will go unchecked.

According to the act and procedures laid down with respect to a variety, the denomination assigned to such by the applicant has to be specified and supported by an affidavit. Complete passport data of the parental lines from which the variety has come and the geographic location from which the genetic material has been taken should also be specified. Regarding denomination, unacceptable denominations have been listed and the penalty for applying for false denomination is also specified.

Given the multiple objectives of the act, it is obvious that its intended impacts are not in favor of one stakeholder at the cost of other stakeholders. As

---

m. See Spielman et al. (2011).

TABLE 10.1 A Comparison Among UPOV 1991, TRIPS, Patents Act, and PVPFR Act

Subject	UPOV 1991	TRIPS	Patents Act	PVPFR Act
Patent protection	Yes	Can be provided but depends on implementation of Article 27.3(b)	Not possible	Not available
Dual or more modes of protection	Yes	Optional	No	No
FR	Optional and limited in scope	Optional subject to implementing Article 27.3(b)	Not applicable	Yes
PBR	Mandatory	TRIPS does not directly provide for PBR but possible through Article 27.3(b)	Not applicable	Yes
Criteria for protection	DUS	Depends on implementation of TRIPS 27.3(b)	Not applicable	DUS
Duration of PBR	Not less than 20 years from the date of grant; for trees and vines, not less than 25 years.	Patents, 20 years	Not relevant for plants or plant varieties	Plant variety protection: (1) for trees and vines, 18 years,[a] (2) for extant varieties, 15 years,[b] and (3) for others, 15 years[c]
Strength of rights	Very strong with dual protection	Depends on implementation of Article 27.3(b). Could be from very strong to moderate or weak	Not applicable	Moderate as it provides for FR and exemption for research

DUS, Distinctness, uniformity, and stability; FR, farmers' rights; PBR, plant breeders' rights; TRIPS, Trade Related Intellectual Property Rights.
[a]From the date of registration of the variety.
[b]From the date of the notification of that variety by the Central Government under Section 5 of the Seeds Act, 1966.
[c]From the date of registration of the variety.

the act allows application for and granting of registration for different types of varieties, how well equipped the stakeholders are at using the act and benefitting from it is an important issue. Given the time lag between application, testing, and granting of a PVP certificate, the number of applications alone cannot be used as a criterion to assess the effectiveness of the act in promoting innovation. The available information from the authority can be found at http://plantauthority.gov.in/List_of_Certificates.htm.

For the period from 2009 until December 1, 2015, 1948 varieties were registered, of which 841 were rice varieties. The breakdown is presented below. Farmers' varieties have not been classified into private, public institution, and SAUs as these varieties are registered in the name of farmers or are registered through civil society organizations representing them.

Type	Total	Private	Public institution	SAUs
Extant varieties	183	40	102	41
New varieties	61	55	5	1
Farmers' varieties	507			

SAU, State agricultural university.

As of March 31, 2016, the total number of certificates issued by the authority was 2193 and, of these, 1021 were certificates for rice.[n] As of June 10, 2016, the breakdown of applications received for registering rice varieties was as follows:[o]

Crop	Total	Public sector	Private sector	Farmers
Rice	5116	274	365	4477

Prima facie, the data reveal that farmers' varieties and extant varieties constitute more than 90% of the registered rice varieties. The number of new varieties is much less (61) and this indicates that innovation with respect to new rice varieties is lower. The same is also true for other crops. According to the ICRA (known earlier as Investment Information and Credit Rating Agency of India Limited), although 4237 applications were filed with the authority during 2007–15, only 1773 certificates were granted up to March 2015. Of these, new varieties accounted for 227 and extant ones accounted for 1546. ICRA also points out that the private sector in the Indian seed industry invests much less in R&D (4%–5% of sales) than the leading global companies, which invest 8%–10% of their sales in R&D (ICRA, 2015). Other studies have pointed out that the public sector dominates in terms of the number of certificates granted but most of the certificates are for extant varieties; extant varieties are also preferred by the private sector but applications for new varieties are more in proportion compared with the public sector. The extensive reliance on extant varieties seems to indicate that the public sector prefers to cash in on the varieties already developed

---

n. www.plantauthority.gov.in/List_of_Certificates.htm
o. www.plantauthority.gov.in/pdf/Status%20Crop%20wise%20Application1.pdf

and tested for plant variety protection while its pipeline of new varieties is not large. One other plausible explanation is that the authority sets a deadline for filing applications under this category and this could have resulted in a rush of applications. However, as Venkatesh and Pal (2014) point out, ICAR could protect 1700 varieties under extant varieties but only 50% of these have been protected under the act. These authors contend that the analysis of crops and varieties protected by ICAR indicates that the public sector focuses on low-value, high-volume crops (Venkatesh and Pal, 2014). They point out that, in the case of cereals, the number of public–private partnerships (PPPs) is increasing (105).

For farmers' varieties, rice applications constitute the lion's share with 3436 out of 4349 applications in 2006–14. The year-wise breakdown in Table 10.2 indicates that the number of applications has increased dramatically in the past few years.

However, the conversion rate, that is, the number of certificates granted against the number of applications received, is low. Although 4262 applications (for all crops) had been received by the end of 2014, the number of certificates granted was just 511 (Hanchinal, 2015). Within rice, applications for farmers' varieties came mostly from Odisha. Hanchinal points out the various issues with granting PVP certificates to farmers and takes the view that if the national agricultural research and extension systems and NGOs work together, much could be achieved to ensure that farmers benefit maximally from the act.

According to Venkatesh et al. (2015), there is a positive relationship between high-value crops and benefits from plant variety protection. Returns from PVP, as seen through higher seed prices, were maximum for cotton, followed by maize and rice (Venkatesh et al., 2015). However, the benefits of PVP are also linked with the seed replacement rate. In the case of crops for which this rate is low, the availability of PVP does not guarantee higher returns. For hybrids, as farmers will obtain diminished yields by planting their own output, the seed replacement rate will tend to be higher.

An assessment of the PVPFRA is beyond the scope of this chapter. However, a few observations can be made in light of the literature and data provided. The clamor for protection of extant varieties has not diminished while protection applications for new varieties has remained low, indicating that the act has

TABLE 10.2 Number of Applications Received by the Authority Under Farmers' Varieties

Crop	2007	2008	2009	2010	2011	2012	2013	2014	Total
Rice	2	4	116	0	938	278	812	1286	3436

Source: Data extracted from Hanchinal, R.R., 2015. Providing intellectual property protection to farmers' varieties in India under the Protection of Plant Varieties & Farmers' Rights Act, 2001. J. Intellect. Prop. Rights 20(1), 11.

not significantly incentivized the development and registration of new varieties of rice. Although the public sector seems to have used the option to register for extant varieties, it is doubtful whether it has used this option to the maximum. Despite the availability of protection for farmers' varieties, farmers have not been able to benefit from this act at least in terms of the number of varieties protected.

It may not be out of place to point out that, although the PVPFRA is necessary, it alone is not enough to create sufficient incentives for innovation. Much-needed reform through the Seeds Act and clarity in policy regarding GMOs in agriculture are necessary to provide a conducive milieu for the development and registration of new varieties. In terms of the registration and categorization of varieties, the numbers alone do not tell us much about the economic value of these varieties as more data and research are needed to assess the impact of plant variety protection in appropriating the value of the varieties by farmers, the public sector, and the private sector.

HYBRID RICE IN INDIA

According to the Directorate of Rice Research, Patna, 78 hybrid rice varieties have so far been notified or released but they are cultivated on less than 10% of the land under rice. The yield of hybrid varieties declines after the first generation and, to maintain yield, farmers have to buy seeds. This is not the case for OPVs. Although seed prices are high, farmers tend to prefer hybrids as they can be economically advantageous due to enhanced yields. Many studies have pointed out that although hybrid rice has substantial potential, its use is low. According to Sarkar and Ghosh (2013), farmers in West Bengal need hybrid varieties that are input efficient and better quality. Despite generation of higher yields from hybrid varieties, farmers saw few advantages due to the higher cost of production and lower price realization. Another study in Bihar pointed out that the cost of irrigation, seeds, and other inputs was higher for hybrids than for high-yielding traditional varieties and access to hybrid rice technology was poor. Another issue with hybrid rice is that of the quality of the rice and its suitability for cooking (Choudhary, 2013).

A study in Tamil Nadu found that the diffusion of hybrid rice technology is low even in traditional rice-growing areas. Although a section of farmers with small and large landholdings are aware of and prefer hybrid rice varieties despite the high costs and issues in adopting new technology, marginal farmers prefer traditional high-yielding varieties on account of the low cost of cultivation and ease in using traditional technology (Sivagnanam, 2014). According to Pray (2011), the diffusion of hybrid rice is confined to a few states while its diffusion in southern India is negligible. The author points out that private firms have a market share of 95% in hybrid rice while hybrids are grown on ~6% of the area cultivated under rice. Although the public sector has developed 28 varieties, the private sector (with 15 varieties) leads in hybrid rice. In terms of yield, there is a 17% increase in average yield over traditional high-yielding varieties and 90% of the gain accrues to the farmers (Pray, 2011). This underscores the importance of hybrids for increasing rice production in India. The Central Rice Research Institute states:

> *Although rice hybrids developed by the public and private sectors have made some progress in irrigated areas of semi-arid to subhumid regions, the yield realization needs to be enhanced through increased heterosis as well as introgressing known pest and disease tolerance genes in the parents. There is also a need to develop hybrids for rainfed lowlands of high rainfall areas. The yield potential of hybrids for such situations needs to be enhanced by increasing the degree of heterosis through the development of suitable parental lines (CRRI, 2011).*

(CRRI has been renamed the ICAR-National Rice Research Institute and is based in Cuttack.)

Spielman et al. (2013) point out that the constraints in scientific discovery, technology development, and product delivery adversely affect the development and diffusion of hybrid rice in India despite its potential. They suggest that more investment by the public and private sectors is needed along with a stronger IP regime with better resolution of infringement issues. They also suggest that more attention should be paid to issues such as germplasm diversity and grain quality, which could help in increasing the use of hybrids. They propose that hybrid rice be promoted as a platform for pro-poor GM crop development. As the seed prices of hybrid varieties tend to be higher, and since farmers need to buy seeds for replanting every season, subsidizing hybrid rice seeds for poor and marginal farmers could enable faster and wider diffusion.

Much needs to be done to realize the potential of hybrid rice in India, including increasing investment in developing appropriate varieties and incentivizing diffusion and adoption across different agroecological regions. The current PVP regime need not be revised for hybrid rice but additional incentives to promote R&D in hybrid rice can be thought of. For example, the state can support private-sector R&D in cases for which the private sector can develop suitable hybrid varieties and can promote PPPs for their development and diffusion.

GENETICALLY MODIFIED RICE

The potential of GM rice was put forth through the development of Golden rice. Ironically, the development of Golden rice highlighted the issues in IP management and access to technologies protected by IP.[p] Irrespective of Golden rice becoming available sooner or later, GM technology can be used to develop rice varieties suitable for India. Since the government has invested

---

p. See Fischer (2013).

heavily in S&T infrastructure for GM technology, GM rice is a technical option that should be explored fully. However, there are policy and regulatory issues with respect to GM crops that constrain this. Given the controversies in *Bt* cotton over the high technology fees charged by Monsanto–Mahyco and on the access to technology, heavy reliance on the private sector in GM rice varieties should be avoided. A better approach would be to develop GM rice in the public sector on a priority basis and allow the private sector to also develop GM rice. It is important that the public sector take up the hard and important task of developing GM rice and become involved in technology transfer. Other options could include developing varieties of GM rice with traits sought and needed by farmers and stacking genes in GM rice so that more than one trait is embedded in that rice variety.

The public sector should use the PVP regime effectively and obtain patent protection for key technologies and processes. Similarly, the use of OPVs for the diffusion of GM rice would make it attractive to farmers, although farmers may not choose OPVs if they do not confer distinct advantages over the varieties in use. Almost all varieties of *Bt* cotton are hybrids, and therefore farmers have no choice except to buy the seeds every season to maintain yield. Despite this they prefer hybrids as the gains are worth the higher price paid for seeds. However, in the case of cereals, such as rice, such an approach should be avoided if the high price for seeds may constrain adoption. Instead, GM rice should be made available as hybrids and as OPVs. However, farmers' pragmatic response would be to prefer GM rice hybrids if they find them to be very advantageous in terms of yield, costs, and income over OPVs.

The effective use of competition law and policy, and facilitating access to and sharing of technology, can help in overcoming problems associated with monopoly control through IP rights.

Jikun et al. (2015) argue that GM rice could result in the reduced use of pesticides and improvements in farmers' health. GM rice is yet to take off in Asia although Asia led the world in the Green Revolution. Although there are many issues to consider, such as consumer acceptance and impacts on trade, for India a major issue is whether the large-scale adoption of GM rice would affect its rice exports, particularly to Europe. Another issue is whether the adoption of GM rice would adversely affect exports of basmati rice. With respect to primary or secondary centers of genetic diversity in rice, the cautious view proposed by the Task Force on Application of Agricultural Biotechnology headed by M.S. Swaminathan should be taken into account (www.frontline.in/static/html/fl2113/stories/20040702002809400.htm).

Since India uses geographical indications to promote basmati rice and protect it, it is essential that the cultivation of GM rice not be encouraged in regions where basmati rice is grown extensively. Given the potential of GM rice, it is critical that the policy for GM rice be synergized with the overall rice strategy for India. However, this will not be possible until there is clarity on the policy on agricultural biotechnology in India.

ACCESS TO GERMPLASM, THE RICE GENOME, AND IP RIGHTS

Germplasm was once considered as a common heritage of humankind and free flow of germplasm was possible as there were hardly any restrictions on its use for crop development. Since the acrimonious debates in the 1980s and the coming into force of the CBD, the situation has changed. National governments now decide on access to germplasm and it is often combined with formal ABS. With respect to germplasm in CGIAR centers, it is held in trust by FAO and access and sharing are mostly done through material transfer agreements (MTAs). ICAR, SAUs, institutes under CGIAR, and others had previously freely shared germplasm but now the procedures have become complex and the use of germplasm comes with the conditions indicated in MTAs and other transfer documents. Although germplasm per se cannot be patented, the products developed through its use can be protected through IP. As India is a party to the CBD and ITPGRFA, and with the setting up of the NBA and the authority, accessing germplasm has become complicated. India has ratified the Nagoya Protocol of the CBD. Globally, not all countries that have ratified the CBD have become parties to the ITPGRFA. Some germplasm is held in countries that are not parties to the CBD (e.g., the United States) and some is held by the private sector.

Whether an MTA provides for freedom to operate (i.e., it allows use of germplasm for research and development of new varieties and other uses), or whether it restricts such use through conditions or through reach-through licenses is an important concern. Although some studies exist on the sharing of germplasm by CGIAR centers, we have not come across many studies that look into issues in the sharing of rice germplasm and associated IP issues. It can therefore be presumed that Indian researchers do not face significant problems in sharing germplasm and accessing it from sources within India. However, with the enactment of the PPVFRA and the coming into force of ABS, there is a need to perform empirical studies on the flow of germplasm and whether IP is a constraint to this. Since there is an increased emphasis on PPPs in crop and product development, such studies can help in formulating guidelines on the sharing and use of IP in such PPPs. We should understand that both data and material (germplasm) are important for plant breeding.[q]

A study on India's contribution to global germplasm exchange shows that India has sent rice germplasm to 50 countries. Pointing out this and the issues in sharing and use of germplasm in the context of post-CBD and ITPGRFA, Jacob et al. (2015) state:

> *One of the major concerns is whether, with the foregoing of the case-to-case basis transaction mechanism, as done under the post-CBD era, India is going to lose direct control over utilization and benefit acquirement of its indigenous germplasm accessions. There is similar concern over the utilization of the germplasm provided by India in the past, which is currently held by multiple international*

---

q. See Rimmer (2005).

organizations, including CG centers, under the purview of their respective legal clauses that are outside India's legal ambit. The norms for sharing of such germplasm accessions are currently not uniform across institutions as has been cited in the case of several mega-hybrid consortium projects involving multinational companies, where the CG genebanks are sharing germplasm with the private sector on payment basis. The concept of minimizing cost of germplasm transactions by centralizing the conservation strategy also requires a more objective analysis since accumulation of resources at a single point invites more issues than mere cost benefits that are often cited.

A 2008 study by CAMBIA pointed out that patents have been granted on various components of the rice genome and broad patent claims on different components of the rice genome have been made. Another study by Jefferson et al. (2015) stated: "We identified 3,956 sequences overlapping across all organisms examined (maize, rice, soybean, and human) and subjected their corresponding 763 patents to further characterization, including claim analysis."[r] The authors noted that post-Myriad, the claims have been modified.[s] More studies are needed to understand the scope of the claims made and the claims granted in patents. An extensive study of claims and patents on components of the rice genome will enable mapping of the patent landscape and identify areas with freedom to operate.

As part of the rice strategy for India, there is an urgent need to make in-depth studies on the transfer and use of rice germplasm from India and the rules governing such transfer and use. These studies should also take into account IP issues and examine whether the current practices, including MTAs/SMTAs, are adequate to protect India's interest and to prevent misappropriation of rice germplasm from India by third parties. Access to rice germplasm and data on the rice genome cannot be treated as separate issues. Given the importance of using genomics for plant breeding and other purposes, the rice strategy for India should include a plan to benefit the most from access to germplasm and studies on the rice genome.

GEOGRAPHICAL INDICATION AND RICE IN INDIA: BASMATI AND BEYOND

Geographical indications (GIs) are a form of IP that are well suited for agricultural products that are associated with a specific location or region with distinct qualities or characteristics. In India basmati rice is a well-known example of a

---

r. The 2008 study was done by CAMBIA (www.cambia.org/daisy/RiceGenome/3648.html).

s. In *Association for Molecular Pathology, et al. v. Myriad Genetics, Inc., et al.* 133 S. Ct. 2107 the Supreme Court of the United States held that "A naturally occurring DNA segment is a product of nature and not patent eligible merely because it has been isolated, but cDNA is patent eligible because it is not naturally occurring." This judgment raised the bar for patenting genes and their components.

rice variety that is known for its distinct qualities with a huge export market. In 2013–14, basmati exports from India were ~INR 29,000 crores. GIs are part of the TRIPS Agreement but they are different from other IPR in many ways. One important difference is that GI is closely linked or associated with a specific territory, locality, or region and hence derives its value from this and the specific qualities or characteristics associated with that product. The protection for GI is justified on many grounds, including protection of consumers against fraud; protection of the interests of the producer(s) of the good; regional, local, and territorial development and conservation; and protection of biological, cultural, and biocultural diversity. GI is often associated with localized products that are traded widely or consumed for their specific traits.

GI is considered as a collective right in India and grant of ownership to the registered owner (often a collective or a legal entity or an agency) represents the interests of the producers.[t] The rights of use for GI are hence provided to the producers who are registered as authorized users. As GI is a collective right and since often it is used to protect the rights of producers, technically, there is no bar to the state becoming involved in this as an applicant either directly, through any of the agencies of the state, or as a representative of the producers.

To give effect to the provisions of TRIPS, the Geographical Indications of Goods (Registration and Protection) Act of 1999 was passed and it came into effect in September 2003. The implementing rules were specified in Geographical Indications of Goods (Registration and Protection) Rules 2002. GI is defined in the Act, Section 2(e), as:

> *an indication which identifies such goods as agricultural goods, natural goods, or manufactured goods as originating or manufactured in the territory of a country, or a region or locality where a given quality, reputation, or other characteristic of such goods is essentially attributable to its geographical origin. For manufactured goods, one of the activities of either the production or of processing or preparation of the goods concerned takes place in such territory, region, or locality, as the case may be.*

Under Section 11 of the Act, an applicant shall be any association of producers or any organization or authority established by or under any law representing the interests of the producers of the concerned goods. Furthermore, the act states that an applicant of the GI can be the registered proprietor while an authorized user can be any person claiming to be the producer of the goods in respect of which a GI has been registered as under Section 17. Thus, an entity A can be the registered proprietor of the GI while X, Y, or Z who are producers of the good for which the GI has been granted can be producers. In other words, individual ownership of GI is not possible under the act but the collective rights can be used by individuals as producers. The government of India has identified different agencies for applying for and obtaining GI for different products. For

---

t. See generally Das (2006), Marie-Vivien (2008, 2010), and Chandola (2006).

example, a well-known GI, "Darjeeling Tea," is registered in the name of the Tea Board and the Tea Board is the agency that will oversee the use of this GI and protect it within India and elsewhere. Although the authorized tea growers can use this GI, they do not own it, nor can they enter into any litigation on the GI (i.e., Darjeeling Tea). For registering GIs on agricultural products, the Agricultural and Processed Food Products Export Development Authority (APEDA) is the authorized agency to file applications.

APEDA filed an application for GI in 2008 for registering basmati. In February 2016, the Intellectual Property Appellate Board (IPAB) directed the GI registry to go ahead with GI for basmati as per the demarcation conducted by APEDA, which was the original applicant for GI. Thus, farmers in 77 districts in 7 states (Punjab, Haryana, Himachal Pradesh, parts of Uttar Pradesh, Uttarakhand, Delhi, and Jammu and Kashmir) will benefit from this and can use basmati as a GI for the rice they produce. Contesting the order by IPAB, Madhya Pradesh filed an appeal in the Madras High Court. On August 16, 2016, the Madras High Court ordered the state of Madhya Pradesh to file a fresh application for GI. Meanwhile, Pakistan has declined to be a joint applicant for basmati. The fight among different states in India over basmati and GI protection indicates that GI protection can be a contentious issue involving protracted litigation.

In recent years GI protection has been extended to rice varieties from other regions. For example, Kaipad rice, a rice variety cultivated by a unique method in the northern parts of Kerala, has been registered under the Geographical Indications of Goods (Registration and Protection) Act of 1999. In this case the applicants were Malabar Kaipad Farmers' Society of Kannur and the Centre for IP Protection of Kerala Agricultural University. GI had earlier been obtained in Kerala for Navara rice, Pokkali rice, Palakkadan Matta rice, and Jeerakasala and Gandhakasala, two rice varieties grown in the Wayanad region. There seems to be some more initiatives for the granting of GI for different varieties of rice grown in other parts of India.

Although these initiatives indicate that farmers and producer collectives are keen to apply and use the GI tag to differentiate their rice varieties and gain a premium in the market, we have not come across any study that quantifies the economic gains and costs of applying for and using GI. The nature of GI is such that it is useful for products that can be identified with a specific region, but large-scale and wider production of that product could result in dilution of the GI and thereby affect the goodwill and premium. For example, although Kaipad rice growers will benefit from the GI granted for that variety, they cannot expand the area of cultivation indiscriminately as that would go against the very idea of GI and affect the name and reputation of that variety. In other words, although GI can be used to obtain a premium in the market, there is a limit to the quantity producible or area under cultivation of that variety. This results in a dilemma for producers as GI can be a boon but it is not an unlimited bounty. Even if demand increases, the producers may not be able to meet that demand

and this in turn can result in like products or products with false claims and false association(s) with a region. Challenging such products and producers can be a costly affair for many producers who are most likely to be users but not owners of that GI. Therefore, they have to depend on the owner of the GI to enforce and fight against infringement.

For traditional rice varieties that qualify for GI, although every attempt should be made to benefit the most from GI, additional protection through trademarks and registered marks can be sought and used. A combination of GI and trademark will be a good marketing strategy that can be used to protect IP rights.

The potential for using GI for rice varieties, particularly traditional rice varieties, should be explored further and used as part of the rice strategy for India. Such a strategy for traditional rice varieties that are organically grown will benefit the conservation of those varieties. In the long run the use of GI for rice varieties should be integrated with the rice strategy for the promotion and protection of products, the interests of producers, and regional development. Rice varieties protected under GI can be used for the production of value-added food products that will have a premium on account of the GI for the rice variety. Such products may not qualify for GI but can be protected through trademarks.

Although basmati is the prime example for GI protection in India, the GI component in the rice strategy need not be confined to basmati. Instead, the strategy should examine the potential and scope of GI for other types of rice varieties, including traditional, indigenous, and organic rice varieties, and explore the role for IP in this.

ALTERNATIVE APPROACHES IN IP AND MANAGEMENT OF IP

Although patents and plant variety protection are well-known modes of IP for rice varieties, in recent years there have been discussions on open-source and open innovation approaches to plant variety development and the sharing of such varieties.[u] This should be understood in the context of discussions and initiatives on open-source biotechnology and open-source drug discovery.[v] In software, open-source models and licenses have worked well and these initiatives aim at using open-source models, licensing, and approaches in plant breeding, sharing of varieties, and the use of and sharing of germplasm. Srinivas (2006) suggests that open-source approaches can be combined with participatory plant breeding for developing new varieties. There have been few initiatives like the open-source seed initiative in the United States. Although no specific projects are using open-source models in rice breeding in India, there is at least one project, the Genomic Open-Source Breeding Informatics Initiative (GOBII) with participation of IRRI and ICRISAT, whose objective is "to work closely with CGIAR centers to develop open-source computational

---

u. See Kloppenburg (2013), Deibel (2012), Luby et al. (2015), and Srinivas (2006).

v. See Adenle et al. (2012), Marden and Godfrey (2012), and Tsioumani et al. (2016).

infrastructure and analysis capabilities, enabling the implementation of genomic and marker-assisted selection as part of routine breeding programs for staple crops in the developing world" (http://cbsugobii05.tc.cornell.edu/wordpress/). Such projects combine open-source tools and bioinformatics in breeding programs. As part of the rice strategy, a few initiatives on open-source seed and plant breeding could be started and supported.

However, one contentious issue is licensing, particularly enforcing rights over germplasm and varieties and enforcing licenses. The general public license developed in the context of software could be modified and, using the approach taken in the Open-Source Drug Discovery Project, could be used for material sharing, access to, and contribution to the project. The Open-Source Seed Project in the United States opted for a simple approach instead of going for a complicated licensing model. In this, when the buyer opens the packet, they accept the conditions imposed.

IP rights can be obtained as a defensive strategy and to prevent misappropriation. Institutions and breeders should not shun IP and make varieties and germplasm freely available, as happened in the days of the Green Revolution. Instead, they should adopt an IP policy that encourages innovation, facilitates diffusion, and encourages sharing. With this alternative, licenses and licensing mechanisms can play an important role.

CONCLUSIONS

The increasing importance of IP in agriculture dictates that it must be an important component in the rice strategy for India. Our brief analysis in this chapter indicates that, in the post-Green Revolution agricultural scenario, the challenge lies in balancing the use of IP for incentivizing innovation and ensuring that IP does not adversely affect the public interest and access to public goods.[w] The data for rice varieties registered under the PPVFRA indicate that the number of new varieties registered is low. We do not know whether this indicates that breeders from the public sector and private sector do not have much interest in developing new varieties and registering them or whether the PPVFRA is not sufficient for incentivizing innovation. For hybrid rice, the potential is well recognized, but the constraints are many and these need to be overcome to tap the full potential. The lessons from *Bt* cotton are important for the promotion and diffusion of hybrid rice and GM rice. Although central and state governments should not weaken market forces through arbitrary price fixation for seeds, they should play a key role in fostering innovation in the public sector for hybrid seeds and transgenic rice, ensuring that competition takes place in the market, and promoting OPVs over hybrids if necessary. GIs are good modes of protection and the controversy over GI for basmati indicates that contentious issues exist in identifying and demarcating areas eligible for classification into areas in

---

w. For a discussion on public interest in plant breeding, see Prifti (2014).

which basmati is grown. The scope for using GI for promoting traditional rice varieties should be explored.

Concerns that deserve attention are climate change, access to technologies, and IPR and their implications for rice.[x] We suggest that more research be undertaken on these areas.

Similarly, given the increasing relevance of and thrust in organic farming and sustainable agriculture in rice, questions relating to organic rice varieties, access to germplasm and revival of traditional varieties, and what the role of IPR is in this can be studied. Another issue that deserves attention is IP and access to genome editing, the application of synthetic biology in rice, and how IP can affect access to and use of technologies. Broadly speaking, the role of IP in emerging technologies in the context of rice has to be examined in terms of empirical research, including patent landscaping and policy choices.

Finally, we suggest that a comprehensive IP strategy for rice be formulated as part of the rice strategy for India. Such a strategy developed after consultation with stakeholders can play an important role in realizing the objectives of the rice strategy for India.

ACKNOWLEDGMENTS

The authors thank the external reviewers for their extensive comments that helped in revising this chapter. Research assistance provided by Dr. Amit Kumar is acknowledged also.

REFERENCES

Adenle, A.A., et al., 2012. Analysis of open source biotechnology in developing countries. Technol. Soc. 34, 256–269.

Agarwala, S., et al., 2012. Adaptation and Innovation: An Analysis of Crop Biotechnology Patent Data. OECD, Paris.

Aoki, K., 2008. Seed Wars: Controversies and Cases in Plant Genetic Resources and Intellectual Property. Carolina Academic Press, Durham, NC.

Blakeney, M., 2011. Trends in Intellectual Property Rights Relating to Genetic Resources for Food and Agriculture. CGRFA: FAO, Rome.

Blakeney, M., 2013. Climate change and intellectual property. In: Kole, C. (Ed.), Genomics and Breeding for Climate Resilient Crops. Springer, Berlin.

Chandola, H., 2006. Basmati rice: geographical indication or mis-indication. J. World Intellect. Prop. 9 (2), 166–188.

Choudhary, R., 2013. Spread of New Varieties of Hybrid Rice and Their Impact on Overall Production and Productivity in Bihar. Agro-Economic Research Centre for Bihar & Jharkhand, Bhagalpur University, Bhagalpur.

Central Rice Research Institute, CRRI, 2011. Vision 2030. CRRI, Cuttack.

Das, K., 2006. International protection of geographical indications with special reference to "Darjeeling" Tea. J. World Intellect. Prop. 9 (5), 459–494.

Deibel, E., 2012. Open variety rights. J. Agrar. Change 13 (2), 282–309.

---

x. See Agarwala et al. (2012), Blakeney (2013), and Srinivas (2013).

Drexl, J., Ruse-Khan, H.G., Nadde-Phlix, S. (Eds.), 2014. EU Bilateral Trade Agreements and Intellectual Property. Springer, Berlin.

Dunwell, J.M., 2010. Patents and intellectual property rights issues. In: Kole, C. (Ed.), Transgenic Crop Plants. Springer, Berlin.

Dutfield, G., 2008. Turning plant variety into intellectual property: the UPOV convention. In: Tansey, G., Rajotte, T. (Eds.), The Future Control of Food. Earthscan, London.

Fischer, G., 2013. The Golden Rice Project: use of patent analytics for non-commercial activities. In: WIPO, Regional Workshop on Patent Analytics, August 26–28, 2013, Rio de Janeiro.

Fowler, C., Mooney, P., 1990. Shattering: Food Politics and the Loss of Genetic Diversity. University of Arizona Press, Tucson, AZ.

Frison, C., Lopez, L., Esquinas, J.T. (Eds.), 2011. Stakeholder Perspectives on the International Treaty on Plant Genetic Resources for Food and Agriculture. Earthscan, London.

Halewood, M., Norigea, I.L., Louafi, S. (Eds.), 2013. Crop Genetic Resources as Global Commons. Routledge, London.

Hanchinal, R.R., 2015. Providing intellectual property protection to farmers' varieties in India under the Protection of Plant Varieties & Farmers' Rights Act, 2001. J. Intellect. Prop. Rights 20 (1), 7–18.

ICRA, 2015. Seeds Industry. ICRA, Mumbai.

Jacob, S.R., Tyagi, V., Agrawal, A., Chakrabarty, S.K., Tyagi, R.K., 2015. Indian plant germplasm on the global platter: an analysis. PLoS ONE 10 (5), e0126634.

Jefferson, O.A., et al., 2015. The ownership question of plant gene and genome intellectual properties. Nat. Biotechnol. 11 (11), 1138–1143.

Jikun, H., et al., 2015. Impact of insect-resistant GM rice on pesticide use and farmers' health in China. Sci. China Life Sci. 58 (5), 466–471.

Kloppenburg, J., 2004. First the Seed: The Political Economy of Plant Biotechnology 1492–2000. University of Wisconsin Press, Madison, WI.

Kloppenburg, J.R., 2013. Re-purposing the master's tools: the Open Source Seed Initiative and the struggle for seed sovereignty. In: Food Sovereignty: A Critical Dialogue, International Conference, September 14–15, 2013, Yale University.

Llewelyn, M., Adcock, M., 2006. European Plant Intellectual Property. Hart Publishing Co., Oxford.

Luby, C.H., et al., 2015. Enhancing Freedom to Operate for plant breeders and farmers through open source plant breeding. Crop Sci. 55, 2481–2488.

Marden, E., Godfrey, R.N., 2012. Intellectual property and sharing regimes in agricultural genomics. Drake J. Agric. Law 17, 369–411.

Marie-Vivien, D., 2008. From plant variety definition to geographical indication protection. J. World Intellect. Prop. 11 (4), 321–344.

Marie-Vivien, D., 2010. The role of the state in protection of geographical indications. J. World Intellect. Prop. 13 (2), 121–147.

Narasimhan, S.M., 2008. Towards a Balanced "Sui Generis" Plant Variety Regime. UNDP, New York, NY.

Oguamanam, C., 2014. Farmers' rights and the intellectual property dynamic in agriculture. In: David, M., Halbert, D. (Eds.), Sage Handbook of Intellectual Property. Sage Publications, London.

Pray, C.E., 2011. Impact of private R&D returns and productivity in hybrid rice in India. In: ICABR Meeting, Rome.

Prifti, V., 2014. The role of public interest in plant-related patents. Available from: http://papers.ssrn.com/sol3/papers.cfm?abstract_id=2542640

Rimmer, M., 2005. Japonica rice: intellectual property, scientific publishing, and data-sharing. Prometheus 23 (3), 325–347.

Santilli, J., 2012. Agbiodiversity and the law. Routledge, London.

Sarkar, D., Ghosh, J.K., 2013. Spread of New Varieties of Hybrid Rice and Their Impacts on Overall Production and Productivity in West Bengal. AERC, Visva-Bharati, Shantiniketan.

Sastry, K.R., Rashmi, H.B., Badri, J., 2011. Research and development perspectives of transgenic cotton. J. Intellect. Prop. Rights 16, 139–153, March.

Sivagnanam, K.J., 2014. Spread of New Varieties of Hybrid Rice and Its Impact on Overall Production and Productivity in Tamil Nadu. AERC, University of Madras, Chennai.

Spielman, D., et al., 2011. The Seed and Agricultural Biotechnology Industries in India. International Food Policy Research Institute, Washington, DC.

Spielman, D.J., Kolady, D.E., Ward, P.S., 2013. The prospects for hybrid rice in India. Food Security 5 (5), 651–665.

Srinivas, K.R., 2006. Intellectual property rights and bio commons. Int. Soc. Sci. J. 58 (188), 319–334.

Srinivas, K.R., 2013. Climate change, technology development and transfer: an exercise in thinking outside the box. In: Brown, A.E.L. (Ed.), Environmental Technologies, Intellectual Property and Climate Change: Accessing, Obtaining and Protecting. Edward Elgar, Cheltenham (UK).

Srinivas, K.R., 2015. Intellectual property rights and politics of food. In: Jerring, R.J. (Ed.), The Oxford Handbook of Food, Politics, and Society. Oxford University Press, New York, NY.

Tsioumani, E., et al., 2016. Following the open source trail outside the digital world. Triple C 14 (1), 145–162.

Venkatesh, P., Pal, S., 2014. Impact of plant variety protection on Indian seed industry. Agric. Econ. Res. Rev. 27 (1), 91–102.

Venkatesh, P., Sangeetha, V., Pal. S., 2015. India's experience of plant variety protection. In: AAEA WAEA Joint Annual Meeting.

Yadav, D., et al., 2014. Intellectual property rights in plant biotechnology: relevance, present status, and future prospects. In: Barh, D. (Ed.), OMICS Applications in Crop Science. CRC Press, Boca Raton, FL.

Chapter 11

Postharvest Management and Value Addition of Rice and Its By-Products

Arindam Samaddar*, Mohammed Mohibbe Azam**,
Kunjithapatham Singaravadivel†, Natarajan Venkatachalapathy†,
Braja Bandhu Swain‡, Purnananda Mishra§
*International Rice Research Institute, New Delhi, India; **Indian Institute of Rice
Research, Hyderabad, Telangana, India; †Indian Institute of Crop Processing Technology,
Thanjavur, Tamil Nadu, India; ‡International Livestock Research Institute, New Delhi,
India; §Indian Institute of Rice Research, ICAR, Cuttack, Odisha, India*

This chapter is divided into two broad sections. The first section discusses postharvest management and value addition of rice and the second section deals with value addition of rice by-products.

RICE POSTHARVEST MANAGEMENT AND VALUE ADDITION

Postharvest management is an integral component in improving and stabilizing India's food supply chain. Best practices ensure the stable supply of hygienic and high-quality food grains for the country. Rice postharvest management is a sequence of technical activities designed to reduce losses and to increase the value of the harvested paddy. These technical activities include harvesting, threshing, drying, cleaning, storing, processing, and packaging and have to take place alongside economic activities, such as transporting, marketing, and controlling quality and administration.

Approximately 9% of paddy is lost through the traditional methods of drying and milling and improper methods of storage, transportation, and handling (DMI, 2004). It has been estimated that total postharvest losses of paddy represent 2.72% of total production at the level of the producer. Threshing alone accounts for the maximum loss of 0.89%; other losses occurring in different operations are tabulated in Table 11.1.

Data collected from six states indicate that postharvest losses were lowest in Punjab (1.82%) and maximal (8.29%) in Tamil Nadu (Kannan, 2014). However, losses to each operation varied between states (Table 11.2).

The Future Rice Strategy for India. http://dx.doi.org/10.1016/B978-0-12-805374-4.00011-7

301

TABLE 11.1 Estimated Postharvest Losses of Paddy at the Producers' Level

Operations	Losses (% of total production)
Transportation from field to threshing floor	0.79
Threshing	0.89
Winnowing	0.48
Transportation from threshing floor to storage	0.16
Storage	0.40
Total	2.72

Source: Marketable Surplus and Postharvest Losses of Paddy in India, 2002, Directorate of Marketing and Inspection, Nagpur.

The millennium study conducted by the Ministry of Agriculture reported paddy postharvest losses of 11% (Gaikwad et al., 2006). A recent study on postharvest losses of major agricultural commodities conducted by the Indian Council of Agricultural Research (2005–07) reported that 5.2% of the 150 million tons of rice annually produced in India is lost after harvesting (CIPHET, 2010).

Postharvest Management in India

Postharvest management practices in India vary depending on the agroecological conditions. In the kharif season, harvesting occurs during the monsoon season when the paddy's moisture content is 18%–25%. This moisture causes germination, heat development, and discoloration of the paddy. If the paddy is left with residual moisture, then milling will also cause further losses in the outturn

TABLE 11.2 Postharvest Losses in Six Different States Across India

Loss particulars (%)	Assam	Karnataka	Punjab	Tamil Nadu	West Bengal	Uttar Pradesh
Harvest	0.62	1.90	1.52	3.10	0.78	2.71
Threshing	1.27	0.20	0.04	2.11	0.32	1.28
Winnowing	0.98	0.08	—	0.18	0.13	0.40
Transportation	1.97	0.57	0.06	0.56	0.55	0.48
Handling	0.71	0.28	0.20	—	0.31	0.30
Storage	2.08	3.83	—	2.34	1.78	0.40
Total	7.63	6.86	1.82	8.29	3.87	5.57

Source: From Kannan, E., 2014. Assessment of Pre- and Postharvest Losses of Important Crops in India. In Research Report: IX/ADRTC/153A. Published by Agricultural Development and Rural Transformation Centre Institute for Social and Economic Change, Bangalore.

(Raj and Singaravadivel, 1980). This problem occurs in coastal areas, such as Karnataka, Kerala, Tamil Nadu, Andhra Pradesh, Odisha, and West Bengal.

Farmers retain part of their harvest for seeding purposes and home consumption. Local practices of storing in gunny bags, bamboo bins, and wooden structures cause moisture absorption and lead to rat infestations. In addition, significant losses during transportation can occur due to improper stitching of bags, multiple handling, and spillage.

Manual threshing damages grains and causes breakage during milling. Contact with soil also paves the way for microorganism contamination. Immature grains pose problems during milling and storage if not properly dried, as they generate heat and, if left unattended, can result in discoloration.

Proper drying practices before storage are therefore necessary to prevent a decline in grain quality. Impurities also cause rapid deterioration of grains. The high-moisture paddy causes the development of hot spots, which leads to mold and insect attack. Farmers and traders store grains using various methods with farmers using bags for short-term storage while traders procure huge quantities in bulk. Paddy is transported from the fields to the drying sites and then moved to warehouses. Traditionally, transportation was performed in gunny bags on bullock carts or tractors with trailers; nowadays, trucks and rail wagons are used for the long-distance transportation of paddy to markets. Lack of proper roads and inclement weather can cause delay in paddy transportation, which can again affect grain quality. These traditional practices need to be replaced with improved methods and technologies to prevent such needless loss at the producers' level.

Improved Methods

Harvesters, combine harvesters, mechanical threshers (Fig. 11.1A), and winnowers were introduced to reduce labor and increase the timeliness of operations (Fig. 11.1). These power tiller-operated mowers or harvesters were later replaced by tractor-drawn harvesters. Threshing of the harvested sheaves can be performed either manually or mechanically; however, harvesting and threshing

(A)   (B)

FIGURE 11.1   (A) Paddy thresher and (B) combine harvester.

activities in waterlogged fields still prove to be difficult even with technologically advanced machines.

The scarcity of agricultural laborers led to the development and popularity of combine harvesters. Tractor-operated and self-propelled combine harvesters (Fig. 11.1B) are commercially available for dry and wet-puddled conditions. Small cleaners that run on electrical energy and diesel engine and power tiller-/ tractor-coupled cleaners are available and are feasible for cleaning the produce at the farm level. A suitable combine machine should also be designed for simultaneously drying grains or segregating the green grains. If the harvest is done manually, as well as by harvesters, then the development of an efficient winnower-cum-grader will also help in obtaining good-quality paddy. A mechanical bulk handling system—with containers and pneumatic conveyors— should be developed for easy transportation.

Drying is the most critical operation after harvesting. Delayed, incomplete, or ineffective drying will reduce grain quality and result in losses. Drying and storage are related processes and can sometimes be combined in one piece of equipment (i.e., in-bin drying). Improper drying may cause cracks or fissures to develop in paddy, which ultimately leads to breakage during milling. Therefore, the drying process must be adjusted so that cracks do not develop. The most common method of drying is sun drying whereby the paddy is spread on yards or paved floors (or sometimes on a road or a mud floor) in 3–5 cm thick layers and turned every now and then to achieve uniform drying (Fig. 11.2A). However, during inclement weather, drying is not possible; moreover, yard drying is unhygienic and causes losses due to scattering or consumption by pests. On the other hand, delays in drying also cause quality deterioration from microorganisms and discoloration. Crack formation should be avoided during drying, especially for raw milling.

The alternative to sun drying is mechanical drying, in which paddy is held in a drying chamber and hot air is blown through it. Among the various dryers, LSU-type dryers (Fig. 11.2B) were traditionally used in modern rice mills, but nowadays fluidized bed dryers (Fig. 11.2C) are being implemented. The paddy is tempered for uniform distribution of moisture and temperature of dried grain

(A)   (B)   (C)

FIGURE 11.2 (A) Yard drying, (B) Louisiana State University (LSU) dryer, and (C) fluidized bed dryer.

during and after drying; however, as the mechanical dryers are fabricated mostly by local artisans without any scientific principle, this can cause nonuniform drying and reduce drying efficiency.

In most modern parboiled rice mills in India, LSU dryer inverted "V"-port cross-flow dryers are available. Nowadays, fluidized bed dryers are being erected in many rice mills. In some mills coolers are used before drying to reduce the paddy temperature and avoid yellowing of rice kernels. Steamed paddy loading into dryers will take about half an hour, but if the paddy temperature is above the GT, the intensity of yellow to reddish color will increase.

Solar energy is available in many agroecological zones in India. To date, no commercial-model solar dryers are available for parboiled paddy drying in rice mills. The mechanism of solar drying, which is widely available for other crops, could be standardized for use in rice mills in the future and would circumvent the problems associated with yard drying. Development of on-farm mechanical dryers using biomass as fuel will also help small farmers to reduce their losses. A mobile mechanical dryer using a tractor power take-off shaft would be useful as it could be taken to different places during the paddy season. A farm-level, in-bin, drying-cum-storage facility could also be developed. Solar drying and mechanical drying have proven to be equally good during storage, especially in terms of germination (Bhardwaj and Sharma, 2015). Harvested paddy could be dried in two stages to obtain good milling yield with fewer broken grains, although in practice this could become costly. However, in conventional rice milling with yard drying, two stages of drying are being used even now.

Mechanical drying is faster, it provides uniformity, and more importantly it can be done round the clock even during inclement weather. However, standard practice in rice mills generally involves loading the steamed paddy directly into the dryer and directly milling without any rest period.

Different receptacles can be used for the bulk storage of grains, including godowns (rural warehouses made of bricks with cement flooring) or silos (towers made of metals, concrete, and bricks) (Fig. 11.3A). In tropical conditions,

(A)   (B)

FIGURE 11.3   (A) Paddy storage silo (2000 tons) and (B) cover and plinth (CAP) storage.

moisture migration and condensation cause problems in silo storage, especially during an extended storage period. Respiratory activities of high-moisture paddy increase the temperature, which leads to moisture migration and discoloration within the storage structure. Cover and plinth (CAP) storage (Fig. 11.3B), an economical way of large-scale storage, is also used across the country for large-scale paddy storage. However, spillage loss and condensation in top-layer bags can cause deterioration. Nowadays, rice millers tend to store their procured paddy in silos for 3–8 months; however, in tropical conditions, it is difficult to store for a longer period without aeration.

In Tamil Nadu six 1000-ton-capacity silos are available in the Tamil Nadu Civil Supplies Corporation (TNCSC) rice mill in Thiruvarur. To overcome the problems associated with condensation, five silos are used for storage and one is kept as a service silo, with paddy regularly interchanged from one to another. Some private rice millers are now moving toward metal silos for paddy storage, some of which have 2000 ton capacity, sensors, and the capacity of aeration.

Processing/Milling

After storage, paddy is milled to obtain rice for direct consumption or to produce value-added products. The processing technologies vary depending on the end-product requirement, but most include the basic steps of cleaning, drying, parboiling, drying, and milling. Moisture content of paddy plays a major role during processing and high-moisture paddy should be sufficiently dried before storage. Parboiling improves the milling quality and nutritional content of paddy. More than 60% of the Indian population consumes parboiled rice, which is prevalent in Kerala, Tamil Nadu, Odisha, West Bengal, Bihar, Chhattisgarh, and Karnataka.

Rice Processing Patterns and Milling Technology

Rice milling is the oldest and largest agroprocessing industry in the country with a current turnover of more than INR 25,500 crore per annum. Approximately 85 million tons of paddy are milled every year, providing staple food and other valuable by-products to more than 60% of the population. Single hullers, of which there are more than 82,000 registered in the country, were traditionally used to mill paddy grain either in its raw condition or after parboiling. A high proportion of these single hullers (60%) are also linked with parboiling units and sun-drying yards. Most of the tiny hullers (250–300 kg/h capacity) are employed for custom milling of paddy. Other types of machines present in the country include double-hulling units (>2600 units), under-runner disc shellers-cum-cone polishers (5000 units), and rubber-roll shellers-cum-friction polishers (>10,000 units) (Fig. 11.4). Over the years there has been a steady growth in the number of modern rice mills in the country, most of which have a capacity of 2–10 t/h (DMI, 2004).

FIGURE 11.4   Double-huller mill.

Two hullers are used for two-stage milling, one for partial shelling and the second for continual polishing. The huller mills are modernized with sheller, horizontal, or cone emery polishers; silky or humidified polishers; and color sorters.

Parboiling of Paddy

Parboiling is a hydrothermal treatment in which paddy is soaked in water and steamed to obtain gelatinized rice. Parboiling seals any cracks within the rice, causing the rice to become harder and resulting in a higher milling yield with less breakage. Different methods of parboiling are in vogue, including household parboiling (Singaravadivel and Raj, 1986); single steaming (Pillaiyar, 1977); double steaming, hot-water soaking, and steaming (CFTRI, 1960); pressure parboiling (Pillaiyar and Singaravadivel, 1993); soaking at 80°C (Ali and Ojha, 1976); and short soaking-cum-tempering (Pillaiyar et al., 1993). Modern parboiling methods, specifically soaking methods, are used in modern rice mills to reduce processing time and quality and quantity losses (Fig. 11.5). When compared with raw rice with the same degree of milling, parboiled rice demonstrated higher nutrient status, less cooking loss, more swelling when cooked to the desired softness, easy digestibility, and a higher protein efficiency ratio. Higher oil levels in bran with enhanced stability were also observed in parboiled rice. This might be because of the internal migration of water-soluble molecules from the aleurone layer into the endosperm during soaking and disruption of fat globules and movement toward the outer aleurone layer during the steaming phase of the parboiling process (Kik, 1955; Sondi et al., 1980; Subrahmanyan et al., 1938).

FIGURE 11.5 Parboiling tanks and LSU dryer.

Modern Rice Mills

Rice milling systems range from traditional to large, complex, modern rice-processing installations (Fig. 11.6). Home-systems include hand-pounding in a pestle and mortar, *chakki*, custom-hiring of hullers, and using emery under-runner disc sheller-cum-huller mills, under-roller emery disc sheller-cum-cone polisher mills, and modern rubber-roller rice mills. Rubber-roller sheller mills give the highest outturn of rice from paddy (2%–3% more than emery shellers and 6%–8% more than hullers). Raw milling outturn occurs more in the modern rice milling process. It is also worth noting that the quality of rice bran obtained in modern rice mills is higher with increased oil content (NPCS, 2007).

After the paddy is dried, it should be tempered before milling. Modern rice milling removes the husk and bran from paddy to produce polished rice. Unit operations involve cleaning, dehusking, husk separation, paddy–brown rice separation, polishing, grading, sorting, weighing, and bagging. Paddy varieties that are crack resistant or nonpasty are best for raw milling, while improved parboiling methods can further reduce losses. Commercialization of available lab-scale parboiling-cum-drying technologies, such as hot humid air, microwave, radio frequency, and ohmic heating, is required for quicker processing. Improved machinery, such as rubber-roller shellers, vertical and humidified polishers, and color sorters, should be used for better quality control.

Grading of paddy is not a traditional practice; however, to obtain better head rice recovery, grading is recommended. Rice grading after milling is usually done by mechanical devices, including rotating graders, plansifters, trieurs, circular purifiers, color graders/sorters, etc. During the process of milling, one-eighth of grains are broken down and further results in reduction in the milling yield. Breaking 10% of grains results in a 1% loss in milling outturn.

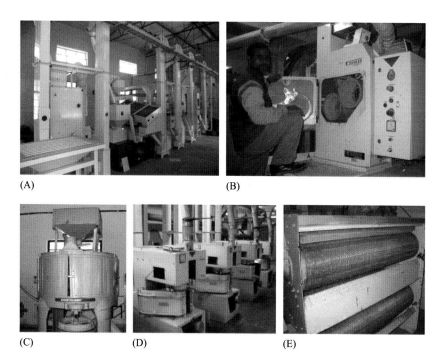

FIGURE 11.6   (A) Modern rice mill, (B) rubber-roller sheller, (C) cone polisher, (D) vertical polishers, and (E) thickness grader.

Brown rice is polished to obtain white-color rice, which is preferred by most consumers. Unfortunately, polishing reduces the nutritional content of rice. Government agencies, such as the Food Corporation of India and other state civil supply corporations procuring rice from millers, follow Fair Average Quality (FAQ) norms. It is recommended to fix the degree of milling to 5%, but in practice rice is milled as high as 8%–10% to obtain white color.

New Developments and Future Scope

The locations of rice mills are confined to a few selected production centers. The absence of a village-level rice milling unit means that farmers have to travel long distances to mill their rice, leading to increased transportation and handling losses. Thus, there is a need to develop village-level agroprocessing units with technical efficiency so as to obtain a higher quantity and quality of rice. Value addition and the generation of gainful and sustainable employment opportunities are other possible benefits arising from this production catchment agroprocessing industry.

Designing a smaller capacity sheller-cum-paddy separator could help to produce enough brown rice to meet the growing demand, especially for organically produced traditional varieties. The ultimate cause of rice breakage during milling is defective grains; therefore, maintaining good grain quality through proper harvesting, drying, and storage practices is of prime importance for reducing rice breakage. The type of mill, moisture content, and grain size and shape are also important. Crack-resistant paddy with reduced husk content gives greater yield; such varieties could be developed genetically.

Value Addition of Rice

Rice can also be converted into different forms for consumption (Fig. 11.7). The major value-added products from rice in almost all the states of India are flaked rice (Fig. 11.7A) and puffed rice (Fig. 11.7B). In addition, many rice-based products are available in different states, including *laiya*, a roasted rice produced in Uttar Pradesh and used on special occasions. Rice in Assam is generally waxy in nature and different traditional products, such as *tilpitha*, *komol chaol*, and *joha* (cooked *joha* rice with pigeon meat is a special preparation in the rural areas of Assam), are consumed. In Punjab *rauh di kheer* (rice cooked for a long time in sugarcane juice) is a special dish. In Tamil Nadu and Kerala rice is consumed as cooked rice, cold rice, and modified products, such as *puttu* (Fig. 11.7C), *idli* (Fig. 11.7D) and *dosa*. In Maharashtra traditional rice products available include *pej* for children, *poha*, *bhadang*, *kurmura*, *papad*, *mirgund*, and *laddu*.

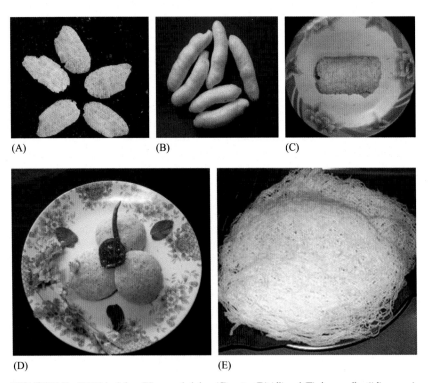

(A)   (B)   (C)

(D)   (E)

FIGURE 11.7   (A) Flaked rice, (B) expanded rice, (C) *puttu*, (D) *idli*, and (E) rice noodles (*idiyappam*).

Extruded rice products, such as rice vermicelli and rice noodles (*idi-yappam*) (Fig. 11.7E), are consumed in southern India. The vermicelli is prepared by extruding the precooked rice flour and drying it before sale (Fig. 11.8A–C). To prepare rice noodles, the rice flour is first pregelatinized, roasted, and dried, and that flour is then extruded and steamed. For use at a later date, the noodles are dehydrated at 55–60°C and packed. Analog rice with fortification by a hot extrusion process is a new approach for value-added rice product preparation. For this extrusion process, low-cost broken rice can be used as the raw material, and the appearance and texture as well as the nutrient content of the kernels can be modified to the specific requirements of target consumer groups. The fortified extruded kernels look and behave exactly like normal rice, and the nutrients embedded are efficiently protected from leaching during washing and cooking. As the extruded rice cooks instantly, various ready-to-eat (RTE) rice preparations, such as tamarind, *sambar*, and lemon, can be prepared. For the preparation of ready-to-cook (RTC) rice products, quick-cooking rice is added with the respective taste-garnishing materials, such as lemon powder, roasted mustard seed, bengal gram *dhal*, curry leaf, chili, etc., before packaging.

Rice Vermicelli (Fig. 11.8)

(A) (B)

(C) (D)

FIGURE 11.8 (A) Rice extrusion, (B) drying, (C) vermicelli from *pisini* rice, and (D) Uruli roaster for roasting rice flour.

Flaked rice is prepared by roasting the soaked paddy in patties or in an electrical (Fig. 11.9A) or sand roaster (Fig. 11.9B). Flaking is then performed in hot conditions in an edge runner (Fig. 11.9C) or roller flaker and broken rice is separated by sieving (Fig. 11.9D). To prepare expanded rice, the soaked paddy should be sand-roasted (Fig. 11.10A–B) and then immersed immediately in water (Fig. 11.10C). The water should then be drained and the rice tempered overnight in a gunny bag (Fig. 11.10D). The dried milled rice is then mixed with salt, sugar, and sodium bicarbonate solution (Fig. 11.10E–F), roasted in sequences in two patties, and finally sand-roasted (Fig. 11.10G–H). It is cooled and packed in airtight polyethylene bags.

Flaked Rice Production (Fig. 11.9)

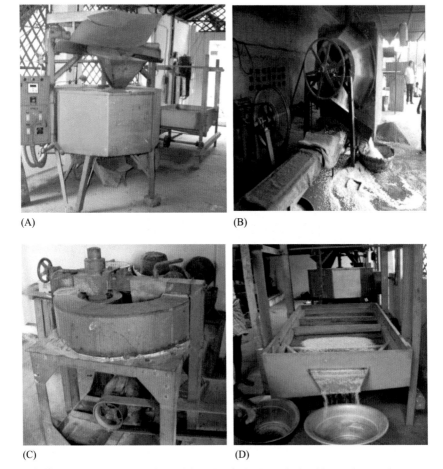

(A)   (B)

(C)   (D)

FIGURE 11.9 (A) Soaked paddy roasting (electrical), (B) soaked paddy roasting (sand-roaster), (C) flaking in edge runner, and (D) sieving.

Expanded Rice Production (Fig. 11.10)

FIGURE 11.10 (A) Raw paddy roasting in stages, (B) sand-roasting in third stage, (C) roasted paddy immersed in water, (D) soaked paddy tempering, (E) milling, (F) rice mixed with salt–sugar solution, (G) rice roasting in pans, (H) expansion in sand roasting, and (I) mechanical sand roaster.

Dehydrated products (e.g., *fryums* and *vattals*) are prepared by drying the cooked extruded rice flour in different shapes. They are consumed after deep-frying. Fermented rice products, such as *idli* and *dosa*, are prepared by grinding the soaked rice and black gram and leaving to fermentation overnight. The resulting batter is then steam-cooked in *idli* mold for *idli* or pan-fried for *dosa*. Instant *idli* mix is prepared by mixing dry rice and black gram flour with salt, sodium bicarbonate, and citric acid. The *idli* batter can be prepared in 5 min by mixing the dry mix with water. Instead of chemicals, patented microbial cultures are also used for the preparation of instant *idli* mix.

Value-Added Products and Modernization

Primary processing, such as cleaning and grading, can ensure a higher price for paddy, as well as lead to the development and commercialization of many value-added products, such as brown rice, germinated brown rice, extruded food, RTE, and RTC rice. Ready-made mixes (i.e., *dosa* and *idli*) and noodles are just some examples of products that have huge potential in the export market. Rice milk has economic importance, especially if prepared from traditional colored rice. Even though laboratory-level technology is available, commercial technology to produce rice milk with a longer shelf life is needed. Suitable modern machinery for value-added products in both small scale and industrial scale is required, as well as standardization of technologies for the production of higher nutritional and functional foods.

Policy Implications

Proper drying of paddy grain can prevent many problems during storage and processing. Government organizations can provide small dryers to farmers on an individual basis and community drying yards or solar tunnel dryers can be provided in villages.

Procuring agencies need to have dryers either at the procuring center or at a storage point so that high-moisture paddy can be dried before storage. The TNCSC requires farmers to use cleaners at the paddy procurement center before storage, which has resulted in better-quality paddy grains.

FAQ norms adopted by government organizations need a few modifications. For example, there needs to be a reduction in the percentage of broken grains obtained during parboiling. The current percentage allowed is 15%, but this could be reduced to 5% as technology improves and machinery becomes available. Cracks in paddy result in broken grain, which in turn reduces the milling yield. Therefore, for the procurement of raw paddy, a new norm of <5% cracks could be stipulated. A policy could be formed for the complete replacement of hullers with modern rice mills and rubber-roller shellers.

Considering the importance of rural storage in the marketing of agricultural produce, the Directorate of Marketing and Inspection initiated a Rural Godowns Scheme in collaboration with the National Bank for Agriculture and Rural Development and the National Cooperative Development Corporation. Its objective was to construct scientific storage godowns with allied facilities in rural areas and to establish a network of rural godowns in the states and union territories. Cooperative storage facilities are provided to the producer at cheaper rates, therefore reducing storage costs. These cooperatives also provide a pledge loan against the produce and storage is more systematic and scientific than traditional methods. Financial assistance and subsidies are provided by government organizations/banks to build cooperative storage.

A bulk handling system from procurement to transportation and storage has to be developed in both government organizations and the private sector. The possibilities of using hermetic storage cocoons and modified atmospheric storage in silos could be explored. The government should also frame policies to give incentives to farmers who adopt an improved bulk handling system and produce better-quality produce.

VALUE ADDITION OF RICE BY-PRODUCTS

Rice Milling By-Products

Depending on the variety, paddy rice is generally composed of 20% rice hull or husk, 11% bran layer, and 69% starchy endosperm (also referred to as the total milled rice). In an ideal milling process this would result in 20% husk, 8%–12% bran (depending on milling degree), and 68%–72% milled rice or white rice. The minor components of rice milling, such as rice husk, rice germ, and bran, are treated as waste products by the rice milling industry instead of useful by-products. The economic stability of the rice milling industry largely depends on commercial use of its by-products and these by-products can be used in a better and more profitable manner for both industrial and feed purposes. Rice straw is another agricultural waste that is abundantly available. The properties and the effective use of these-by-products are discussed in this section.

Rice Bran

Paddy is milled to remove the husk and this yields brown rice. During the second stage of milling, the outer brown layer is removed to produce white or polished rice. The removed brown layer is called rice bran, and the major components of which are edible oil, high-quality proteins, and dietary fibers. Rice bran accounts for only 8%–12% of the total weight of rough rice, but contains about 90% of its nutrients and nutraceuticals (Houston, 1972). It is composed of carbohydrate (33%–36%), crude fat and oil (18%–21%), crude protein (14%–16%), dietary fibers (9%–15%), and ash (8%–10%) (Heli Roy and Shanna Lundy, 2005). Rice bran is rich in B-complex vitamins (Tadera and Orite, 1991) and contains several minerals, including iron and zinc (Lu and Luh, 1991).

Potential Dietary Uses of Rice Bran

Rice bran is a rich source of hypoallergenic protein, edible oil, dietary fiber, and several nutrients essential to life. The addition of bran improves the storage stability of food because of its antioxidant properties (Kim and Godber, 2001; Nanua et al., 2000) and decreases fat absorption during frying (Hammond, 1994). It is suitable for baked products (Sharif, 2009) and can be used to substitute up to 20% of wheat flour without adversely affecting its quality and taste (Premakumari et al., 2012; Sharif, 2009).

Challenges in Using Rice Bran

Rice bran is underused as a value-added food product because of the presence of various antinutritional factors, such as lipases, trypsin inhibitor, hemagglutinin–lectin, and phytates. If fresh rice bran is not stabilized, its lipase enzyme causes very rapid hydrolysis of the oil, converting it to free fatty acid (FFA). Rice bran becomes unfit for human consumption when the FFA concentration surpasses 5%, and it becomes unfit even for cattle feed once the concentration exceeds 12%.

Stabilization of Rice Bran

Several methods have been developed for rice bran stabilization. These methods include dry or moist heat treatment (Lakkakula et al., 2004; Kim et al., 1987), extrusion cooking (Sayre et al., 1982), fluidized bed drying (Fernando and Hewavitharana, 1993), parboiling (Narisullah and Krishnamurthy, 1989), and microwave heating (Malekian et al., 2000). The drawbacks of most of these methods are the severe processing conditions that are capable of damaging valuable components of bran and incompletely inactivating enzymes, as well as the high operational costs. Microwave heating is the most effective and economical method (Malekian et al., 2000); it requires less processing time, and has few adverse effects on nutritional value (Yoshida et al., 2003). Ohmic heating is another potential method to stabilize rice bran. Dhingra et al. (2012) from CIPHET, Ludhiana, have developed a protocol through which 10 kg of bran/batch can be stabilized by ohmic heating; stabilized bran was found to be stable even after 75 days.

Rice Bran Protein

The protein content in rice bran is 10%–15%. Rice bran protein (RBP) is composed of water-soluble, salt-soluble, alcohol-soluble, and alkali-soluble storage proteins (Fabian and Ju, 2011) and has unique nutritional value and nutraceutical properties (Saunders, 1990). Among cereals, rice protein has the highest nutritional value owing to its high content of limiting essential amino acids, such as lysine and threonine, which are generally deficient in cereals (Juliano, 1985; Mawal et al., 1987). At the same time, RBP is well known as one of the lowest allergy-generating proteins for pure protein products, protein supplements, or the development of infant formula (Gnanasambandam and Hettiarachchy, 1995). RBP concentrate and isolate are not commercially produced because of the lack of commercially feasible extraction methods (Fabian and Ju, 2011).

Extraction of Rice Bran Protein

Several methods have been developed for the extraction of rice bran protein, which include extraction with alkali, with multiple solvents or agents, through enzymes, and through subcritical water extraction.

Alkaline extraction is the most common method with the highest protein yield, but it converts protein to a toxic product, lysinoalanine (Cheftel et al., 1985). Multiple extraction methods are possible, but are not commercially feasible. For example, enzymatic extraction (Ansharullah et al., 1997) yields 88% high-quality protein suitable for use as a nutritional food ingredient but the high cost of the enzymes makes this a nonviable method (Hamada, 1999). High protein yields (84%) are also obtained when subcritical water is used (Watchararuji et al., 2008), but the high temperature involved in the extraction could possibly denature the proteins. A viable commercial extraction technique for separating RBP is yet to be developed. Extraction remains a challenge because of poor solubility (Hamada, 1997) and the bran's phytate and fiber content (Juliano, 1985) makes the protein bodies very hard to separate from other components.

Application of Rice Bran Protein

RBP hydrolysates can be used as nutritional supplements, functional ingredients, and flavor enhancers in foods, coffee whiteners, cosmetics, personal care products, and confectionary, and in the fortification of soft drinks and juices. They are also used in soups, sauces, gravies, meat products, and other savory applications (Giese, 1994; Weir, 1986).

With growing interest in natural food ingredients, such as plant proteins, the demand for rice protein should also increase and should play a major role as a plant protein for food and pharmaceutical industries. Extraction of RBP is difficult because of its complex nature. This complexity makes it more difficult to find a single suitable solvent for extraction. The use of enzymes and subcritical water treatment showed promising protein yields; however, the relatively expensive cost of the enzyme should be addressed, while the exact yield and quality of protein extracted at relatively high temperatures for subcritical water treatment needs further studies. More research is needed to develop a more efficient and economically viable method for RBP extraction.

Rice Bran Oil

Rice bran oil (RBO) is a cooking oil with a very appealing nut-like flavor. Its high smoke point (254°C) also makes it suitable for high-temperature cooking methods, such as stir frying and deep-frying. It offers several unique properties that make it interesting as a specialty oil in niche markets. The most notable feature is its high content of components with nutraceutical value, such as gamma-oryzanol and tocotrienols (Table 11.3) (Rogers et al., 1993; Sayre and Saunders, 1990). Moreover, the fatty acid composition of RBO closely matches that recommended by various organizations, such as the American Heart Association(AHA), World Health Organization (WHO), and the ICMR. Consumption of RBO normalizes blood cholesterol by reducing total plasma cholesterol and low-density lipoproteins. It also shows a significant reduction in cholesterol absorption. The cholesterol-reducing power of RBO is better than that of coconut, canola oil, corn oil, and peanut oil (Ausman et al., 2005; Wilson et al., 2000). Most of the medicinal properties of RBO are due to oryzanol, which has the ability to reduce plasma cholesterol, reduce cholesterol absorption, and decrease early atherosclerosis (Chou et al., 2009; Lichtenstein et al., 1994). RBO has a good shelf life, good oxidative stability, and good thermal stability (Usha and Premi, 2015). Ideal fatty acid composition, high content

TABLE 11.3 Comparison of Natural Antioxidants in Edible Oils

Oil type	Vitamin E tocopherol (ppm)	Vitamin E tocotrienol (ppm)	Oryzanol (ppm)	Total natural antioxidants (ppm)
Rice bran	81	336	2000	2417
Olive	51	0	0	51
Canola	650	0	0	650
Peanut	487	0	0	487
Soybean	1000	0	0	1000
Grape seed	256	149	0	405

ppm, Parts per million.
Source: Available from: www.californiariceoil.com/nutrition.htm

of antioxidants and nutraceuticals, high smoke point, and its low absorption in prepared food make this oil superior to all other cooking oils.

RBO Production Potential

India produced ~157.197 million tons of paddy during 2014–15, which is equivalent to 104.798 million tons of rice (67% recovery), which could yield ~8.488 million tons of rice bran (if 90% of the paddy is used for consumption and bran recovery is 6%). In theory, this amount of bran has the potential to yield 1.44 million tons of RBO (17% recovery); however, in reality the actual production of RBO was only 0.93 million tons, just 63.3% of the potential. The untapped potential for RBO was 0.54 million tons (Table 11.4).

One of the reasons for this wide production-potential gap is the scattered presence of rice mills as transportation time is crucial to maintain the quality of rice bran required to prepare edible oil. Another reason is the practice of one-stage milling, which results in a mixture of hulls and bran with an oil content too low for economic oil extraction. The third reason is the lipolytic enzyme system in the bran, which causes rapid deterioration of RBO by releasing FFAs.

These problems can be overcome if a two-stage milling process is adopted, which will separate bran from the husk and oil extraction will be economical. Control of the rapid release of FFA can be done through heat stabilization or immediate extraction of oil from the bran. The latter is possible if milling and oil extraction units are on the same campus.

This practice will reduce the cost of stabilization. Before adopting this practice, it is desirable to know the content of FFA and peroxide value in the stored bran. Deterioration of bran takes place because of the presence of enzymes in the bran and not because of the atmospheric oxygen.

TABLE 11.4 Production of Rice Paddy and Its By-Products in 2014–15 (in million tons)

Study number	State	Rice	Paddy (3/2 = 1.5)	90% Paddy (paddy × 0.9)	Husk (90% paddy × 0.2)	Bran (90% paddy × 0.06)	RBO (bran × 0.17)
1	Andhra Pradesh	11.56	17.34	15.606	3.1212	0.93636	0.159181
2	Assam	4.86	7.29	6.561	1.3122	0.39366	0.066922
3	Bihar	6.38	9.57	8.613	1.7226	0.51678	0.087853
4	Gujarat	1.64	2.46	2.214	0.4428	0.13284	0.022583
5	Haryana	4.01	6.01	5.408	1.0816	0.32449	0.055163
6	Karnataka	3.66	5.49	4.941	0.9882	0.29646	0.050398
7	Chhattisgarh	6.02	9.03	8.127	1.6254	0.48762	0.082895
8	Madhya Pradesh	3.63	5.45	4.901	0.9801	0.29403	0.049985
9	Maharashtra	2.93	4.40	3.956	0.7911	0.23733	0.040346
10	Odisha	8.29	12.44	11.192	2.2383	0.67149	0.114153
11	Punjab	11.12	16.68	15.012	3.0024	0.90072	0.153122
12	Jharkhand	3.32	4.98	4.482	0.8964	0.26892	0.045716
13	Tamil Nadu	5.83	8.75	7.871	1.5741	0.47223	0.080279
14	Uttar Pradesh	12.22	18.33	16.497	3.2994	0.98982	0.168269
15	West Bengal	14.71	22.07	19.859	3.9717	1.19151	0.202557
16	India	104.80	157.20	141.477	28.2955	8.48864	1.443068

RBO, Rice bran oil.

Bridging of the Gap in the Demand–Supply of Edible Oil

In India the demand for both edible and nonedible oils is increasing due to contributing factors, such as rising income, the growing population, and expanding urbanization. The result is an overall decline in the per capita availability of edible oils. The aggregate consumption of edible oils rose to 19 million tons in 2014–15 from around 14 million tons in 2009–10. Domestic demand for vegetable oils and fats has been rising at the rate of 3%–6% per annum. Per capita consumption of vegetable oil increased from 14.4 kg/year in 2009–10 to 16.7 kg/year in 2014–15, which is still very low in comparison with the world average of 27.6 kg/year for 2014–15 (Mehta, 2016). If the oil consumption growth rate were to hit 3% per year, consumption could reach 26.8 million tons by 2025. If the consumption growth rate were 4%, there would be a demand for 30 million tons of vegetable oil in India (Mehta, 2016). Domestic edible oil production from traditional sources was 7.2 million tons in 2014–15 and therefore to meet additional requirements an additional 14.4 million tons of edible oil were imported. India needs an additional million tons of edible oil every year to meet the growing requirements. The use of RBO will not only reduce the gap between demand and supply of edible oils but will also save valuable foreign currency. According to the Solvent Extraction Association of India, India imports edible oil worth INR 55,000 crore every year. With full exploitation of rice bran oil, the country could reduce its edible oil import bills by approximately INR 3,000 crore every year.

Processing Potential

There is a need to process all the rice bran produced in India to recover RBO, which could make a significant contribution to the vegetable oils basket. At the same time, the large number of high-value by-products mentioned earlier also needs to be recovered during RBO processing, which will change the economy of the entire process. The Indian edible oil industry is composed of 15,000 oil mills, 600 solvent extraction units, 250 *vanaspati* units, and about 400 refining units.

Promotion of Rice Bran Oil

RBO is indigenous to India and can be easily produced in the country. Its promotion will help to reduce the imports of edible oil and at the same time decrease the gap between demand and supply. The health benefits of RBO should be highlighted and it should be promoted as healthy oil. People are becoming more health conscious and taking more steps to stay healthy and therefore the health benefits of RBO will attract health-conscious consumers. Moreover, India's disposable income is rising and people are ready to pay a premium if they find valuable products.

Some major constraints need to be overcome in the promotion of RBO. One is that people are skeptical about RBO's health benefits because of the lack of consumer awareness. The task is therefore to educate people about RBO and its health benefits. Marketing of RBO has to be more compelling to persuade consumers to try a new product and perceptual barriers about the color and taste of the oil need to be addressed to dispel the myths that dark-colored refined oils are impure and unhealthy. Value-added products of rice and rice by-products should also be exploited, as these will generate additional income and reduce the overall cost of RBO.

Codex standards for RBO should be finalized. The Codex, also known as Codex Alimentarius, is a collection of internationally recognized standards, codes of practice, guidelines, and other recommendations relating to foods, food production, and food safety. The Codex Alimentarius is recognized by the World Trade Organization as an international reference point for the resolution of disputes concerning food safety and consumer protection. All rice-producing countries should come under the umbrella of the International Association of Rice Bran Oil (IARBO). It will help in sharing information and statistical data on rice bran oil and a dedicated website for RBO under IARBO will further promote this healthy oil.

Rice Husk

Rice husk accounts for ~20% of the weight of paddy. The main components of rice husk are silica (18%–22%), cellulose (28%–38%), hemicellulose (28%), and lignin (9%–20%). Upon burning, rice husk leaves a residue called rice husk ash (RHA) containing 87%–97% silica (Worasuwannarak et al., 2007).

Rice husk requires large volumes for storage and transportation because of its low bulk density (102–106 kg/m^3). It has low specific heat (0.29 kcal/kg °C) and low thermal conductivity (0.035 kcal/h m °C) values, thereby making it suitable for use in thermal insulation (Mishra et al., 1986). Its caloric value is around 13.5 MJ/kg (Wan et al., 2010).

Use of Rice Husk in India

Rice husk is mostly used as boiler feed for the generation of steam in rice mills that produce parboiled rice. This is the largest single source of rice husk consumption in India. It is also used as fuel in brick kilns, roadside *dhabas*, or eateries; in the generation of producer gas for running IC engines; in the generation of electricity; in the running of pump sets; etc. (Yahaya and Ibrahim, 2012; Yusuf et al., 2008). Rice husks are used for the manufacturing of particleboards, soil mulch, animal feed, and poultry litter (Ajiwe et al., 1998; Bhatnagar, 1994; Ebaid and El-Refaee, 2007). They are also used for producing sodium silicate, activated carbon (Granados and Venturini, 2008), and for furfural, which is widely used in the pharmaceutical industry, oil refineries, and resin manufacturing (Ren et al., 2012). Rice husk has also been used for producing bioethanol (Srivastava et al., 2014). Husks generated from raw rice mills and boiler ash generated from parboiled rice mills are sold to other industries or dumped in low-lying areas.

Pure amorphous silica can be produced from rice husk (Chakraverty et al., 1988, 1990). Amorphous silica is used in the rubber industry as a reinforcing agent, in cosmetics, in toothpaste, and in the semiconductor industry (Bose et al., 2015).

RHA obtained from rice husk also has various industrial applications, such as insulation cover for steel ingots, abrasives in metal cleaning, a floor sweeping aid, a carrier for fungicides, and for pest control. It can be used for manufacturing refractory bricks (Adylov et al., 2005; Kumar et al., 2012) and pozzolanic concrete blocks (Faiziev, 2003), and for the treatment of wastewater generated from industry (Giddel and Jivan, 2007).

The most promising and profitable use of rice husks is in the generation of electricity. Power plants in the range of 1–10 MW can become commercially viable. For a 1-MW power plant, approximately 24 tons of rice husk will be required, and every day 4.32 tons of RHA will be obtained. Most of the rice husk-/biomass-fired power plants follow the combustion route just like coal-fired plants. These plants are usually off-grid type and are most suitable for medium-to-large commercial enterprises. Akshya Urja (December 2014) reports that Satia Paper Mills Ltd. in Muktsar, Punjab, produces 12.5 MW of power using rice husk. Another paper mill in Saila Khurd near Chandigarh has rice husk–based power plants of 1, 5, and 10 MW capacity each. Similarly, OSWAL Woolen Mills has a 3.5-MW power plant that is based on rice husks only.

Potential Availability of Rice Husk in India

In 2014–15, rice lands all over India harvested ~157 million tons of paddy (the equivalent of 104.79 million tons of milled rice). Assuming that 90% of the paddy produced is processed into rice in modern mills, the potential husk availability has been computed by taking an average value of 0.2 husk-to-paddy ratio (Table 11.4). In 2014–15, total husk production in India was 28.295 million tons. West Bengal was the number one contributor and produced 3.972 million tons, followed by Uttar Pradesh (3.299 million tons), Andhra Pradesh (3.121 million tons), and Punjab (3.002 million tons).

Potential Use of Rice Husk: Policy and Strategy Required

Despite having so many well-established uses, only a small portion of rice husks is used in a meaningful way. The remaining husks are burned or dumped as solid waste. Many reasons are associated with the underuse of rice husks (Giddel and Jivan, 2007), including (1) lack of awareness of their potential for farmers and industry persons, (2) insufficient information about proper use, (3) socioeconomic problems, (4) penetration of technology, (5) lack of interest, (6) lack of awareness of environmental impact, and (7) inefficiency of information transfer.

The complete exploitation of the potential of rice husks to produce electricity is lacking, and three valuable products, such as silica, activated carbon, and sodium silicate, have numerous industrial applications.

For the effective use of rice husks, the following strategic measures need to be taken:

- To promote rice husk use in technologies for energy and other products, greater awareness will have to be created among entrepreneurs. Demonstration-cum-training centers can be used to effectively disseminate these technologies among the youth.
- Technologies can be popularized through exhibitions, TV programs, the establishment of technology demonstration centers, etc. The National Research Development Corporation can play a vital role in technology promotion.
- A capital subsidy of 30%–40% should be provided for all rice husk–based technology promotion and adoption. Research institutions can lead in developing low-cost rice by-product technologies or equipment. However, the subsidy needs to be provided only at the beginning for infrastructure development. The business model should be developed keeping in view the sustainability of the project.
- A rice husk–based small electricity generation unit (50-kW capacity) is ideal for isolated rural areas. This could be promoted by government agencies. These systems along with electricity generation will create employment and save diesel oil, etc. Rice husk–based electricity generation and the supply of up to 30-kW capacity have been popularized in India by Husk Power System (HPS) and Decentralized Energy System India (DESI), two enterprising units that have successfully provided electricity access using this resource. HPS has successfully used rice husk to provide decentralized electricity in rural areas and has so far installed more than 80 plants in 300 villages. On the other hand, DESI power placed emphasis on production uses of husk-based systems to displace the diesel-based electricity supply to microenterprises.
- Cooperation between India and other rice-producing countries is essential for the development of rice husk use technologies, updating of technologies based on experiences, technology transfer, etc. International organizations, such as FAO, UNIDO, UNDP, IRRI, ICAR, etc., can promote greater international cooperation.

Rice Straw

Asian countries contribute 52% of crop residue worldwide. Rice straw alone makes up 34.3% of the residue supply. It is a rich source of carbon with a high content of cellulose (32%–47%), hemicellulose (19%–27%), and lignin

(5%–24%) (Lim et al., 2012). It is an important source of livestock feed for large ruminants in South Asia (~90% are dependent on it) and Southeast Asia (30%–40% are dependent), including China and Mongolia (Devendra and Thomas, 2002). Rice straw contributes 40%–50% of total feed intake by large ruminants (Rao and Hall, 2003); however, this varies from region to region and state to state. The production and use pattern of rice straw in India varies across states and agroecology. Table 11.5 gives an overview of rice straw production and its major uses in the major rice-producing states across India.

The use of rice straw depends on four major interacting factors: farmers' preferences, crop/straw production amounts, access to alternative biomass resources, and straw demand in a particular region/agroclimatic zone (Erenstein et al., 2011). Demand for rice straw is higher in states with high livestock density, a high number of human inhabitants, and good access to biomass production. For instance, rice straw has a higher demand in eastern India because of its high human and livestock density than in Punjab and Haryana. However, the demand for straw is increasingly coming from various sectors, such as packaging, pot- and brick-making industries, mushroom-growing farmers, etc.

The patterns of rice straw usage in different states have changed over the years because of the mechanization of harvesting activities and land-use patterns. For example, combine harvesters that are being used in many states leave all rice residues in the field and farmers are not left with any choice but to burn the remaining residue.

Potential Use of Rice Straw

Rice straw is the single most important dry fodder for dairy animals in several states of India. Farmers in most of the major rice-producing states feed rice straw to their dairy cattle and draft animals. The states where farmers consume rice as the main staple also use rice straw as the main dry fodder for their large ruminants.

Due to the rapid population growth, urbanization, and increasing income, demand for livestock products is increasing day by day (Gulati and Reardon, 2008; Maltsoglou, 2015). However, agricultural and grazing land have been declining because of urbanization and growth in infrastructure. This development reduces the availability of feed, especially dry fodder for livestock. It has been estimated that the availability of dry fodder for livestock will decline by 26% by 2025 (GOI, 2011). By 2020 India will require 526 million tons of dry fodder, mainly crop residue, to meet the demand for livestock feed (Dikshit and Birthal, 2010). With increasing demand for rice straw for different purposes (i.e., mushroom cultivation and pot packaging), the future availability of rice straw will decline for livestock feed (GOI, 2011). Over time the dependency on dry fodder as livestock feed will keep on increasing. In the future smallholder farmers will shift to cereal residues (Herrero et al., 2009;

TABLE 11.5 Straw Production and Use Pattern in Major States in India

State	Rice straw production (paddy × 1.75) (in million tons)	Straw use patterns
Punjab	29.190	The majority of the straw (70%–80%) is burned in the field, 7% is used for livestock feed, and 1%–2% is used for roof construction (Government of Punjab, 2013). Swain et al. (2012) observed 50%–60% burned in the field in Haryana and 10%–15% used for livestock feed
Haryana	10.516	
Uttar Pradesh	32.078	Around 30% of the straw is used as fodder for livestock, 22% is burned in situ, 19% is incorporated in soil, and 16% is sold in the market (Singh et al., 2007)
Bihar	16.748	60% of the rice straw is used as livestock feed, 5%–10% is used as household fuel, and the rest is used for other purposes (Singh and Singh, 2012)
West Bengal	38.614	Rice straw is considered as the most important dry fodder for livestock feed, while 60%–70% is used for this purpose and 10%–15% is used as household fuel. The rest of the rice straw is used for roof making and packaging
Assam	12.758	
Odisha	21.761	
Madhya Pradesh	9.529	More than 50% of the straw is burned in fields, while 20% is used as dry fodder for livestock feed and the rest for household fuel or for roofing
Andhra Pradesh	30.345	Rice straw is mainly used for livestock feed in rice-growing areas (i.e., mostly in coastal region). Although farmers in Tamil Nadu use a combine harvester, none of them burn in the field. Farmers prepare straw bricks and sell them to neighboring districts and the paper industry, and also to neighboring states, such as Kerala. A very small proportion of the straw is used for roofing
Karnataka	9.608	
Tamil Nadu	15.304	
Kerala	1.330	Rice straw is usually used for mulch or is burned. The predominance of small and fragmented land parcels, lack of mechanization, and high labor costs make the collection of rice straw difficult (Energy Report Kerala, 2013)
Gujarat	4.305	Some 20%–30% of the rice straw is used for animal feed and 30%–40% is used to produce porcelain, glass, and pottery, and also paper pulp
Maharashtra	7.691	
Rajasthan	0.820	Most of the rice straw produced is used for feed; however, some straw left in the field is harvested by a combine harvester
India	275.095	The use of straw is not uniform across states; it varies across the states and is even mixed within a state

Straw production estimated by following the conversion rate of Bhattacharya et al. (1993) and Koopman and Koppejan (1997).

Valbuena et al., 2012), as the availability of dry fodder declines and pasture and grazing areas are taken over by agriculture or housing (Hegde, 2010; ICAR, 2012).

Rice straw as dry fodder has a huge potential to contribute to the growth of the dairy sector as one of the major cereal dry fodder sources for dairy cattle and buffalo. The digestibility of rice straw is comparable with that of other cereal residues (Jackson, 1978), although there is wide variation in the digestibility of different varieties of rice straw (Pattanaik et al., 2004). To increase livestock productivity and the availability of better varieties of rice straw as fodder, there is scope to improve rice breeding programs by developing varieties that will serve as dual-purpose crops (i.e., higher grain production and good-quality straw for feeding dairy animals). Rice straw digestibility is one of the least preferred traits considered in most breeding programs for new varietal improvements. This would require a lot of advocacy and explaining of the economic benefits of better digestibility both in the scientific programs and to the farming community.

The efficiency of rice straw feeding could be enhanced through chopping and adding minerals. Besides chopping, the nutritional value of rice straw can be enriched through physical and chemical treatment. The nutritional value of straw can be enriched by treating with urea and molasses (Abate and Melaku, 2009). The treated straw increases the intake and palatability and helps improve cattle health. Hart and Wanapat (1992) indicated that urea-treated (5%) rice straw improved overall intake, nutrient digestibility, and VFA production, and increased the passage rate of particles in the rumen. Urea treatment is the most suitable for small livestock holders. To improve the storage mechanism and make the feed available during flood and drought, rice straw can be densified through baling.[a] Densification has at least two advantages: (1) improving handling and efficiency throughout the supply chain system, and (2) controlled particle size distribution for improved feedstock uniformity and density. Densification of paddy straw makes it available for commercial use by increasing the energy content and reducing the transportation cost and storage space. State departments should make an effort to organize awareness programs to educate farmers on the use of chaff cutters and enriching straw as high-quality complete food for dairy cattle (Kurup, 2011).

Paddy straw can also be used for mushroom farming. It is recommended to use paddy straw for the cultivation of button and oyster mushrooms in winter and paddy-straw mushrooms in summer. The state governments should mobilize farmers to adopt mushroom cultivation for additional income generation. The market for mushrooms is already established and it can be further enhanced through subsidies, incentives, or soft loans for making

---

a. Compression baling and roll-compression can reduce biomass volume to one-fifth of its loose bulk.

mushroom huts. The leftover paddy straw after harvesting mushroom can be used as manure (after composting) for other crops, which would save expenses on chemical fertilizers.

Rice Straw as an Alternative Source of Energy

Rice straw has immense potential to create bioenergy, an alternative renewable source of power. Biomass can be directly used as fuel for combustion and for heat and electricity generation. Biochemical processes convert biomass into value-added products, such as ethanol, hydrogen, and methane. Revitalization of the biomass sector for the next 5–10 years could attract INR 300 billion in investments and provide labor wages for 16.8 million person-days per year (Khandelwal, 2015).

A fair amount of rice straw is wasted because of inefficient and improper harvesting and collection processes. Multiple cropping systems (such as the rice–wheat system) leave a very short time window for rice straw collection and land preparation for the next crop. The lack of labor during the harvesting season has also led to higher labor costs. Most farmers burn the remaining rice straw or stalks to clear their field and sow the wheat crop. This problem is compounded by the fragmented nature of agricultural holdings, the absence of storage facilities, and the need to further process the straw (drying) for some applications, which adds cost. Biomass power plants need a ready and accessible supply of fuel throughout the year to avoid fluctuating costs. The biomass power plants have huge operational and maintenance costs that are tied to the prices of feedstock.

Due to the high content of alkali, potassium, chloride, and silica in rice straw, the low ash fusion temperature causes deposition and shuts down boilers. The plant load factor (or power plant energy output) decreases and maintenance costs rise because of frequent stoppages, cleaning, and plant shutdowns. At present most power plants mix rice straw with other fuels to prevent damaging the boilers. Improvements in plant performance and machinery are still needed in the biomass sector.

Cohesive state-level policies in addition to innovative financial mechanisms (i.e., green bonds) need to be in place for the biomass sector to be reenergized. States must have a clear target for biomass cogeneration and a tariff system aligned with both state and central electricity regulatory commissions. Best practices that encourage the sector include single-window clearance procedures, exemption from electricity duty, and open access for the sale of generated power. Support for land acquisition along with linkages and development of the fuel supply chain through various government schemes, and the development of targeted rural straw collection centers with all the facilities and machinery are key components of the policy to boost the use of biomass for energy generation. Greater synergy and goal setting between R&D and the biomass power industry need to be fostered, especially for improving operational performance.

Policy Requirement for Effective Rice Straw Management

Apart from the Ministry of New and Renewable Energy, other ministries linked to rural development, agriculture, health and family welfare, environment, and climate change need to be involved to realize the direct and indirect benefits of rice straw and its by-products. Links between the fuel-supply chain (especially rice straw collection) and MGNREGA or with the village panchayat need to be strengthened, along with collaboration with the MNRE, developers, and farmers. Community engagement or communication with self-help groups can help overcome biomass procurement problems. In addition, states need to provide support for the collection of paddy straw waste from farms. Machines, such as spreaders, rakes, and balers, could be made available after the assessment of various members involved in the rice straw supply chain based on straw production through various schemes (Box 11.1).

Multiple options for straw management must be made available locally instead of on a state or national platform. Solutions can be tailored depending on the existing market opportunities, such as areas with the potential for mushroom farming or where livestock fodder is in high demand. Another way of encouraging diversification lies in the availability of biomass power plants or industrial zones. For example, in Haryana, Uttar Pradesh, and Punjab, paddy straw is in surplus and underused, and the presence of biomass power plants can change the practice of burning of residue as a strategy to clear and prepare the land for a wheat crop. Aside from bioenergy, farming communities near processing industries or manufacturing plants (e.g., paper and board) can earn by supplying straw as a raw material.

BOX 11.1 Case Study: Zamindara Farm Solutions

Zamindara Farm Solutions is an agribusiness enterprise based in Fazilka, Punjab. It has readily grown into a prominent social enterprise by using the radio-taxi model for the rental of farm machinery. As part of its marketing strategy, it has developed schemes, such as *"chalak bane malik"* (driver becomes owner), to encourage dedicated farmers to become involved by operating machinery and providing services to other farmers.

Apart from providing "operate now/own later" schemes, the organization links farmers to biomass power plants in the area. Through its efforts, farmers can invest in hay balers, collect straw from various farms, and sell it to the power plant. This has become a successful business model and has considerably reduced rice straw burning in the vicinity of biomass power plants in the districts of Fazilka and Muktsar in Punjab. Coordinated aggregation of biomass created multiple employment opportunities and used surplus rice straw for energy production.

A holistic approach should be applied in assessing different rice straw management options in-field (i.e., incorporation and mulching), off-field (i.e., energy and nonenergy), and in return to field (i.e., compost). Removing the straw removes nutrients, processing the straw (drying, compacting, and chopping) requires energy, and all options produce greenhouse gases at different steps in different quantities. Burning has positive aspects, such as the control of some diseases. Life cycle and sustainability assessments should be conducted for the different options, including rice production, processing, and by-product management, to obtain a clearer understanding of their sustainability to feed into policy recommendations.

Rice straw management within a value-addition framework can be tailored to available livelihood options. For example, rice straw that has been used for mushroom farming could be later used for livestock feed or used to produce biogas. The biogas could then be used in the household for cooking. Leftover material from the biodigesters could also be added as nutrient on the farm.

Farmers must obtain financial and agency support to engage in alternative livelihood systems that may yield better returns than farming. Awareness campaigns can be used to attract and encourage farmers to prioritize rice straw and other by-products.

REFERENCES

Abate, D., Melaku, S., 2009. Effect of supplementing urea-treated barley straw with lucerne or vetch hays on feed intake, digestibility and growth of Arsi Bale sheep. Trop. Anim. Health Prod. 41, 579–586.

Adylov, G.T., Kulagina Mansurova, E.P., Rumi, M.K., Faiziev, S.A., 2005. Lightweight dinas refractories based on rice husk ash. Refract. Ind. Ceram. 46 (3), 187–188.

Ajiwe, V.I.E., Okeke, C.A., Ekwuozor, S.C., Uba, I.C., 1998. A pilot plant for production of ceiling boards from rice husks. Bioresour. Technol. 66, 41–43.

Ali, N., Ojha, T.P., 1976. Parboiling. In: Araullo, E.V., de Padua, D.B., Graham, M. (Eds.), Rice Postharvest Technology. International Development Research Centre, Ottawa, Canada, pp. 163–204.

Ansharullah, Hourigan, J.A., Chesterman, C.F., 1997. Application of carbohydrases in extracting protein from rice bran. J. Sci. Food Agric. 74, 141–146.

Ausman, L.M., Rong, N., Nicolosi, R.J., 2005. Hypocholesterolemic effect of physically refined rice bran oil: studies of cholesterol metabolism and early atherosclerosis in hypercholesterolemic hamsters. J. Nutr. Biochem. 16, 521–529.

Bhardwaj, S., Sharma, R., 2015. Recent advances in on-farm paddy storage. Int. J. Farm Sci. 5 (2), 265–272.

Bhatnagar, S.K., 1994. Fire and rice husk particleboard. Fire Mater. 18, 51–55.

Bhattacharya, S.C., Pham, H.L., Shrestha, R.M., Vu, Q.V., 1993. CO_2 emissions due to fossil and traditional fuels, residues and wastes in Asia. AIT Workshop on Global Warming, Issues in Asia. 8–10 September 1993, AIT, Bangkok, Thailand.

Bose, D.N., Haldar, A.R., Chaudhuri, T.K., 2015. Preparation of polysilicon from rice-husk. Curr. Sci. 108 (7), 1214–1216.

Central Food Technological Research Institute (CFTRI), 1960. Circ. No. 7 (revised). CFTRI, Mysore, India.

Chakraverty, A., Banerjee, H.D., Mishra, P., 1990. Production of amorphous silica from rice husk in a vertical furnace. AMA Agric. Mech. Asia Afr. Lat. Am. 21, 69–75.

Chakraverty, A., Mishra, P., Banerjee, H.D., 1988. Investigation of combustion of raw and acid-leached rice husk for production of pure amorphous white silica. J. Mater. Sci. 23, 21–24.

Cheftel, J.C., Cuq, J.L., Lorient, D., 1985. Amino acids, peptides and proteins. In: Fennema, O.R. (Ed.), Food Chemistry. Marcel Dekker, New York, NY, 245p.

Chou, T.W., Ma, C.Y., Cheng, H.H., Chen, Y., Lai, M.H., 2009. A rice bran oil diet improves lipid abnormalities and suppresses hyperinsulinemic responses in rats with streptozotocin/nicotinamide-induced type 2 diabetes. J. Clin. Biochem. Nutr. 45 (1), 29–36.

Central Institute of Post-Harvest Engineering and Technology (CIPHET), Indian Council for Agricultural Research (ICAR), 2010. Estimate of economic value of harvest and post-harvest losses in India. Estimation of quantitative harvest and postharvest losses of major agricultural produce in India (for 2005–07). Ludhiana, India.

Devendra, C., Thomas, D., 2002. Crop-animal interactions in mixed farming systems in Asia. Agric. Syst. 71, 27–40.

Dhingra, D., Chopra, S., Rai, D.R., 2012. Stabilization of raw rice bran using ohmic heating. Agric. Res. 1 (4), 392–398.

Dikshit, A.K., Birthal, P.S., 2010. India's livestock feed demand: estimates and projections. Agric. Econ. Res. Rev. 23, 15–28.

Directorate of Marketing & Inspection (DMI), 2004. Post-harvest profile of paddy/rice. Available from: http://agmarknet.nic.in/rice-paddy-profile_copy.pdf

Ebaid, R.A., El-Refaee, I.S., 2007. Utilization of rice husk as an organic fertilizer to improve productivity and water use efficiency in rice fields. Afr. Crop Sci. Conf. Proc. 8, 1923–1928.

Energy Report Kerala, 2013. Government of Kerala.

Erenstein, O., Samaddar, A., Teufel, N., Blümmel, M., 2011. The paradox of limited maize stover use in India's smallholder crop–livestock systems. Exp. Agric. 47, 677–704.

Fabian, C., Ju, Y., 2011. A review on rice bran protein: its properties and extraction methods. Crit. Rev. Food Sci. Nutr. 51 (9), 816–827.

Faiziev, S., 2003. Synthesis of ceramic compounds utilizing woody waste materials and rice husk. Constr. Build. Mater. Mater. Sci. Forum 437, 411–414.

Fernando, W.J.N., Hewavitharana, L.G., 1993. Effect of fluidized bed drying on stabilization of rice bran. Drying Technol. 11, 1115–1125.

Gaikwad, V.R., Sambrani, S., Prakash, V., Kulkarni, S.D., Murari, P., 2006. State of Indian Farmer: A Millennium Study: Post-Harvest Management, vol. 16. Ministry of Agriculture, New Delhi.

Giddel, M.R., Jivan, A.P., 2007. Waste to wealth: potential of rice husk in India, a literature review. International Conference on Cleaner Technologies and Environmental Management PEC, Pondicherry, India (vol. 2, pp. 4–6).

Giese, J., 1994. Proteins as ingredients: types, functions, applications. Food Technol. 48, 50–60.

Gnanasambandam, R., Hettiarachchy, N.S., 1995. Protein concentrates from unstabilized and stabilized rice bran: preparation and properties. J. Food Sci. 60, 1066–1069.

Government of India (GOI), 2011. Report of the Sub Group III on Fodder and Pasture Management, Planning Commission of India, GOI. Version 1.5, 21 September.

Government of Punjab, 2013. Policy for management and utilization of paddy straw in Punjab.

Granados, C.D., Venturini, R., 2008. Activated carbons obtained from rice husk: influence of leaching on textural parameters. Ind. Eng. Chem. Res. 47, 4754–4757.

Gulati, A., Reardon, T., 2008. Organized retail and food price inflation: opening the 'black box.' Business Line, May 24.

Hamada, J.S., 1997. Characterization of protein fractions of rice bran to devise effective methods of protein solubilization. Cereal Chem. 74, 662–668.

Hamada, J.S., 1999. Use of protease to enhance solubilization of rice bran proteins. J. Food Biochem. 23, 307–321.

Hammond, N., 1994. Functional and nutritional characteristics of rice bran extracts. Cereal Foods World 39, 752–754.

Hart, F.J., Wanapat, M., 1992. Physiology of digestion of urea-treated rice straw in swamp buffaloes. Asian-Aus. J. Anim. Sci. 5, 617–622.

Hegde, N.G., 2010. Forage resource development in India. Souvenir of IGFRI Foundation Day, November.

Heli Roy, R.D., Shanna Lundy, B.S., 2005. Rice bran. Pennington Nutrition Series Pennington Biomedical Research Center, Baton Rouge, LA, (p. 8).

Herrero, M., Thornton, P.K., Notenbaert, A.M., Msangi, S., Wood, S., Kruska, R., Dixon, J., Bossio, D., van de Steeg, J., Freeman, H.A., Li, X., Rao, P.P., 2009. Drivers of change in crop–livestock systems and their potential impacts on agroecosystems services and human well-being to 2030. Nairobi, Kenya.

Houston, D.F., 1972. Rice bran and polish. In: Houston, D.R. (Ed.), Rice: Chemistry and Technology. American Association of Cereal Chemists, Inc., St. Paul, MN, p. 272.

Indian Council for Agricultural Research (ICAR), 2012. Crop residue management with conservation agriculture: potential, constraints, and policy needs.

Jackson, M.G., 1978. Rice straw as livestock feed. Ruminant Nutrition: Selected Articles from the World Animal Review. FAO.

Juliano, B.O., 1985. Rice bran. In: Juliano, B.O. (Ed.), Rice Chemistry and Technology. American Association of Cereal Chemists, St. Paul, MN, pp. 647–680.

Kannan, E., 2014. Assessment of Pre- and Post-Harvest Losses of Important Crops in India. Research Report: IX/ADRTC/153A. Agricultural Development and Rural Transformation Centre Institute for Social and Economic Change, Bangalore.

Khandelwal, A., 2015. State level policy cohesion key to momentum. Bio Power India 1 (4), 3–8, Available from: http://biomasspower.gov.in/document/Magazines/Biopower/BioPower%20Issue%204%20April%20June%202015.pdf/.

Kik, M.C., 1955. Influencing of processing on nutritive value of milled rice. J. Agric. Food Chem. 3, 600.

Kim, J.S., Godber, J.S., 2001. Oxidative stability and vitamin E levels increased in restructured beef roast with added rice bran oil. J. Food Qual. 24, 17–26.

Kim, C.J., Byun, S.M., Cheigh, H.S., Kwon, T.W., 1987. Optimization of extrusion of rice bran stabilization process. J. Am. Oil Chem. Soc. 64, 514–516.

Koopman, A., Koppejan, J., 1997. Agricultural and forest residues generation, utilization and availability. Paper presented at the Regional Consultation on Modern Applications of Biomass Energy. 6–10 January 1997, Kuala Lumpur, Malaysia.

Kumar, A., Mohanta, K., Kumar, D., Parkash, O., 2012. Properties and industrial applications of rice husk: a review. Int. J. Emerg. Tech. Adv. Eng. 2 (10), 86–90.

Kurup, M.P.G., 2011. Technologies and practices for improving livestock feeding in India. Successes and failures with animal nutrition practices and technologies in developing countries. In: Makkar, H.P.S. (Ed.), Proceedings of the FAO Animal Production and Health Proceedings. No. 11. FAO Electronic Conference, 1–30 September 2010, Rome, Italy.

Lakkakula, N.R., Lima, M., Walker, T., 2004. Rice bran stabilization and rice bran oil extraction using ohmic heating. Bioresour. Technol. 92, 157–161.

Lichtenstein, A.H., Ausman, L.M., Carrasco, W., 1994. Rice bran oil consumption and plasma lipid levels in moderately hypercholesterolemic humans. Arterioscler. Thromb. Vasc. Biol. 14, 549–556.

Lim, J.S., Manan, Z.A., Alwi, S.R.W., Hashim, H., 2012. A review on utilization of biomass from rice industry as a source of renewable energy. Renew. Sust. Energ. Rev. 16, 3084–3094.

Lu, S., Luh, B.S., 1991. Properties of the rice caryopsis. Lu, B.S. (Ed.), Rice Production, vol. 1, second ed. AVI Publishing Co., Westport, CT, pp. 389–419.

Malekian, F., Rao, R.M., Prinyawiwatkul, W., Marshall, W.E., Windhauser, M., Ahmedna, M., 2000. Lipase and lipoxygenase activity, functionality, and nutrient losses in rice bran during storage. Agricultural Center, Louisiana Agricultural Experiment Station, Baton Rouge (Bulletin No. 870, p 1–68).

Maltsoglou, I., 2015. Household expenditure on food of animal origin: a comparison of Uganda, Vietnam, and Peru. Available from: www.fao.org/ag/againfo/projects/en/pplpi/docarc/wp43.pdf

Mawal, Y.R., Mawal, M.R., Ranjekar, P.K., 1987. Biochemical and immunological characterization of rice albumin. Biosci. Rep. 7, 1–9.

Mehta, B.V., 2016. Overview of Indian vegetable oil industry and future outlook. Malaysia–India Palm Oil Seminar. 19 May 2016, Hyderabad.

Mishra, P., Chakraverty, A., Banerjee, H.D., 1986. Studies on physical and thermal properties of rice husk related to its industrial application. J. Mater. Sci. 21 (6), 2129–2132.

Nanua, J.N., McGregor, J.U., Godber, J.S., 2000. Influence of high-oryzanol rice bran oil on the oxidative stability of whole milk powder. J. Dairy Sci. 83, 2426–2431.

Narisullah, Krishnamurthy, M.N., 1989. Effect of stabilization on the quality characteristics of rice bran oil. J. Am. Chem. Oil Soc. 66, 661–663.

NPCS Board of Consultants and Engineers, 2007. Handbook on Rice Cultivation and Processing. NIIR Project Consultancy Services, New Delhi (544 p.).

Pattanaik, A.K., Dutta, N., Ramakrishna Reddy, C., Viraktamath, B.C., Sharma, K., Blümmel, M., 2004. Evaluation of genetic variability of rice straw cultivars for better straw quality by in vitro gas production. Nutritional Technologies for Commercialization of Animal Production. Abstracts of Eleventh Animal Nutrition Conference. 5–7 January 2004, Jabalpur (p. 27).

Pillaiyar, P., 1977. Improvements in conventional parboiling of paddy: elimination of foul odor and increase in yield of rice. Report of the Action Oriented Field Workshop for Prevention of Postharvest Rice Losses. Food and Agriculture Organization of the United Nations, Malaysia.

Pillaiyar, P., Singaravadivel, K., 1993. Pressure parboiling of wet season paddy. Madras Agric. J. 80 (12), 713–716.

Pillaiyar, P., Singaravadivel, K., Desikachar, H.S.R., 1993. Low moisture parboiling of paddy. J. Food Sci. Technol. 30, 97–99.

Premakumari, S., Balasasirekha, R., Gomathi, K., Supriya, S., Jagan Mohan, R., Alagusundram, K., 2012. Development and acceptability of fiber enriched ready mixes. Int. J. Pure Appl. Sci. Technol. 9, 74–83.

Raj, S.A., Singaravadivel, K., 1980. Influence of soaking and steaming on the loss of simpler constituents in paddy. J. Food Sci. Technol. 17 (3), 141–143.

Rao, P.P., Hall, A.J., 2003. Importance of crop residues in crop-livestock systems in India and farmers' perceptions of fodder quality in coarse cereals. Field Crops Res. 84, 189–198.

Ren, S., Xu, H., Zhu, J., Li, S., He, X., Lei, T., 2012. Furfural production from rice husk using sulfuric acid and a solid acid catalyst through a two-stage process. Carbohydr. Res. 1 (359), 1–6.

Rogers, E.J., Rice, S.M., Nicolosi, R.J., Carpenter, D.R., Muclelland, C.A., Romanczyc, Jr., L.R., 1993. Identification and quantitation of gamma-oryzanol components and simultaneous assessment of tocopherols in rice bran oil. J. Am. Oil Chem. Soc. 70, 301–307.

Saunders, R.M., 1990. The properties of rice bran as a food stuff. Cereal Foods World 35, 632–662.

Sayre, B., Saunders, R., 1990. Rice bran and rice bran oil. Lipid Technol. 2, 72–76.

Sayre, R.N., Saunders, R.M., Enochian, R.V., Schultz, W.G., Beagle, E.C., 1982. Review of rice bran stabilization systems with emphasis on extrusion cooking. Cereal Foods World 27 (7), 317–322.

Sharif, M.K., 2009. Rice Industrial By-Products Management for Oil Extraction and Value Added Products. PhD thesis. National Institute of Food Science and Technology, University of Agriculture, Faisalabad, Pakistan (206 p.).

Singaravadivel, K., Raj, S.A., 1986. An appraisal of commonly adopted parboiling methods. Agric. Rev. 7 (2), 98–104.

Singh, D.K., Singh, V.K., 2012. Fodder production and utilisation patterns in disadvantaged areas: a study of Eastern India. Tropentag Annual Conference on Resilience of Agricultural Systems Against Crises. 19–21 September 2012, Gottingen.

Singh, J., Erenstein, O., Thorpe, W., Varma, A., 2007. Crop–livestock interactions and livelihoods in the Gangetic plains of Uttar Pradesh, India. Crop–Livestock Interactions Scoping Study: Report 2. Research Report 11. International Livestock Research Institute, Nairobi, Kenya (88 p.).

Sondi, A.B., Mohan Reddy, I., Bhattacharya, K.R., 1980. Effect of processing condition on the oil content of parboiled rice bran. Food Chem. 5, 277.

Srivastava, A.K., Agrawal, P., Rahiman, A., 2014. Delignification of rice husk and production of bioethanol. Int. J. Innov. Res. Sci. Eng. Technol. 3 (3), 10187–10194.

Subrahmanyan, V., Sreenivasan, A., Das Gupta, H.P., 1938. Studies on quality of rice. I. Effect of milling on the chemical composition and commercial qualities of raw and parboiled rices. Indian J. Agric. Sci. 8, 459.

Swain, B.B., Nils, T., Meera, M., Surabhi, M., 2012. Optimizing Livelihood and Environmental Benefits from Crop Residues in Smallholder Crop–Livestock Systems: South Asia Regional Case Study. Project Report. International Livestock Research Institute, New Delhi.

Tadera, K., Orite, K., 1991. Isolation and structure of a new vitamin B_6 conjugate in rice bran. J. Food Sci. 56, 268–269.

Usha, P.T., Premi, B.R., 2015. Rice bran oil: nature's gift to mankind. Available from: https://www.nabard.org/pdf/issue7td-13.pdf

Valbuena, D., Erenstein, O., Homann-Kee Tui, S., Abdoulaye, T., Claessens, L., Duncan, A.J., Gérard, B., Rufino, M., Teufel, N., van Rooyen, A., van Wijk, M.T., 2012. Conservation agriculture in mixed crop-livestock systems: scoping crop residue trade-offs in sub-Saharan Africa and South Asia. Field Crops Res. 132, 175–184.

Wan, A.K., Ghani, W.A., Abdullah, M.S.F., Matori, K.A., Alias, A.B., da Silva, G., 2010. Physical and Thermochemical Characterisation of Malaysian Biomass Ashes. The Institution of Engineers Malaysia, Petaling Jaya, Malaysia, (vol. 71 (2), pp. 9–18).

Watchararuji, K., Goto, M., Sasaki, M., Shotipruk, A., 2008. Value-added subcritical water hydrolysate from rice bran and soybean meal. Bioresour. Technol. 99, 6207–6213.

Weir, G.S.D., 1986. Protein hydrolysates as flavorings in development. In: Hudson, B.J.R. (Ed.), Food Protein. Elsevier, London, pp. 175–217.

Wilson, T.A., Ausman, L.M., Lawton, C.W., Hegsted, D.M., Nicolosi, R.J., 2000. Comparative cholesterol lowering properties of vegetable oils: beyond fatty acids. J. Am. Coll. Nutr. 19 (5), 601–607.

Worasuwannarak, N., Sonobe, T., Tanthapanichakoon, W., 2007. Pyrolysis behaviors of rice straw, rice husk and corncob by TG-MS technique. J. Anal. Appl. Pyrol. 78, 265–271.

Yahaya, D.B., Ibrahim, T.G., 2012. Development of rice husk briquettes for use as fuel. Res. J. Eng. Appl. Sci. 1 (2), 130–133.

Yoshida, Y., Niki, E., Noguchi, N., 2003. Comparative study on the action of tocopherols and tocotrienols as antioxidant: chemical and physical effects. Chem. Phys. Lipids 123, 63–75.

Yusuf, M., Farid, N.A., Zainal, Z.A., Azman, M., 2008. Characterization of rice husk for cyclone gasifier. J. Appl. Sci. 8, 622–628.

Chapter 12

Institutional Innovations in Rice Production and Marketing in India: Experience and Strategies

Sukhpal Singh

Centre for Management in Agriculture (CMA), Indian Institute of Management (IIM), Ahmedabad, Gujarat, India

INTRODUCTION

Rice is the most important crop worldwide as it is the staple food for more than half of the world's population. It provides employment to one billion people globally and is therefore crucial for food security; however, this crop also has a large physical and environmental footprint in terms of area devoted to it. In India a large proportion of rice farmers and farm workers suffer from a lack of food security and poor working conditions (Gathorne-Hardy et al., 2016).

Low yields, increasing costs of cultivation, and low prices obtained in smallholder rice farming have been and continue to be important concerns in India. This applies to all stakeholders, including the private sector, involved in the marketing of agricultural inputs and purchasing of farm produce. The only way to help such farmers is to help decrease their costs of production and marketing, provide stable and remunerative market access, improve prices obtained, or increase yield. However, returns to labor in small-scale farming are increasingly determined outside agriculture through more integrated produce and labor markets, opportunity costs of farm labor, and the aspirations of farm family youth (Gatzweiler and von Braun, 2016). Therefore, there is a role for institutional innovations (i.e., changing the ways of doing things), which involves rules, norms, organizations, and organizing mechanisms. It has been recognized that institutional innovations are as important for sustainable growth and development as technological innovations (Gatzweiler, 2016). These innovations could be in the form of new institutional mechanisms for the provision of farm inputs and services, new platforms for the marketing of rice

The Future Rice Strategy for India. http://dx.doi.org/10.1016/B978-0-12-805374-4.00012-9

or linking farmers with markets, or new credit institutions for achieving inclusive (including marginal and small farmers, women farmers and farm workers, and farmers and farm workers in remote, resource-constrained, and dry-land regions) and sustainable agricultural development in a context, such as that of Indian agriculture.

The need for institutional innovations in the rice sector arises on account of the need for increased yield [with the yield gap ranging from 4% in West Bengal to 66% in Jharkhand (Singh, 2012)], reduced production costs, and an improved quality of output. This is in addition to achieving sustainability in economic, environmental, and social dimensions. On the marketing side price realization is poor, that is, "producer remuneration" is low as farmers may have to resort to selling below the minimum support price (MSP) in some states. This occurs when there is no public procurement system in place (Niti Aayog, 2016) or when public agencies procure rice from rice mills and not from farmers. In the states of Gujarat, Maharashtra, West Bengal, and Madhya Pradesh, a majority of the farmers are not even aware of the MSP. In Bihar and West Bengal the farmers do not sell to the government procurement agencies and in these and many other states, a significant proportion of farmers sell below the MSP (Niti Aayog, 2016). High marketing (transaction) costs or a high cost of market access may also be experienced, which leads to poor marketing margins for primary producers and a low share of consumers' prices for producers. There is also a lack of quality standards and incentives and a lack of market infrastructure. In addition, various input and output markets are interlocked, which leads to overpricing of farm inputs and underpricing of farmers' output. This is further complicated by a lack of producer collectivization, with only 2.2% of the farmers being members of any farmer association and only 4.8% having a member belonging to a self-help group (Witsoe, 2006). It is also argued that institutional innovations in smallholder market linkage are needed in terms of partnerships, the use of information and communication technology (ICT), leveraging networks, value chain financing, smallholder policies, and even in contracts that can promote both efficiency and inclusiveness of the linkage (Mendoza and Thelen, 2008).

This chapter examines the various types of institutional innovations in the rice agribusiness sector from the global literature and the Indian smallholder agricultural context, where other ways of reaching small farmers or linking them with markets have not worked as small producers have been marginalized or excluded from institutional frameworks (Gatzweiler, 2016). It takes a value chain approach to identifying and analyzing the institutional innovations at different stages of the chain. The next section provides a theoretical underpinning to the chapter in terms of an analytical framework. The section "The Indian Experience" provides an analysis of a few cases of success and failure in this domain, wherein it also examines the inclusiveness of various institutional innovations. The cases of institutional innovations in paddy production and marketing in India range across the value chain, such as inputs (e.g., bioinput production, machinery rental, and water), land (e.g., *Kudumbshree*), marketing (design and practice of contract farming, public–private partnerships, PPPs, and organic certification), Agricultural Produce Market Committee (APMC) markets, the role of Primary Agricultural Cooperative Societies (PACS), and new aggregation mechanisms, such as producer companies (PCs). The section "Conclusions" concludes with major lessons for strengthening and leveraging institutional innovations for a more sustainable rice sector in India and infers the evolution and role of such institutional innovations and the implications for policy and practice.

INSTITUTIONAL INNOVATIONS: A FRAMEWORK

Institutions and institutional context are important determinants of development. Institutions refer to the "rules of the game" in a society or, more formally, the socially designed constraints and enablers that shape human interactions within action situations (Gatzweiler, 2016). They are made up of formal and informal constraints and opportunities, such as rules and laws, norms of behavior or codes of conduct, and their enforcement characteristics. They altogether define the incentive structure of the society and, more so, economy.[a] These institutions are further embedded in local social and cultural systems, which leads to institutional "thickness." This refers to the dense presence of organizations in a local area, their strong interactions in the local area, their domination due to this high level of interaction, and their shared commitment to a common cause: though all of this need not be formal (Neilson and Pritchard, 2009).

Innovation is the implementation (use) of something new or improved (whether technology or otherwise) in products (goods or services), processes, marketing, or organizational methods. In other words, it means applying ideas, knowledge, or practices that are new to a particular context with the purpose of creating positive change that will provide a way to meet needs, take on challenges, or seize opportunities. Such novelties and useful changes could be substantial (a large change or improvement) or cumulative (small changes that together produce a significant improvement) (IICA, 2014, p. 3).

A novel idea implemented in a particular way can be considered an innovation if it is new in the context, even though it may not be new to the world (IICA, 2014, p. 3; Raffaelli and Glynn, 2015).

There are many types of innovations, such as technological, social, product, process, marketing, and organizational; institutional innovation is just one type (IICA, 2014). There could be path dependence in institutions

---

a. Institutions are also different from organizations, the former being the rules of the game and the latter the players in the game. Both of them influence each other in terms of which organizations come up, and how they evolve is determined by the institutional framework (rules of the game) and they in turn influence how the institutional framework itself evolves.

(Ebbinghaus, 2005) versus innovations in institutions. Path dependence refers to the recognition in institutional theory that the past shapes the future or that history matters. Thus, path dependence could be about an unplanned "trodden trail" that emerges due to repeated use of the path by an institution, which leads to diffusion of the persistence of the same institution; or about a "road juncture," which is a branching point at which one of the available paths needs to be chosen to continue the journey, which is about institutional change (Ebbinghaus, 2005). In the Indian context, the various amendments in the Cooperative Societies' Act could be more related to path dependence while the Producer Company Act could be more about institutional change. Institutional innovations could occur in land systems, labor systems, social systems, or the organization of activity, production, and marketing (including market and policy reforms). Innovations can also take place in a top-down or bottom-up manner. Institutional innovations entail a change of policies, standards, regulations, processes, agreements, models, methods of organization, institutional practices, or relationships with other organizations. This creates a more dynamic environment that encourages improvements in the performance of an institution or system to make it more interactive and competitive (IICA, 2014, p. 4). It can be a continuous process of incremental change, a response to a crisis or a failure (Shiller, 2006), or a process of creative destruction (Gatzweiler, 2016). Therefore, institutional innovation can be the creation of a new institution or the change in an existing institution (Raffaelli and Glynn, 2015). The emergence of second-hand tractor markets in the Indian Punjab can be seen as an institutional innovation in response to a crisis of overtractorization and nonviable use of the machine in the post-Green Revolution period (Singh, 1999).

However, bringing about institutional change in terms of innovations is not easy. Major concerns about institutional innovations include the following: they generally take place outside the formal system; there is very little policy support before they are proven (Ruttan, 1989); they could be market, social, or environmental driven; there could be exclusion from and inclusion in institutional innovation depending on the type of crop, place, technology, market, and/or type and nature of the organization; the sustainability of such innovations; and the scale-up of such innovations. On the other hand, barriers to such innovations can include (1) infrastructural barriers, relating to the knowledge infrastructure made up by departments of R&D, universities, research centers, and all related regulations, and also the physical infrastructure, consisting principally of roads and telecommunications; (2) hard and soft institutional barriers, relating to formal rules and regulations (hard) and relating to symbols, values, and norms (soft); (3) network barriers, calibrated by the strength of connectivity, whereby strong interactions cause blindness toward new ideas from outside and weak interactions hinder members from combining their forces to work for change; and (4) market structure barriers, relating to the position of and relations between market parties along the value chain (Totin et al., 2012).

THE INDIAN EXPERIENCE

This section examines the various institutional innovations in primary production, commodity selling, processing and value addition, and marketing links in the rice value chain. The aim is to specifically highlight the role of each innovation in terms of addressing the issues of yield, cost of production and marketing, and remuneration for the primary producer and farm worker as a part of the rice value chain domestic, or global.

Innovations in the Production Sector

There are cases of custom rentals of agricultural machinery for paddy, such as PACS in Punjab, which rent out laser land levelers to increase irrigation efficiency and reduce costs. There are 1200 such Cooperative Agricultural Machinery Service Centers (CAMSC) in Punjab, each of which has a capital investment of INR 16–40 lakh and annual expenses of INR 9–20 lakh. The net returns of such centers range from INR 16,000 to INR 6.3 lakh, with 25–40 machines in each CAMSC. The CAMSC hiring costs are lower than privately owned machines and equipment. They are mostly used by marginal and small farmers who have a higher percentage of income from the nonfarm sector and a higher percentage of family labor than those in tractor-owning households. Timely availability was the only negative issue and was caused by the small number of machines at the PACS (Sidhu and Vatta, 2012).

This is similar to the innovation in agricultural mechanization in China where the average farm size is only 0.34 acre and the lease of land to farmers is for 30 years. With only 5% of farm power being animal based (as against 9% in India), the custom-hiring farmer cooperative companies operate across provinces, with one combine service enterprise (CSE) harvesting 200 farms, 133 hectares, or 2 farms per day with 100 days of work. These cooperatives adopt a strategy to not compete with each other, but they choose to access lower cost spare parts together for a group of 5–10 CSEs that are part of the cooperative. These are all private initiatives that were initially supported by the state with harvest calendars across regions. The CSEs have begun to manage themselves owing to their personal experiences across provinces (Yang et al., 2013). However, the Chinese cooperatives have been able to scale up and reap economies of scale in buying, as well as in selling their services. This is very different from the PACS in Punjab, which are more localized and have to compete with private entrepreneurs.

In the Indian Punjab, Zamindara Farm Solutions (ZFS), a farm machinery entrepreneur, is another interesting innovation for decreasing the cost of paddy production. Originally established in 2005, ZFS today owns 170 machines that have been used by 6000 farmers over 7 years across 4 districts with a 300 km radius from the original center. It is run as a business model in an environment of overtractorization of the farm sector, where affordability for such costly machines is an issue and the crisis of mechanization is seen in the presence of second-hand tractor markets,[b] which are held weekly or fortnightly across many *mandi* towns and large villages. Zamindara's investment of INR 1 million in 2005 led to a turnover of INR 60 million by 2011–12. It uses a "library model" and a "taxi model" for the custom hiring of machines and tractors, with the "library model"[c] providing machines and "taxi model"[d] providing tractors along with drivers. This model (franchising) was adopted along with the distribution of tractors by the parent company (Zamindara Distributors). The franchisee invests INR 30,000 and obtains a 10% share in rentals, pays the cost of the tractor in equalized monthly installments from revenue generated, pays lower rent for self-use, and ends up owning the tractor, which is promoted as the scheme called *Chalak Bane Maalik* (driver becomes owner) (primary survey).

There is also a private paddy–wheat custom-hiring service sector in India, where the owners are mostly graduates or diploma holders and the customers are medium landowners and operators (~15 acres). This also includes some landless and marginal farmers in Maharashtra, mostly with electrical tubewell (multiple) irrigated lands on which they are growing traditional crops, such as wheat and paddy. Most harvesting machines are tractor-driven, except in Maharashtra, and are mostly of Standard and John Deere brands due to the brand reputation and other farmers' experiences. These brands have been purchased since 1990 in Punjab, and since 2005 in Gujarat and Maharashtra. They were bought from a company, dealer, or another farmer using 100% credit. Replacement sales occurred only in Punjab. Use varied from 90 days in Maharashtra to only 50 days in Punjab and Gujarat and 600–800 h annually. They were mostly used in rabi season in Gujarat and Maharashtra and in both seasons in Punjab. Custom-hiring service was provided across states like in China (Singh, 2010).

There have also been well-known cases of irrigation access and development in the form of informal tubewell companies in Gujarat, which led to transformation of the farm sector in dry-land conditions (Shah, 1996).[e] In Andhra Pradesh, informal groups worked in irrigation in low-investment activities but not in high-risk and high-investment activities as farmers preferred individual investments due to the high transaction costs (Aggarwal, 2000). Regulation of private tubewell water prices by the village council (*panchayat*) in West Bengal is also a noteworthy institutional innovation for providing equitable access to water to small and marginal farmers. West Bengal also

---

b. These markets can also be seen as a case of institutional innovation as they try to resolve the issue of overtractorization in the state, whereby, in regions where these markets are held, every third farming household owned a tractor. For details, see Singh (1999).

c. This refers to machines and equipment being made available in one place from which farmers take it on their own.

d. This refers to tractors being rented out along with drivers as full service on a per hour basis.

e. For details, see Singh (1999).

successfully implemented cooperative tubewells by S&M farmers, which improved efficiency (lower cost) and equity in water access, and reduced reverse tenancy (Rawal, 2002; Mukherjee, 2007).

Land leasing and tenancy innovations. In case of land leasing and pooling, a project in Kerala (*Kudumbshree*) tried land lease banks. In these regions absentee lands have been leased through land and labor banks by labor groups and this has been successfully running for the past 3 years. *Kudumbshree*, a Kerala government project started in 1998, involves 4 million women below the poverty line, who account for 50% of the households in Kerala. The program integrates itself into the decentralized planning process in the state, allowing local communities greater opportunity to determine their own priorities and to implement their own solutions. The organizational structure of *Kudumbshree* includes neighborhood groups (NHGs) of 15–40 families each, area development societies at the ward level, and a community development society (CDS) at the village or municipal level. The NHGs serve as the basic unit of planning, along with trust and credit societies to fund micro and group enterprises, and the group has democratically elected members. The government agency (*Kudumbshree* project) has a budget and paid staff to provide support, training, and coordination and it is integrated into the local decentralized planning process. One-third of the total development funds are set aside for CDS plans. The basic premise of *Kudumbshree* was that the poor needed to be active agents in their own development (Mukherjee-Reed and Reed, 2013). It had 45,611 groups (joint liability groups) doing farming on leased land in 2014 (*Down to Earth,* December 1–15, 2015).

The group farming initiative under *Kudumbshree,* which has a membership of 250,000 poor women, took up farming with 44,225 small collectives whose members now cultivate more than 10 million acres across the state and earn INR 15,000–20,000 per year depending on the crops grown. The benefits of *Kudumbshree* include greater social inclusion, especially for women; transformation of wage laborers into independent producers; and higher production of food crops to promote food sovereignty (Mukherjee-Reed and Reed, 2013).

However, despite all the policy talk and liberalization of land laws at the state level, leasing for the landless remains a problem. This could also be in the form of reverse leasing to such communities as was the case in Africa, where an innovative paddy-processing company in Malawi worked with 7000 farmers who were provided with inputs and various other services besides the purchase of their output. The company leased out 70 hectares of its own irrigated land to 220 farmers for cultivating paddy twice a year, with the average area cultivated by a farmer being 0.4 hectare. The company also rented out machines and equipment to the farmers, for which the farmers paid USD 60 per hectare. Since the company itself was not involved in paddy production, it allowed many existing and new players to work with the company and profit from the value chain. The business model of the company centered on sourcing from small-scale farmers (Chitka, 2015).

In the Philippines the shift from share tenure to lease tenure and subtenancy (50:50 shares of input and output) during the 1970s can also be seen as a case of institutional innovation, although subtenancy was illegal under the land reform code. Subtenancy emerged because of the disequilibrium between leasehold rent and the opportunity cost of labor due to the surplus amount of labor and the higher yields of rice due to new technologies. It was a mechanism for landowners to avoid the law, which prevented them from shifting their land to more efficient tenants. Thus, the share to land was the lowest and the operator's surplus the highest for land under leasehold tenancy, followed by that under shared tenancy. On the other hand, the share to land was the highest and no surplus was received by the operator under the subtenancy arrangement as the surplus was shared as land rent by the leaseholder and the landlord. Although this arrangement (subtenancy) opened a new route for some tenants to become farmers, it was at the cost of equity that farm efficiency was promoted by this institutional innovation as the operator's surplus for the real farmer (subtenant) was negative compared with that under shared tenancy or leasehold systems (Ruttan, 1989).

Another institutional innovation from the Philippines in the 1970s was the *gamma,* in which participation in harvesting rice crops was limited to workers who had performed weeding operations without receiving wages. Earlier systems of wage work had paid harvesting and threshing laborers one-sixth of the harvest. Thus, *gamma* was introduced to reduce the wage rate for harvesting to an amount equal to the marginal productivity of labor, which was being violated due to higher rice yields in the presence of an abundant supply of labor. This can be seen as an innovation that was negative as it tried to maintain the existing distribution of surplus from paddy production and was therefore antilabor. However, if this had not happened, the farmers would have resorted to mechanization of harvesting and threshing operations, which would have led to even lower employment opportunities for labor and lower earnings (Ruttan, 1989).

Similar to what happened as a market-based institutional innovation in the Philippines, in Gujarat, there is a system of *bhagidaari* (labor tenancy) under which local or migrant workers manage the entire farm on behalf of the owner. The owner provides all the farm inputs, including machinery service, and an advance for the purchase of inputs. The tenant provides only labor or labor and farm supervision. A study conducted by Bhatt (2008) in Anand District in central Gujarat found that the sharecropper contributes the labor and obtains one-fourth of the produce although other systems of tenancy also exist, such as an equal share of both input and output between the landowner and sharecropper; a 25% share of output to the landowner; and one-tenth of the output (*Pavda Bhaag*) to the sharecropper in which the sharecropper provides only his own labor, with additional labor cost along with all other input expenditure being borne by the landowner. This system falls between cash leasing, in which a lump-sum, prenegotiated, cash payment is being made to the farmer at the end of the year; and sharecropping, in which the produce is shared between the landlord and the tenant in a fixed proportion. Here, a percentage of the output is being taken by the *Bhagidaar* in return for the labor input given by him. There is a slight variation in percentage shares across crops and regions, varying between 15% and 30% of the produce depending on the crop, as well as the type of land. This system has been prevalent for more than 40 years (Rutten, 1986), and is expanding. It is so popular that it accounts for more than 95% of the labor engaged in some crops and areas.

The *bhagidaars* live on the farms with their family and take care of the crop (whose sale is in the hands of the landowner), receiving 15%–30% of the value of the produce. This kind of sharecropping is very different from other systems prevalent in India, such as fixed cash rent leasing, fixed input and produce share, fixed produce, or mortgage. This is close to the permanent attached labor (e.g., *sanjhi* or *siri* in Punjab) system prevalent in some parts of India a few decades ago, in which the workers shared the cost of production and produce depending on their relative contribution to the farm enterprise of the owner. This is pure labor tenancy as the *bhagidaar* contributes only the labor component and obtains a share in produce (15%–30%). This is an important strategy as landowners can manage labor issues by having permanent labor on the farm through this arrangement.

This system has been prevalent since the 1970s and Rutten (1986) called it a "putting-out system" or "contract-type tenancy" and reported a share for the tenant on the order of 12%–16% of the produce depending, among other things, on the crop at that time in central Gujarat. He also reported that this system was used by large and medium farmers to manage only a part of their farmland, usually a few acres allotted for each *bhagidaar,* which required intensive labor and supervision. As the author writes:

> *Contract tenancy is therefore nothing but a convenient arrangement under which a middle-large farmer who leases out in part relieves himself from the burden of labor management while still performing the major entrepreneurial functions. Since he still does exercise control over agricultural production and since the contract is on a crop sharing basis for the duration of only one season at a time, he can be rest assured that the yield would be maximized (Rutten, 1986, p. A-18).*

Thus, as Rutten concludes:

> *By way of organizing the agricultural work in the field and the recruitment of labor in an entrepreneurial manner, these middle-large farmers have solved the problem of fluctuations in supply and demand of labor while still exercising control over the process of agricultural production. This, however, is entirely at the expense of the casual laborers working in their fields who most of the year do not earn more than INR 6 to 8 per day and that only on days they are able to find work…. Thus creation of an intermediate stratum has become an important aspect of the entrepreneurial way of farm management in which responsibility for the welfare of laborers working in the fields is abdicated to an increasing extent to a lower layer in rural society (Rutten, 1986, p. A-19).*

Institutional Innovations in Marketing of Rice

In addition to their role of supplying farm inputs, PACS procure rice for the government agencies from member farmers at the MSP in MP and Bihar. This is a result of policy innovation at the state level, especially in times when commission agents have been legally removed from APMC *mandis* in one state (MP) and the APMC Act has been abolished altogether in another state (Bihar). This is quite an innovative institutional arrangement as PACS are best positioned to do this and help their members realize the MSP at the door-step level (Krishnamurthy, 2012; primary survey). This was also found to be the case in Turkey where credit cooperatives turned out to be better than the commission agents and traditional marketing cooperatives in linking small farmers with modern market arrangements that required quality supply and therefore investment by the cooperative, as well as the farmers. They were able to do so as they introduced certain changes in the governance structure of the producer union (created within the cooperative, facilitated by the legal provision to allow it to obtain a producer union certificate for the sale of fruits and vegetables), such as member screening to ensure homogeneous membership, voting rights in proportion to patronage, zero purchase of low-quality produce, and authority for managers to make business decisions and thus these become more like new-generation cooperatives (NGCs) (Lemeilleur and Codron, 2011).

The removal of commission agents *(kacha arthiya)* from the APMC system in MP is another innovative institutional policy change that is not talked about as part of the agricultural market reforms in India. This is because of political economy reasons wherein *arthiyas* have become a powerful stakeholder in agricultural markets, although in MP this was undertaken way back in the 1980s (Krishnamurthy, 2012). This has led to improvements for both buyers and sellers of crops, such as rice, as sellers do not have to pay any implicit charges and interlocking is eliminated (at least formally), and buyers save commission on purchases as they buy directly.

It is important to realize that for farmer incomes to improve they need to have a presence in value-added products that have higher margins. For this to happen, farmers need to own or control some of the downstream links in the value chain, which requires capital and capability. It is here that the role of PCs comes in as they are more like business entities with no interference from the state or any other external entity (Mahajan, 2015). These are new, more market-oriented, cooperative companies that can help farmers buy and sell more effectively. They gained currency across India during the last decade when the amendment to the Companies Act (2003) made this possible and India now has hundreds of PCs, with many of them being supported by state agencies (Singh and Singh, 2014). In rice, there are 13 such PCs in Uttarakhand, 5 each in Telangana and Madhya Pradesh, and 2 each in UP, Chhatissgarh, Tamil Nadu, and West Bengal (SFAC website). These agencies are the most suitable for changing the market scenario as they are businesslike, free from any external (government or other) control, and work like any other corporate entity. However, issues of access to capital and market orientation still remain (Singh and Singh, 2014).

The PCs suffer from a lack of finance in the formative years as the kind of support provided by the MP government has not been provided by either the other state governments or the union government until recently. However, their own mobilization from members remains low in most cases (Singh and Singh, 2014). In fact, the idea of a cooperative (read PC) is not only for pooling of produce but also for pooling of members' capital (Shah, 2016). Singh and Singh (2014) found that most of the MPDPIP PCs are into a seed production business, which is not a very good business as it involves a small number of members and is high cost. Therefore, it does not create member centrality and the large patronage needed for the PC to scale up. The viability of the PCs was dependent on different factors in different contexts, including scale, nature of farmers, crops or produce handled, professional management of the value chain, and women members' involvement.

Contract farming has been studied as an institutional innovation in agribusiness (Vande Velde and Maertens, 2014). There have been interesting cases of organic basmati rice contract farming wherein private companies organized basmati rice growers to undertake more sustainable and market-driven production through contract farming. Almost 80% of the basmati production in India is exported. Although the volume of basmati rice exports (11.61 lakh tons) is not even one-third of the nonbasmati exports (39.05 lakh tons), it accounts for as much as three-quarters of the value of the nonbasmati rice exports. Basmati prices are subject to speculation and determined by international market prices. Basmati rice is grown only on the Indian subcontinent in the foothills of the Himalayas, responding well to the climatic conditions of this area. Being the largest producer and exporter of basmati rice, India commands a premium over its traditional rivals in terms of price and quality. There is no MSP for basmati rice, and therefore the market operation is quite flexible and favorable for processors and exporters. Being an export crop, basmati also enjoys a well-established high-value market, which in turn has created a lot of space for exporters. Thus, it has been one of the more penetrated commodities among organic crops and has not seen any farmer defaults in contracting in terms of side selling. Two examples of institutional innovation are Agrocel and Kohinoor Foods Ltd. In Haryana, Agrocel organized the farmers into organic and fair trade associations for the certification of basmati rice under a group contract farming program. As such, farmers could obtain a premium price and market for both organic and fair trade in a commodity that did not even have MSP protection from the state (Fig. 12.1). Agrocel also worked as an organizer of contract farming with basmati growers for another exporter (Fig. 12.2) (Singh, 2009b).

In Uttarakhand, Kohinoor Food Ltd. (KFL), formerly known as Satnam Overseas Limited, one of India's leading companies in the organized marketing of rice, including basmati rice, attempted a PPP in organic basmati rice. In 2004 KFL

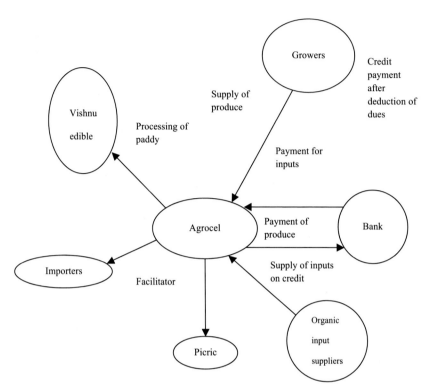

FIGURE 12.1   **Agrocel organic basmati rice supply chain.** *(Adopted from Singh, S., 2009. Organic Produce Supply Chains in India: Organization and Governance. Allied, New Delhi.)*

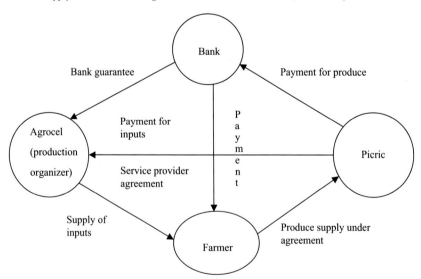

FIGURE 12.2   **Picric organic basmati rice supply chain.** *(Adopted from Singh, S., 2009. Organic Produce Supply Chains in India: Organization and Governance. Allied, New Delhi.)*

made contact with a basmati farmers' federation in Dehradun District to promote organic farming and allied sectors throughout the state. This was organized by the Uttarakhand Organic Commodity Board (UOCB), a state government agency. UOCB took responsibility for the internal control system and organic certification, which enabled KFL to avoid the preoperational work of motivating farmers to adopt organic cultivation. Since the majority of the farmers in Uttarakhand have small holdings, one federation was not sufficient for KFL's requirement. KFL and the first farmer federation therefore identified seven other UOCB-promoted farmer federations and a total of eight federations (four each in Dehradun and Udham Singh Nagar districts) were organized to participate in the program. Formal contracts between each federation and KFL were signed with UOCB as the organic certification service provider and mediator. Technical support to the farmers was provided by KFL. The farmer federations procure paddy from farmers as they have a *mandi* license, pay *mandi* (APMC) charges, receive payments and service charges (2.5%) from KFL, and pay to individual farmers (Singh, 2009a).

Compared to the mandi system, the farmers earned approximately INR 235 per metric ton. The organic yield was higher, production cost lower, and price higher than conventional basmati. KFL also gained INR 245 per metric ton from this and one-quarter of this was spent on extension support to farmers. A subsidy of INR 250 per farmer (~INR 10 per kg) was provided by the UOCB as a part of its support for organic certification. The farmers were also able to make some more money by weighing and bagging their produce themselves. Moreover, the 1.5% commission to the federation not only covered its operation costs but also served as a cash reserve that could be used to make emergency cash loans to the members. Starting with only 190 farmers and 119 acres under the organic project in 2005, the project expanded to cover 864 farmers and 748 acres in 2007. As the Organic Basmati Export Program of the UOCB was a government-supported project, it has placed emphasis on including small and marginal farmers. This meant that a large number of farmers had to be covered to produce sufficient quantities of paddy. With time, the confidence of the farmers in KFL and in organic farming in general increased (Singh, 2009a).

This is similar to a paddy rice contract farming project in Cambodia. Under the umbrella of the agricultural association, a private rice mill company aided by farmer associations has provided input supply, including credit and extension, to more than 50,000 contract growers. This is an interesting case of institutional innovation. A similar project was attempted in the rice-growing region during the late 1990s when the aromatic Thai rice (Hom Mali) was in demand across Asia. The company promoted a similar variety from Cambodia for export and that too was organic. Since the cost of organic certification was high, the contract farmers were asked to produce a noncertified crop using organic methods. Community association officials were trained by the company in the basics of producing organic aromatic rice. The associations monitor farmer members and report to the company and the company provides its services through the association, and technical services through the agent.

The economics of households producing this crop was robust and it led to improvements in children's schooling, housing, and assets, such as two-wheelers and televisions. The intervention also led to some farmers opting out of the contract farming scheme and going for the open local market within the country. There were also reports that farmers felt that the associations were not independent of the control of the company and had little bargaining power to negotiate better terms for the farmers. The inclusiveness of its operations was also doubtful as the contracted farmers had larger landholdings than the noncontracted farmers. This was because the company specified that a minimum of 1 hectare of land was needed to become a contract grower (Roy, 2015).

These examples from India and Cambodia show that it is possible to attend to high-quality export markets if partnerships with small farmers are innovatively built and managed in a participatory manner as a multistakeholder initiative.

Integrated Solutions for Farmers: Franchising as an Institutional Innovation

Mahindra Shubhlabh Services Limited (MSSL), a farm input and output facilitation company that wanted to provide a one-stop solution to farmers, had 57 such outlets in 10 states in northern, western, and southern India, and only 3 of them were company owned and run. The rest were all run by franchisees. Generally, one district had one franchisee and this was exclusive license and business format franchising. Each franchisee had 15–25 spokes (village cluster-level outlets). The franchising system composed 2.5% of the MSSL's business. Franchisees were selected based on their agricultural input and/or output business volumes and experience in the local area. A franchisee was typically an *arthiya* (a commission agent) or an agroinput dealer. A franchisee employed 5 field staff, each one managing 100 farmers or 500 acres of a crop (each farmer growing at least 5 acres) in a village or cluster of villages, and all of them were supervised by one supervisor. For the farm advisory service, a fee of INR 50 was charged in cash from the farmer and the remainder (INR 100) of the credit was recovered at the time of delivery of the crop. The crop was monitored regularly by the field staff. The equipment was owned by the franchisee. The franchise contract was for 3 years initially but extendable. The franchisee obtained a commission as a distributor of inputs.

MSSL provided support, such as business planning for the outlet, training, business rollout, a territory manager, all input supplies, and bank linkages, as well as promotional support and accounting packages. The franchisee was evaluated for its performance in terms of planned and achieved targets. The franchisee was also expected to carry out some local promotional activity. The franchisee owned or leased the outlet and did not have to pay any security deposit. For franchise location and franchisee selection, an intensive business feasibility study was conducted to determine the business prospects for the

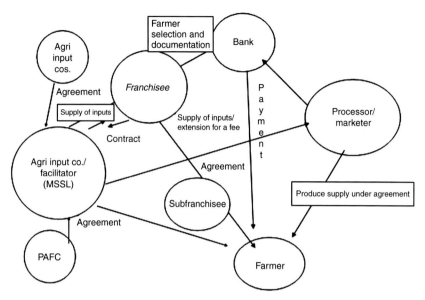

FIGURE 12.3   The MSSL franchisee model (constructed by the author).

district in terms of selection of crops, business planning for the center, and business rollout and launch plan. Crop selection included corporate involvement in the crop, nature of the crop, yield improvement potential, and scalability. Selection of the area for operation involved examining input usage levels. Operational aspects included agricultural marketing laws, credit issues, and government support. Fig. 12.3 shows the various linkages in the franchise system for inputs and output of the company. As can be seen, the franchisee route became too lengthy and complex to reach farmers effectively and efficiently and therefore failed.

Most of the failure was due to the selection of incorrect locations, crops, and franchisees, alongside the business model itself. Most of all, the company did not have much presence in fast-moving inputs, such as seeds, fertilizers, and pesticides, and in output markets in terms of processing, marketing to make volume, and being in regular contact with the franchisees and farmers. In the one instance where it introduced its own pesticide, the performance of the system improved. There were also cases of channel conflict for MSSL as the franchisees and dealers of the input companies vied for the same market. The input companies were also not interested in joining hands with the company anymore. The major problems in managing the farmer interface were defaults by farmers, reluctance of buyers of produce to pay a premium price, and recovery of loans (Singh, 2014). This experience of MSSL confirms that for an innovation to be called an institutional innovation, it should be novel, useful, and legitimate, which are its salient characteristics (Raffaelli and Glynn, 2015).

However, one of its franchisees, Bhuvi-care Private Limited, did succeed. The company was originally established in Tirunelveli District of Tamil Nadu in 2002 by three agricultural engineers who had run a business of selling and installing drip irrigation systems (Irrigation Engineering Services Private Limited). Bhuvi-care entered into an agreement with MSSL in May 2002 to act as a franchisee. Bhuvi-care Mahindra Krishi Vihar (BCMKV) started its services for paddy farmers in Tirunelveli District in the 2002-I season. Its farmer base as well as contracted acreage base grew in 2003 with farmers registering for a rate of INR 500 (approximately USD 10) per acre per season for the integrated set of services. In 2003 BCMKV also included maize under its services and during the 2003-I season 163 farmers registered a total area of 500 acres at the rate of INR 250 (approx. USD 5) per acre. The registration fee in maize decreased to INR 150 (approx. USD 3) per acre from the next season onward (2003-II season). The fees declined as the firm realized that maize requires less attention than rice and a supervisor can manage 250 acres of the crop (for rice, only 75–100 acres could be managed by one supervisor). Moreover, the franchisee also realized that they could obtain more remuneration by way of input sale commissions, shelling charges, and trading commission for maize and so the registration fee could be decreased. This resulted in attracting more farmers under their arrangement and the increased participation compensated for the reduction in the fee. The shelling of maize cobs is fully arranged by BCMKV, which procures the maize for Godrej Agrovet, with whom MSSL has arranged a buyback agreement (Sulaiman et al., 2004).

BCMKV provided all other services as envisaged by the franchisor in the model. For instance, the field supervisor of the franchisee visited farmer fields once a week in the case of rice and once a fortnight for maize. For credit, it established links with ICICI Bank and the State Bank of India. All rice farmers did not require the procurement service; however, the franchisee compulsorily procured paddy of those who did not have a good repayment record so that the loan repayment was not affected (Sulaiman et al., 2004).

The franchisee succeeded as the franchise system evolved at the local level through a series of experimentations and several failures. It is quite interesting to note that the franchisee succeeded in developing this system into a viable business model by making a series of innovations to the original model proposed by the franchisee, which failed in its (franchisor) directly managed center. Besides the extension-plus feature of the MSSL model, and the partnership with other players being an important element, partnership with a contact farmer was an innovation introduced by the franchisee to the MSSL approach (Singh and Bhagat, 2004; Sulaiman et al., 2004). Furthermore, the franchisee innovated on the original model proposed by MSSL through a process of experimentation, reflection, and learning. Realization that the doorstep delivery of inputs was not viable led to experimentation with contact farmers in each village. MSSL failed to make any arrangement for the buyback of paddy in Tirunelveli and through this the franchisee lost heavily as it could not recover some of the loans provided to farmers. This led it to set up the seed company Tamirabarani Seeds to procure from selected progressive farmers and market seed-quality paddy to those who had registered 30 acres to produce quality seeds of paddy varieties ASD-16, ADT-39, and ADT-43. The Korean paddy transplanter supplied by MSSL was not appropriate for the type of rice cultivation in Tirunelveli and those farmers who tried it were very unhappy with the results. The franchisee compensated the affected farmers for this lapse and later withdrew this machine from its equipment rental portfolio. It also developed a transparent credit delivery system wherein each farmer was provided with a passbook that contained details of all transactions. These types of innovations to the original model helped the franchisee to expand its operations (Sulaiman et al., 2004).

Another agribusiness start-up aimed at helping farmers with better inputs, extension, and markets [Green Agrevolution Private Limited (GAPL), Bihar, India] has also used a franchising model under which it runs nine outlets or centers (Dehaat) across four districts. These outlets serve a total of 4000 farmer members (who pay INR 200 annually each) located within a 10–12 km radius with services such as soil sample analysis, crop selection, technical support during the season, and marketing of produce. Green Agrevolution, established in February 2012, undertakes marketing and processing of farm output (Kumar, 2013) and has a nonprofit arm that works with farmers and development agencies for extension.

The farmers in Bihar are generally smallholders, with 92% operating less than 2 hectares. However, Dehaat farmers are typically larger than their non-Dehaat counterparts in both owned and operated landholdings. Whereas overall owned land on average is 3.33 acres, it is 3.71 acres for Dehaat farmers and 2.78 acres in the case of non-Dehaat farmers. Operated holdings are 3.63 acres on average but 3.89 acres and 3.27 acres for Dehaat and non-Dehaat categories, respectively. This also represents some amount of leasing in practice, which is about 9% of total operated land. In general, Dehaat farmers cultivate more area under high-value crops, such as fruits, vegetables, potato, and maize, than their non-Dehaat counterparts. Furthermore, small farmers have a larger proportion of their area under vegetables than other categories though their absolute average area is smaller than that grown in other categories and this holds across districts (Singh, 2016).

Dehaat farmers tend to have a reduced cropping intensity compared to their non-Dehaat counterparts across both districts. One reason for this could be the higher area under fruit crops, which are perennial or annual crops. However, across both categories, marginal and small farmers have a higher cropping intensity than other categories. This is expected as small farmers are more intensive cultivators of their smaller plots. Only 10.7% of the farmers face a shortage of agriinputs in Dehaat and the major shortage is of seeds. More than 93% of the farmers reported an increase in yields, and, in most cases, this increase was up to 15%. The prevalence of this phenomenon was more common among marginal and small farmers than among semimedium and medium farmers. Approximately one-fifth of the farmers confirmed that Dehaat could help them in crop selection and this help was more successful in kharif crop selection. Very few farmers (9%) reported that they could cut the cost of cultivation through the intervention of Dehaat extension. Approximately 42% of the Dehaat farmers received marketing or sales support from Dehaat, with smallholders being the largest group, followed by marginal and semimedium farmers (in equal numbers) (Singh, 2016).

These findings on franchise operations and their farmer-level impact show that the franchise model is working but needs improvement for more effective farmer-level impacts, especially on small farmer livelihoods. The extension contribution of Dehaat is noteworthy as extension is more by default than by design in mainstream agriinput marketing channels (Kaegi, 2015). On the other hand, in the context of abolition of the APMC Act in the state, Dehaat is making an important contribution by facilitating a new and more direct market linkage for small farmers in new and high-value crops that need prompt handling.

The functioning of the Dehaat centers and their farmer uptake rates show that new channels can lead to more informed farmer-level input use and the realization of higher prices in a smallholder context. However, as revealed by the GAPL case study, the shortage of capital to scale up such innovative initiatives remains an issue. It is here that investment support for agri-start-ups is needed and the start-up fund should be channeled to such innovative agencies. Furthermore, as the Ministry of Agriculture recently made a degree in agricultural sciences mandatory to obtain a farm input distribution license, such agencies can fill the space and step up in larger numbers to provide more effective and timely extension backed by farm input supply and output handling services.

CONCLUSIONS

The experience of institutional innovations using formal and informal contracts and partnerships shows that it is possible to link smallholders with markets and enable them to produce quality products. This is provided that these institutions are designed and managed well, which depends on their governance structure and supportive policies and legal environment. Such institutions and organizations are able to successfully deal with modern markets and benefit member farmers.

The experiences of franchising in agribusiness in India show that it is possible to use this strategy for better farmer interface and resolution of some supply chain and value chain issues, including last-mile reach, lower cost, better relationships, and scale. The commodities range from just input supply to extension-plus and even output handling. This shows that all agribusiness activities and functions are amenable to franchising. GAPL and ZFS have used franchising to create value for smallholders at the local level through local enterprise promotion and growth, which is more sustainable as the franchisor and the franchisee grow together. ZFS has created a new institution of custom service-providing entrepreneurs for agricultural machinery and equipment. This will help small farmers to become viable by decreasing the cost of cultivation for which overtractorization and undermechanization are the prevalent paradox.

Further franchising evolves over time in developing country agribusiness situations. The various models analyzed show that it is becoming an emerging model for agribusinesses, whether through established companies or start-ups, and it does deliver the goods. It was found to be more cost effective than the company owned–company operated outlets model and also better in delivery and quality. This can be seen by the movement of many players to this route over time in India. What seems missing is that, in some cases, franchisees may need more support as they themselves are small and hard pressed to make a living out of this venture. The franchisees need to be treated as more than just distributors. The franchisors need to pick up local entrepreneurs and invest in them for the long term and bring in value-added services. In many instances local agencies, such as farmer companies or cooperatives, can also be brought in as franchisees instead of individuals to enable faster reach and scaling up. The policy for agriculture and agribusiness can also encourage this channel as it promotes local entrepreneurship and builds local capacity for undertaking value-added services.

Interestingly, all of these innovations have happened without any presence or support from the state in local areas. However, over time, some of these innovations may not remain inclusive if not supported by the state and other development agencies, as they may compromise on this inclusivity due to pressures of scale-up and viability. Therefore, there is a need for external policies and mechanisms to promote and support such institutional innovations to leverage their inclusiveness and sustainability advantages. Furthermore, incubating such ideas and handholding them through public–private partnerships can help achieve better scaling up and enhanced impact of such institutional innovations.

The use of a warehouse receipt system is another institution that is innovative and available but not fully used as of now. It can help remove the interlinking of factor (including credit) and product markets for small farmers, which many other initiatives have found difficult to do. Some of the PCs in MP (such as Ram Rahim) have successfully used a warehouse receipt (WR) system to enable farmers to withhold sale of their produce to obtain higher prices. The WR effectively provided them with a loan against their produce (stored in a warehouse) that was recognized by the banks under the WR Act (Raghunathan, 2016). This will also help farmers remove credit constraints and therefore distress sales immediately after harvest as they can obtain a loan against the WR for their produce. This can help in obtaining higher prices, as well as providing access to lower-cost finance as part of value chain finance.

Furthermore, although PCs generally focus on small producers to achieve inclusiveness, there is some merit in mixed-member PCs (in terms of farmer base) as this helps achieve scale and mobilize more equity. It is also argued that, if they are composed of only small and marginal producers, they find it difficult to break even sooner. Due to the small scale of their operations, bringing in larger farmers at a later stage can create problems of governance as PCs were originally designed with patronage cohesiveness. However, scale is only one factor and others, such as differentiation, meeting buyer preferences, or serving niches, are also important as part of the strategy (Penrose-Buckley, 2007).

The PCs also need to choose their activity portfolio carefully, keeping in mind the member centrality. They should also do adequate value chain mapping of the relevant commodity, sector, or service before undertaking any intervention (Mahajan, 2015). It is possible to identify new activities in local areas that are valuable for small farmers (e.g., custom hiring of farm machinery and equipment that they can't afford to buy but can rent). This is being done in some parts of India viably by private entities and PACS. Furthermore, like the NGCs, PCs can also discourage, if not restrict, large membership using by-laws. This will allow them to remain viable and not become unwieldy in terms of membership size as has been seen with PCs promoted by NDDB. The PCs need to identify other businesses that can bring in a larger number of members to reach scale and cater to the diverse interests of members. The NDDB experience also suggested that a large member base and involvement was crucial as it crossed states to achieve economies of scale and scope and obtain member centrality and patronage (Shah, 2016). Furthermore, the NDDB-promoted PCs are asset light (e.g., no dairy cooperative society at the village level but a *sahayak* to manage procurement) and have a high asset turnover ratio. These provisions have differentiated these PCs from the erstwhile dairy cooperatives promoted by NDDB in India under Operation Flood (Shah, 2016; Singh, 2015).

Regarding worker interest in the rice value chain, going by the experience of *Kudumbsree* in Kerala and the *bhagidaari* system in Gujarat, there is a need for more innovation as far as access to land for landless farm workers is concerned. The creation of land banks at the village or cluster level can help match demand and supply for leasing. Owners desirous of leasing out can deposit their land parcels with this bank, which can act as a clearing house (Agarwal and Sharma, 2012; Niti Aayog, 2015). The bank would also be responsible for passing on the lease rent to the owner. But, given the present scenario of land leasing and agrarian distress in most parts of India, there is also a need to cap the land rent so that there is no exploitation of small leasees in a competitive market.

Finally, in smallholder agricultural situations, institutional innovations need to proactively involve various stakeholders, especially workers, in agribusinesses for more creative work and production, for example, in paddy farm mechanization that creatively involves workers instead of casting them aside.

REFERENCES

Agarwal, B., Sharma, P., 2012. An idea to bank on. The Times of India, 12 January 2012, New Delhi, Ideas Page.

Aggarwal, R., 2000. Possibilities and limitations of co-operation in small groups: the case of group owned wells in southern India. World Dev. 28 (8), 1481–1497.

Bhatt, S., 2008. Nature of agricultural land lease contracts: a study of three villages in Anand district. Agricultural Situation in India, May 2008, pp. 63–69.

Chitka, R., 2015. Rice: smallholder farmers in Malawi can be profitably included. In: Harper, M., Belt, J., Roy, R. (Eds.), Commercial and Inclusive Value Chains: Doing Good and Doing Well. Practical Action, Warwickshire, United Kingdom, pp. 89–96.

Ebbinghaus, B., 2005. Can path dependence explain institutional change? Two approaches applied to welfare state reform. MPIfG Discussion Paper No. 05/2. Max Planck Institute for the Study of Societies, Cologne, Germany.

Gathorne-Hardy, A., Reddy, D.N., Venkatanarayana, M., Harriss-White, B., 2016. Systems of rice intensification provide environmental and economic gains but at the expense of social sustainability: a multidisciplinary analysis in India. Agric. Syst. 143, 159–168.

Gatzweiler, F.W., 2016. Institutional and technological innovations in polycentric systems: pathways for escaping marginality. In: Gatzweiler, F.W., von Braun, J. (Eds.), Technological and Institutional Innovations for Marginalised Smallholders in Agricultural Development. Springer Open, London, pp. 25–40.

Gatzweiler, F.W., von Braun, J., 2016. Innovation for marginalised smallholder farmers and development: an overview and implications for policy and research. In: Gatzweiler, F.W., von Braun, J. (Eds.), Technological and Institutional Innovations for Marginalised Smallholders in Agricultural Development. Springer Open, London, pp. 1–21.

IICA, 2014. Innovation in agriculture: a key process for sustainable development. Institutional Position Paper. Inter-American Institute for Cooperation in Agriculture (IICA), San José, Costa Rica.

Kaegi, S., 2015. The experience of India's agricultural extension system in reaching a large number of farmers with rural advisory services. Background Paper to the SDC Face-to-Face Workshop "Reaching the millions!", March 2015, SDC, Hanoi, Vietnam.

Krishnamurthy, M., 2012. States of wheat: the changing dynamics of public procurement in Madhya Pradesh. Econ. Polit. Wkly. 47 (52), 72–83.

Kumar, V., 2013. Farms and farmers: fertile gains. Business Outlook. 31 August 2013.

Lemeilleur, S., Codron, J.-M., 2011. Marketing cooperative vs. commission agent: the Turkish dilemma on the modern fresh fruit and vegetable market. Food Policy 36, 272–279.

Mahajan, V., 2015. Farmers' producer companies: need for capital and capability to capture the value added. Access Development Services State of India's Livelihoods Report 2014. Oxford University Press, New Delhi.

Mendoza, R.U., Thelen, N., 2008. Innovations to make markets more inclusive for the poor. Dev. Policy Rev. 26 (4), 427–458.

Mukherjee, A., 2007. Implications of alternative institutional arrangements in groundwater sharing: evidence from West Bengal. Econ. Polit. Wkly. 42 (26), 2543–2551, 30 June.

Mukherjee-Reed, A., Reed, D., 2013. Taking solidarity seriously, analysing Kudumbshree as a women's social and solidarity economy experiment. Paper presented at UNRISD Conference– Potential and Limits of Social and Solidarity Economy. 6–8 May 2013, Geneva, Switzerland.

Neilson, J., Pritchard, B., 2009. Value Chain Struggles: Institutions and Governance in the Plantation Districts of South India. Wiley-Blackwell, Chichester, West Sussex.

Niti Aayog, 2015. Raising agricultural productivity and making farming remunerative for farmers. An Occasional Paper, Niti Aayog. December 2015, Government of India, New Delhi.

Niti Aayog, 2016. Evaluation report on efficacy of minimum support prices (MSP) on farmers. PEO Report No. 231. Niti Aayog, January 2015, Government of India, New Delhi.

Penrose-Buckley, C., 2007. Producer Organisations: A Guide to Developing Collective Rural Enterprises. Oxfam, Oxford.

Raffaelli, R., Glynn, M.A., 2015. Institutional innovation: novel, useful, and legitimate. In: Shalley, C.E., Hitt, M.A., Zhou, J. (Eds.), The Oxford Handbook of Creativity, Innovation and Entrepreneurship. Oxford University Press, Oxford, (Chapter 22).

Raghunathan, R., 2016. The case of Ram Rahim Pragati Producer Company Ltd. An unpublished note.

Rawal, V., 2002. Non-market interventions in water-sharing: case studies from West Bengal. J. Agrar. Change 2 (4), 545–569.

Roy, R., 2015. Angkor rice: 50000 Cambodian farmers growing for export. In: Harper, M., Belt, J., Roy, R. (Eds.), Commercial and Inclusive Value Chains: Doing Good and Doing Well. Practical Action, Warwickshire, United Kingdom, pp. 97–104.

Ruttan, V.W., 1989. Institutional innovation and agricultural development. World Dev. 17 (9), 1375–1387.

Rutten, M., 1986. Social life of agricultural entrepreneurs: economic behavior of life style of middle-large farmers in central Gujarat. Econ. Polit. Wkly. 30 (13), A15–A22.

Shah, T., 1996. Catalysing Co-operation: Design of Self-Governing Organisations. Sage, New Delhi.

Shah, T., 2016. Farmer producer companies: fermenting new wine for new bottles. Econ. Polit. Wkly. 51 (8), 15–20.

Shiller, R.J., 2006. Behavioral economics and institutional innovation. South. Econ. J. 72 (2), 269–283.

Sidhu, R.S., Vatta, K., 2012. Improving economic viability of farming: a study of co-operative agro machinery service centres in Punjab. AERR 25, Conference Number, 427–434.

Singh, S., 1999. Institutional innovations in Indian agriculture: a case of input markets. Inst. Dev. 6 (2), 3–9.

Singh, A.K., 2009. Basmati rice and Kohinoor Foods Limited. In: Harper, M. (Ed.), Inclusive Value Chains in India: Linking the Smallest Producers to Modern Markets. World Scientific, Singapore (Case Study 6, Chapter 4, pp. 110–127).

Singh, S., 2009b. Organic Produce Supply Chains in India: Organisation and Governance. Allied, New Delhi.

Singh, S., 2010. Agricultural Machinery Industry in India: Growth, Structure, and Buying Behavior. Allied, New Delhi.

Singh, M., 2012. Yield gap analysis in rice, wheat and pulses in India. Ann. Agric. Res. 33 (4), 247–254.

Singh, S., 2014. Agribusiness franchising in India: experience and potential. IIMA WP No. 2014-12-09, IIM, Ahmedabad.

Singh, S., 2015. Promotion of milk producer companies: experience of NDS. Presentation at the Workshop on Governance Issues in Producers' Organizations held at and organized by NDDB, Anand, 8 October.

Singh, S., 2016. Innovative agricultural input marketing models in India: performance and potential. Final report for the MoA, GoI, CMA. IIM, Ahmedabad.

Singh, R., Bhagat, K., 2004. Corporate initiatives in Indian agriculture. Indian Manage. 43 (2), 72–79.

Singh, S., Singh, T., 2014. Producer Companies in India: Organisation and Performance. Allied, New Delhi.

Sulaiman, R.V., Hall, A.J., Suresh, N., 2004. Effectiveness of private sector extension in India and lessons for the new extension policy agenda. Paper Presented at the Workshop on Rural Innovations: Emphasizing the Postharvest Sector, organized by CRISP, Hyderabad. 22–29 November, ICRISAT, Hyderabad.

Totin, E., van Mierlo, B., Saïdou, A., Mongbo, R., Agbossou, E., Stroosnijder, L., Leeuwis, C., 2012. Barriers and opportunities for innovation in rice production in the inland valleys of Benin. NJAS-Wagen. J. Life Sci. 66–63, 53–66.

Vande Velde, K., Maertens, M., 2014. Contract farming in staple food chains: the case of rice in Benin. Bioeconomics Working Paper No. 2014/9. Division of Bioeconomics, Department of Earth and Environmental Sciences, University of Leuven, Geo-institute, Belgium.

Witsoe, J., 2006. India's second green revolution? The sociological implications of corporate-led agricultural growth. In: Kapoor, D. (Ed.), India in Transition: Economics and Politics of Change Series. Centre for the Advanced Study of India, Philadelphia, PA.

Yang, J., Huang, Z., Zhang, X., Reardon, T., 2013. The rapid rise of cross-regional agricultural mechanization services in China. Am. J. Agric. Econ. 95 (5), 1245–1251.

Chapter 13

The Rice Seed System in India: Structure, Performance, and Challenges

Aldas Janaiah*, Behura Debdutt**
*International Rice Research Institute, New Delhi, India; **Odisha University of Agriculture and Technology, Bhubaneswar, Odisha, India

Seed is the most important input in crop production. Therefore, delivering quality seeds of modern varieties (MVs) and hybrids is an important strategy for increasing productivity and promoting agricultural growth in many developing countries. The seed market has started to expand and show true competitiveness, with increasing levels of private-sector participation in the seed production and delivery system (Pal et al., 2000; Pray and Deininger, 1998; Selverajan et al., 1999; Tripp and Pal, 2001). Despite this, farmers' own harvests and exchange with neighbors remain the major sources of seed for nonhybrid crops, such as rice (Pal et al., 2000). The regular use of quality seeds, even for nonhybrid crops, such as rice, increases yields without substantial incremental costs.[a] However, continuous saving of seeds from own harvests (without proper cleaning) would seriously affect seed health and lead to lower rice yields (Mew, 1997).

In Vietnam, 60% of the rice farmers change their seeds every year or season (Khoa et al., 1996), while most rice farmers in Thailand change their varieties (seeds) every 3–5 years (Somkid et al., 1996). A study in Bangladesh has also documented that the use of clean seeds from a farmer's own harvest can raise paddy yield by 8%–10% (Diaz et al., 2000). Experiences in the state of Andhra Pradesh, where the highest rice seed replacement rate (SRR) is registered, suggest that regular use of purchased seed can increase paddy yield by 12%–17% (Pal et al., 2000). It is generally noted that average paddy yields are high in regions where the SRR is high.

---

a. Plant breeders and geneticists generally suggest that the genetic purity of rice seed would be lost if farmers used their own harvest as seed without proper cleaning for three consecutive seasons/years.

The Future Rice Strategy for India. http://dx.doi.org/10.1016/B978-0-12-805374-4.00013-0
359

Keeping in mind the increasing importance of seed quality in rice production, an efficient seed-provisioning system is considered an essential element of India's overall rice strategy. The principal objective of this chapter is to analyze the current structure and performance of various players in the current seed system to draw a meaningful outlook on India's rice seed sector. The chapter discusses four key aspects of the seed system: (1) structure, roles, and current status of key participating entities in the seed system; (2) hybrid rice technology and its seed system; (3) the current policy environment; and (4) key challenges and the future outlook.

STRUCTURE, ROLES, AND CURRENT STATUS OF KEY ENTITIES OF THE SEED SYSTEM

An ideal seed system in a developing country, such as India, ought to serve farmers by providing an *adequate* supply of *quality* seeds of *MVs* at *affordable prices* at the *right time* (Hossain et al., 2001) (Fig. 13.1). Four prospective participating entities in the current seed system—the public sector, the private sector, farmers' cooperatives, and NGOs—could cater to all of these

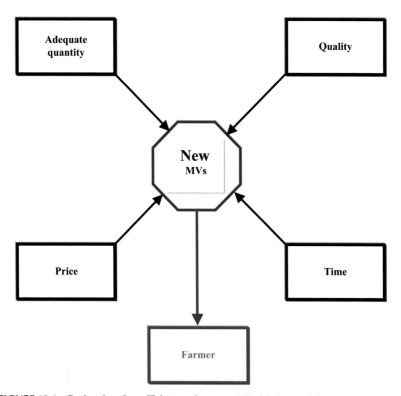

FIGURE 13.1   Basic roles of an efficient seed system. *MVs*, Modern varieties.

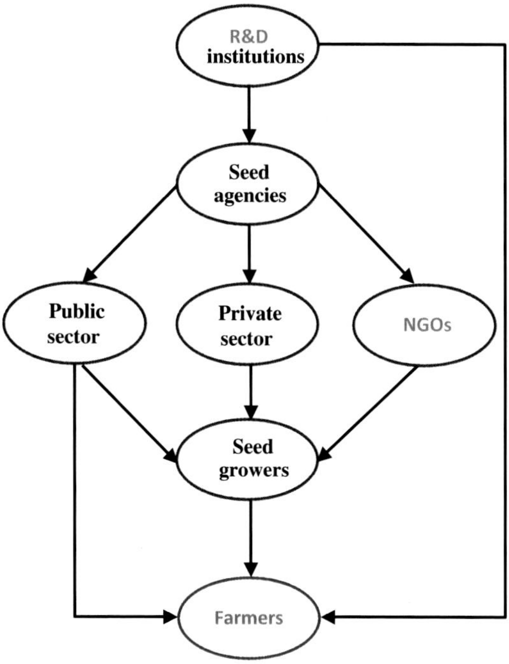

FIGURE 13.2   **Key players in an ideal seed system.**

basic needs (Fig. 13.2). However, an ideal seed system needs institutional support, favorable policy conditions, and strengthening of the public–private interface to enable active participation of all key entities in an efficient manner. The different entities of an ideal and efficient seed system should play a wide range of roles, such as development and delivery of modern varieties, production of source seed, seed multiplication, quality enforcement, and marketing. The performance of these different entities in these different roles is discussed further.

Development and Delivery of Modern Varieties

Plant breeders are the foremost players in an ideal seed system. Therefore, the effectiveness and strength of a plant breeding program (varietal development activity) can ultimately determine how effective the delivery of seeds is.

With the exception of a few large companies in India (Pal et al., 2000), private seed companies in many developing countries have not been actively involved in research for new varieties of self-pollinated crops, such as rice and wheat (Pray and Deininger, 1998). It is important to note that the private sector in India is appreciably participating in the delivery of rice seeds of publicly bred MVs because of the growing public–private sector interface and the flow of information and materials (Pal et al., 2000).

Agricultural research in India has been largely undertaken in the public-sector domain (a few exceptions to this rule exist, such as the recent research efforts into hybrid crops in the private sector). The key institutions involved in the development of improved crop varieties and hybrids include state agricultural universities (SAUs), central research institutions under the Indian Council of Agricultural Research (ICAR), public-sector international research centers under CGIAR, and a few large private seed companies. SAUs, ICAR institutions, and CGIAR centers form the basis of a strong Indian-wide research network for crop improvement; 120 national agricultural R&D centers (ICAR system) and 70 SAUs are currently involved in a varietal development program in India.

Approximately 4200 MVs and hybrids, including ~1000 for rice, were developed and released by SAUs and the ICAR over the past 5 decades. Many of these MVs and hybrids have been widely adopted by farmers in the different states of India. In addition, several improved crop varieties were developed by CGIAR centers and were again widely adopted in India. For instance, improved breeding materials from the International Rice Research Institute covered about 12% of the total rice area in India (Janaiah et al., 2006). A sizable number of MVs and hybrids were also developed and released by the private sector, which commercially marketed a number of its own hybrids.

The public R&D system has contributed substantially to the delivery of new MVs and hybrids to the farmers of India through transfer of technology (TOT) programs, such as on-farm verification trials, frontline demonstrations, the minikit program, etc. Farmers who participated in these TOT programs have become sources of new MV seeds for fellow farmers in their area. Furthermore, publicly bred MVs and other improved germplasm were also made accessible to the private sector, farmers' cooperatives, and NGOs, especially for nonhybrid crops, such as rice. The delivery of publicly bred MVs of rice by the private sector and farmers' cooperatives has increased over the years (Pal et al., 2000). Following the new seed policy of 1998 and 2002, a considerable public–private interface emerged that allowed the public sector to make its improved germplasm freely accessible to the private sector for both the development and delivery of MVs.

Source Seed Production

Breeder seed (BS) is the progeny of the nucleus seed for a particular variety and is produced by the originating breeder or by a sponsored breeder and a research center. It is the responsibility of research institutions (SAUs and ICAR) to produce and supply the source seed (BS) of a notified crop variety to both public and private seed companies. These companies in turn produce foundation seed (FS) from which certified seed (CS) is produced and distributed to farmers. Thus, the growth of the seed industry primarily depends on the demand for and supply of source seed. Liberalization of the seed policy in recent years has sparked a growing demand for BS from private companies, cooperatives, and NGOs, especially for nonhybrid crops. However, private-sector participation in BS production and distribution remains to be expanded.

Indents from various seed-producing agencies are collected by the State Departments of Agriculture (SDA) and submitted to the Department of Agriculture and Cooperation (DAC), Ministry of Agriculture, and Government of India. The DAC compiles all information on the crop and sends it to the crop project coordinator at ICAR, who will then perform the final allocation of production responsibility to the appropriate SAU/ICAR institutions. Indents are compiled and forwarded to ICAR at least 18 months in advance. Production of the BS is reviewed every year by ICAR–DAC in the annual seed review meeting and the actual production of the BS by different research centers is suggested to DAC by ICAR. On receipt of information from ICAR, the available BS is allocated to all the indenters (seed agencies) in an equitable manner for further multiplication to CS.

There has been a steady increase in the production of rice BS over the years. Table 13.1 depicts the indent and production of rice BS in India from 2004–05 to 2013–14, and it is clear that the production has outstripped the indent. The production of rice BS has increased by about threefold over the past 10 years: a clear indication of the fast-growing rice seed industry for nonhybrid rice seeds.

TABLE 13.1 Indent and Production of Breeder Seed (BS) of Rice in India, 2004–05 to 2013–14

Year	Indent (tons)	Production (tons)
2004–05	119.9	257.2
2005–06	157.5	285.0
2006–07	207.6	340.5
2007–08	249.1	392.3
2008–09	302.8	433.3
2009–10	388.0	538.7
2010–11	460.4	609.5
2011–12	577.2	682.8
2012–13	665.8	798.6
2013–14	768.5	845.8

Source: Department of Agriculture and Cooperation, Ministry of Agriculture, Government of India (2001–14 reports).

Seed Multiplication

State seed development corporations (SSDCs), national seed corporation (NSC), state farms corporation (SFC), private seed companies, farmers' cooperatives, and NGOs multiply FS from BS on their own seed farms. FS is required to meet the seed certification standards prescribed in the Indian Minimum Seed Certification Standards, in both field and laboratory testing. Using FS, seed agencies then produce CS in farmers' fields through contractual arrangements. It is worth mentioning here that two-thirds of India's total CS production of modern rice varieties takes place in the state of Telangana alone. There are two reasons for this. First, the most popular MVs of rice, such as Swarna, BPT 5204, Vuijetha, Cotton Dora, etc., which were developed and released by the local SAU, are widely grown in different states of India. So much so that nearly one-fourth of India's total rice area is cultivated with these varieties (Janaiah et al., 2006). Second, ~400 seed companies that engage in seed production are located in Telangana. Thus, Telangana is widely known as the seed capital of India. Both the public (SSDCs, NSC, and SFC) and private sectors have a huge network for CS production through "contractual arrangements" with progressive seed farmers in Telangana.

CS is the progeny of FS and must also meet the standards of seed certification prescribed in the Indian Minimum Seed Certification Standards (1988). In the case of self-pollinated crops, CS can also be produced from CS provided it does not go beyond three generations from FS stage-I.

CS production is organized by various seed agencies (SSDCs, departmental agricultural farms, private companies, cooperatives, NGOs, etc.) through respective contractual seed farmers. Approximately 300,000 farmers are engaged in CS production for different seed agencies in Telangana alone. FS is supplied to selected farmers (both contract and registered seed growers) for the multiplication of CS. The farmers who undertake CS production have to meet certain standards, including isolation of the fields, appropriate cultivation practices, and removal of plants of undesirable varieties from the plots (roguing). The seed certification agencies closely supervise such plots. The seed is then purchased from contracted seed farmers by the respective seed agencies, tested in laboratories for purity and germination, graded, and packed with a certification mark before it is released for sale in the market.

The number of private seed companies engaged in the production and distribution of FS and CS has increased significantly over the past decade. Approximately 600 private seed companies are participating in the seed business in India, of which ~400 are based in Telangana. Total CS production in India (from all sources) increased from 0.24 million tons during 2001–02 to 0.72 million tons in 2013–14, a threefold increase in 15 years (Fig. 13.3). The private sector continued to dominate with the lion's share of CS production for hybrid crops. However, the share of the private sector and farmers' cooperatives in CS production has increased tremendously over the last decade, especially in major

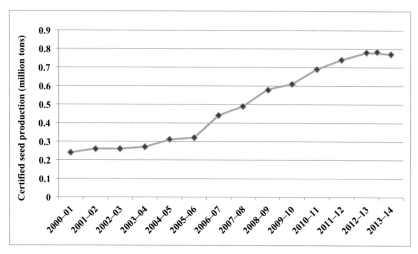

FIGURE 13.3   Year-wise distribution of certified rice seeds in India (2000–01 to 2013–14).

nonhybrid rice crops. The share of the private sector, cooperatives, and NGOs in total rice CS production was nearly two-thirds in Andhra Pradesh and Telangana together (Janaiah, 2003). The increased access of public crop varieties to the private sector after the liberalized seed policies of 1998 has contributed to increasing participation of the private sector in CS production in nonhybrids, such as rice (Janaiah, 2003).

The Department of Agriculture and Cooperation closely monitors the demand for and supply of CS in consultation with state governments. Fig. 13.4 indicates the demand and supply ratio of quality certified rice seeds in different

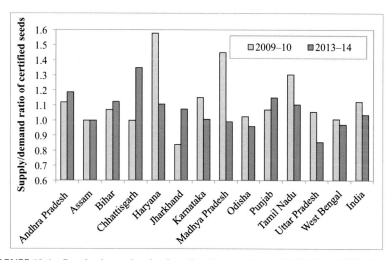

FIGURE 13.4   Supply–demand ratio of quality rice seeds during 2009–10 and 2013–14.

TABLE 13.2 Requirement, Availability, and Supply of Certified Seeds of Rice in India, 2009–10 to 2014–15

Year	Requirement (millions tons)	Availability (million tons)	Supplied (million tons)
2009–10	0.66	0.73	0.61
2010–11	0.72	0.86	0.69
2011–12	0.83	0.92	0.74
2012–13	0.78	0.80	0.72
2013–14	0.84	0.89	0.78
2014–15	0.85	0.86	0.77

Source: Department of Agriculture and Cooperation, Ministry of Agriculture, Government of India (2010–15 reports).

Indian states during 2009–10 and 2013–14. With the exception of a few states, such as Uttar Pradesh, Odisha, West Bengal, Assam, and Madhya Pradesh, the supply of rice CS outstrips the demand. However, quality and timely availability remain major concerns.

The total volume of CS produced is always more than the requirement (Table 13.2), while the actual quantity of CS supplied to farmers is 10%–20% lower than the requirement. Table 13.2 reveals the requirement, availability, and actual supply of certified seed in India over the past 6 years. There has been an impressive growth in the production and distribution of rice CS over the past 15 years (Fig. 13.3 and Table 13.2). This may be because of the increased participation of the private sector and cooperatives in rice CS production due to free access to improved germplasm from public-sector R&D. Increased awareness of the role of quality seed among farmers, due to the marketing strategy of the private sector, is also an important factor in the increased demand for CS. As a result, the SRR (the percentage of area under CS in the gross cropped area of that crop) for rice has increased significantly in various states (Table 13.3). The National Seed Plan-1998 envisaged a 33% SRR for self-pollinated crops (i.e., rice) and 50% for cross-pollinated crops (i.e., maize). Among all states in India, Andhra Pradesh (including Telangana) has the highest SRR for rice. The SRR has also increased markedly over the period in the eastern states of India (Table 13.3).

Seed Certification, Quality Control, and Storage

Seed quality is the most important element of a seed-provisioning system. There are ~120 national seed testing laboratories and 25 state seed certification agencies (SSCAs) that ensure adequate seed quality. SSCAs are quality enforcement agencies that maintain, certify, and regulate seed quality issues in

TABLE 13.3 State-Specific Seed Replacement Rate (SRR)% of Rice in Major Rice-Producing States of India, 2001–13

| | | | | | | | States | | | | | | | |
Year	AP	Assam	Bihar	Chhat-tisgarh	Haryana	Jharkhand	Karna-taka	MP	Odisha	Punjab	TN	UP	WB	All India
2001	42.00	—	6.33	—	11.25	—	22.00	3.30	9.59	11.00	17.00	14.28	22.00	19.22
2002	62.00	—	6.87	—	14.12	—	25.00	3.91	5.57	17.00	15.00	15.90	23.00	19.31
2003	49.00	—	6.80	2.49	14.49	—	25.00	3.90	6.13	23.00	14.00	17.25	25.00	19.16
2004	52.00	3.01	10.00	5.89	14.05	—	30.30	3.39	4.73	15.00	12.00	17.90	25.30	16.27
2005	61.00	6.82	12.00	7.15	17.00	—	29.00	6.41	6.83	19.00	55.00	20.29	25.50	21.33
2006	60.00	10.72	12.00	8.50	17.50	—	34.00	3.96	6.40	21.00	56.00	22.67	26.00	22.41
2007	73.00	17.30	15.00	7.65	18.00	—	36.00	8.85	12.04	24.00	65.00	25.00	26.50	25.87
2008	82.00	23.00	19.00	13.81	17.86	14.25	33.00	10.06	14.80	21.00	67.00	28.57	28.00	30.05
2009	81.75	32.42	35.19	20.39	18.52	23.42	38.85	15.60	19.07	27.83	76.65	28.40	30.00	33.60
2010	86.45	46.45	41.70	28.41	21.69	25.13	40.90	13.31	20.64	40.08	62.70	37.77	32.00	37.47
2011	87.21	46.82	38.02	34.33	23.54	17.12	41.04	16.85	21.65	52.78	67.98	31.61	33.67	40.42
2012	88.10	47.40	42.30	29.40	26.12	22.60	43.20	18.20	23.40	56.80	69.30	33.20	36.10	44.30
2013	91.20	48.30	43.60	33.10	29.30	24.10	45.60	20.40	26.10	58.20	70.30	35.10	37.80	45.30

Source: Department of Agriculture and Cooperation, Ministry of Agriculture, Government of India (2001–14 reports).

their respective state. The technical staff of SSCAs supervise both FS and CS production in farmers' fields (contract seed growers) of SSDCs, NSC, SFC, cooperatives, and also private seed companies. The private sector has its own set of technical staff to supervise its contract seed growers' fields.

Inadequately trained staff and poor infrastructure led to many SSCAs not being able to meet the growing certification demand, especially for nonhybrid crops, such as rice. Thus, the new seed policy of 2002 allowed the private sector, cooperatives, and NGOs to undertake self-certification, and label CS with "truthful labeling (TFL)." Although TFL is a welcomed step in promoting private-sector participation, an inadequate "law enforcement mechanism" may not protect farmers' interests in the case of private-sector misuse of TFL provisions and distribution of poor-quality seeds.

Adequate modern storage facilities are yet to be created to ensure retention of seed quality during storage. For example, Andhra Pradesh and Telangana together produce two-thirds of India's CS of rice, but have the capacity to store only 120,000 tons of seed in both the public and private sectors. This accounts for only 30% of the CS currently produced in these states. This calls for the expansion and modernization of seed production and postproduction infrastructure for maintaining adequate seed quality.

Seed Distribution and Marketing

Three formal marketing channels exist for the distribution of publicly produced CS. First, SSDCs and NSC have their own marketing network with many sales centers nationwide. Farmers can purchase TFL-tagged CS from a sales outlet of public-sector seed agencies. The second marketing channel is through licensed private seed dealers. The number of seed sales outlets (seed dealers) in India has increased significantly over the past few years (Table 13.4), with ~200,000 seed outlets now engaging in marketing of seed (of both the public and private sectors) to farmers in India. The third channel for the marketing of seed is through the SDA under various government-funded development programs (mostly subsidized seed distribution). Such programs primarily target small and marginal farmers to support them with the use of quality seeds of MVs.

The private sector is allowed to freely fix the retail price of seeds with adequate incentives to seed dealers. As the profit margin is less for the marketing of publicly produced CS, seed dealers prefer to participate in the marketing of private-sector seed. In practice, seed dealers charge a higher price to farmers during periods of scarcity because of the absence of adequate regulatory mechanisms.

Community-Level Seed Production Programs and Projects

In many developing countries, NGOs and cooperatives usually promote community-level seed production, often through farmers' organizations (Almekinders et al., 1994; Wiggins and Cromwell, 1995). A number of large cooperatives

TABLE 13.4 State-Specific Number of Seed Sales Outlets in India
(2009–10 to 2012–13)

States/UTs	2009–10	2010–11	2011–12	2012–13
Andhra Pradesh	9,480	9,590	10,670	13,550
Assam	41	41	41	43
Bihar	5,633	5,633	5,633	5,685
Chhattisgarh	270	274	272	273
Delhi	176	176	176	176
Gujarat	7,827	10,981	11,832	13,856
Haryana	5,453	6,699	8,231	9,800
Himachal Pradesh	232	271	324	440
Jammu and Kashmir	1,170	1,178	1,094	1,190
Jharkhand	400	400	400	400
Karnataka	4,815	5,069	5,216	5,616
Kerala	1,076	1,076	1,076	1,076
Madhya Pradesh	11,328	11,350	11,369	12,120
Maharashtra	31,061	30,678	31,363	41,850
Odisha	1,362	1,362	1,362	1,362
Punjab	7,234	7,234	7,879	8,450
Pondicherry	50	45	50	68
Rajasthan	8,622	8,622	8,622	8,622
Tamil Nadu	9,053	9,100	9,122	9,245
Uttar Pradesh	25,774	25,898	26,038	27,176
Uttarakhand	752	1,323	1,323	1,384
West Bengal	34,000	34,000	35,000	36,175
Other states	60	70	81	94
All India	165,869	171,070	177,174	198,651

UTs, Union territories.
Source: Department of Agriculture and Cooperation, Ministry of Agriculture, Government of India (2001–14 reports).

and NGOs in India actively promote community-level seed production, especially for nonhybrid crops, such as rice. In Andhra Pradesh and Telangana, the public sector (mostly SDA, SSDCs, and the local SAU) is directly involved in promoting community seed production through programs, including the Seed

Village Program. Established in the early 1990s to promote seed production by entrepreneurial farmers at the village level, the main objective of this program is to make each village self-sufficient in seed requirement. The SSDC and local SAU supply FS to the participating farmers, while SDA staff provide technical support. The participating farmers (seed producers) are allowed to sell their seed produce freely to fellow farmers within the village and even outside the village. The SDA also procures seeds from the Seed Village Program and supplies them to small and marginal farmers at a subsidy. Although this program contributed substantially to the availability of seeds locally, the maintenance of adequate quality is questionable because of the absence of quality regulatory and maintenance mechanisms at the village level.

The total quantity of CS distributed in India through all channels increased from 50,000 tons in 1970–71 to ~1 million tons in 2000–03 (Janaiah, 2003).

HYBRID RICE TECHNOLOGY AND THE SEED SYSTEM

Hybrid Rice in India: An Overview

India was the first country to initiate hybrid rice research at the Central Rice Research Institute, Cuttack, in the early 1950s (Sampath and Mohanty, 1954). Inspired by Chinese success, collaborative research efforts with IRRI, Philippines, into hybrid rice started in the early 1980s. ICAR began a focused hybrid rice R&D program in India in 1989 with the objective of developing and releasing indigenous rice hybrids to farmers (DRR, 1997). The rigorous efforts in India over the past 25 years resulted in the development and release of ~70 rice hybrids from both the public and private sectors (DRR, 2014). India was the second country after China to develop and release a rice hybrid, whereas other countries, such as Vietnam and Bangladesh, imported their rice hybrids from China (Janaiah and Hossain, 2003). It was widely reported that many farmers who grew hybrid rice for one or two seasons had started to drop out of hybrid rice cultivation trials in India (Janaiah, 1995, 2002; Janaiah et al., 1993) and Bangladesh (Hossain et al., 2001). Therefore, the rate of hybrid rice adoption by farmers was limited and scattered until 2004.

Since the release of the first-generation hybrids in the early 1990s, hybrid rice R&D in India has encountered major challenges, such as nonacceptable grain and cooking quality, pest and disease susceptibility, higher seed costs, etc. (Janaiah, 2002). It is important to note that the hybrid rice R&D strategy being adopted is dynamic. The program is continually refined by taking farmers' feedback and constraints into consideration and amending research priority settings to meet emerging challenges. Thus, there has been a considerable improvement in the recently released hybrids in terms of grain quality, yield gain, pest and disease resistance, seed yield, etc., when compared with the first- and second-generation hybrids of the 1990s. Several farm-level impact studies carried out during the 1990s and early 2000s (Janaiah, 2002; Janaiah and Hossain, 2003) provided useful feedback on and insights into the hybrid rice R&D program. This allowed the program to reorient its strategy toward farmers' preferences, resulting in the development of some farmer-acceptable hybrids after 2004. However, it has been reported that even the rice hybrids released after 2004 were not acceptable to farmers and consumers in terms of long-term sustainability of the technology. One positive aspect of the new hybrids is that there have been considerable improvements in F_1 seed yield of rice hybrids in seed farmers' fields over the past 10 years (Janaiah and Xie, 2010).

The Nature and Extent of Hybrid Rice Adoption in India

In India the hybrid rice R&D strategy initially aimed to reverse the decelerating yield trend observed under the intensive rice–rice systems of southern India and the rice–wheat systems of northern India. Favorable environments in eastern India, especially boro rice lands, were later targeted for expansion of hybrid rice cultivation when it became evident that the first-generation hybrids were not suited to the irrigated rice lands of southern and northern India (DRR, 1997; Rao et al., 1998).

Expectations rose and ambitious targets were fixed during the early 1990s for the expansion of hybrid rice cultivation. It was projected that hybrid rice would cover nearly 5%, 25%, and 60% of India's total rice area by 2000, 2010, and 2020, respectively (Barwale, 1993). Based on these projections, an ex ante evaluation study estimated that hybrid rice would need to contribute 35%–40% to meet the additional rice demand by 2020 (Janaiah et al., 1993). Furthermore, it was projected that this technology would generate huge employment opportunities for female workers through hybrid seed production in rural India. An ex ante assessment of hybrid rice potential in India based on on-farm trial data (1992–93 and 1993–94) reported 12% yield gains of hybrids over inbred varieties (Janaiah, 1995).

However, farmers' perceptions during on-farm testing indicated that the poor grain and cooking quality of the tested rice hybrids would constrain large-scale acceptance of this technology by both farmers and consumers in India (Janaiah, 2002; Janaiah et al., 1993). Many of the first-generation rice hybrids released during the 1990s were those tested during on-farm trials in 1992–93 and 1993–94, when farmers' perceptions during the prerelease testing period were ignored. Many commercial farmers in the irrigated ecosystem began dropping out from hybrid rice cultivation after one or two crop seasons. In spite of the efforts of seed companies to move hybrid rice from one state to another to obtain a market, area expansion remained much below the projected level until early 2000. Thus, the adoption rate of hybrid rice in India was meager (less than 1% of total rice area) until 2003 (Table 13.5).

In 2004 the hybrid rice R&D strategy in India was reoriented. The multilocation testing approach was expanded to on-farm evaluation of a wide range of rice hybrids under diverse production environments across the country. The

TABLE 13.5 Area Planted to Hybrid Rice Cultivation in India, 1996–2014

| Year | Gross rice area (million ha) | Area planted to hybrid rice[a] | |
		Thousand ha	% of gross rice area
1996	42.84	50	0.12
1997	43.43	90	0.21
1998	43.45	100	0.23
1999	44.80	150	0.33
2000	45.16	175	0.39
2001	44.71	180	0.40
2002	44.90	200	0.45
2003	41.18	275	0.67
2004	42.59	570	1.34
2005	41.91	750	1.79
2006	43.66	1000	2.29
2007	43.81	1100	2.51
2008	43.91	1400	3.20
2009	45.54	1467	3.22
2010	41.92	1800	4.29
2011	42.86	2000	4.67
2012	44.01	1733	3.94
2013	42.75	2000	4.68
2014	43.95	2667	5.92

[a]*Estimated based on seed production during preceding year considering seed rate at 15 kg/ha.*
Source: For seed data: DRR (1996–2015 reports).

strategy was also refocused toward favorable rainfed areas where rice is grown under groundwater, mainly in the eastern parts of the country. The current base yield of different inbred rice varieties under rainfed uplands is low (3.0–3.5 t/ha), with the yield of hybrid rice under on-farm testing reported at 4–4.5 t/ha, a yield gain of 30%–35%. Furthermore, large private seed companies also started engaging in R&D and seed production after 2004, anticipating a huge business as rice was a widely cultivated crop (Ramesha et al., 2009). Some state governments began to subsidize hybrid rice seed during the initial years to promote this hybrid rice. Thus, next-generation hybrids (post-2004) started picking up in the eastern and northern states, and these hybrids outperformed the hybrids of the 1990s. Hybrid rice varieties (released after 2004) are appreciably superior to inbred rice varieties in terms of yield in states such as Chhattisgarh and

Eastern Uttar Pradesh. Hybrid rice outyielded the existing inbred varieties under farmers' field conditions by ~36% in Chhattisgarh and by 24% in Uttar Pradesh (Janaiah and Xie, 2010). However, grain and cooking quality remain a major challenge even with rice hybrids released after 2004.

The adoption rate of hybrid rice, which was less than 1% during the first decade after release, increased to 3.2% and 6% in 2008 and 2014, respectively (Table 13.5). The area planted to hybrid rice during 2014 was estimated at 2.7 million ha of the total rice area, largely concentrated in the eastern parts of the country. The contribution of hybrid rice to total rice production in India was 5.6%, although its share of total rice area was only 3.2% during 2008 (Janaiah and Xie, 2010). The long-term adoption of hybrid rice will depend on the sustainability of the technology in farmers' fields due to continuous problems of poor grain and cooking quality, susceptibility to pests and diseases, higher seed cost, etc.

The Hybrid Rice Seed System

The availability of quality hybrid seed at a reasonable price is crucial to the success of any hybrid technology. The success of hybrids in crops, such as maize, pearl millet, *jowar*, and sunflower, has clearly shown the necessity for an economical and efficient production and distribution of seed for the large-scale adoption of hybrid crop technology in any country like India (Janaiah, 2003).

India has a strong infrastructure for the seed sector. The private sector took the lead in the production and marketing of hybrid rice seed from the start of the hybrid rice program, as it has with other crop systems. More than 30 private seed companies have taken up large-scale seed production and ~10 of these possess their own R&D setup. It is noteworthy to mention that ~85% of the total hybrid rice seed in India is produced in Northern Telangana (Karimnagar, Warangal, and Nizamabad districts). The maximum area planted to hybrid rice seed production is in Karimnagar and Warangal districts, where all the leading seed companies are involved in large-scale seed production. Seed yields are higher in the dry season than in the wet season; hence large-scale seed production is generally done in the dry season because it offers congenial conditions for successful seed production.

It is worthy to note that the average farm size of a hybrid rice seed producer is only 1.2 ha (i.e., small farmers are in seed production), and these smallholders largely have assured irrigation; 300,000 small farmers in Telangana are engaged in hybrid seed production of various crops, including rice. The private seed companies purposively opt to engage small farmers who have an assured irrigation facility for this work, as they usually produce seeds using their entire crop areas (seed companies also insist on the same) and they engage fully in seed production without shifting to other farm activities. Thus, almost the entire cropped area of small seed producers is devoted to seed production (Janaiah and Xie, 2010).

All seed producers have a contract agreement with a middleman, an "organizer," who in turn enters into an agreement with the seed companies. Some seed companies also have direct contract agreements with seed farmers through seed organizers. The organizer facilitates the identification and mobilization of seed producers, supply of parental line seeds, arrangement for procurement of seed by seed companies, payment of the correct amount to seed producers from the companies, etc. The contractual arrangement between seed growers and the organizer is for various roles, such as the seed price to be paid by the seed companies to the growers, the supply of parental line seeds and gibberellic acid (GA_3), and payment of a risk allowance in case of crop failure. It is most important that seed companies do not enter into any direct agreement with the seed producers. The agreements between seed growers and the organizer, and between the organizer and seed companies, do not have any legal entity; they are like private agreements. Thus, seed producers will not have any legal rights to claim compensation in the event of seed production failure due to natural disasters (as was experienced by 25,000 *Bt* hybrid cotton seed producers in Gadwal, Telangana, during 2015–16). Although the seed industry in India is the largest source of employment for rural people, it is yet to be organized and institutionalized to obtain social security benefits from industry.

The private sector procures hybrid rice seed from contract seed growers through organizers at about USD 1.00 per kg. At this price, hybrid rice seed production is significantly more profitable than the alternative economic activity (inbred rice cultivation). Hybrid rice seed production generated a net profit of USD 2,007 per hectare, which is substantially more profitable than inbred rice cultivation (Janaiah and Xie, 2010). Therefore, hybrid seed production will not be a limiting factor once demand for hybrid seed is created among farmers, once grain and cooking quality improves, and once the rice hybrids have wider adaptability.

Growth in the Hybrid Rice Seed Industry

Approximately 30 seed companies entered into the hybrid rice seed industry, including a few big MNCs, as they anticipated a huge boom in business in view of rice being grown on about 45 million ha nationwide. It is estimated that, during the 2013 DS, hybrid rice seed production was carried out on nearly 30,000 ha with ~40,000 tons of hybrid seed being produced, largely in Northern Telangana. Among the public-sector seed agencies, the NSC and the state seed corporations of Maharashtra, Karnataka, and Uttar Pradesh are undertaking hybrid rice seed production on a small scale.

Hybrid rice seed production in India started with less than 200 tons of total production in 1996, but, surpassed 40,000 tons in the 2013 DS (Table 13.6). Seed yields obtained were initially very low (0.3–0.5 t/ha), but with experience over the years, average seed yields of 1.5 t/ha are now being obtained (Table 13.6).

TABLE 13.6 Area and Production of Hybrid Rice Seed (F_1) in India, 1996–2013 DS

Year	Area (ha)	Seed production (tons)	Average seed yield (kg/ha)
1996	195	200	1,026
1997	1,075	1,200	1,116
1998	1,485	1,800	1,212
1999	1,630	2,200	1,350
2000	1,660	2,500	1,506
2001	1,630	2,700	1,656
2002	1,625	2,900	1,785
2003	1,635	3,100	1,896
2004	2,865	4,000	1,396
2005	4,350	8,600	1,977
2006	6,800	12,500	1,838
2007	15,000	20,000	1,333
2008	18,000	22,000	1,222
2009	18,000	27,000	1,500
2010	20,000	30,000	1,500
2011	18,000	26,000	1,444
2012	21,000	30,000	1,429
2013	30,000	40,000	1,333

Note: 1995–96 to 2012–13 DS is shown as 1996–2013 DS.

The latest generation of rice hybrids has considerably outperformed existing inbred rice varieties in terms of yield gain in eastern India. Although there has been a considerable improvement in grain quality and consumer acceptance over the period, the future large-scale adoption of hybrid rice will largely depend on further improvement in grain quality comparable to that of popular inbred varieties. Hybrid rice seed production would not be a constraint to the large-scale adoption of acceptable hybrid rice, as F_1 seed production is highly profitable for seed producers. The key challenges for hybrid rice R&D, however, are the development of new rice hybrids with a competitive and comparable grain and cooking quality, wider adaptability with pest and disease resistance, suitability for irrigated areas, a further increase in yield potential, a reduction in retail seed price, etc.

Hybrid rice seed production is considered a highly knowledge-intensive process. The risks of obtaining a low yield from poor synchronization of parental lines, weather changes, etc., are high in hybrid seed production. Farmers would not engage in it unless it were more profitable than the alternative activity (inbred rice cultivation) and unless the additional profit compensated for the risks and skills involved. Therefore, it is essential to understand the economics of F_1 seed production in farmers' fields with respect to the costs involved, contractual arrangements between seed growers and companies, profitability, and farm-level constraints.

THE CURRENT POLICY ENVIRONMENT

In view of the importance of quality seed for agricultural development, policymakers in India have consistently made efforts to formulate appropriate seed policies since the Green Revolution period. The first Seed Act-1966 was enacted by the Government of India to streamline the seed system in inbred varieties under the public domain, and to ensure a timely supply of quality seeds at an affordable price to popularize high-yielding crop varieties. The first seed review team was set up during 1967 to review seed provisioning and suggest measures to improve the seed delivery system. However, during this review, the expert group recommended a national seed policy with adequate provisions for private-sector participation. As a result, a liberalized seed policy came into implementation in the early 1990s with many incentives and attractive provisions for expanding private seed sector participation. A seed policy review group (in 1995) made several recommendations to further improve India's seed system that resulted in the formulation of the National Seed Policy Act of 2002.

Until the early 1980s, seed provision in India was largely in the public domain, especially under the purview of the SSDCs (public-sector agencies for the states), NSC, and SFC (the public-sector agencies at the central level). The liberalization of seed policies in the early 1990s stimulated the involvement of the private sector and cooperatives in the seed business. As in many countries, private-sector participation in India was historically confined to the production and marketing of hybrid seeds of various crops, such as *jowar*, maize, cotton, sunflower, and vegetables. Since the 1980s, there has been a visible shift in the seed delivery structure with considerable participation by the private sector and cooperatives.

There have been significant policy initiatives in the seed sector, which entailed noteworthy implications for the industry, as well as the market. Amendment of the Industrial Licensing Policy in 1987 removed the restrictions that applied to private investment by foreign companies, monopoly houses, and large domestic firms, resulting in their participation in the Indian seed market.

Liberalization of seed laws through the New Seed Policy-1988 initiated a major change in the import and export of seeds. The importation of vegetable, flower, and fruit seeds was placed under an Open General License; however, the import of wheat and rice seeds was not allowed under the new policy. The importation of seeds of oilseeds, pulses, and coarse cereals was initially allowed for 2 years for companies that had collaborative agreements, provided that the foreign supplier agreed to supply the parental lines, nuclear seed, or BS technology to its Indian client.

Other incentives for the development and production of new seed varieties included a huge cut in the import duty on equipment used for seed production from 100% to 15%, an income tax rebate on R&D expenditure, and the availability of easier and cheaper financing sources for the organizations involved in seed research and production. These policy initiatives have resulted in a boom in private-sector activity in the seed industry.

Trade Related Intellectual Property Rights and Plant Breeder Rights are the other major changes in seed policy and law. It is expected that developers of new varieties and seeds will have greater rights on the process and product developed and will have greater control over the subsequent use of the developments as a direct consequence of these policy changes. Increased protection offered to those who develop new seeds will provide great incentive for firms to invest in R&D. They will also actively try to develop new seeds either independently or in collaboration with research institutions and agricultural universities.

Government plans to bring about changes in the Land Ceiling Act to allow firms to hold larger amounts of land for agricultural purposes will also have a bearing on the seed industry, as firms will have larger amounts of land dedicated to seed multiplication and production. With the IPR and import restriction changes, farmers are likely to have increased choice and access to high-quality seeds of different varieties. This will most probably result in an increase in yield. As firms will have greater control over the marketing of the varieties they develop, the prices of these seeds are likely to rise and farmers might view this unfavorably. However, if the increase in seed prices is matched by an increase in yield, there is no cause for worry.

The New Economic Policy, the New Seed Policy, and changes to IPR-related policies have triggered general public apprehension that the Indian seed industry will be overrun by imported seeds and dominated by transnational corporations with very little chance for domestic firms to survive. However, this fear is not founded, as seeds will be imported only if they result in higher yield, improve the quality of the crop, or are less expensive than the local seed. Furthermore, the imported seeds need to be suited to the various agroclimatic zones of the country, which are very different from those of the countries from where these seeds are likely to be imported. Consumers' taste regarding the agricultural output also acts as a barrier to importing seed. Therefore, even MNCs must be able to adapt and modify their seeds to local conditions and the local environment for which they need to have established domestic R&D. Competing with the local industry in terms of cost will also be difficult for the MNCs if they do not have development and production bases in the country. Local companies will be competitive as long as they devote some energy to R&D.

Seed growers are not coming forward to take up seed production because of the risks involved in the various stages of seed production from selection of fields, harvesting, transporting, grading, treating, and packing. The Government of India approved the implementation of a pilot scheme on seed crop insurance during 1999–2000 that aimed to cover the risk involved in seed production so as to increase the area under BS, FS, and CS. The seed crops of red gram, *bajra*, cotton, gram, groundnut, *jowar*, maize, paddy, soybean, sunflower, and wheat were included under the scheme. The scheme covers all the seed-producing entities under government or private control that are involved in the production of notified varieties. The scheme covers risk of seed crop failure due to natural fire, storms, cyclones, landslides, drought, dry spells, excessive rain, and large-scale incidence of pests and diseases. The compensation is 40% of the sum assured if the crop fails within 1.5 months of sowing and 80% after that period and until harvest. The minimum area of crop failure has to be 0.5 acre. The sum insured is equivalent to the past 3–5 years' average seed yield in the unit area multiplied by the sale price of the seed.

On the whole, there is a positive policy environment for expanding the rice seed sector. The private sector can freely access improved crop germplasm through a mutually agreed memorandum of understanding and PPP-mode. The private seed sector can also liberally access capital markets to strengthen the industry.

CONCLUSIONS, KEY CHALLENGES, AND FUTURE OUTLOOK

Seed provisioning is a complex process with various entities involved in different channels, starting from the development of new crop varieties at the R&D level to seed retailers who deliver the seed to farmers at the bottom level. India's seed system has been developing alongside institutional arrangements over the past 4 decades to meet the seed-provisioning requirements of the farming community. Farm-level seed demand has been growing rapidly in India because of the increased awareness of the importance of quality seed, frequent advertisements through the press and electronic media, the government's special seed promotional activities, etc.

Nearly 120 national agricultural R&D centers under the crop-specific ICAR system, CGIAR centers, and 70 SAUs are currently involved in varietal development programs for Indian farmers. The liberalized seed policy under the new economic regime facilitated the entry of large-scale private-sector members, including MNCs, into agricultural R&D. The public and private sectors together developed and released ~4200 crop varieties and hybrids, including ~1000 for rice in India. The production of rice BS has increased threefold over the past 10 years, while the CS supply has increased by fourfold over the past 15 years. The SRR for rice in many states of India has outstripped the recommended 33%, with almost half of India's rice area planted with new seed every year. To capture the growing seed market for rice, ~400 seed companies in both the private and public sectors (including small companies, cooperatives, and NGOs) are now engaged in rice seed provisioning.

The development and release of hybrid rice during the early 1990s brought new impetus to the rice seed industry. Nearly 30 seed companies working in the Indian hybrid rice seed business have developed and commercially released ~70 rice hybrids over the past 20 years. The public and private sectors together produced ~40,000 tons of F_1 seed during 2013, of which 85% was produced in Telangana alone due to favorable climate and availability of skilled seed producers.

The latest generation of rice hybrids has considerably outperformed existing inbred varieties in yield gain in eastern India. Although there has been a considerable improvement in grain quality over the period, the large-scale adoption of hybrid rice in the future largely depends on further improvement of grain and cooking quality comparable with that of popular inbred varieties. Hybrid rice seed production would not be a constraint to the large-scale adoption of acceptable hybrid rice as F_1 seed production is highly profitable for seed producers and the industry. The key challenges for hybrid rice R&D, however, are the development of new rice hybrids with a competitive and comparable grain quality; wider adaptability and suitability for irrigated areas; a further increase in yield potential; a reduction in retail seed price; etc.

Key Challenges and Future Outlook

Although the seed system in the country is fairly well streamlined, a few policies and institutional challenges confront its future growth. The key challenges and future perspectives are summarized here.

Inadequate Infrastructure

A major issue is whether the current seed system has an adequate modern infrastructure base for providing efficient seed provisioning in the country. For instance, Telangana and Andhra Pradesh are far ahead of many other states in terms of institutional and infrastructure facilities for seed marketing. However, the present storage capacity in these two states is still only 50% of total CS production. Moreover, the modern infrastructure facilities, such as cold storage, modern seed processing units, etc., are inadequate. Modern infrastructure is essentially required to maintain adequate seed quality and to make seeds available at the right time. Poor and outdated infrastructure in many public- and private-sector seed companies is one of the main reasons why farmers are provided with poor-quality seeds. Competitive growth of the seed sector will strongly depend on the level of modern infrastructure; thus public-sector financial institutions and large-scale private-sector firms should provide financial support to the private seed sector and seed cooperatives at a low rate of interest to enable them to establish modern infrastructure.

Willful Supply of Spurious Seeds

It is a common practice in some Indian states, especially during the peak season, for a few small companies (or a small group of seed producers) to market spurious seeds to make a quick profit. The current seed laws have inadequate provisions for taking action quickly against such willful offenses and these small companies tend to disappear within a year or so. The implementation mechanism for protecting farmers' interests against the seed-related offenses by some players in the current seed system is also weak. The legal framework to handle seed-related offenses is mainly guided by Seed Act-1966. Most of the provisions under this act are primarily meant for public-sector-related issues. Therefore, a new Seed Bill-2004 was drafted by the Government of India with a wide range of provisions to deal with all related issues of seeds, technology access, varietal and hybrid release mechanism, quality norms, etc., for hybrid seeds, GM crops, etc. This Seed Bill has been pending for approval since 2004 because of strong opposition from some farmers' groups. Thus, a transparent and vibrant legal framework with a regulatory system is needed to handle the complex issues arising out of today's seed sector. There should be a proper mechanism to implement various provisions under the current seed laws, as well as future laws, that allow farmers to claim and obtain due compensation from the seed companies in case of crop failures on account of poor seed quality.

Violation of the Scientific Norms in the Marketing of Seeds of New Crop Varieties

Some seed companies are in a hurry to deliver seeds from new crop varieties and hybrids to reap the market benefits as quickly as possible in view of the growing competition among seed companies. In doing so, a few seed companies have violated scientific norms in the testing of new varieties and hybrids before commercial release. A few seed companies (and even some public institutions) have courted publicity through various mass media outlets for varieties and hybrids that are still in scientific evaluation trials. The classic example of such a violation of scientific norms is the way the private sector introduced the seeds of *Bt* cotton in Andhra Pradesh during 2002–03 (Qayum and Sakkari, 2003). Similarly, the seeds of rice hybrids were marketed by both the public and private sectors during the mid-1990s without establishing adaptability, yield superiority, and consumer acceptance of the rice hybrids, something that was not accepted by farmers at a later date (Janaiah, 2002). Thus, the state should have a proper regulatory mechanism to force both the public and private sectors to follow scientific procedures in the evaluation of a new variety or hybrid before commercialization (marketing).

Labeling the Source of a Variety/Hybrid

Public–private interface has been growing rapidly after the liberalized seed policy in the early 1990s. The private sector, cooperatives, and NGOs now have free access to the BS of publicly bred crop varieties to enable the large-scale production and marketing of CS to farmers. However, private seed companies often do not label CS with the correct source of the variety. A few companies have been found to label CS with their own name, even though the BS of that variety was received from a public-sector research institution. In view of increasing concerns about plant variety protection, plant breeders' rights, patents, etc., there should be a transparent regulatory mechanism that would enforce the compulsory "labeling" of seeds with the original source of the variety or hybrid, and a few basic features, such as potential yield, year of release, parentage, recommended ecosystem, and resistance to biotic and abiotic stresses, if necessary.

The rice seed sector has a bright future if these key challenges are met.

REFERENCES

Almekinders, C.J.M., Louwaars, N.P., Bruijin, G.H., 1994. Local seed systems and their importance for improved seed supply in developing countries. Euphytica 78, 207–211.

Barwale, B.R., 1993. Hybrid Rice Food Security in India. Macmillan India Limited, Madras, India.

Diaz, C., Hossain, M., Bose, M.L., Merca, S., Mew, T., 2000. Effect of seed quality on rice yield. A paper Presented at the Review and Planning Meeting of Seed Health Project. 25–26 November, 2000, Bangladesh Institute of Development Studies, Dhaka, Bangladesh.

Directorate of Rice Research (DRR), 1997. Development and Use of Hybrid Rice Technology in India: Project Report. DRR, Hyderabad.

Directorate of Rice Research (DRR), 2014. Development and Use of Hybrid Rice Technology in India (1996–2014 Reports). DRR, Hyderabad.

Hossain, M., Janaiah, A., Husain, M., Firdousi, N., 2001. The rice seed delivery system and seed policy in Bangladesh. Bangladesh J. Dev. Stud. 27 (4), 1–40.

Janaiah, A., 1995. Economic Assessment of Hybrid Rice Potential in India: An Ex-Ante Study. Ph.D. Thesis. Institute of Agricultural Sciences, Banaras Hindu University, Varanasi, India.

Janaiah, A., 2002. Hybrid rice for Indian farmers: myths and realities. Econ. Polit. Wkly. 37 (42), 4319–4328.

Janaiah, A., 2003. The seed delivery system in Andhra Pradesh: institutional and policy issues. Indian J. Agric. Market. 17 (2).

Janaiah, A., Hossain, M., 2003. Can hybrid rice technology help productivity growth in Asian tropics? Farmers' experiences. Econ. Polit. Wkly. 38 (25), 2492–2501.

Janaiah, A., Xie, F., 2010. Hybrid Rice Adoption in India: Farm-Level Impacts and Challenges. Technical Bulletin 14. International Rice Research Institute, Los Baños, Philippines.

Janaiah, A., Ahmed, M.I., Viraktamath, B.C., 1993. Hybrid rice: an opportunity for food security in India. Yojana 37 (20), 7–9.

Janaiah, A., Hossain, M., Otsuka, K., 2006. Productivity impact of the modern varieties of rice in India. Dev. Econ. 44 (2), 190–207.

Khoa, N.T., Lan, P.T.P., Dau, H.X., Dinh, H.D., Hach, C.A., Luat, N.V., Paris, T., 1996. Farmers' current rice production practices and seed grain quality in Cantho and Tien Giang Provinces, Vietnam. In: Mew, T.W., et al. (Eds). Proceedings of Seed Health for Disease Management Conference. IRRI, Los Baños, Philippines, pp. 26–29.

Mew, T.W., 1997. Seed health testing: progress towards 21st century. In: Hutchins, R. (Ed). Developments in Rice Seed Health Testing Policy, pp. 129–138.

Pal, S., Tripp, R., Janaiah, A., 2000. Public–private interface and information flow in the rice seed system of Andhra Pradesh, India. Policy Paper 12. National Centre for Agricultural Economics and Policy Research (NCAP), New Delhi, India.

Pray, C.E., Deininger, D.U., 1998. The private sector in agricultural research systems: will it fill the gap? World Dev. 26, 1127–1148.

Qayum, A., Sakkari, K., 2003. Did Bt cotton save farmers in Warangal? Report of a season long impact study of Bt cotton, Kharif 2002 in Warangal District of Andhra Pradesh. AP Coalition in Defense of Diversity, Deccan Development Society, Hyderabad, India.

Ramesha, M.S., Hari Prasad, A.S., Revathi, P., Senguttuvel, P., and Viraktamath, B.C., 2009. Rice hybrids released in India. Technical Bulletin No. 40. Directorate of Rice Research (ICAR), Hyderabad, India.

Rao, N.G.P., Singh, A., Sivasubramanian, V., Murthy, K.S., Mukhopadhyay, A.N., Abraham, C.C., 1998. Rice research and production in India: present status and a future perspective. Quinquennial Review Report. Directorate of Rice Research, Hyderabad.

Sampath, S., Mohanty, H.K., 1954. Cytology of semi sterile rice hybrid. Curr. Sci. 23, 182–183.

Selverajan, S., Joshi, D.C., O'Toole, J.C., 1999. The Indian Private Sector Seed Industry. Rockefeller Foundation, Manila, Philippines.

Somkid, D., Kamoisak, K., Somsak, T., Kannika, P., 1996. Farmers' clean seed practices and seed health testing in Thailand. In: Mew, T.W., et al. (Eds). Proceedings of Seed Health for Disease Management Conference. IRRI, Los Baños, Philippines, pp. 23–26.

Tripp, R., Pal, S., 2001. The private delivery of public crop varieties: rice in Andhra Pradesh. World Dev. 29, 103–117.

Wiggins, S., Cromwell, E., 1995. NGOs and seed provision to smallholders in developing countries. World Dev. 23, 413–422.

Chapter 14

Export Competitiveness of Indian Rice

A.V. Manjunatha*, Lalith Achoth**, Mamatha N.C.*

*Agricultural Development and Rural Transformation Centre, Institute for Social and Economic Change, Bengaluru, Karnataka, India; **Dairy Science College, Bengaluru, Karnataka, India

INTRODUCTION

The area under rice cultivation in India was 44.13 million ha during 2013–14, with production of 106.65 million tons and yield of 2416.2 kg/ha. Rice production had increased from 95.97 million tons during 2010–11 (DES, 2016). The major rice-growing states in India are Uttar Pradesh, Madhya Pradesh, West Bengal, Andhra Pradesh, Odisha, Punjab, and Bihar, with these states together accounting for ~70% of the total rice production in the country. The triennium average of area and yield of rice in the major rice-growing states of India is presented in Table 14.1. Uttar Pradesh, with 6,195,000 ha of area under rice cultivation, is followed by Madhya Pradesh and West Bengal, with 5,612,000 and 5,463,000 ha, respectively. Yield per hectare is highest in Uttar Pradesh, followed by Punjab, Haryana, Madhya Pradesh, and Andhra Pradesh. The triennium average yield in Uttar Pradesh was 4627 kg/ha, followed by 3897 kg/ha in Punjab, 3191 kg/ha in Haryana, 3133 kg/ha in Madhya Pradesh, and 3081 kg/ha in Andhra Pradesh (DES, 2016).

India was the largest exporter of rice in 2013–14, followed by Thailand, Vietnam, Pakistan, and the United States (Fig. 14.1). India is the second-largest consumer after China and hence plays an important role in the global rice economy. India views rice as a strategic commodity for food security and the government intervenes in the market through grain procurement, price support, and export subsidies. Asian countries account for two-thirds of total world rice production; however, other major non-Asian rice-producing countries, including Brazil, the United States, Egypt, Madagascar, and Nigeria, also produce rice. In Africa, rice is the fastest growing staple food with annual total cereal production rising steadily from 9.3% in 1961 to 15.2% in 2007 (Muthayya et al., 2014).

The Future Rice Strategy for India. http://dx.doi.org/10.1016/B978-0-12-805374-4.00014-2

TABLE 14.1 Triennium Average of Area and Yield of Rice in the Major Rice-Growing States of India (2011–12 to 2013–14)

State	Area (in 1000 ha)	Yield (kg/ha)
Uttar Pradesh	6195.0	4627.0
Madhya Pradesh	5611.0	3132.7
West Bengal	5463.8	2745.1
Odisha	4069.2	1695.0
Andhra Pradesh	4026.7	3080.6
Bihar	3251.2	2065.3
Punjab	2838.0	3897.0
Tamil Nadu	1707.5	3243.4
Karnataka	1344.7	2697.1
Maharashtra	1571.0	1913.0
Haryana	1226.0	3190.6
Other states	4780.1	2075.1

Source: DES (Directorate of Economics and Statistics), 2016. Department of Agriculture, Cooperation and Farmers' Welfare, Ministry of Agriculture and Farmers' Welfare, Government of India. Available from: http://eands.dacnet.nic.in/.

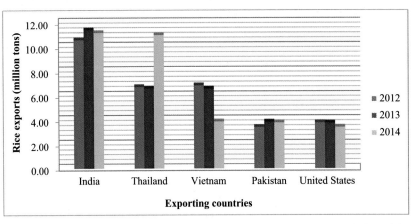

FIGURE 14.1 Major rice exporting countries. (*International Trade Statistics, 2016. Available from: www.intracen.org/default.aspx.*)

Before 1994, India exported relatively small quantities of rice because only exports of basmati were permitted and this was subject to a stipulation of a minimum export price. After 1994, exports of basmati rice were liberalized by the elimination of minimum export prices and the ban on exports of nonbasmati rice was removed. These bold steps, taken to liberate rice production from government controls, led to a dramatic improvement in export volume. Over the next few years, when domestic prices were far lower than the international prices of export competitors, India became a major rice exporter.

A commodity may be export competitive, an import substitute, or neither. To evaluate competitive advantage in trade, various methods exist to empirically map the effective rates of protection of the relevant commodities (Hoda and Gulati, 2008). These methods capture the price competitiveness of specific crops, whereby competitiveness is defined as the ability to supply (potential of export) a product or service on a sustainable (long-term) and viable (profitable) basis (Davar and Singh, 2013). World trade has been facing stiff competition since the emergence of the WTO, due to liberalized policies of governments and the openness of world markets (Davar and Singh, 2013). In spite of their notable export performance, Indian rice exports face problems, such as a lack of consistency in export performance. Domestic problems for rice exports include the following:

- Multiple taxations, such as a purchase tax (on indirect exports), market fees, the Rural Development Fund, and administrative charges. These taxes render rice internationally uncompetitive, making Indian rice costlier in international markets than that of competing countries.
- Lack of proper infrastructure. Rice mills are not fully modernized to ensure high milling recovery.
- Absence of pure seed varieties to meet the quality requirements of grain size.
- The major rice-producing countries usually decrease their prices to capture international markets; however, prices are inelastic for Indian rice because of the relatively high cost of production, thus making it uncompetitive in international markets (Acharya et al., 2012; Datta, 2001).

This chapter focuses on estimating the export competitiveness of Indian rice. The specific objectives are (1) to study the trends in basmati and nonbasmati rice, and (2) to document the changing direction of rice exports and domestic production. These problems relate to the global competitiveness of India's rice and therefore a separate section on measuring competitiveness is presented. The subsequent section deals with the revealed comparative advantage (RCA) of Indian rice exports, which has been decomposed into three components—world trade effects, market distribution effects, and competitiveness effects—that can explain the actual gain or loss in exports. An attempt is also made to study the prospects and challenges for rice trade as well as the export value chain of Indian rice. The final section offers conclusions and policy suggestions.

ANALYTICAL FRAMEWORK

The empirical methods used to achieve the objectives of the study are given in the following sections.

Growth Rate

The growth rate of basmati and nonbasmati rice has been computed using the exponential growth function as in the following equation:

$$Y_t = a \times b^t \times e^u Y \tag{14.1}$$

where Y_t = basmati/nonbasmati rice exports, a = intercept, b = regression coefficient, t = time variable, e = error variable, and u = disturbance term.

The growth rate coefficients (b's) were computed by transforming Eq. (14.1) in logarithmic form as:

$$\text{Log } Y_t = \ln a + \ln b + e \tag{14.2}$$

The compound growth rate in percentage has been computed using Eq. (14.3):

$$G = \{(\text{antilog of } \ln b) - 1\} * 100 \tag{14.3}$$

Markov Chain Model

The direction of trade and changes in exports were examined by employing a first-order Markov chain model. It is a stochastic process (X_i) that describes the finite number of possible outcomes S_i (i = 1, 2, 3...r) that a discrete random variable X_t (t = 1, 2, 3...t) can assume with the passage of time. This is said to have first-order Markovian property if the conditional probability distribution of X_i is dependent only on the state the system is in at step i and not in steps 0, 1, 2, 3, $i - 1$ (Lee et al., 1965). Mathematically, the stochastic process (X_i) has Markovian property if

$$P\left(X_{i+1} = \frac{S}{X_0} = T_0, X_1 = T_1 \ldots X_{i-1} = T_{i-1}, X_i = r\right) = P\left(X_{i+1} = \frac{S}{X_i} = r\right) \tag{14.4}$$

where $P\left(X_{i+1} = \dfrac{S}{X_i} = r\right)$ is the one-step transitional probability of going from state r at step i to state s at step $i + 1$. This represents the conditional probability of X_{i+1} given X_i. If, for each r and s, $P\left(X_{i+1} = \dfrac{S}{X_i} = r\right) = P\left(X_i = \dfrac{S}{X_0} = r\right)$ for all i, then the one-step transitional probability remains stationary. One-step stationary transition probability takes into consideration only one state at each point of time, which will be useful for estimating the market share for one step.

As the present study uses annual export data to predict future exports of rice to different countries and the changing direction of rice area in India from n-step (year) from now, then an n-step stationary transitional probability property will be appropriate. The n-step transitional probabilities are defined as:

$$P_{rs}^{(n)} = p\left(X_{i+n} = \frac{S}{X_i} = r\right)P\left(X_n = \frac{S}{X_0} = r\right) \tag{14.5}$$

where $P_{rs}^{(n)} > 0$ for all states r and s; and $n = 1, 2\ldots$; $\sum_{s=0}^{n} P_{rs}^{(n)} = 1$ for all states r.

Eq. (14.5) assumes that there are $n + 1$ possible states. If the system is currently in state r, it must be in some state n steps from now. In general, the n-step stationary transition probabilities have been estimated using Eq. (14.6):

$$P_{rs}^{(n)} = \sum_{j=0}^{n} P_{rj} P_{js}^{(n-1)} \tag{14.6}$$

where the possible states are 0, 1, 2, 3...n. Therefore, the probability of going from state r to state s in n steps is the probability of going from state r to state s in one step times the probability of going from state j to state s in $n - 1$ steps, summed over all $j = 0, 1, 2, 3\ldots n$ (Murthy and Subramanian, 1999).

Specification of the Markov Chain Model

The share of exports of rice from India (X_{it}) to a particular country and area of rice grown under a particular state (jth) at time t is considered as a random variable and this depends only on its past exports to the country. Following the first-order stationary Markovian property as discussed earlier, the model can be specified as follows:

$$X_{jt} = \sum_{i=1}^{n} X_{jt-1} P_{ij} + e_{jt} \tag{14.7}$$

where X_{jt} represents the exports of rice from India to jth country during the year t, X_{jt-1} are the exports to jth country during the year $t - 1$, P_{ij} is the probability that the exports will shift from ith country to jth country, $e = e_{jt}$ is the error term independent of X_{it-1}, and n is the number of other importing countries.

The transitional probability (P_{ij}), which is central to the Markov chain model, can be arranged in a (columns * rows) matrix and has the following two properties: (1) $0 < P_{ij} < 1$, and (2) $\Sigma P_{ij} = 1$, for all i. The transitional probability P_{ij} indicates the possibility that exports will switch over from country i to country j with the passage of time. The probability P_{ij} for $i \neq j$ indicates the gains or losses in exports of each of the importing countries. The probability P_{ij} for $i = j$ (diagonal probability) indicates the probability of retention of an importing country.

Policy Analysis Matrix

The policy analysis matrix (PAM) is an organizational framework that is used to represent the effects of policies and policy changes on farming and agribusiness. A PAM is a budget-based method for quantitative economic policy analysis and it allows for the evaluation of public investment projects and government policies in the agricultural sector (Akramov and Malek, 2012; Pearson et al., 2003). It is a system of double-entry bookkeeping that provides complete and consistent coverage for all policy influences on returns and costs of agricultural production. It addresses practical matters, such as the impact of policy on competitiveness and farm-level profits, the influence of investment policy on economic efficiency and comparative advantage, and the effects of agricultural policy on changing technologies.

The PAM is composed of two sets of identities (Monke and Pearson, 1989): one set defines profitability and the difference between private and social values, and the other measures the effects of divergences (distorting policies and market failures).

Private values are those prices that are observed in goods and services actually being exchanged, namely, the price of the crop; the cost of seed, fertilizer, farmyard manure, and pesticide; and the wage rate. These are used in budgets and are also called market or financial prices (Rani et al., 2014). Social values are the prices that would prevail in the absence of any policy distortions (i.e., taxes or subsidies) or market failures (i.e., monopolies). They reflect the value to society as a whole rather than to private individuals, and are the values used in an economic analysis when the objective is to maximize national income.

Determination of social values is one of the main tasks of economists, since these values offer the best indication of optimizing income and social welfare. For internationally traded goods, free on board (FoB) is used for exports and cost insurance freight is used in the case of domestic factors that are not traded in international markets, making social costs difficult to estimate. Social costs are estimated for these goods using the value of marginal product approach, in which the share (S_i) of various inputs (X_i) is assumed to be the elasticity coefficients in line with the logic of Euler's theorem together with the mean values of inputs and outputs (Y) and prices (P_y).

Private profits reflect the private profitability of the rice cropping system given existing technologies, output values, input costs, and the policy environment in the country ($D = A - B - C$). The second row measures the social profitability of the same cropping system in terms of social prices ($H = E - F - G$). The social profitability is calculated at shadow prices for inputs and outputs that are calculated by taking into account such market-influencing policies as taxes, subsidies, tariffs, and import duties. The columns of the PAM illustrate revenues, costs, and profits. Costs are divided into two components: tradable inputs and domestic production factors. Tradable inputs are those inputs that are traded in the market and they are further divided into four subcomponents: seeds, fertilizers, depreciation of implements, and insecticides. The domestic cost factors are those inputs that are not traded in the international market: capital, labor, and contracted services. The third row in the PAM table shows divergences, which reflect transfers in the economy due to policy distortions. This allows for capturing the differences between private and social profitability. The valuations of tradable inputs as well as domestic cost factors can also be affected by government taxes and subsidies.

It should be noted that, for a given commodity system, costs and profits represent an aggregate for all activities from farm to wholesale. Private profits (D) are the aggregate measure of net returns for all activities in the system and a high value would suggest a system that is competitive from a financial point of view. A negative value would be a strong indication that the system is unsustainable since there are no incentives for individual farmers to continue cultivating rice. Social profits (H) represent the foreign exchange saved by reducing imports or earned by expanding exports of a unit of this commodity. A positive value means that production is adding to national income, whereas a negative value suggests that the country as a whole would be better off (in terms of national growth) by not producing the commodity. It is therefore an indication of international comparative advantage (Souza and Revillion, 2013). The coefficient gives an insight into whether a country has a comparative advantage in the export trade of rice in a free-trade condition or otherwise. Cell L is the difference between D and H. Disaggregation of the transfer shows whether each policy provides positive or negative transfers to the system due to distortions. Positive transfers indicate that policies are favoring domestic industry and negative transfers indicate discrimination.

The PAM framework is used to estimate important indicators for policy analysis. The nominal protection coefficient (NPC), a simple indicator of the incentives or disincentives in place, is defined as the ratio of domestic price to a comparable world (social) price. The domestic price used in this computation could be either the procurement price or minimum support price, while the world reference price is the international price adjusted for transportation, marketing, and processing costs.

Other indicators that are calculated from the PAM include effective protection coefficient (EPC) and domestic resource cost (DRC). EPC is the ratio of value added in private prices (A − B) to value added in social prices (E − F). If the EPC value is greater than 1, this indicates that government policies provide positive incentives to producers, whereas values less than 1 indicate that producers are not protected through policy interventions. The DRC measures the relative efficiency between agricultural commodities. It is defined as the shadow value of nontradable factor inputs used in an activity per unit of tradable value added (G/E − F). The DRC indicates whether the use of a domestic factor is socially profitable (DRC < 1) or not (DRC > 1).

The data requirements for constructing a PAM include the value of the main product, input requirements, and market prices for inputs and outputs. Transportation costs, port charges, and exchange rate are also required to arrive at social prices. In this chapter a PAM was compiled for rice for the years 2010–11, 2011–12, and 2012–13 and a triennium average was considered. The PAM for basmati rice was calculated for 2012–13. The data on basmati rice have been considered from the cost of cultivation of basmati rice in Haryana, which had the highest area (60%) in India. The years considered for the current study have the latest available data and the years correspond to the period when the Thai rice paddy pledging policy was implemented. Data on cost of cultivation published by the Directorate of Economics and Statistics (DES) of the Ministry of Agriculture and Farmers' Welfare, Government of India, have been used for this purpose.

Revealed Comparative Advantage

RCA measures the change in the comparative advantage of a country's exports. It is the ratio of P_{ij} divided by Q, where P stands for the export value of a commodity (i) of a country or region (j) divided by the total export value of all commodities of a country or region; and Q stands for the total (i) export value of the commodity (i) in the world divided by the total (i) export value of all commodities in the world (Gong, 2011). RCA was first applied by Balassa (1965). He illustrated that two of the factors (relative size and the value of its aggregate exports) determine the value of a country's export share for a particular product, thus causing the export share to be a conceptually unsatisfactory measure of its RCA. Eliminating these factors results in an adjusted export share for each country and a set of these shares will correlate perfectly with their matching set of RCA indices. This chapter uses constant market share (CMS) analysis. This method examines the competitiveness attributable to world trade effect, market distribution effect, and competitiveness effect (Leamer and Stern, 1976).

World Trade Effect

World trade effect measures the hypothetical increase in the focus of a country's exports if its exports grow at the same rate as world exports. The magnitude of this effect would show the potential increase in India's exports if the major importing countries were able to maintain their share. Hence, the world trade effect may alternatively be viewed as the increase or decrease in a country's exports due to expansion in world trade under the assumption that initial market share is maintained. Thus, given a constant overall market share in individual markets, a country's export volume may increase as a result of a general expansion in the total market size (Haque et al., 2014). In other words, a country may gain from the world demand share if it is able to maintain its market share.

Market Distribution Effect

Market distribution effect measures the extent to which a country's exports are concentrated in markets where demand is growing either faster or slower than total world demand (in those markets). It is the deviation in the growth of a particular market for a specific commodity from the average growth rate of world exports for that commodity. Market distribution depends on trade policies and income growth of foreign countries. Growth rates of exports may deviate in the case of changing income elasticity of demand from commodity to commodity;

the income elasticity of demand for the same commodity fluctuates from one market (region) to another and the real income may not grow at the same rate in all regions (Haque et al., 2013). A positive distribution effect suggests that exports are concentrated in relatively expanding areas in world trade, while a negative distribution effect suggests that exports are concentrated in a market where demand is growing slower than world demand. It is usually assumed that this effect is independent of other effects discussed earlier and it largely reflects the role of domestic factors of the exporting country (Muhammad and Yaacob, 2009).

Competitiveness Effect

Competitiveness effect measures the difference between the actual increase in the focus of a country's exports and the increase that would have occurred if the focus country maintained its market share in those markets. This residual term indicates the improvement or deterioration in the competitiveness of exports depending on whether the term has a positive or a negative sign (Haque et al., 2013). A negative sign indicates that the country fails to maintain the market share because of a lack of competitiveness (Haque et al., 2014). However, this residual may provide a biased measure of general competitiveness due to interacting effects of commodity composition and market distribution. In rapidly emerging markets, the country may experience a diminishing share in the world market if it cannot cope with growth to that extent. Furthermore, the net competitiveness effect is to be reflected in the negative sign because of favorable market and commodity growth.

RESULTS AND DISCUSSION

Trends in Basmati and Nonbasmati Rice Exports

Rice exports from India are dominated by basmati rice, with nearly two-thirds of the basmati rice produced in India being exported. Basmati rice is a leading aromatic fine-quality rice of world trade and it fetches a relatively higher price than nonbasmati rice in the international market. It is mainly grown on the Indo-Gangetic plains. The meaning of basmati is derived from "bas," which means aroma, and "mati," meaning sense. Thus, the word basmati represents ingrained aroma. This aroma gives basmati its uniqueness unparalleled by any other rice grain in the world. Many scented varieties of rice have been cultivated on the Indian subcontinent for a long time but basmati differentiates itself from all other aromatic rice because of its typical aromatic characteristics coupled with the silky texture of its long grain.

Basmati rice has a large grain size, unique aroma, and a cooking quality that are preferred by consumers in the Gulf region; therefore, the export market is large and is growing steadily. The major destinations for basmati rice exports from India are Saudi Arabia, Iran, the United Arab Emirates, Iraq, and Kuwait.

TABLE 14.2 Major Importers of Basmati Rice From India During 2014–15

Importing country	Quantity (tons)	Share (%)
Saudi Arabia	966,931	26.1
Iran	935,568	25.3
United Arab Emirates	278,601	7.5
Iraq	235,448	6.4
Kuwait	166,469	4.5
Other countries	1,119,267	30.2
Total	3,702,284	100.0

Source: APEDA (Agricultural & Processed Food Products Export Development Authority), 2016. Ministry of Commerce & Industry. Government of India. Available from: http://apeda.gov.in/apedawebsite/index.html.

Saudi Arabia has traditionally been the largest market for Indian basmati rice (Table 14.2).

During the 1990s India exported 0.232 million tons of basmati rice valued at USD 164 million. Exports increased to 0.85 million tons by 2000–01, with a value of USD 474 million. However, during 2013–14 basmati rice exports from India witnessed a quantum leap to 4.686 million tons, valued at USD 520 million. The growth in exports of Indian basmati rice from 1990–91 to 2013–14 displayed an exponential increase (Fig. 14.2). The growth in quantity and value was ~11.3% and 13.7%, respectively, per annum.

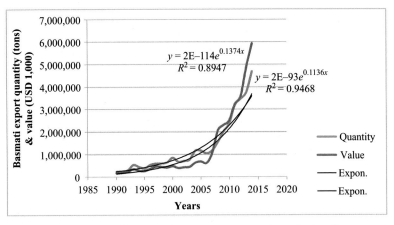

FIGURE 14.2   Growth trend in basmati rice exports in quantity and value. First exponential line (above) pertains to Expon. (Qty.) and y & R^2 (below) pertains to Qty. Second exponential line (below) pertains to Expon. (Value) and y & R^2 (above) pertains to Value. *(Authors' estimates based on data collected from APEDA.)*

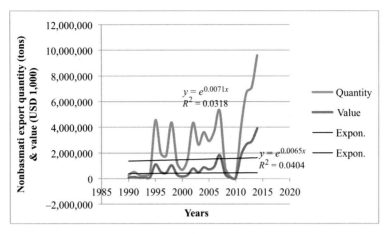

FIGURE 14.3   Growth trend in nonbasmati rice exports in quantity and value. First exponential line (above) and y & R^2 (above) pertains to Expon. (Qty.). Second exponential line (below) and y & R^2 (below) pertains to Expon. (Value) *(Authors' estimates based on data collected from APEDA.)*

The major destinations for India's nonbasmati rice exports were the developing countries, namely, Bangladesh, the United Arab Emirates, South Africa, Nepal, Saudi Arabia, and Sri Lanka. During 1990–91 exports of nonbasmati rice were 0.272 million tons, which increased gradually over the years to 0.682 million tons during 2000–01 and to 9.59 million tons in 2013–14. To assess the growth in exports of Indian nonbasmati rice in terms of value, the compound growth rate was computed for the period 1990–91 to 2013–14. The overall exports of nonbasmati rice recorded a compound growth rate of 4.47% in terms of value and 3.35% in terms of quantity, which was lower than the growth in basmati rice (Fig. 14.3). Thus, it can be inferred that the demand for Indian basmati rice is higher than for nonbasmati rice in the international market.

The Shifting Focus of Rice Cultivation in the States of India

The changing pattern of rice area in rice-growing states was estimated by obtaining the transitional probability matrices for the period 1968–69 to 2012–13 (Table 14.3). The analyses were carried out for states located in the various zones of India. The northern zone comprises Haryana, Punjab, Uttar Pradesh, Himachal Pradesh, and Jammu and Kashmir. The western zone contains Rajasthan, Gujarat, Maharashtra, and Goa. The central zone involves Madhya Pradesh and the southern zone comprises Andhra Pradesh, Karnataka, Kerala, and Tamil Nadu. The eastern zone covers Bihar, West Bengal, Odisha, and the northeastern states.

The results of the transitional probability matrix for the rice-growing states are presented in Table 14.3. It is evident from the table that the northern and eastern states are the stable rice-growing states compared with the western, southern, and northeastern states. This implies that Punjab, Haryana, and Uttar

TABLE 14.3 Transition Probability Matrix for Rice Area Across Different States of India

States	Haryana	Punjab	Uttar Pradesh	Himachal Pradesh	Jammu and Kashmir	Madhya Pradesh	Rajasthan	Gujarat	Maharashtra	Goa	Andhra Pradesh	Karnataka	Kerala	Tamil Nadu	Bihar	West Bengal	Odisha	Northeastern states
Haryana	0.741	0.000	0.186	0.000	0.000	0.000	0.000	0.073	0.000	0.000	0.000	0.000	0.000	0.000	0.000	0.000	0.000	0.000
Punjab	0.067	0.917	0.013	0.000	0.000	0.000	0.000	0.002	0.000	0.000	0.000	0.000	0.000	0.000	0.000	0.000	0.000	0.000
Uttar Pradesh	0.000	0.011	0.504	0.001	0.000	0.369	0.000	0.000	0.000	0.000	0.000	0.000	0.000	0.000	0.000	0.044	0.000	0.071
Himachal Pradesh	0.000	0.000	0.000	0.372	0.000	0.000	0.000	0.000	0.000	0.000	0.000	0.000	0.000	0.628	0.000	0.000	0.000	0.000
Jammu and Kashmir	0.000	0.000	0.000	0.000	0.000	0.000	0.000	0.000	0.000	0.000	1.000	0.000	0.000	0.000	0.000	0.000	0.000	0.000
Madhya Pradesh	0.000	0.000	0.093	0.002	0.007	0.108	0.000	0.017	0.053	0.002	0.103	0.045	0.000	0.043	0.052	0.254	0.128	0.092
Rajasthan	0.000	0.000	0.000	0.000	0.099	0.000	0.523	0.000	0.218	0.006	0.000	0.000	0.000	0.000	0.000	0.000	0.000	0.153
Gujarat	0.133	0.111	0.044	0.000	0.000	0.000	0.000	0.375	0.000	0.000	0.000	0.288	0.000	0.000	0.000	0.049	0.000	0.000
Maharashtra	0.000	0.000	0.000	0.002	0.081	0.232	0.000	0.000	0.643	0.000	0.000	0.000	0.000	0.000	0.000	0.000	0.000	0.043
Goa	0.000	0.000	0.000	0.000	0.000	0.000	0.000	0.000	0.000	0.000	1.000	0.000	0.000	0.000	0.000	0.000	0.000	0.000
Andhra Pradesh	0.000	0.017	0.181	0.001	0.009	0.018	0.010	0.008	0.047	0.001	0.448	0.004	0.000	0.000	0.000	0.091	0.000	0.166

Karnataka	0.000	0.000	0.189	0.000	0.000	0.368	0.000	0.140	0.000	0.000	0.000	0.295	0.000	0.000	0.000	0.008	0.000	0.000
Kerala	0.000	0.000	0.000	0.000	0.000	0.000	0.000	0.000	0.000	0.000	0.000	0.000	0.941	0.000	0.059	0.000	0.000	0.000
Tamil Nadu	0.000	0.000	0.000	0.010	0.000	0.059	0.000	0.000	0.008	0.001	0.014	0.000	0.000	0.593	0.186	0.000	0.130	0.000
Bihar	0.000	0.000	0.000	0.003	0.005	0.000	0.007	0.000	0.002	0.003	0.033	0.000	0.003	0.100	0.838	0.004	0.000	0.003
West Bengal	0.000	0.000	0.000	0.000	0.000	0.083	0.000	0.000	0.000	0.002	0.189	0.065	0.000	0.000	0.000	0.520	0.027	0.113
Odisha	0.000	0.000	0.000	0.000	0.000	0.043	0.000	0.000	0.000	0.002	0.000	0.019	0.000	0.038	0.000	0.145	0.753	0.000
Northeastern states	0.000	0.000	0.363	0.000	0.010	0.269	0.000	0.003	0.013	0.000	0.000	0.015	0.000	0.000	0.000	0.019	0.000	0.308

Source: Authors' estimates based on data collected from DES (Directorate of Economics and Statistics), 2016. Department of Agriculture, Cooperation and Farmers' Welfare, Ministry of Agriculture and Farmers' Welfare, Government of India. Available from: http://eands.dacnet.nic.in/.

Pradesh have been the stable rice-growing states in the northern zone, as reflected by the high probability of retention of 0.917, 0.741, and 0.504, respectively. Similarly, in the eastern zone, Bihar, West Bengal, and Odisha have been the stable rice-growing states, as reflected by the probability of retention of 0.838, 0.520, and 0.753, respectively. The transitional probabilities for the rice-growing states of the western zone (Rajasthan, Gujarat, Maharashtra, and Goa) were 0.522, 0.374, 0.643, and 0, respectively. The probability value of Goa clearly indicates great instability in rice cultivation. Similarly, among the southern states, Kerala has a higher probability of retention (0.9407) than Tamil Nadu (0.5926) and Andhra Pradesh (0.4481) and the probability is lowest in Karnataka (0.2952). However, the central zone (Madhya Pradesh) has a low retention (0.1082) as do the northeastern states (0.3077), indicating instability in rice growing and a high probability of shifting from rice to other crops. It is interesting to note that the major gainers in rice-growing states are the northern and eastern zones. Tamil Nadu, which has a retention probability of 0.5926 of area under rice, is likely to gain from Himachal Pradesh (0.62). Similarly, Andhra Pradesh is likely to gain from Jammu and Kashmir and Goa, while Karnataka is likely to gain from Gujarat (0.28) and West Bengal (0.06). Moreover, it can be noted that the rice-growing areas of the southern states are likely to gain from the northern and western zones of India.

Changing Direction of Rice Exports From India

The shifting pattern of rice exports from India was estimated in the Markov chain framework by approximating the transitional probability matrices for the annual export shares of rice (in terms of value) for the period 2001–14. The major importers of rice from India are Saudi Arabia, Iran, the United Arab Emirates, Bangladesh, and Kuwait. From Figs. 14.4 and 14.5, it is interesting

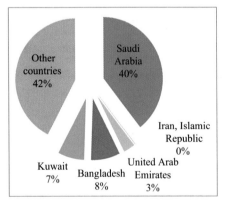

FIGURE 14.4 Rice exports in 2001. *(APEDA (Agricultural & Processed Food Products Export Development Authority), 2016. Ministry of Commerce & Industry. Government of India. Available from: http://apeda.gov.in/apedawebsite/index.html.)*

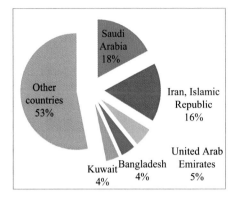

FIGURE 14.5   Rice exports in 2014. *(APEDA (Agricultural & Processed Food Products Export Development Authority), 2016. Ministry of Commerce & Industry. Government of India. Available from: http://apeda.gov.in/apedawebsite/index.html.)*

to note that Saudi Arabia's share of India's total exports declined drastically from 40% to 18% between 2001 and 2014. At the same time, Iran improved its imports to 16% from almost zero during 1990.

Major rice importers were considered for the Markov chain analysis while rice exports to the remaining countries were pooled under other countries. The results of the transitional probability matrix are shown in Table 14.4. It is evident that the United Arab Emirates and Iran are stable rice importers, having a probability of retention of 0.6025 and 0.5593, respectively. The transitional probability for Saudi Arabia is 0.3388 and for Bangladesh and Kuwait it is low

TABLE 14.4 Markov Chain for Rice Exports From India

Country	Saudi Arabia	Iran, Islamic Republic	United Arab Emirates	Bangladesh	Kuwait	Other countries
Saudi Arabia	*0.3388*	0.0000	0.0000	0.0000	0.0001	0.6611
Iran, Islamic Republic	0.2001	*0.5593*	0.0000	0.0000	0.0437	0.1970
United Arab Emirates	0.0000	0.1883	*0.6025*	0.0000	0.2093	0.0000
Bangladesh	0.8577	0.0000	0.0250	*0.0062*	0.1111	0.0000
Kuwait	0.0000	0.0000	0.2288	0.0000	*0.1386*	0.6326
Other countries	0.1263	0.0000	0.0389	0.2201	0.0215	*0.5932*

Source: Authors' estimates based on data collected from APEDA.

(0.0062 and 0.1386, respectively). However, the probability of retention among other countries is 0.5932, indicating loyal markets for Indian rice. Other markets are gaining in popularity, which goes to show that markets are becoming diversified.

Although Saudi Arabia is the major rice importer from India, it is more likely to gain from Bangladesh (0.8577) and Iran (0.2001). The United Arab Emirates has a high probability of retention of its own share of imports of Indian rice and is likely to gain 22% from Kuwait. Similarly, Bangladesh is likely to gain 22% from other countries. The other countries that are importing Indian rice, in addition to having a reasonably good retention probability, are likely to gain from Saudi Arabia (66%), Kuwait (63%), and Iran (19%).

Comparative Advantage of Indian Rice Cultivation

The results of the PAM shown in Table 14.5 indicate that private profits (INR/ha) were positive, signifying that profits are being generated through rice production in India. The social profits (the foreign exchange saved by reducing imports, or earned by expanding exports of a unit of this commodity) were also positive. The negative social profits in rice cultivation indicate that, if the input industry of resources is made more efficient and inputs are priced at their optimum levels, the profits from rice cultivation in India could be further enhanced. As such, this is an indication of international comparative advantage. The coefficients give an insight into whether a country has a comparative advantage in the export trade of a commodity in a free-trade condition or otherwise.

Divergences in revenues are caused by distortions in output prices; divergences in tradable input costs are caused by distortions in tradable input prices; divergences in domestic factor costs are caused by distortions in domestic factor prices; and the net transfer effect arises from the total impact of all divergences. The effects of divergences are found by applying the divergences identity.

TABLE 14.5 Policy Analysis Matrix (Triennium Average: 2010–11 to 2012–13)

| | | Cost (INR) | | |
Item	Revenue (INR/ha)	Tradable inputs	Nontradable inputs	Profit
Private price	1585.71	138.71	828.03	724.43
Social price	2544.22	109.86	917.00	1517.36
Divergence	−853.04	28.85	−88.96	−792.92

Source: Authors' estimates based on data collected from DES (Directorate of Economics and Statistics), 2016. Department of Agriculture, Cooperation and Farmers' Welfare, Ministry of Agriculture and Farmers' Welfare, Government of India. Available from: http://eands.dacnet.nic.in/.

It is evident that the output price of rice is distorted as the landed price of rice in India is far higher than the price prevailing in the international market. This could be because Indian rice cannot compete perfectly in the international market because of aspects regarding quality, the government's protectionist policy due to the staple nature of the crop, and so on. Divergences in tradable inputs are positive, suggesting that tradable input prices are distorted and higher than those prevailing in the international market. In fact, the prices of inputs are higher than at the international level, thus penalizing rice farmers. Nontradable input prices are lower due to high-productivity inputs, which are in fact helping rice farmers. The same inferences were drawn in the study on rice production done in Mercosur by Souza and Revillion (2013).

The major source of divergence is market failure. A market fails to discover competitive prices that reflect social opportunity cost and this leads to an inefficient allocation of products or factors. Three basic types of market failures create divergences: First, when the seller or the buyer controls market prices. Second, in the event of negative or positive externalities for which the benefits cannot be realized as compensation. Third, factor market imperfections due to the inadequate development of institutions to provide competitive services and full information. The high competitiveness witnessed in Indian rice production could be because prices are not being transmitted adequately from the international market to domestic markets due to asymmetric information, trade restrictions, and varietal and quality differences. Policy should be designed to align production to international standards so that India can make greater inroads into the world rice market.

Results of the PAM for basmati rice (2012–13) are presented in Table 14.6. The first row provides the measure of private profitability (Rs. 184.31). This demonstrates the competitiveness of basmati rice given current technologies, prices for inputs and outputs, and policy. The social profit (Rs. 1421.48) indicates that the state uses scarce resources efficiently and has a static comparative advantage in the production of basmati rice. However, Table 14.6 demonstrates

TABLE 14.6 Policy Analysis Matrix for Basmati Rice (2012–13)

Items	Revenue (INR/ha)	Cost (INR)		Profit
		Tradable inputs	Nontradable inputs	
Private price	1280.00	145.85	949.83	184.31
Social price	2469.59	76.70	971.41	1421.48
Divergence	−1189.59	69.16	−21.58	−1237.16

Source: Authors' estimates based on data collected from DES (Directorate of Economics and Statistics), 2016. Department of Agriculture, Cooperation and Farmers' Welfare, Ministry of Agriculture and Farmers' Welfare, Government of India. Available from: http://eands.dacnet.nic.in/.

that private revenues were lower than social revenues, evidencing that public policies are negatively affecting rice production. In the last column the difference between social and private prices was negative, which indicates that prices prevailing in India are lower than the prevailing world price.

The competitiveness of cultivation was further analyzed in the context of protection coefficients. The competitiveness for rice was measured using ratios: NPC, DRC, and EPC. Detailed estimation of the NPC of rice is given in Table 14.7 and the reference of calculation is based on Gulati et al. (1990). The summary result on the protection coefficients of rice is reported in Table 14.8. NPC is a direct measure of the price competitiveness of a country, DRC is required to save or to earn a unit of foreign exchange through the production or export of the commodity, and EPC captures transfers due to distortions in input as well as output price on the product's value added.

The NPC value of rice was 0.65 in 2010–11 and 0.64 in 2011–12. These values are less than 1, which implies that Indian rice was unprotected by the policy

TABLE 14.7 Variables Used for Estimating Export Competitiveness

	Particular	Unit	2010–11	2011–12	2012–13
1.	FoB price at Bangkok in January	USD/ton	516.80	542.00	564.20
2.	Freight (January)	USD/ton	9.00	12.50	9.50
3.	Exchange rate (January)	USD 1 = Rs.	45.50	51.19	54.24
4.	Price of rice [(1 + 2) * 3]	INR/ton	23,923.90	28,384.85	31,117.49
5.	Port clearance charges (1.5%)	INR/ton	358.85	425.77	466.76
6.	Landed price	INR/ton	24,282.76	28,810.62	31,584.25
7.	MSP	INR/ton	14,285.71	15,428.57	17,857.14
8.	Marketing cost and distribution (margins at 5% of procurement price)	INR/ton	714.28	771.42	892.85
9.	Transportation cost from Kolkata to Hyderabad	INR/ton	3,146.66	3,343.33	3,343.33
10.	Reference price (6 − 8 − 9)	INR/ton	20,421.81	24,695.86	27,348.06

International prices are of Thai (milled) white 5% broken variety and belong to those months (October to January in different years) that correspond to the peak marketing season of rice in India; marketing costs and traders' margins (except for transportation cost) are estimated at 5% of the procurement price.

TABLE 14.8 Protection Coefficients for Rice (Including Both Basmati and Nonbasmati) and Basmati Rice

Years	NPC	EPC	DRC
Rice (basmati and nonbasmati)			
2010–11	0.59	0.63	0.36
2011–12	0.62	0.61	0.39
2012–13	0.65	0.67	0.38
Basmati			
2012–13	0.52	0.47	0.41

Source: Authors' estimates based on data collected from DES (Directorate of Economics and Statistics), 2016. Department of Agriculture, Cooperation and Farmers' Welfare, Ministry of Agriculture and Farmers' Welfare, Government of India. Available from: http://eands.dacnet.nic.in/.

framework, competitive in the world market, and worth exporting. Other studies in Indian states also reported an NPC value of less than 1 (Rani et al., 2014).

DRC allows a comparison of efficiency and thus produces different outputs. It is analyzed in a similar way to private profitability, that is, minimizing DRC is the same as maximizing social profits (Souza and Revillion, 2013). The DRC coefficient compares the opportunity costs of using domestic primary resources (namely, land, labor, and capital) and traded inputs in domestic production with the value added by that production at border prices. All DRC estimates for rice were less than 1 (0.36, 0.39, and 0.38) during 2010–11, 2011–12, and 2012–13. This reveals that the country had a comparative advantage in rice production, thus indicating high production efficiency and positive effects due to higher net revenues. This also signifies that domestic rice production is efficient in the use of resources.

EPC is a more reliable indicator than the NPC, even when these values are positive and less than 1. The EPC of rice was 0.63, 0.61, and 0.67 during 2010–11, 2011–12, and 2012–13, respectively. This indicates that value added at domestic prices is higher than the value added at border prices and rice in India is effectively protected through the combination of domestic output and input price policy; in other words, the domestic output is effectively protected by the domestic policy regulating input and output prices. This specifies that India is an efficient producer of rice, which could be due to the emergence of proficient production technology and policy impacts in the country.

The NPC for basmati rice was 0.52, indicating competitiveness in the world market for rice exporting. The estimated DRC was less than 1 (0.41), indicating either production efficiency or that it is desirable to produce basmati rice and expand its production to obtain higher net revenue. The EPC was also less than 1 (0.47), indicating that public policy interventions in basmati rice production are reduced and that there is an increasing rate of competitiveness of basmati rice, mainly because of the adoption of improved production technology.

Revealed Comparative Advantage of Indian Rice Exports

The principal components that have contributed to the growth of rice exports are the world trade effect, market distribution effect, and competitiveness effect (Poramacom, 2002). These principal components explain the changes in a country's share of trade in world markets. CMS identifies the cause of the extent to which a country's export growth differs from that of the rest of the world. In this study the CMS of India's rice exports during 2001–14 was estimated. Decomposition of the export gain or loss of Indian rice is presented in Table 14.9.

It is evident from Table 14.9 that the world trade effect was positive in terms of both quantity and value of rice. The results were 4.88% for quantity and 2.50% for value, meaning that the increase in world exports had a positive impact on the increase in India's rice exports on the whole. The market distribution effect of rice was negative for both quantity and value, that is, −137,157 tons (−5.88%) and USD −97,559 (−2.79%), respectively: this is minimal. This distribution effect suggests that exports were concentrated in a market where demand was not growing in line with world demand. The competitiveness effect of rice exports of India was positive from 2000 to 2013. This indicates an improved position of exports in terms of competitiveness and that India has the ability to respond to changing environments and adapt its supply situation to world conditions. In addition, the rice pledging scheme in Thailand during 2011 boosted the share of Indian rice exports in the world market. However, Thailand faced fierce competition from the rice exporting countries of Vietnam and India, whose rice appears to be a very close substitute to Thai rice (Mahathanaseth and Pensupar, 2014). It is very significant that most of the growth of rice exports from India is due to the competitiveness of Indian rice, specifically basmati rice.

TABLE 14.9 Decomposition of Export Gain or Loss of Indian Rice

	Quantity		Price	
	(tons)	(%)	(USD thousand)	(%)
World trade effect	113,744	4.88	87,507	2.50
Market distribution effect	−137,157	−5.88	−97,559	−2.79
Competitive effect	2,355,251	101.00	3,508,756	100.29
Total change	2,331,837	100.00	3,498,704	100.00

Source: Authors' estimates based on International Trade Statistics, 2016. Available from: www.intracen.org/default.aspx.

Prospects and Challenges for Rice Trade

Rice is a major staple food and it is central to food security for most of the world's population. Indian rice exports face major problems that need to be tackled if India is to emerge as a steady and reliable exporter of rice. Even though the recent trend of rice exports in both basmati and nonbasmati has been positive, the dominant characteristic appears to be fluctuations in exports. Nonbasmati rice was a new entrant into exports during the 1990s and fluctuations in its export were large and clear.

Figs. 14.6 and 14.7 show the significant amount of fluctuation reflected in quantity as well as in total value of rice exports. Indian basmati rice exporters also face problems, especially with respect to price, as basmati rice from Pakistan is 20%–25% cheaper than Indian basmati rice in the international market.

Fig. 14.8 shows that India was the largest global exporter of rice in 2014–15, with a share in world trade of 27%. Thailand was second with a share of 26.5%. Thailand has consistently been the largest exporter of rice for the last decade; however, trade decreased after 2012 because of its paddy pledging program, which has subsequently been discontinued. Vietnam and Pakistan were the third- and fourth-largest exporters of rice, with shares of 9.38% and 9.15%, respectively.

Various taxes are imposed on rice exports by the state governments, such as a purchase tax on indirect exports, market fees, the Rural Development Fund, administrative charges, and so on. These taxes affect the pricing of rice internationally and as a consequence Indian rice becomes costlier than that of other competing countries. An increase in the cost of inputs used for paddy

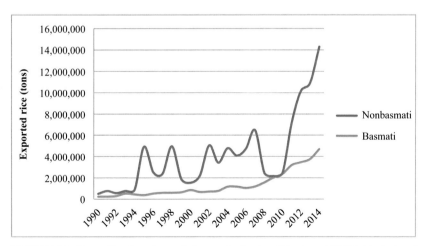

FIGURE 14.6 **Basmati and nonbasmati rice production in million tons.** *(APEDA (Agricultural & Processed Food Products Export Development Authority), 2016. Ministry of Commerce & Industry. Government of India. Available from: http://apeda.gov.in/apedawebsite/index.html.)*

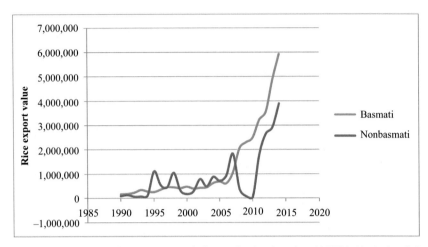

FIGURE 14.7 Basmati and nonbasmati rice production in value. *(APEDA (Agricultural & Processed Food Products Export Development Authority), 2016. Ministry of Commerce & Industry. Government of India. Available from: http://apeda.gov.in/apedawebsite/index.html.)*

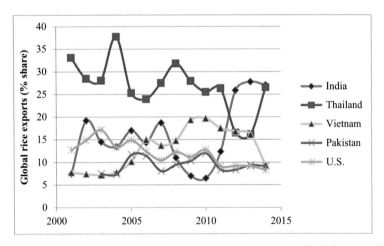

FIGURE 14.8 Percentage share of exporting countries. *(International Trade Statistics, 2016. Available from: www.intracen.org/default.aspx.)*

cultivation results in increasing production costs, which become even costlier when the paddy is converted to rice. The Indian government has increased the minimum support price for paddy every year to safeguard the interests of growers. However, major rice-producing nations have started to decrease the price of rice and offer large subsidies in order to capture the international market. Indian rice prices, however, are inelastic because of the relatively high costs of production, leading Indian rice to become internationally uncompetitive. Many prospects of basmati rice export have been lost in recent years to other competing countries (i.e., Pakistan) because of these high prices. This has to be offset by achieving higher productivity through increasing the outlay for research and development.

Other challenges facing India include the lack of proper arrangements for the production of the quantity of quality seeds needed for the cultivation of rice for export purposes. In the absence of genetically pure basmati varieties, a variation with different plant height, grain size, and maturity is found in the majority of basmati rice fields. This is likely to be one of the major reasons for the poor quality of some Indian basmati rice. Lack of proper infrastructure also affects exporters as they are unable to store their stocks properly and safely at seaports, thus adding additional expenditure.

World rice consumption outstripped production from 2011 to 2014, although the supply and demand gap was met by liquidating stocks. Population growth and the associated complication of increased land for habitation, roads, and growing urbanization have gradually eaten into rice lands. Nearly 25 million acres of fertile rice land in South Asia, constituting about 10% of the world's irrigated rice land, are showing signs of soil fatigue due to overuse of chemical fertilizer-based intensive monoculture (Datta, 2001). Recent expansion suggests that yields are stagnating in countries such as Japan, South Korea, China, Indonesia, the Philippines, and northwestern parts of India. Thus, rice exporting countries, including India, may not worry about a shortfall in demand for rice exports or a decline in the international price of rice. As a consequence of GATT negotiation, several rice-consuming countries, such as Japan, South Korea, and Indonesia, have started opening up their domestic markets for which India can become a potential exporter (Datta, 2001). However, meeting this future rice demand could be challenging given the increasing competition for land and water and the environmental degradation. Although protectionist trade policies have proved beneficial to India in the short term, the effect could prove to be unsustainable. Furthermore, India's production-oriented aid measures could become fiscally impossible to sustain. Public-sector debt is very high, which undermines capital investment through heavy debt service burdens (Coface, 2009). Moreover, in India, long-grain rice would gain significantly but the gains would be almost offset by losses to consumers. This is due to the consistent price change related to large volumes and characterizes moderate impacts per producer or consumer (Wailes, 2005).

Kumar and Rosegrant (1996) reported that the national yield rate of rice was fluctuating substantially. This fluctuation was found to be more evident in those states that had been pursuing high-yielding technology and that had obtained higher yield than the national average. The absence of appropriate rice technology for resource-poor and adverse agroclimatic conditions has also been a persistent feature that is not certain to be resolved satisfactorily in the near future. Moreover, the sudden upsurge in nonbasmati exports in recent years was achievable because of the accumulation of buffer stocks for the purpose of the Targeted Public Distribution System (TPDS). Launched in 1997, the TPDS aimed to provide subsidized food and fuel to the poor through a network of ration shops, as a sizable part of the population lacks sufficient purchasing power to procure rice at fair prices. The National Food Security Act (passed in 2013) gives statutory backing to the TPDS, and this marked a shift in the right to food as a legal right rather than a general entitlement. The percentage of the total rural and urban population that is classed as poor declined by 23.4% between 1993–94 and 2011–12, suggesting that the number of poor households that are eligible for assistance [as they exist below the poverty line (BPL)] has declined. However, the government did not minimize the estimated number of BPL households and continues to provide BPL allocations based on 1993–94 poverty estimates. In this situation, genuine fears are being expressed against committing exports of nonbasmati rice in the future (Balani, 2013; Datta, 2001). It is a wise strategy for India to export basmati rice rather than nonbasmati rice as it generates significantly higher returns per unit of output (0.97 times for basmati and 0.31 times for nonbasmati). In pursuing this strategy, some of the major problems, such as the fear of an increasing nonbasmati rice price and groundwater depletion, could be decreased.

The Export Value Chain

More than 480 million tons of milled rice are used every year, with more than 85% of this for human consumption (FAO, 2013). China and India collectively account for nearly 50% of global rice consumption, with daily per capita rice consumption greater than 300 g (>110 kg per capita annually in Asia). Unlike other cereal foods, paddy gives a wide range of products: fine rice, cut rice, derived products (from large-sized varieties), stuffed rice products, and by-products, such as husk and rice bran. By-products, such as husks, are used as fuel in steam or processing industries or as a burning agent in brick manufacturing. Rice bran contains 18%–20% oil content, which can be extracted as rice bran oil (RBO) in solvent extraction plants (Nagaraj and Krishnegowda, 2015). The production and trade of rice have been one of the major concerns of the Indian economy; however, no proper supply chain framework has yet been developed, which normally leads to unfulfilled demand, stockouts, overstocking, and distribution problems.

Analyzing a value chain in a particular region or in a particular consumer/producer segment requires an understanding of the different activities of stakeholders. This is essential to allow planning for the present and future, to reduce transaction costs, and to recognize each partner's competitive advantage. However, this requires demand and supply estimates under alternate scenarios with and without a value chain interview (Reddy, 2013). As markets develop, the value chains become more complex, with additional channels competing for both inputs and outputs. This market map helps in understanding policy issues that affect the functioning of the chain and also the institutions and organizations that provide the services (e.g., market information and

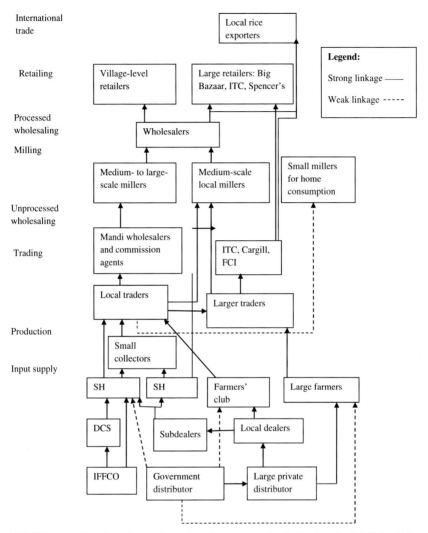

FIGURE 14.9   The rice value chain. *(Adapted from McCarthy, S., Singh, D.D., Schiff, H., 2008. Value Chain Analysis of Wheat and Rice in Uttar Pradesh, India. ACDI/VOCA for World Vision, Lucknow, India.)*

quality standards). Fig. 14.9 presents the members involved in the rice value chain in India. Input suppliers include major chemical companies, government distributors, small wholesalers/retailers, and even small retail shops that sell seed, fertilizer, and pesticide to farmers at the village level. Producers include small, medium, and large farmers. The spectrum of aggregators and traders handles rice quantities from small scale to large scale. Millers are involved in cleaning, polishing, and packaging according to standards required by major retailers and local exporters.

CONCLUSIONS AND POLICY SUGGESTIONS

India constitutes the major share of global rice exports, particularly basmati rice. Almost two-thirds of the basmati rice produced in India is exported and the compound growth rate of basmati rice exports from India (1990–91 to 2013–14) was ~11.3% in quantity and 13.7% in value per annum. For nonbasmati rice, the compound growth rate was a mere 4.47% in terms of value and 3.35% in terms of quantity. This is because the demand for Indian basmati rice is comparatively high in international markets when compared with that for nonbasmati rice. Even though the trend of rice exports in both basmati and nonbasmati is positive, the dominant characteristic appears to be wide fluctuations in exports, for which basmati rice yields, more value than nonbasmati rice. It is imperative to look for technological and institutional factors rather than mere price incentives to augment supplies and sustain exports of rice in the coming years. In addition, virtual water transfer through trade is becoming an important component of water management in India, where water is a scarce resource. Considering the sustained decline in oil prices and thereby the earnings of the Gulf countries (India's major export destinations), there is a need to analyze the impact of these developments on Indian rice exports and develop suitable strategies accordingly.

The changing patterns of the rice-growing states in India were estimated for the period of 1968–69 to 2012–13 using Markov chain analysis. It was evident that northern, southern, and eastern zones were stable rice-growing states compared with northeastern and western states. In the northern zone, Punjab, Haryana, and Uttar Pradesh were stable rice-growing states with a probability of retention of 0.9171, 0.7413, and 0.5043, respectively. In the eastern zone, Bihar, West Bengal, and Odisha were stable rice-growing states. It was interesting that Jammu and Kashmir and Goa had zero probability of retention of rice and were therefore not important rice-growing states. The shift in rice cultivation from south to east, where water resources are more abundant, is a good sign. Rice production in those areas can efficiently use the resources.

With regard to the changing direction of rice trade analyzed by the Markov chain framework, the top five importers of rice from India were selected. It is evident from the result that the United Arab Emirates and Iran are stable rice importers, having a probability of retention of 0.6025 and 0.5593, respectively. Bangladesh was seen as having the least probability of retention at 0.0062. However, the probability of retention among other countries was 0.5932, indicating reliable markets for Indian rice in minor markets.

The competitiveness in Indian rice analyzed through the PAM revealed that private profit was positive in the triennium period of 2011–12 to 2013–14 and, as such, rice cultivation in the country was profitable and was not affected by price fluctuations in other commodities because of globalization and the WTO. Social profitability analysis, for which costs and prices at the social optimum values were considered, also showed that rice cultivation was profitable in India even if India were to liberalize trade for rice even further. There is vast scope for augmenting the exports of superfine quality rice to many countries. However,

Indian exports are bogged down by quality problems. Efforts could be directed to further reduce production costs in rice cultivation through resource-saving technologies, which would lead to increased competitiveness. India should ensure that developed countries reduce their domestic support measures and tariff and nontariff barriers and to sustain exports India should remove a large number of domestic constraints. Indian rice is fairly price competitive and there is the potential to further increase competitive strength through concerted by-products.

The government should intervene to correct market failure and thus offset divergences, especially with respect to input supplies. For example, successful regulation of a monopoly would increase seller prices, cause private and social prices to equalize, and increase income for farmers. The second source of divergence is distorting government policy with regard to pricing and the cost of inputs. This prevents the efficient allocation of resources, leading to the inefficient realization of objectives and creating divergences. A tariff on rice imports, for example, could be imposed to raise farmers' income to foster the equity objective and increasing domestic rice production could protect the interests of farmers.

If the government enacted efficient policies to offset market failures and if the government were to decide to remove distorting policies, this would greatly streamline rice production in India. If these actions could be carried out, divergences would be removed or minimized.

Based on the results of decomposition of export gains or losses of Indian rice, even though world trade effects are positive, India might gain from its share of world demand if it is able to maintain its market share. When market distribution effects are negative, this infers that Indian exports are concentrated in a market where demand is growing slower than world demand. This should be changed so that export performance will improve. However, the competitiveness effect is 100%, which favors India's exports and the increase that has happened by maintaining its market share.

The rice value chain in India faces significant inefficiency at various stages of the chain. The traditional Indian agricultural value chain is crowded by many players and the layers of intermediaries who operate all along the chain result in weak farm–firm linkages (Ghosh et al., 2015). Hence, there is a need to redesign procurement, distribution, collaboration, and the logistics system. A framework that suggests a mechanism for tracing and visibility of inventory in the system, at each level of supply and the value chain, is essential in order to make Indian rice efficient and globally competitive.

REFERENCES

Acharya, S.S., Chand, P.R., Birthal, S.K., Negi, D.S., 2012. Market Integration and Price Transmission in India: A Case of Rice and Wheat With Special Reference to the World Food Crisis of 2007–08. FAO, Rome, Italy.

Akramov, K., Malek, M., 2012. Analyzing profitability of maize, rice, and soybean production in Ghana: results of PAM and DEA analysis. Ghana Strategy Support Program Working Paper 0028.

Balani, S., 2013. Functioning of the Public Distribution System. PRS Legislative Research, New Delhi.

Balassa, B., 1965. Trade liberalization and revealed comparative advantage. Manchester Sch. Econ. Soc. Stud. 33 (2), 99–122.

Coface, 2009. Risk assessment: India. Available from: www.coface.com/Economic-Studies-and-Country-Risks/India

Datta, S.K., 2001. Problems and prospects of India's rice trade in a WTO regime. In: Datta, S.K., Deodhar, S.Y. (Eds.), Implications of WTO Agreements for Indian Agriculture. Oxford IBH and Co., Delhi.

Davar, S.C., Singh, B., 2013. Competitiveness of major rice exporting nations. Indian J. Ind. Relat. 48 (3), 513.

DES (Directorate of Economics and Statistics), 2016. Department of Agriculture, Cooperation and Farmers' Welfare, Ministry of Agriculture and Farmers' Welfare, Government of India. Available from: http://eands.dacnet.nic.in/

FAO, 2013. Food and Agriculture Organization, Rome, Italy. Available from: http://faostat3.fao.org/home/E

Ghosh, M.D., Ray, A.K., Biswas, M.C., 2015. Linking small and marginal farmers with the growth of India. Int. J. Adv. Res. 3 (4), 1191–1193.

Gong, X., 2011. A study on trade of complementarity among Xinjiang and its neighboring countries. Asian Soc. Sci. 7 (1), 128–133.

Gulati, A., Hanson, J., Pursell, G., 1990. India—Effective Incentives in India's Agriculture: Cotton, Groundnuts, Wheat and Rice. No. 332. The World Bank, Washington, DC.

Haque, A., Sultana, S., Kedah, Z., Yasmin, F., Momen, A., 2014. Gaining of competitive advantage of Malaysian telecommunication products: measure of competitiveness. Int. Rev. Bus. Res. Pap. 10 (2), 27–45.

Haque, A., Yasmin, F., Anwar, N., Ibrahim, Z., 2013. Export of furniture product from Malaysia: market prospects and challenges. J. Econ. Behav. Stud. 5 (7), 406.

Hoda, A., Gulati, A., 2008. WTO Negotiations on Agriculture and Developing Countries. International Food Policy Research Institute, Washington, DC, Issue Brief 48.

Kumar, P., Rosegrant, W.M., 1996. Projections and policy implication of medium and long term rice supply and demand in India. Keynote paper presented to IRRI-NCAP-sponsored seminar on "Vision on India's Rice Trade" held in New Delhi, 25 April.

Leamer, E.E., Stern, R.M., 1976. Quantitative International Economics. Transaction Publishers, Piscataway, NJ.

Lee, T.C., Judge, G.G., Takayama, T., 1965. On estimating the transition probabilities of a Markov process. J. Farm Econ. 47 (3), 742–762.

Mahathanaseth, I., Pensupar, K., 2014. Thai agricultural policies: the rice pledging scheme. International Workshop on Collection of Relevant Agricultural Policy Information and Its Practical Use, Taipei, Taiwan.

Monke, E., Pearson, S., 1989. The Policy Analysis Matrix for Agricultural Development. Cornell University Press, Ithaca, NY.

Muhammad, N.M.N., Yaacob, H.C., 2009. Export competitiveness of Malaysian Electrical and Electronic (E&E) product: comparative study of China, Indonesia and Thailand. Int. J. Bus. Manag. 3 (7), 65.

Murthy, D.S., Subramanian, K.V., 1999. Onion exports markets and their stability for increasing India's exports: Markov chain approach. Agric. Econ. Res. Rev. 12 (2), 118–128.

Muthayya, S., Sugimoto, J.D., Montgomery, S., Maberly, G.F., 2014. An overview of global rice production, supply, trade, and consumption. Ann. NY Acad. Sci. 1324, 7–14.

Nagaraj, B.V., Krishnegowda, Y.T., 2015. Value chain analysis for derived products from paddy: a case of Karnataka state. Int. J. Manag. Value Supply Chains 6 (1), 47.

Pearson, S., Gotsch, C., Bahri, S., 2003. Applications of the policy analysis matrix in Indonesian agriculture. Available from: http://stanford.edu/group/FRI/indonesia/newregional/newbook.htm

Poramacom, N., 2002. Revealed comparative advantage (RCA) and constant market share model (CMS) on Thai natural rubber. Kasetsart J. Soc. Sci. 23, 54–60.

Rani, U., Reddy, G.P., Prasad, Y.E., Reddy, A.A., 2014. Competitiveness of major crops in post-WTO period in Andhra Pradesh. Indian J. Agric. Econ. 69 (1), 125–141.

Reddy, A.A., 2013. Training Manual on Value Chain Analysis of Dryland Agricultural Commodities. ICRISAT, Patancheru, India.

Souza, Â.R.L.D., Revillion, J.G.C., 2013. Policy analysis matrix applied to rice production in Mercosur. J. Agric. Policy 22 (1), 55–71.

Wailes, E.J., 2005. Rice: global trade, protectionist policies, and the impact of trade liberalization. In: Aksoy, M.A., Beghin, J.C. (Eds.), Global Agricultural Trade and Developing Countries. The World Bank, Washington, DC, pp. 177–194.

Chapter 15

Extension Policy Reforms in India: Implications for Rice Production Systems

Suresh Chandra Babu
International Food Policy Research Institute, Washington, DC, United States

INTRODUCTION

The agricultural sector is a key contributor to national income in most developing countries and it often employs a major proportion of the population. Hence, for policymakers, the productivity of the agricultural sector is of great importance. Among the many factors that contribute to the productivity of the agricultural sector, assistance in the form of extension services is well recognized. The provision of extension includes several services, including exposing farmers to new methods to increase yields, ensuring the delivery of good-quality inputs, assisting in the adoption of climate-smart agricultural practices, exposing farmers to techniques for postharvest protection of produce, providing ways to reduce information asymmetry in the market, and training farmers in techniques to build resilience (Babu et al., 2013). The aims of extension services are to increase the productivity of the entire food system and improve yields, income, and welfare of farmers by translating research results into tangible gains. The recent definition of extension services by the Global Forum for Rural Advisory Services (GFRAS) and the Food and Agriculture Organization (FAO) of the United Nations encompasses a broad set of activities that go beyond production agriculture and include strengthening technical, organizational, and managerial skills of farmers and connecting research, education, and extension systems (Christoplos et al., 2012). In the context of India, extension systems have played an important role in realizing agricultural productivity gains. This was particularly true at the time of the Green Revolution, during which extension workers played a key role in conducting field demonstrations of high-yielding varieties of seeds and ensuring timely delivery of inputs to farmers (Glendenning et al., 2010).

Rice is one of the most important crops in India and it is currently cultivated on 44 million ha. In fact, India has the largest rice-harvesting area in the world.

The Future Rice Strategy for India. http://dx.doi.org/10.1016/B978-0-12-805374-4.00015-4

Rice cultivation is also the primary source of income and employment for more than 50 million households.[a] Approximately 65% of the total population in India eats rice; therefore increasing the productivity of rice farmers is essential for both improving livelihoods and achieving food security. Apart from its economic importance, rice is deeply ingrained in the tradition and culture of India, being used in religious ceremonies, marriages, and on special occasions.

Increasing rice production has been a priority for policymakers due to the central importance of the rice crop to Indian life. The technology- and knowledge-intensive nature of rice has meant that agroeconomic practices have continuously changed. Farmers require different informational support preseason, during the season, and postseason depending on the agricultural zone in which they cultivate. In some areas rice farmers face the problem of drought, whereas in others they face the problem of flooding; hence the development of proper countrywide rice extension systems is important. The rice extension system is currently of a pluralistic nature with provision from the public sector, private sector, and civil society. However, despite the existence of different extension systems, rice farmers continue to use outdated production methods, particularly in remote areas. Farmers cultivating rice still remain vulnerable to sudden weather changes and continue to be at the mercy of moneylenders and middlemen. A well-functioning extension system can strongly help overcome these challenges. It is therefore important to understand the current issues and challenges facing rice producers before suggesting changes to improve the efficiency of the rice extension system in India.

This section "Introduction" reviews extension systems in the rice-growing regions of India and identifies challenges and constraints in linking research to farmers' needs. The section "Conceptual Framework" presents the conceptual framework used to analyze the rice extension system in India. The section "Current Challenges in the Rice Extension System" presents the evolution of extension reforms in the country with regard to rice production. The section "Private Extension Services" uses examples of major extension programs to highlight challenges that public, private, and nonprofit providers of extension systems face. The section "Conclusions and Policy Implications" presents the conclusions of the study and identifies the policy implications of extension reforms with reference to rice producers.

CONCEPTUAL FRAMEWORK

This chapter traces the evolution of extension reforms in the rice-growing areas of India using the conceptual framework presented in this section. The framework acts as an anchor for the analysis and breaks down complex elements of extension provision, studying them in detail to understand their suitability in a given context. It is similar to approaches followed in Birner and Anderson (2007), Birner et al. (2009), Ekboir et al. (2009), and Swanson and Rajalahti

---

a. http://ricestat.irri.org:8080/wrs2/entrypoint.htm

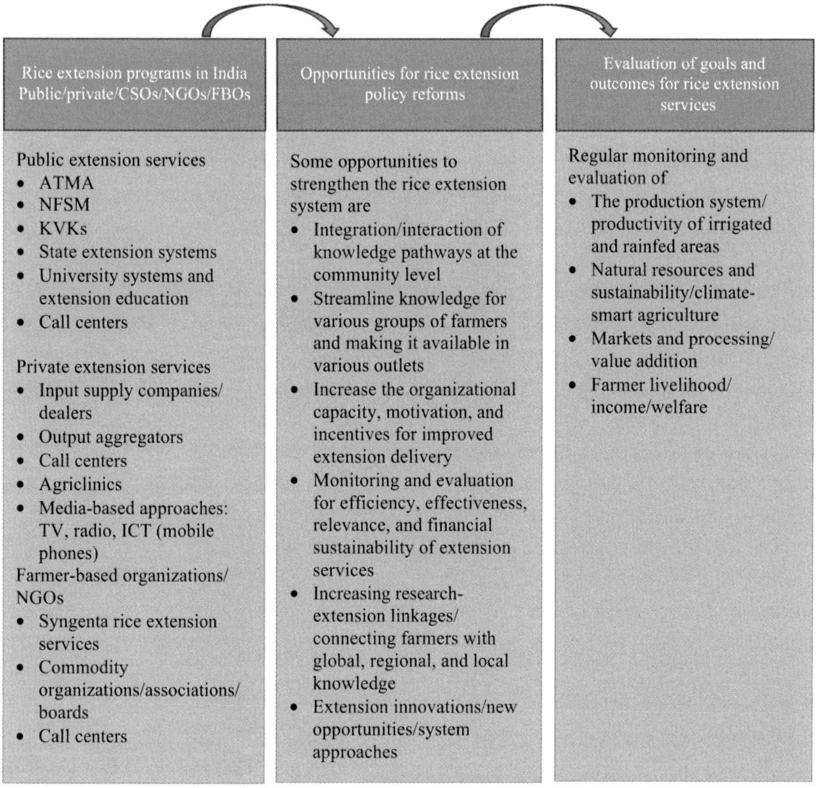

Rice extension programs in India Public/private/CSOs/NGOs/FBOs	Opportunities for rice extension policy reforms	Evaluation of goals and outcomes for rice extension services
Public extension services • ATMA • NFSM • KVKs • State extension systems • University systems and extension education • Call centers Private extension services • Input supply companies/ dealers • Output aggregators • Call centers • Agriclinics • Media-based approaches: TV, radio, ICT (mobile phones) Farmer-based organizations/ NGOs • Syngenta rice extension services • Commodity organizations/associations/ boards • Call centers	Some opportunities to strengthen the rice extension system are • Integration/interaction of knowledge pathways at the community level • Streamline knowledge for various groups of farmers and making it available in various outlets • Increase the organizational capacity, motivation, and incentives for improved extension delivery • Monitoring and evaluation for efficiency, effectiveness, relevance, and financial sustainability of extension services • Increasing research-extension linkages/ connecting farmers with global, regional, and local knowledge • Extension innovations/new opportunities/system approaches	Regular monitoring and evaluation of • The production system/ productivity of irrigated and rainfed areas • Natural resources and sustainability/climate-smart agriculture • Markets and processing/ value addition • Farmer livelihood/ income/welfare

FIGURE 15.1 Tracing pathways to assess rice extension policy reforms in India. *ATMA*, Agriculture Technology Management Agency; *ICT*, information and communication technology; *KVK*, Krishi Vigyan Kendras; *NSFM*, National Food Security Mission. *(Adapted from Babu, S.C., Joshi, P.K., 2015. Future of agricultural extension reforms in developing countries: lessons from India. e-J. Econ. Complex. 1(1). Available from: www.cespi.it/E-journal/Babu&Joshi-June-2015.pdf)*

(2010). The conceptual framework used is presented in Fig. 15.1, and has been adapted from Babu and Joshi (2015).

Reviewing the services that currently exist in extension provision is a good starting point to understand the pathways of rice extension services in India. It is important to look at the various knowledge pathways through which rice farmers receive information related to production and distribution of their crops. It is also worth noting that the kinds of information and advisory services needed may vary before, during, and after the rice production season. Some extension services provide information and advisory services on production technology and others provide information on markets or prices. Keeping this in mind, the first column in Fig. 15.1 shows the different members of the rice extension service provision system in India. This helps to understand the role of different service providers and the sources of information available to rice farmers. It is important to examine the current system to identify the challenges that it is facing. For instance, extension services in rice production have seen a rise in private service providers since the early 1990s. This changes how services are provided, who has access to them, and the funding channels that make service provision financially sustainable. Similarly, the rise in popularity of information and communication technology (ICT) tools has changed the capacity development needs that are required for maximum use of available resources.

Understanding the existing services helps to identify the current gaps in rice extension. This leads to the middle column, which lays out the opportunities and potential strategies for extension service provision in rice cultivation. It is important to look at the different opportunities to formulate a strategic pathway to complement and synergize efforts in agricultural extension services (Babu and Joshi, 2015). For instance, integration of various knowledge pathways at the community level helps to tailor the information disseminated to farmers' needs and maximize the benefits of information outlets. Similarly, creating and improving knowledge-sharing platforms, such as the Rice Knowledge Management platform, to strengthen links between extension researchers and farmers helps to test and increase the impact of extension research.

Lastly, monitoring and evaluation to track progress toward goals and outcomes, and to provide feedback on the current extension programs, need to improve. There needs to be a rigorous framework to regularly monitor rice extension systems with a wide array of indicators and reliable data sources on issues such as the productivity of rice farmers, their livelihood sustainability, and the adoption of climate-smart agricultural practices. It is important for evaluation exercises to move beyond merely being an implementation tool to being a credible source for tracking progress and providing evidence-based feedback to existing programs (Babu and Joshi, 2015).

This section has provided the broad anchoring framework that will be used throughout the chapter to analyze the implications of rice extension reforms in India. The next section traces the evolution of extension reforms for rice producers, giving a brief historical account.

EVOLUTION OF EXTENSION REFORMS IN INDIA

Extension reforms in India span more than 6 decades and this section aims to provide a historical background on the evolution of reforms in rice-growing areas over this time period. Extension systems in India traditionally only conducted village-level crop demonstrations. These demonstrations used state-funded resources and provided very limited information to rice farmers in a particular agroecological zone. Table 15.1 gives a snapshot of the evolution timeline, marking this as the first phase in extension and future developments thereafter.

With the advent of the Green Revolution, information demands of farmers increased because of the introduction of new technology and inputs for rice production. Extension services during this period focused on conducting demonstrations of high-yielding variety (HYV) seeds, irrigation, price support, and ensuring the

TABLE 15.1 Timeline of Rice Extension in India

Timeline of extension reforms in rice extension

Broad categories of extension	Traditional extension	Green Revolution extension	T&V system	Pilot ATMA + agriclinics	ATMA rollout and private extension	Pluralistic extension system
Specific description	Crop demonstrations in villages, limited linkage between research and extension	Demonstrations of HYV seeds, irrigation, price support, and timely delivery of subsidized inputs; established the role of adaptive trials	Frequent training and visits of extension workers to selected farmers; technical training and updating sessions for extension workers	Customized extension services, encouraging private-sector participation	Integrate various extension services, link research with extension provision, decentralize decision-making	Provision of extension services by public, private, and NGO sector
Broad timeframe	1947–65	1965–80	1980–2000	2000–05	2005–11	2011–present
Phases of reforms	Phase 1	Phase 2	Phase 3	Phase 4	Phase 5	Phase 6

HYV, High-yielding variety; T&V, training and visit.

timely delivery of subsidized inputs, such as quality seeds, fertilizers, and pesticides. Extension services were largely provisioned by the public sector during this phase and were fairly standardized in different regions. Inputs were handed over to farmers through state agricultural depots and subsidies were disbursed through various agricultural development programs. Even though this was successful at the time, by the late 1970s such extension services had reached their limit and there was a need to move beyond increasing the adoption of HYVs.

Sustaining the large rice agricultural extension system without contributing to increases in rice productivity was a big challenge for the government. This situation demanded a revamp of the existing system; hence with the help of World Bank funding, a new training and visit (T&V) system was piloted in the state of Rajasthan in 1974 (Ameur, 1994). The program involved extension workers regularly visiting selected farms to conduct training and was also accompanied by technical training sessions for extension workers to keep them updated (Feder et al., 1987). The program was scaled up and introduced in other states by 1977. Although evaluation of the program showed favorable results, its long-term sustainability remained a serious question because of its high staffing and funding requirements (Anderson and Feder, 2004). The system was effective in high-potential irrigated areas, but had little effect on the productivity and incomes of farmers in rainfed areas (Singh et al., 2013). The T&V system continued until the early 1990s when the World Bank funding ended, after which the government stopped the recruitment of new staff. The T&V system, which was plagued with low-quality staff, insufficient infrastructure, and low funding, further deteriorated (Anderson et al., 2006). During this phase, extension services for the rice crop were provided primarily by the State Department of Agriculture through its state, district, and block-level arms, as other ministries (i.e., Ministry of Soil and Water Conservation), research centers, and agricultural universities played a very limited role. This phase did see a rise in informal private extension providers.

Looking at these underused avenues for extension provision, the World Bank started the National Agricultural Technology Project (NATP), under which it introduced the Agricultural Technology Management Agency (ATMA). The idea was to integrate extension programs across departments, link research and extension activities within each district, and decentralize decision-making. The aim of this was to increase the involvement of farmers and the private sector at the block and district level. With support from the World Bank, a pilot study was conducted from 1998 to 2003 across 28 districts in 7 states. The pilot was successful, mainly because of the Bank's funding, and it became popular as one could track the district-level implementation of the project online. The availability of funds from the central government and external pressure led to the scaling up of the ATMA. In 2002 the government helped to form agribusiness and agriclinic centers, which provided support to trained agricultural graduates, allowing them to set up centers, rent and sell farm inputs, and provide advisory services in the form of soil testing and pest management. They were also able to provide information on weather forecasts, prices, and credit in these centers. This was the first time in the history of extension that there was an institutionalization of the provision of customized extension services. This initiative was particularly useful for rice producers because of the introduction of crop-specific services that were tailored to the farmer's needs.

In 2007 the ATMA program was implemented throughout the country and in 2010 revised guidelines were issued to incorporate lessons learned so far. However, the ATMA still faces several capacity and institutional constraints (Babu and Glendenning, 2011). Even after 5 years of scaling up, with limited evaluation, it is not able to encourage innovation by researchers in the public sector and has not been able to establish strong links between farmers and researchers. This is part of the reason for the low productivity and capacity of rice producers in India, particularly smallholder rice farmers who rely on government channels of extension provision.

Several private organizations, NGOs, and farmer associations have emerged on the extension system landscape since the establishment of the ATMA, making it a pluralistic extension system. This change has led to the emergence of new challenges for the system, especially for rice production. Fig. 15.2 summarizes the transitions in extension reforms and shows where Indian rice extension services stand (Babu et al., 2015). In the figure the horizontal axis represents the

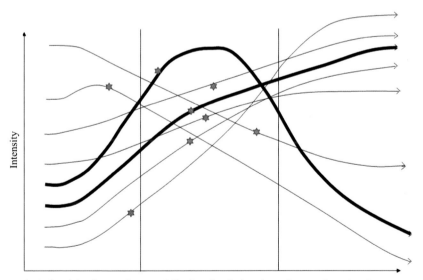

Agricultural development/commercialization of agriculture/extension transformation ✳

FIGURE 15.2   Pluralistic extension in rice and the status of India. *(Adapted from Babu, S.C., Ramesh, F N., Shaw, C., 2015. The current status and role of private extension: a literature review and conceptual framework. In: Zhou, Y., Babu, S.C., (Eds.), Knowledge Driven Development: Private Extension and Global Lessons. Academic Press, London. Available from: http://ifpri.worldcat. org/oclc/911268908)*

level of agricultural development and extension transformation with regard to rice production. This axis has three segments showing low, medium, and high progress in agricultural development of the country. The vertical axis shows the intensity of different extension initiatives for rice production. Furthermore, the stars indicate the position of the Indian rice production and extension system and the level of the extension services on this continuum. For most extension systems, India currently has low-to-medium efficiency and effectiveness, a decreasing role of the public sector, an increasing role of private extension providers, and greater use of ICT in extension provision. For example, the role of farmer-based organizations is still at its nascent level for rice extension, but has shown some promise in organic basmati rice production organized under contract farming systems (Nanda et al., 2015).

The next section will provide details of the specific extension services that currently exist. It describes the existing reform and extension programs in the context of rice cultivation, further presents the challenges facing different extension systems, and suggests ways to overcome them.

CURRENT CHALLENGES IN THE RICE EXTENSION SYSTEM

There have been a number of efforts in the provision of extension services in India since independence. This section describes the existing programs and channels through which farmers receive information. The first subsection describes the public extension systems in India and their effects on rice producers. The next subsection describes private extension systems, explaining a few cases of private extension providers that are continuing to gain popularity. Finally, the last subsection covers extension provision by farmer associations. The challenges facing each of the programs and the different ways to overcome them are presented from the perspective of the rice-producing areas.

Public Extension Services

Public extension services have been very popular in India; in fact, a major portion of extension service provision has come from the government. Even though the public extension system has a wider reach and tends to be more inclusive, it suffers from several organizational and institutional capacity constraints.

Agricultural Technology Management Agency

The ATMA is currently the flagship program for providing extension services in India. It exists at the district level where it directs extension activities based on a Strategic Research and Extension Plan (SREP) prepared using the participatory rural appraisal (PRA) technique by each district. The ATMA project director chairs the ATMA management committee (AMC), which is responsible for coordinating the extension activities in the district. The AMC includes the heads of all line departments and research organizations in the district. At the block level, the Farm Information and Advisory Centre (FIAC) is the primary physical platform where the Block Technology Team (BTT) and Farmer Advisory Committee (FAC) meet to prepare the Block Action Plan (BAP) and implement extension activities. At the state level, an Interdepartmental Working Group (IDWG) formulates a State Extension Work Plan (SEWP) to consolidate the district SREPs. ATMA is yet to establish itself as an autonomous institution at the district level. In 2010 a block to village extension link was formally institutionalized through the concept of a "farmer friend" (FF) for every two villages. A FF is a progressive farmer who has the minimum qualification of a pass in matriculation or intermediate examination and is directly engaged by the block technology manager. Overall, the ATMA provides a decentralized bottom-up structure for the provision of extension services.

ATMA has been successful in bringing recognition to the importance of extension systems in India, which can be seen through the increased funding and human resources. For the case of rice cultivation, ATMA has been able to expand the range of extension activities and be more responsive to various stakeholders, thus lending credibility to the system of extension provision. Furthermore, ATMA has managed to achieve convergence among different programs implemented by the Department of Agriculture (DoA) and improve the relationship of the DoA with other line departments (i.e., agricultural engineering, animal husbandry, and fisheries). This has been particularly beneficial for rice farmers engaging in income-earning activities beyond rice cultivation. Through its bottom-up model, ATMA has expanded the reach of public extension systems to rural areas.

Variations in leadership and commitment are the main factors that drive differences in ATMA's impact. Ensuring that staff positions are filled at all times and the provision of adequate incentives to retain staff are common problems faced by ATMA. The linkages with Krishi Vigyan Kendras (KVKs, Farm Science Centers) need to be proactively strengthened to make better use of the existing research capacity. Delays often occur in the release of funds from the center to the states and this affects the implementation of the SREP and SEWP. Some of the central government schemes do not have sufficient resources for knowledge support, hence presenting an untapped opportunity for ATMA to fill the knowledge gap. ATMA also needs to develop collaborative crop-specific research programs to bring increased focus onto the productivity of important crops, such as rice. Lastly, farmers' influence on decision-making at the block level remains low despite the FF initiative and therefore capacity building of the village-level representatives and farmer interest groups needs more investment.

National Food Security Mission

In 2007 the Government of India launched the National Food Security Mission (NFSM) to increase the production of rice, wheat, and pulses. The aim was to increase rice production by 10 million tons by 2011–12, which was later increased by another 10 million tons for the next 5 years. In coordination with the Directorate of Rice (an agency established to popularize the use of HYV seeds through training and demonstrations at the time of the Green Revolution), the policy included several reforms and subsidies for rice farmers. Some measures included demonstrations on cropping systems; distribution of HYV and hybrid seeds; subsidies for farm machinery such as power weeders, seed drills, and laser land levelers; subsidies for plant protection; subsidies for soil micronutrients; and subsidies for local initiatives that helped to increase farmer productivity (NFSM, 2015).

A midterm evaluation of the NFSM project suggested that productivity gains remained limited (AFC India, 2014). The scope to increase the area under rice cultivation remains limited; therefore it is all the more important to increase productivity per unit area. However, rice yield also recorded an overall negative growth in the first 3 years of the program and the productivity gains that were observed were mainly due to adoption of the system of rice intensification (SRI) practices and hybrid seeds (AFC India, 2014). Although the private sector has been playing an important role in reducing the knowledge gap, it is important to encourage greater synergy between public and private initiatives. Over the past 5 years, there has been an increase in the productivity of rice, adoption of subsidized inputs, and capacity of farmers through farmer field schools (FFS) (AFC India, 2014). However, productivity remains uneven across districts with the real potential of the program still to be realized.

State Extension System

Agricultural extension reforms at the state level are implemented by the State Department of Agriculture (SDoA) through extension staff at the block and district levels. The specific programs vary across states but have some common characteristics. Most of the programs are a continuation of the T&V system of extension provision; thus demonstrations remain a key medium through which information needs of farmers are met. These demonstrations encourage farmers to use equipment and inputs from only a few vendors; thus they are not exposed to different options. In addition, the information disseminated by extension workers remains restricted to crop production (Sulaiman and van den Ban, 2003). Although this is good for rice production, to increase productivity extension services need to expand their focus to include information on postharvest methods. The state departments also continue to function under capacity in terms of human resources, with no new recruitment since 1998. This overburdens the existing staff and prevents them from increasing their capacity and updating their knowledge (20% of staff are university graduates). The ratio of staff to farmers varies widely across the country (1:300 in Kerala to 1:2000 in Rajasthan) and is particularly low in remote areas (Sulaiman and Holt, 2002). The links between the Indian Council of Agricultural Research (ICAR)—an important information source for extension workers—and the state departments remain weak. Monitoring of the staff is top-down and there is no scope for farmers to provide feedback, thus keeping the accountability structure of extension staff vague.

The state extension system has changed little over time and has several overlaps with existing reforms and initiatives. This is particularly true for extension systems for rice producers. It is important for the government to examine its functioning and overhaul the system in a systematic manner with a focus on the adoption of new technologies, building capacity of staff, and building linkages with other initiatives with an aim to increase farmer productivity.

Mass Media Approaches

In 2000 the policy for agricultural extension articulated the scope of mass media and ICT in delivering extension services (MAI, 2015). The government currently employs a variety of information dissemination methods such as public radio, television channels, call centers, and the Internet to overcome the poor coverage of extension services in the country.

Information dissemination using radio is through All India Radio, a state-run channel that broadcasts half an hour of agricultural programs providing information relevant to the agroecological zone in question. The government also launched a Kisan channel on Doordarshan (state-run television channel) telecasting agricultural programs. Even though these schemes help to provide extension services, they cannot incorporate feedback or address specific queries of farmers. Rice farmers also experience different problems in different agroecological zones; hence standard information is not useful. Moreover, rice farmers often have to spend time listening to the whole program to come to the information most relevant to them.

In 2004 the Government of India established Kisan (farmers) call centers as a method of extension service provision in the form of on-demand information to farmers. Under the initiative the government established a toll-free number on which farmers could receive expert advice on their queries. This initiative was started because of the high telephone and mobile penetration in rural areas (Mittal, 2016). The experts that answer farmer queries are agricultural graduates and they are supported by a second level of operators who answer general queries. However, there is a lack of awareness of agricultural practices at the level of the operators and the linkages between the experts and operators are weak (Glendenning et al., 2010). The service is available in 21 local languages, but the provision of localized and specific information has been problematic (Glendenning et al., 2010). There has been considerable improvement in the provision of services through these centers (GOI, 2012); however, the information is still of limited use to rice farmers. Even though the number of sources through which farmers receive information has increased, the effect of mass media on the productivity and income of rice farmers is not known because of limited evaluation. A crop-specific evaluation would be helpful in informing ways to maximize the use of mass media techniques.

Agriclinics

Agriclinics make advisory services available to farmers as an alternative to information provided by private input dealers. This is done by supporting trained agricultural graduates to set up their own agriclinic laboratories. Agriclinics started in 2002 as part of the Indian national program of Agribusiness and Agriclinic Centers. Tamil Nadu, a rice-producing state in India, has agriclinics providing several services, including soil testing to inform on fertilizer application, testing of irrigation water quality, advisory services for farming practices, credit access, pest and disease management, crop insurance, and marketing (Tamil Nadu Agriculture Department, 2008a,b). As these are services that the farmers pay for, operators of the agriclinics have to ensure that they identify farmer needs correctly and provide good-quality, reliable information.

Currently, 19,105 ventures are established across the country; however, the number of agriclinics within rice-producing states varies greatly.[b] Many studies have evaluated the effect of agriclinics on meeting farmers' informational needs and increasing their productivity and income. In 2008 a midterm evaluation reported that the farmers were "very satisfied" with the services provided (Global AgriSystem, 2008), while another study showed that agrientrepreneurs frequently visited farmers and were successful in persuading them to adopt improved technologies (Chandra Shekara and Kanaka Durga, 2007).

However, the uptake of these advisory services depends on farmers' demand and willingness to pay. A study into agriclinics that provide soil-testing services in Tamil Nadu found that the clinics often struggled to find clients. However, once farmers were aware of the potential benefits of this service, they were willing to pay for it. A system of targeted subsidies to operators to increase engagement with rice farmers could be a way to overcome this problem. The study also found that the extension service was used less by smallholder farmers and those from marginalized groups while membership of a farmer-based organization (FBO) increased the likelihood of using paid advisory services. Creating ways to promote the formation of such groups in rice-producing areas would be an important step in increasing the uptake of services in agriclinics. It is also important to build the capacity of agriclinic operators so that they can make information available to farmers. In addition, to maintain quality standards, a reliable system for monitoring the standards of service delivery should be created (Glendenning et al., 2010). The results of this study in Tamil Nadu are applicable to other rice-producing areas of India that face similar problems in the provision of agriclinic services.

Private Extension Services

The public sector has not been able to completely meet the information demands of farmers and therefore private players have emerged in extension provision. The private sector has played an important role in catering to the needs

---

b. www.agriclinics.net/QuerySheet.asp

of different farmer groups, providing customized advisory and support services. This has proved to be particularly beneficial for large-holding rice farmers who can often afford private services. However, since these services come at a cost, they are exclusive and often overlook the needs of smallholder rice farmers.

System of Rice Intensification

SRI is a unique technique to increase rice productivity by changing the management of plants, soil, water, and nutrients. SRI practices are built around certain principles, such as planting young seedlings to enhance plant growth potential, providing growing plants with sufficient water, spacing hills widely so that roots and canopy have room to grow, and promoting active soil aeration. Hence, SRI is more of a traditional approach to increasing productivity.[c]

Although researchers are divided on the benefits of SRI, the number of farmers using this method has increased and it is now practiced in several states in India. However, in some areas farmers abandoned the innovation after a few years of using it. According to Babu and Jayachandran (2014), knowledge inequality is seen as a major cause of this phenomenon. Extension services play an important role in bridging the knowledge gaps in SRI, but the practices in this method are quite labor intensive and require high knowledge and skills of both farmers and extension workers. Table 15.2 shows the knowledge challenge and knowledge intensity for various packages of practices identified under SRI.

In countries such as India where extension systems often function below capacity, adoption of practices such as SRI at scale is quite challenging. Adusumilli and Laxmi (2011) challenge the fact that SRI is more labor intensive, claiming that SRI actually requires only an initial investment of time to learn new skills and better managerial skills. According to their study, labor costs of farmers decrease substantially (43%) upon adoption of SRI practices. Their evaluation of SRI practices adopted in Andhra Pradesh shows a 90% reduction in seed cost and 52% reduction in field water use.

To realize the gains from the adoption of SRI, additional investments need to be made in the rice extension system. It is important to conduct further evaluation and cost–benefit analysis to test the productivity gains of this method in different rice-producing areas.

Information and Communication Technology Applications

The use of ICT applications for argoadvisory services has expanded greatly since 2007, mainly due to the high penetration of mobile phones in rural areas, especially from private-sector providers. ICT initiatives are transforming the traditional agricultural systems using methods such as community radio, SMS, voice-based messaging, and Internet kiosks (Mittal et al., 2010). As mobile phone penetration is very high in rice-producing areas of India, ICT applications are particularly relevant for rice extension.

---

c. http://sri.cals.cornell.edu/aboutsri/methods/index.html

TABLE 15.2 The System of Rice Intensification: Principles, Practices, and Knowledge Intensity

Serial number	Principles of SRI	Actual practices	Knowledge challenges	Knowledge intensities (rated from 1 to 5)
1	Very young seedlings should be used to preserve the plant's inherent growth potential for rooting and tilling	8- to 15-day-old seedlings with three leaves are grown in a raised-bed nursery	Moving from traditional broadcast method of nursery beds to row sowing of seeds	Raised nursery bed, maintenance of single seedlings, and keeping track of age of the nursery requires knowledge intensity of 3
2	Transplanting a single seedling per hill should be done quickly, carefully, shallowly, and skillfully to avoid any trauma to the roots, which are the key to plants' success	Single seedlings are planted with a minimum time interval between the time they are taken out from the nursery and planted carefully at a shallow depth (1–2 cm)	Teaching farm laborers about trauma-free transplanting requires knowledge of skill transfer on-farm; lead farmers need to train their labor force	Nursery needs to be raised next to the main field; multiple nurseries are needed to avoid time delay; highly skilled labor force needed to maintain the depth of planting requires knowledge intensity of 4
3	Reduce the plant population radically by spacing hills widely and squarely so that both the roots and canopy have room to grow and can have greater access to nutrients, sunlight, etc.	Planting in grids of either 20 × 20 or 25 × 25 cm² (or 30 × 30 cm² or even wider if the soil is very fertile) using a rope or roller marker to achieve precise interplant distances (to facilitate intercultivation)	Change from random approximate bunchy planting to single seedling planting with precise grid distance maintenance	Knowing which type of soil requires what type of grid spacing; maintenance of row spacing for intercultivation requires more labor and skill guided by supervision; requires knowledge intensity of 3
4	Provide growing plants with sufficient water to meet the needs of roots, shoots, and soil biota, but never in excess so that the roots do not suffocate and degenerate	Up to panicle initiation: irrigate to 2.5-cm depth after the water ponded earlier disappears and hairline cracks are formed on the soil surface (heavy clay soils should not be permitted to reach the cracking stage, but still are issued less water than with usual flooding); after panicle initiation: irrigate to a depth of 2.5 cm, 1 day after the water ponded earlier disappears	Grading the soil moisture and water required by farmer/laborers, control of water movement among farmers, varying sowing and planting dates, constant monitoring of water levels in flooded zones; uncertainty of rainfall, electricity availability, and labor availability can reduce water control	Knowledge on water needs in different cropping stages; measuring moisture levels to protect crop from wilting and from overirrigation is simply not part of extension or farm level 4 practice; this is not only new, but the adoption of water management as prescribed by SRI is considered high knowledge intensity activity at 4

5	Active soil aeration improves rice crop growth by benefitting both roots and beneficial aerobic soil organisms	Intercultivation with a mechanical weeder at intervals of 10–12 days, starting 10–12 days after transplanting and continuing until the canopy closes, passing between the rows, and making perpendicular passes across the field	Knowledge and use of mechanical weeder; interaction of weeds and crop depends on the level of water; effective use of mechanical weeder depends on soil and water properties	Weeding and weed management are an additional requirement that is a result of principle 4 recommended previously; weeding is also recommended as a practice that can help in soil aeration. This is again a high knowledge-intensive activity, as weeds are usually killed in the traditional system by submerging them under water
6	Augmenting organic matter in soils as much as possible improves the performance of the rice crop by improving soil structure and functioning and supporting beneficial soil organisms	Application of cattle manure, green manure, biofertilizers, and vermicompost is recommended; chemical fertilizer can be used, but it does not have the same beneficial effects on soil systems	More than knowledge, availability of biomass is a constraint; substituting chemical fertilizers requires soil testing and knowledge to use soil-testing results	Knowledge of what to apply as nutrients and in what form requires understanding of what is in the soil. Soil-testing infrastructure remains poor; even if the results are available and recommendations are known, availability of inputs in time is critical for the effective use of knowledge; farmers need knowledge intensity of at least 5 when using optional rates and choosing the combination of organic and inorganic fertilizers

Source: Columns 1 and 2 from system of rice intensification (SRI) brochures, Uphoff, N., 2011. The system of rice intensification: an alternative civil society innovation. Technol. Assess. Theory. Prac. 20(2), 45–51; columns 3 and 4 from Babu, S., Jayachandran, C., 2014. The role of knowledge in agricultural entrepreneurship, innovation, and transformation in South India. Paper Presented at the International Business Conference on Knowledge Inequality, Rome, Italy.

Since 2007, none of the old ICT initiatives have been replaced by new ones; instead, they have been added onto the existing extension systems (Mittal, 2016) with some being more successful than others. Mittal (2016) describes the basic parameters for successful ICT programs as efficiency, relevance, and actionability. This means that information received through ICT applications should be concise and give all the information to farmers in a timely manner; address farmers' information needs; and should provide a course of action for the problem faced. In these three respects, most current ICT applications have fallen short. For instance, rice farmers in different parts of the country face different problems at any given moment. It is important for information to be customized to make the content useful for farmers in different areas and this will require improvements in extension system capacity with increased accountability for results. Additionally, there is a lack of synergy between the various existing initiatives. ICT initiatives also face a problem of trust because most farmers take face-to-face interaction more seriously. Poor connectivity in remote areas also reduces the inclusivity of these ICT applications. Some of the rice-producing states in India are among the areas with the poorest Internet or mobile connectivity, but these areas require the support of extension services the most. Connectivity problems therefore limit the scope of ICT extension in these areas.

In 2000 Indian Tobacco Company (ITC) launched the e-Choupal initiative with an aim to remove intermediaries and reduce transaction costs (Annamalai and Rao, 2003; Bowonder et al., 2007). This was done by creating kiosks with computers and Internet access in villages to provide farmers with information on local weather, agricultural best practices, feedback on quality of crops, and soil tests. Even though the magnitude and cause of income increase are not known, farmers in villages with e-Choupals saw an increase in productivity and a decrease in transaction costs (Bowonder et al., 2007). e-Choupals were initiated to reduce procurement costs of products directly purchased by ITC to export, such as coffee, soy, and wheat (Annamalai and Rao, 2003; Upton and Fuller, 2004); hence, they have had little benefit for rice farmers. However, e-Choupals still provide a good example of a private extension initiative that can potentially reduce market asymmetries and increase the productivity of rice farmers.

Other Private Rice Extension Services

Rice farmers are using several other private extension services in India. For instance, Syngenta is a unique agribusiness venture that provides farmers with several support services to help increase productivity. It helps farmers by providing end-to-end solutions for rice production, whereby farmers can hire Syngenta to take care of selecting inputs, sowing seeds, and harvesting the rice crop. The rice farmers only have to ensure that the water level is correct. This is a useful service as it frees up rice farmers for other income-generating activities; however, such extension services can be used only by rice farmers who can afford them.

Similarly, Sarveshwar Organic Foods Limited has conducted several training and awareness generating programs to help rice farmers improve quality and increase income through exports. A recent study in Jammu and Kashmir shows how the organization was successful in implementing extension services to increase the productivity and income of farmers (Nanda et al., 2015). The study points out that organizing farmers with a clear goal and proper training can help in realizing these outcomes. The government should promote such training initiatives by private-sector players for other varieties of rice.

Farmer Associations

Despite the number of extension service initiatives, there remains a large number of smallholder farmers whose demand for information remains unmet. The poorly functioning and understaffed public extension system often excludes this category of farmers who often reside in remote areas. Due to their low incomes, smallholder rice farmers cannot afford private extension services; hence, the emergence of farmer-based organizations (FBOs) and self-help groups (SHGs) as important vehicles of extension provision for them. These groups use a participatory method of extension provision that makes it easier to incorporate farmers' feedback and the main information sources are progressive farmers within that area. However, the formation of FBOs and SHGs is very limited throughout the country. Feder et al. (2010) identified some problems with these groups, such as the limited availability of competent service providers, deep-seated cultural attitudes that prevent effective empowerment of farmers, and difficulties in implementing farmer control of service-provider contracts. Despite these reasons, farmer associations are a low-cost and effective medium of information dissemination that can be used to support existing government and private initiatives in reaching end users more effectively. The southern rice-producing states have more groups than the other states in India (Birner and Anderson, 2007). The government, private sector, and NGOs should facilitate the formation of such groups and invest in building their capacity to provide quality information to farmers by building strong links with ATMA and KVKs.

Table 15.3 summarizes the findings from the earlier analyses by presenting the scope of various extension services and their ability to help rice farmers overcome problems. Some of these problems are small-scale production leading to high production costs, shortage of manual labor, ineffective natural resource management, asymmetry in markets and prices, difficulty in making correct varietal choices, and lack of timely availability of inputs. The extension services are ranked from being most applicable (represented by four asterisks) to being least applicable (represented by one asterisk).

As presented in Table 15.3, most extension services have the scope to help those farmers who are producing on a small scale. Problems of a shortage in labor supply are not addressed by most extension systems, although selected collective approaches to combine resources have shown some promise. Natural

TABLE 15.3 Application of Extension to Emerging Challenges in Rice Farming Systems

Type of extension systems	Problems in rice cultivation					
	Small-scale agriculture	Shortage of labor	Natural resource management	Markets and prices	Varietal choices for specific agrozones	Timely availability of inputs
ATMA	****	*	****	*	***	**
NFSM	****	*	***	*	**	***
State extension system	***	**	**	***	****	****
Radio channel	***	*	***	***	*	*
TV: Kisan channel	****	*	****	***	***	*
Kisan call center	**	*	***	***	***	*
Agriclinics	***	*	****	*	****	***
System of rice intensification	***	*	****	****	***	****
Private ICT applications	****	*	****	****	*	*
Private extension services: Syngenta	***	****	****	****	****	****
Farmer associations	***	**	***	***	**	*

Notes: ****, Highly applicable; ***, applicable; **, less applicable; *, least applicable.

resource management comes within the purview of most services except for the state extension system, in which it is less applicable. Apart from the ATMA, NFSM, and agriclincs, all extension services aim to provide information on markets and prices. Technology availability is most applicable to private extension services such as Syngenta because of their ability to provide appropriate technology to rice farmers. The emphasis on timely availability of inputs comes within the mandate of the state extension systems.

CONCLUSIONS AND POLICY IMPLICATIONS

First, as a large proportion of farmers are rice producers, and there is a high level of rice consumption and exports in India, the crop is important from both a livelihood and food security perspective. There have been several initiatives to improve rice extension, but there is a long way to go in improving the productivity and income of rice farmers. It is important to conduct a systematic crop-wise evaluation to see the extent to which extension systems can increase the productivity of farmers, as we have shown that productivity increases due to different extension systems have been differential across regions.

Second, the links between rice extension workers and research activity are weak despite significant investment in research. ATMA and ICAR entities (i.e., KVKs) often function in an isolated manner, each conducting its own extension activities and implementation. The rice programs under both institutions should be combined, aiming to create a complete package of services for rice farmers. There needs to be a clear distinction between recurring costs, such as salaries of extension workers, and operational costs (farm visits) in the budget. Money is often used up only in paying a salary to extension staff (Balasubramanian, 2014). For instance, it is important to include water management (canals, groundwater, and common-pool irrigation sources) in extension services, but most rice extension focuses on crop production problems, although several innovations are tested to address nonproduction issues.

Third, a major constraint to the government's extension efforts has been their top-down bureaucratic nature. Most schemes work inefficiently and are not able to incorporate farmers' feedback in their service delivery. The government is not fully using private and NGO initiatives as tools to increase the reach of its extension programs. Public extension systems should provide more flexibility in functioning in different rice-producing areas. Additionally, officers often spend most of their time performing public duties, such as subsidy delivery, which reduces the time for other extension services. The government's efforts in the delivery of extension services have not fully met the needs of rice-producing areas. Even targeted schemes such as the NFSM (for rice) have yet to show their full potential.

Fourth, most rice extension systems also suffer from a lack of accountability in service delivery. It is important for extension providers to understand farmers' information needs (Babu et al., 2010) to ensure that their demands are met in a timely and efficient manner. Farmers should have multiple channels through which to raise issues and voice their concerns and the government should encourage a more decentralized structure in extension service delivery through the formation of SHGs. It is important to move the existing initiatives from being subsidy driven to being more demand driven in nature.

Fifth, the government also needs to integrate the various channels through which knowledge and information are disseminated to rice farmers. It is important to have a common online platform through which all information can be made available to rice farmers. With ICT applications, the possibility of such knowledge integration is quite high. The government can promote the development of model farms funded privately or by the state as examples of best practices and these farms can be further used to increase tourism in the area.

Finally, it is important to strengthen rice farmers' links to markets. For instance, rice farmers should be informed about the latest consumer preferences and trends in rice consumption at national and global levels.

Investment in reliable monitoring systems to track crop-wise progress on delivery and impact of extension services will be important for increasing the productivity of rice farmers. It is important that the government focus on the provision of extension, especially to smallholder rice farmers. As rice is such an important crop for the nation, it is important to ensure that improving the livelihood and productivity of rice farmers remains a priority of the government. Extension systems need to go beyond production orientation to address markets, prices, climate change, nutrition, and sustainability issues when promoting rice extension approaches. This will require an overhaul of the capacity of frontline extension workers to become social entrepreneurs of the rice farming systems.

REFERENCES

Adusumilli, R., Laxmi, S.B., 2011. Potential of the system of rice intensification for systemic improvement in rice production and water use: the case of Andhra Pradesh. Paddy Water Environ. 9 (1), 89–97.

Agriculture Finance Corporation of India. 2014. Executive Summary of the Report on the impact evaluation of the National Food Security Mission. Mumbai, India. Available from: http://nfsm. gov.in/Publicity/2014-15/Books/NFSM%20IE-EXEC%20SUMM-231114.pdf

Ameur, C., 1994. Agricultural extension: a step beyond the next step. World Bank Technical Paper 247. The World Bank, Washington, DC.

Anderson, J.R., Feder, G., 2004. Agricultural extension: good intentions and hard realities. World Bank Res. Obs. 19 (1), 41–60.

Anderson, J.R., Feder, G., Ganguly, S., 2006. The Rise and Fall of Training and Visit Extension: An Asian Mini-Drama with an African Epilogue. The World Bank, Washington, DC.

Annamalai, K., Rao, S., 2003. What Works: ITCs e-Choupal and Profitable Rural Transformation: Web-Based Information and Procurement Tools for Indian Farmers. Digital Dividend "What Works" Case Study Series. World Resources Institute, Washington, DC.

Babu, S., Glendenning, C.J., 2011. Decentralization of Public-Sector Agricultural Extension in India: The Case of the District-Level Agricultural Technology Management Agency (ATMA) (No. 1067). International Food Policy Research Institute, Washington, DC.

Babu, S., Jayachandran, C., 2014. The role of knowledge in agricultural entrepreneurship, innovation, and transformation in South India. Paper Presented at the International Business Conference on Knowledge Inequality, Rome, Italy.

Babu, S.C., Joshi, P.K., 2015. Future of agricultural extension reforms in developing countries: lessons from India. e-J. Econ. Complex. 1(1). Available from: www.cespi.it/E-journal/Babu&Joshi-June-2015.pdf

Babu, S., Glendenning, C.J., Asenso-Okyere, K., 2010. Review of agricultural extension in India: are farmers' information needs being met? IFPRI Discussion Paper 01048. Washington, DC. Available from: http://sa.indiaenvironmentportal.org.in/files/Review%20of%20Agricultural%20Extension%20in%20India_0.pdf

Babu, S.C., Joshi, P.K., Glendenning, C.J., Asenso-Okyere, K., Sulaiman, R., 2013. The state of agricultural extension reforms in India: strategic priorities and policy options. Agric. Econ. Res. Rev. 26 (2), 159–172.

Babu, S.C., Ramesh, N., Shaw, C., 2015. The current status and role of private extension: a literature review and conceptual framework. In: Zhou, Y., Babu, S.C. (Eds.), Knowledge Driven Development: Private Extension and Global Lessons. Academic Press, London, Available from: http://ifpri.worldcat.org/oclc/911268908.

Balasubramanian, R., 2014. Presidential address. Agricultural research and extension in India: reflections on the reality and a roadmap for renaissance. Indian J. Agric. Econ. 69 (1).

Birner, R., Anderson, J.R., 2007. How to make agricultural extension demand driven? The case of India's agricultural policy. Discussion Paper 00729. International Food Policy Research Institute, Washington, DC.

Birner, R., Davis, K., Pender, J., Nkonya, E., Anandajayasekeram, P., Ekboir, J., Mbabu, A., Spielman, D., Horna, D., Benin, S., Cohn, M., 2009. From "best practice" to "best fit": a framework for analyzing pluralistic agricultural advisory services worldwide. J. Agric. Educ. Extens. 15 (4), 341–355.

Bowonder, B., Gupta, V., Singh, A., 2007. Developing a rural market e-hub: the case study of e-Choupal experience of ITC. Available from: www.planningcommission.gov.in/reports/sereport/ser/stdy_ict/4_e-choupal%20.pdf

Chandra Shekara, P., Kanaka Durga, P., 2007. Impact of agri-clinics and agri-business centers on the economic status of the farmers. IUP J. Agric. Econ. 4 (3), 66–78.

Christoplos, I., Sandison, P. Chipeta, S., 2012. Guide to evaluating rural extension (No. C20-38). GFRAS.

Ekboir, J.M., Dutrénit, G., Martínez, G., Vargas, A.T., Vera-Cruz, A.O., 2009. Successful Organizational Learning in the Management of Agricultural Research and Innovation. The Mexican Produce Foundations, International Food Policy Research Institute. Available from: www.ifpri.org/sites/default/files/publications/rr162.pdf.

Feder, G., et al. 2010. Promises and realities of community-based agricultural extension. IFPRI Discussion Paper 00959. International Food Policy Research Institute, Washington, DC.

Feder, G., Slade, R.H., Lau, L.J., 1987. Does agricultural extension pay? The training and visit system in northwest India. Am. J. Agric. Econ. 69 (3), 677–686.

Global AgriSystem, 2008. AgriClinics and AgriBusiness Centre Evaluation Study. Global AgriSystem, New Delhi.

Glendenning, C.J., Babu, S., Asenso-Okyere, K., 2010. Review of Agricultural Extension in India. Are farmers information needs being met? IFPRI Discussion Paper 01048.

Government of India (GOI), 2012. Report of the Working Group on Agricultural Extension for Agriculture and Allied Sectors for the Twelfth Five Year Plan (2012–17). Government of India, New Delhi.

Ministry of Agriculture, India, 2000. Submission on agricultural extension. Ministry of Agriculture, Department of Agriculture and Cooperation, Extension Division, New Delhi. Available from: http://agricoop.nic.in/imagedefault1/Uploaded%20ATMA%20Guidelines%20%20-%20FINAL.pdf

Mittal, S., 2016. ICT for strengthening extension to reach the last mile in India. New Delhi. Forthcoming publication in print.

Mittal, S., Gandhi, S., Tripathi, G., 2010. Socio-economic impact of mobile phone on Indian agriculture. ICRIER Working Paper no. 246. International Council for Research on International Economic Relations, New Delhi. Available from: www.icrier.org/page.asp?MenuID=24&SubCatId=175&SubSubCatId=691

Nanda, R., Sharma, R., Gupta, V., Tondon, G., 2015. Extension and advisory services for organic basmati rice production in Jammu and Kashmir, India: a case study of Sarveshwar Organic Foods Ltd. In: Zhou, Y., Babu, S.C. (Eds.), Knowledge Driven Development: Private Extension and Global Lessons. Academic Press, London, Available from: http://ifpri.worldcat.org/oclc/911268908.

NFSM, 2015. NFSM Documents. Available from: http://nfsm.gov.in/interventions/2014-15/Intervnetions_rice.pdf

Singh, K.M., et al., 2013. Agricultural Technology Management Agency (ATMA): a study of its impact in pilot districts in Bihar, India. MPRA Paper No. 45544. Available from: http://mpra.ub.uni-muenchen.de/45544/

Sulaiman, R., Holt, G., 2002. Extension, poverty and vulnerability in India: country study for the Neuchatel Initiative. Working Paper 154. Overseas Development Institute, London.

Sulaiman, R., van den Ban, A.W., 2003. Funding and delivering agricultural extension in India. J. Int. Agric. Extens. Educ. 10 (1), 21–30.

Swanson, B., Rajalahti, R., 2010. Strengthening agricultural extension and advisory systems: procedures for assessing, transforming, and evaluating extension systems. Agricultural and Rural Development Discussion Paper 45. The World Bank, Washington, DC.

Tamil Nadu Agriculture Department, 2008a. National Agriculture Development Programme (Rashtriya Krishi Vikas Yojana, RKVY): Implementation of Project Proposals Approved by State Level Sanctioning Committee in the Meeting of 15 December 2007. G.O.M.s No. 82, 2008. Available from: http://tn.gov.in/gosdb/gorders/agri/agri_e_82_2008.pdf

Tamil Nadu Agriculture Department, 2008b. National Agriculture Development Programme (Rashtriya Krishi Vikas Yojana, RKVY): Implementation of Project Proposals for the Year 2008–09. G.O.M.s No. 339, 2008. Available from: http://tn.gov.in/gosdb/gorders/agri/agri_e_339_2008.pdf

Upton, D.M., Fuller, V.A., 2004. The ITC e-Choupal Initiative. Harvard Business School Case N9-604-016. Harvard Business Publishing, Boston, MA, (Revised January 15, 2004).

Chapter 16

Accelerating Impact Through Rice Innovation Systems: Integrating Knowledge, Technology, and Markets

Shaik N. Meera*, Rikin Gandhi**, Rabindra Padaria[†]
*Indian Institute of Rice Research, ICAR, Hyderabad, Telangana, India; **Digital Green, New Delhi, India; [†]Indian Agricultural Research Institute, ICAR, New Delhi, India

Research and development strategies in the rice sector in India have undergone profound transformations and advisory services for smallholder farmers have followed a similar path of change. Most extension programs have started to move away from centralized systems to try to improve links with research and farmers (World Bank, 2007).

Agricultural Extension and Advisory (AEA) is no longer viewed as an agency but as a system that is integral to innovation systems and that focuses on facilitating interaction and learning rather than solely training farmers. New models are more important than ever because agricultural advisory services (extension services) are shifting their focus and changing their roles to improve service provision. CGIAR research on agricultural extension from an innovation systems perspective shows that it has a vital role to play in helping to strengthen capacity, to innovate, and to broker linkages (Spielman et al., 2008).

AEA services are defined as systems that facilitate the access of farmers, their organizations, and other value chain and market members to knowledge, information, and technologies; that facilitate their interaction with partners in research, education, agribusiness, and other relevant institutions; and that help them to develop their own technical, organizational, and management skills and practices, as well as to improve the management of their agricultural activities (Birner et al., 2009; Christoplos, 2010).

The Future Rice Strategy for India. http://dx.doi.org/10.1016/B978-0-12-805374-4.00016-6

WHAT AILS AGRICULTURAL EXTENSION AND ADVISORY IN THE RICE SECTOR IN INDIA

An attempt is made here to extrapolate commodity-specific AEA strategies from those of general AEA strategies. Cornerstones of the Indian AEA system are reaching out to the 119 million smallholders who farm an average of 1.23 ha of land area using a pluralistic system that includes public AEA services (on average, one extensionist per 1200 farmers), a multitude of private extension schemes, 230 million members of agricultural cooperatives, 646 Krishi Vigyan Kendras (KVKs), and several nongovernment organizations (NGOs). It is estimated that public AEA services reach only 6% of the 119 million Indian farmers (Ghimire et al., 2014).

The Indian AEA system is one of the world's largest extension systems, consisting of major streams, such as state departments of agriculture (predominantly extension systems), the Indian Council of Agricultural Research (ICAR), state agricultural universities (SAUs) with the "Agroadvisory and Technology Backstopping System" (Krishi Vigyan Kendra and Agricultural Technology Information Center), Ministry of Rural Development, Ministry of *Panchayat Raj*, Ministry of Information and Broadcasting, Ministry of Natural Resources, Ministry of Water Resources, India Meteorological Department under the Ministry of Earth Sciences, Ministry of Environment and Forest, and private extension agencies, NGOs, and farmers' organizations. Agriculture is largely a state subject, while the central government shares policy, programs, schemes, and funding support, and plays a facilitating role so as to ensure overall national food security.

AEA systems specifically targeted at the rice sector have experienced several challenges when attempting to make a significant impact. Some of these challenges are (1) poor access of farmers to improved seed varieties for enhanced productivity at the farmer level; (2) a lack of sustainable mechanisms for providing improved agronomic skills and enhanced farmers' organizations; (3) weak marketing and business linkages with private-sector members (i.e., agroprocessors) and the input industry (seeds and fertilizers); and (4) a lack of synergies and cooperation with other members performing a facilitating role in the rice sector.

Availability, accessibility, and applicability of rice knowledge and services among AEA systems and farmers are still major challenges facing the rice sector today. Furthermore, existing capacity-building programs and facilities are inadequate to meet the demands of millions of rice stakeholders, including farmers. There is no real-time knowledge and technology exchange among the rice-based institutes, state departments of agriculture, other stakeholders, and farmers. Stakeholders (farmers, scientists, extension officials, the private sector, and other key players) demonstrate a lack of familiarity with up-to-date rice technologies and information sourcing, which suggests that existing information-sharing mechanisms are insufficient.

An excellent review by Evenson (1998) indicated that many extension programs have been highly effective in helping farmers to achieve higher productivity. This review summarizes estimates of economic impact from 57 economic studies undertaken in 7 African countries (Burkina Faso, Côte d'Ivoire, Botswana, Nigeria, Ethiopia, Kenya, and Malawi), 7 Asian countries (Bangladesh, Indonesia, Malaysia, Nepal, the Philippines, South Korea, and Thailand), 3 Latin American countries (Brazil, Paraguay, and Peru), the United States, and Japan. The studies were grouped into two categories: distribution of the estimates by level of statistical significance and, when reported, by level of rate of return to extension. It also appears that some programs have not been so effective.

Transfer of technology models, popular in the last few decades in India, projected rice farmers as passive recipients of knowledge and technology. There is an increasing belief that AEA services for rice should be participatory, demand driven, and community oriented and therefore this disconnect resulted in ineffective targeting of rice varieties, hybrids, and technologies and hence lower adoption, missing forward and backward linkages across the value chain. The complete disconnect of rice farmers with the value chain and hence the innovation system makes extension delivery a challenging job in the years to come.

It is in this context that the proposed framework for rice extension strategies that integrate knowledge, technology, and markets will be developed. The authors opine that such a framework should

- Provide better, faster, and cheaper solutions to reach out to rice farmers.
- Integrate knowledge, technologies, and markets.
- Offer components of advisory services that are empirically proven elsewhere.
- Increasingly harness digital technologies.

IMPACT ACCELERATION IN RICE: CASES

The strategy for technology delivery and impact acceleration should be based on different isolated yet successful cases. In this section a few extension delivery and market linkage methods (non-ICT based) are discussed that could be piloted or scaled up in the Indian context.

Campaigns: Vietnam (Three Reductions, Three Gains)

The Three Reductions, Three Gains (3R3G) campaign in Vietnam aimed to lower the cost of growing rice in irrigated systems (through fewer seeds and less nitrogen fertilizer and pesticides) while maintaining yield, improving farmers' health, and better protecting the environment (through reduced reliance on agrochemicals).

The national program of Vietnam used standard extension activities combined with an elaborate and creative mass media campaign. A multistakeholder participatory planning process was used to develop a campaign package to reach and motivate large numbers of rice farmers in the Mekong Delta (Huan et al., 2005). It consisted of communication media (TV, radio, print, and demonstrations) and materials (soap operas, leaflets, pamphlets, and farmer field days) geared toward increasing the farmers' ability and motivation to modify their resource management practices by adopting a relatively knowledge-intensive technology. The strategy was to change farmers' attitude toward input use from one of "more is better" to one of "less is more sensible" through several information-delivery systems. These included billboards along main roads and soap operas that aired on national radio and television stations. Public amplifiers in the villages were used to replay the radio broadcasts before daybreak just before the farmers went to their rice fields. It became nearly impossible for a farmer not to hear about 3R3G. The 3R3G technology is not a physical good (such as in the case of a new high-yielding rice variety) but rather a package of input management recommendations that farmers can use in their profit-maximizing or input-use decision-making process.

The 3R3G technology capitalizes on the synergistic effects of reducing three inputs together without sacrificing yield. For example, if seed rates are lowered, less fertilizer is required, which individually and jointly makes the crop less attractive to pests, thus reducing the need for insecticides for the whole cropping season. Evidence of the economic impact of 3R3G at the farm level shows an increase in net income from USD 92 to 118/ha.

Extension in Vietnam is based on master farmer training of ~400 people per year per province. It is an example of a well-functioning centralized system like that seen in China. It has excellent rural infrastructure and a very homogeneous rice-growing environment with farmers with average land size of more than 1 hectare. The critical question in adapting this strategy is whether this concept can be transferred to the Indian context. Given the penetration of digital technologies among rice farmers, the concept can be refined to suit the Indian context.

The STRASA Model

While working out a futuristic strategy, there is a need to discuss how project-based extension strategies evolve depending upon the need of stakeholders. While the 3R3G campaign can be the backbone for an AEA framework for the rice sector, the essential project-based strategies can be worked out based on need.

The STRASA project tried to tackle a few of the challenges inherent in the public-sector AEA system.

The Stress-Tolerant Rice for Africa and South Asia (STRASA) project began at IRRI in collaboration with the Africa Rice Center at the end of 2007. The aims of the project were to develop and deliver rice varieties tolerant of abiotic stresses to the millions of farmers in unfavorable rice-growing environments.

STRASA was conceived as a 10-year project with a vision to deliver the improved varieties to at least 18 million farmers on the two continents. The project was also anticipated to have significant spillover effects for nonparticipating countries.

http://strasa.irri.org/

The uniqueness of this model lies in the synergy between technology need and the development-delivery approach. Critical learning from this approach includes

- Adding desirable traits to farmer-preferred varieties by moving a highly desirable trait (e.g., submergence tolerance) into a widely accepted variety to accelerate product delivery.
- Using varietal targeting by leveraging digital technology (e.g., geospatial mapping tools) to improve targeting and to prioritize trait development.
- Developing seed roadmaps to systematically define market opportunities, product profiles, and volumes needed to achieve target adoption; identifying strategic partners for production; and laying out timelines for delivery.
- Using participatory varietal selection that lets farmers select what variety is most appropriate.

The STRASA model focused on organized, large-scale, prerelease seed multiplication and dissemination to ensure seed availability and fulfillment of demand upon commercial release. Toward this end, partnerships in the public, private, and civil society sectors were encouraged. Successfully engaging communities on a small scale (e.g., subprovincial level) can circumvent bureaucracy and get the right partners on board to help scale up. An ability to mobilize communities around new technology depends on addressing an important need. For example, seed roadmaps are needed for prioritizing and targeting trait integration pipelines and for planning seed production and delivery by creating a transparent coordination mechanism among partners.

Zoom-In Zoom-Out Model

AfricaRice attempted open-air video presentations that enhanced learning, experimentation, confidence, trust, and group cohesion among rural people between 2005 and 2009. The videos strengthened the capacity of more than 500 organizations and hundreds of thousands of farmers. AfricaRice's integrated rural learning approach also helped women to access new markets and credit. Learning videos allow for unsupervised learning, facilitate institutional innovations, and improve social inclusion of the poor, youth, and women.

Developing effective and efficient learning material requires considerable human input and is relatively time-consuming, so the need to start by addressing topics of high regional relevance cannot be overemphasized. Hence, an approach called zooming-in zooming-out (ZIZO) evolved. A broad multistakeholder consultation process helped to define regionally relevant learning needs. This was followed by in-depth interactions with a few communities to obtain a better feel about their ideas, their innovations, and the words they used related to the learning subject. Only after zooming-in had happened could the exact content of the video be decided upon. Fine-tuning of the learning tools happened during the scaling-up or zooming-out phase.

Videos on drying and storage triggered changes in practice, even without facilitation. Women tended to store seed in a wide range of receptacles but, after having seen the video, many changed to smaller and more airtight containers. Three percent started to paint their earthen pots when the videos were shown without facilitation but this increased to 32% when follow-up was implemented through group discussions. Irrespective of facilitation, more than 90% had learned how to expel air from their storage container. Facilitation has an overall positive effect on the rate of adoption but, if video is used as a standalone method, it can trigger change in a community, albeit small. Achieving impact through mass media has huge potential. A good video puts ideas into the heads of some community members who will experiment with them. In doing so, the video will have achieved its purpose. Adoption and diffusion will depend on how functional and profitable the technologies are in the given context (van Mele, 2006).

Community-Based Plant Clinics

A CABI initiative supports plant health systems in developing countries. Plant clinics are run by trained plant doctors with support from the Plantwise knowledge bank (PWKB), an online–offline gateway to plant health information, to provide actionable plant health advice to farmers.

The PWKB provides free open-access tools for pest management in more than 12 Plantwise implementing countries across Africa. The database has developed secure data and information tools for collecting, managing, and sharing plant clinic data and information by stakeholders in the plant health system. The robust data collection processes, analysis, and insights have enabled countries to prioritize extension, develop and disseminate key extension messages, and identify training needs of extension providers, farmers' needs, and strengths and weaknesses in the plant health sector. The Plantwise approach is currently running close to 700 plant clinics in 12 African countries. Preliminary experiences indicate that the Plantwise approach is an invaluable opportunity to help farmers manage plant health problems while reducing the risk of pesticide use in agriculture.

Plant health clinics in India have come a long way in recent years. They have carved out a niche in the plant health system whereby they can provide timely diagnosis and prescription to keep pests at bay and thereby mitigate losses. As of 2013, the organizations running such clinics include (1) SAUs; (2) the National Horticulture Mission, Ministry of Agriculture & Cooperation, Government of India; (3) traditional universities; (4) CABI; and (5) agriclinics and the Agribusiness Center Scheme (Srivastava, 2013).

The activities of plant clinics extend beyond the advisory and diagnostic, with emphasis also placed on extension and working more closely with farmers and organizations involved in promoting food production. This approach has been successful in India but it needs to be strengthened with ongoing extension strategies in the rice sector.

Rice Sector Development Hubs

As part of its 2011–20 strategic plan, Africa Rice Center is working with its partners across the continent to set up rice sector development hubs to concentrate research and development efforts, establish a critical mass, connect partners along the rice value chain, and facilitate the spread of innovations. The hubs are testing grounds for new rice technologies and they follow a "reverse-research approach," that is, starting from the market (Africa Rice Center, 2011).

The hubs bring together large groups of farmers (1000–5000) and partners from the whole rice value chain (input suppliers, seed producers, processors, millers, wholesalers, retailers, and consumers) to facilitate change. These represent key rice-growing environments and market opportunities across African countries, and will be linked to major national or regional rice development efforts to speed up wider adoption of rice knowledge and technologies. The geographic positioning of each hub is determined in national workshops convened by the national agricultural research and extension systems. As of 2014, national partners had identified 68 hubs in 24 West, Central, and East African countries.

Even though there is a difference between methodologies, the Cereal Systems Initiative for South Asia (CSISA) tried to recreate the system with differential impacts. The Indian Council of Agricultural Research (ICAR) under the "Farmer FIRST (farm, innovations, resources, science, and technology)" program can deploy these strategies. Furthermore, new IRRI programs in India can focus on the strategies that would be most appropriate in the microclimate change scenarios.

Key Learnings From This Section

The Agricultural Technology Management Agency (ATMA) reflects the new role of public extension institutions in India at the district (subprovincial) level. It coordinates diverse AEA services and facilitates linkages between research, extension, the private sector, NGOs, and farmers. The ATMA identifies local research and extension priorities through consultation with farmers and subsequently develops local problem-solving plans. By establishing the ATMA as a coordinating entity for extension services, the government affirmatively supports pluralistic extension service delivery.

This can act as a starting point for introducing and piloting the concepts discussed in the section "Impact acceleration in rice: cases." In India, yield-increasing and cost-saving technologies can be identified and disseminated through a campaign, synergizing the efforts with national and provincial programs, such as the National Food Security Mission. At the district level, the ATMA can form the core of the Indian AEA innovation framework that is being proposed in this chapter.

The STRASA model focused on organized, large-scale, prerelease seed multiplication and dissemination to ensure seed availability, and fulfillment of demand upon commercial release. Toward this end, partnerships in the public, private, and civil society sectors can be encouraged. This means successfully engaging communities on a small scale (e.g., subprovincial level) to circumvent bureaucracy and get the right partners on board to help scale up. The STRASA model can be integrated into structural or functional components of the ATMA or piloted across identified rice ecosystems dependent on need. Success will depend on how this model can be synergized with ongoing government initiatives supporting farmers through various programs. Although this model provides an opportunity for synergy between technology need and the development-delivery approach, it also needs to integrate with national and provincial projects, such as the National Agricultural Development Project (*Rashtriya Krishi Visan Yojana*). At the provincial level, this will result in shorter product development cycles generating superior technologies (a threefold faster timeline from R&D to on-farm adoption), faster adoption and scaling up by farmers, and faster diffusion across states.

ZIZO can be planned and implemented with a broad multistakeholder consultation process to define regionally relevant learning needs. It can be integrated with rice innovations where in-depth interactions in a few communities have documented farmers' ideas, their innovations, and the words they use related to the learning subject. The exact content of a video can be decided upon once zooming-in has taken place.

Although the major roles of plant clinics are diagnostic and advisory, their activities extend beyond, with emphasis on extension and working more closely with farmers and organizations involved in promoting food production. This approach has been successful in India but it needs to be strengthened with ongoing extension strategies in the rice sector. In addition, ICAR institutes and AICRIP centers could work toward developing rice sector development hubs. ICAR, under the Farmer FIRST program, can deploy these strategies in rice regions while new programs at IRRI can focus on the strategies that would be most appropriate in microclimate change scenarios.

DIGITAL EXTENSION FOR THE RICE SECTOR: TACTICAL TO PRACTICAL

Despite the agricultural extension network reaching from the MoA (central level) to KVKs (local level), the present system serves only 6% of the 119 million farmers in India (Ferroni and Zhou, 2011; Ghimire et al., 2014). Although campaigns, rice sector development hubs, plant clinics, and the STRASA model form the core of the innovations framework, it is essential to discuss how well digital technologies can be integrated with the AEA system. Mobile telephone, community radio, satellite television, video projections, computer networks, etc., may be incorporated when building rice innovation systems to effectively deliver knowledge and services.

Digital Tools and Strategies for the Rice Sector

Mobile phones provide multiple opportunities for improving communication. Mobile phones can be used as a basic (with talk and text), a featured (with talk, text, and some other applications), or a smart technology (as above plus they can access the Internet). There are several examples from India and elsewhere as to how mobile phones could be used for credit, service provision, and market and technical know-how. Some of these include banking through M-Pesa (IFFCO Kisan Sanchar Limited, IKSL), market information (Esoko), and technical information (e-Krishok , iCow, and mKisan). The Kisan call center, a landline-based pull service, has given way to mobile applications for high-end smartphones. Basic text messages in KMAS, mKrishi, etc., have evolved to voice messages (IKSL), to multimedia content (m4agriNEI), and to mobile applications (Videokheti).

Community radio (FM radio owned and operated by the community) has brought much relevance to current Indian agriculture. Prime Minister Modi has called for the establishment of one community radio in each agricultural college that will transform the way local agricultural knowledge and experiences are shared. There are several examples from India and elsewhere as to how community radio could be deployed for agricultural development, including Deccan Development Society, Farm Radio International, Farmer Voice Radio, and Mali Shambani (Kenya). In most cases community radio provides rice-related information that is topical, relevant, and of interest for farmers' problems. Follow-up interviews with farmers who have successfully improved their farms, etc., will speed up modern farming.

Social media can be a powerful communication tool to quickly engage large numbers of people, and they have the potential to reach a broader and (likely) younger audience of farmers and stakeholders. Facebook and Twitter are two well-known social media sites whose usage in agricultural fields will witness radical change in the near future. Crowd sourcing can also revolutionize rice farming. In this media approach, required services, ideas, or content are obtained by soliciting contributions from a large group of farmers. Think about the impacts of having crowd sourcing in the whole agricultural value chain! Similarly, *WhatsApp* can transform problem-solving strategies in agriculture. Kenya Seed Company Facebook and Twitter, Google Baraza Agriculture, Ushahidi, and different Facebook forums in India are a few examples.

Smartphones and other smart devices have the advantages of mobile phones but with greater interactivity and wider access to other media (e.g., Internet). Tablets (e.g., iPad) or "audio computers" can also be used in agriculture. *Nutrient Manager* of IRRI, Philippines, is an example of this, and the Indian Institute of Rice Research (IIRR) under its Rice Portal planned to develop a series of mobile applications for the benefit of Indian farmers. IRRI's *Rice Crop Manager* is another application that can be accessed via a smartphone or a computer with an Internet connection. It allows extension officers to give farmers a specific recommendation on nutrient, pest, weed, or water management, depending on the specific variety they used, their yield from the previous season, and the site-specific conditions of their field.

It is also worth mentioning here the mKisan project of the Government of India. Conceptualized, designed, and developed in-house within the Department of Agriculture and Cooperation, the project uses unstructured supplementary service data (USSD). This has improved the outreach of scientists, experts, and government officers posted down to the block level to disseminate information and provide advisories to farmers through their mobile phones. The SMS Portal was inaugurated in 2013 and since its inception nearly 396 crore messages have been sent to farmers throughout the country. These numbers will continue to rise (http://mkisan.gov.in/).

The power of pictures is very important in effective communication. Video and TV are powerful ways to raise awareness, provide technical information, and enhance training. Two examples are Access Agriculture and Digital Green. Articulating a simple, clear, practical message with five to six (or fewer) key points and an effective storyboard will definitely make a difference in modernizing Indian farming.

The web offers a powerful way to reach the "web-connected" fraction of farmers. Extension and development workers are the primary target group for this intervention given that many others may have access issues. These workers then pass on the information they receive to rural communities. The most comprehensive examples are the Rice Knowledge Management Portal (RKMP), mKisan Portal, Farmers' Portal, TNAU Agritech Portal, Agropedia, and Agrisnet.

Big Data Tools

For rice extension to succeed as a system, there is a need to build huge repositories using GIS and state-of-the-art data management systems. Initiatives for this come from frontline extension efforts, such as KVKs, frontline demonstrations, private input databases, weather-based agroadvisories, markets, and postharvest options. All the integrated data platforms should have interoperability and capability to be accessed with mobile platforms.

Africa Trial Sites (http://africats.org/) is a portal that enables national and international research organizations to electronically pool their extensive information related to trial sites. It also provides numerous tools (based on ICT advances in bioinformatics, GIS, and data management) that help farmers, plant breeders, and agronomists to evaluate new varieties more efficiently in the field and gain more useful data from field trials. Users can search the website for trial sites and data by country, design trials to evaluate cultivars, obtain tools to manage trials (from developing a budget to estimating water stress during the growing season), analyze trial data, view results of spatial analyses, examine data on an interactive Google map, and report results online. The combination of African trial site data and interactive data analysis tools has made valuable information much more widely available for the agricultural research, development, and extension community.

The Rice Knowledge Management Portal (www.rkmp.co.in) has a similar data repository comprising data related to multilocation performance of rice cultivars over the past 49 years (www.rkmp.co.in/data-repository). Frontline demonstration results are also made available at www.rkmp.co.in/FLDs. Another example for mapping submergence-prone areas for better targeting of Swarna-Sub1 varieties was developed by IRRI, Philippines (http://acrs2015.ccgeo.info/proceedings/WE1-1-5.pdf).

Empirical Evidence of Digital Interventions

To maximize the impacts of technology in farmers' fields, synergy between extension advisories (knowledge or information) and availability of technologies must be attained. The private sector was involved in an innovative and integrated extension approach at ICAR-IIRR, Hyderabad, to bring together evidence of the impact of digital technologies.

An IIRR initiative on the critical factors that contribute to maximizing technology impacts in farmers' fields is bringing synergy between extension advisories (knowledge or information) and the availability of technologies. During 2012–13 and 2013–14, small action research and surveys were conducted in two major rice-growing states in India (Andhra Pradesh and Karnataka) involving 180 rice farmers and 45 extension professionals. Indicators such as knowledge and the effects of technology on productivity were studied.

The impact of knowledge interventions was found to be significant when blended with field demonstrations in both states. Out of 32 nonnegotiable adoption points, 15 (46%) information needs were met from various ICT tools. Changes in practices due to ICT interventions were found to be in the range of 11%–12% of total adoption points. The majority of farmers opined that a productivity increase was realized to the tune of 30%. Approximately 20% of the farmers benefited by either reduced cost of cultivation or other such advantages.

The impact of knowledge and technology interventions on rice productivity was also calculated. The digital interventions proved to be influencing the adoption of land preparation (0.241^*), seeds and varietal selection (0.376^{**}), nursery management (0.027), crop establishment methods (0.404^*), nutrient management (-0.295^*), water management (0.271^*), plant protection (0.383^{**}), weed management (0.296^*), and harvest management (0.490^*), which were eventually found to improve the productivity of rice in farmers' fields.

Key Learnings From This Section

The use of digital technologies in agriculture and rural development (hence the rice sector) is being viewed as an investment rather than an expenditure. Empowering rural people using digital technologies is seen as yet one more policy instrument for economic development and hence will form part of the proposed rice strategy.

Evidence of leveraging the potential of investments in Digital India, in which the government is investing considerable resources to establish hardware and connectivity, has been increasing. Individual personalized advisory services recognize the importance of linking digital initiatives to groups, such as farmers, farmer producer organizations, etc. Efforts to undertake farmer segmentation are now happening. ICT is adding value in knowledge management as several players are looking for the same information. Although most of the initiatives are currently independently funded and managed, there is much scope for convergence. Institutionalization of successful ICT initiatives will depend largely on policy-level support and there is a need to bridge the gap between practice and policy.

STRENGTHENING THE RICE EXTENSION AND ADVISORY SYSTEM

It is widely recognized that Rice Extension and Advisory (REA) professionals need to have more capacity to deal with emerging challenges (declining natural resource base, climate change, linking farmers to high-value markets, conforming to new standards and certification norms, etc.). However, very little is known about how this capacity can be developed. The futuristic REA system should provide new opportunities for these professionals to be more efficient and updated on various aspects of rice production and postproduction.

Innovative Capacity Development of Extension Professionals

Overall capacity development of extension professionals and farmers is going to be a crucial dimension of an innovation system. The ability of extension agents to use emerging technologies and processes and to train others in these could be enhanced and developed in a training of trainers approach. Key challenges here will be to meet technical capacity development in rice production and postharvest skills and the softer skills for being effective rural advisory agents.

The capacity of members can be developed through courses that provide understanding and practical skills in the latest rice production technologies from seed to market. Each state will identify issues and gaps in the knowledge to be able to set the capacity-building agenda over the project period. The capacity-building syllabus can be built from the demands coming from farmers who are engaged in different extension programs, ICT tools for large-scale communication, and the analytical tools that enable analysis of extension delivery outcomes. Some of the short-term strategies follow:

- A rice extension agronomy program (possibly led by IRRI/ICAR) for building the next generation of rice extension professionals in India. The development of extension leaders will be practically grounded in rice production and postharvest. This will give them the ability to effectively operate at the farmers' level and to better understand the rice crop in their locality.

- An extension leadership program for rice extension professionals to provide them with strong leadership skills that will enable the creation of win-win situations among local rice industry members. The trained rice extension agents will be far more capable in developing localized extension initiatives that include finding competent stakeholders to support farmers through linking of public and private resources.
- A rice knowledge management capacity program to train and develop the capacity of extension agents to address the particular challenge of effectively reaching large numbers of small farmers and "going to scale." This series may aim to develop competency in using ICT tools, such as precision mobile applications (Nutrient Manager and Crop Manager), knowledge portals (Agrisnet Farmers' Portal, RKMP, and RKB), and other innovative tools.

The outcome of this capacity building will be the production of competent, practical-oriented advisors with access to the latest production knowledge and extension tools and an ability to work in a multichannel extension service. These learnings will be translated into field-level impacts through a series of orientation workshops for field extension officers.

E-Learning and MOOCs in Rice Extension Systems

Learning (formal and informal) is central to all innovation systems, including those for the rice sector, and is essential for sustaining the capacity to innovate over the long term. E-learning enables governments, agricultural advisory services, NGOs, farmer organizations, and private companies—in fact, any member in the innovation system—to reach large numbers of professionals (including farmers).

E-learning does not require the complex online workflows associated with standard learning management systems, but the priority in promoting e-learning in extension systems is to build ICT capacity in personnel at all levels of agricultural education, training, and extension. Digital technologies and virtual interactions are not sufficient to form cohesive learning communities; instead peer-to-peer contact significantly improves learning, and mobile phones can provide useful support.

A study conducted at the IIRR focused on e-readiness and information literacy among rice (extension) workers and tried to assess the feasibility of e-learning strategies for agricultural development in general, and the rice sector in particular. A total of 18 e-courses were developed on rice production, protection, and other technologies and are available online at www.moodle.learnrice.in. Both online and offline training are being imparted using this platform. The online version of the e-learning platform was converted into an offline version (Portable *Moodle*——Poodle).

A massive open online course (MOOC) is another strategy aimed at unlimited participation and open access via the web. In addition to traditional course materials, such as filmed lectures, readings, and problem sets, many MOOCs provide interactive user forums to support community interactions among students, professors, and teaching assistants. MOOCs are a recent and widely researched development in distance education that was first introduced in 2008 and emerged as a popular mode of learning in 2012. Extension systems can revolutionize the way adult and continuing education is carried out and the way skills are developed in the agricultural sector.

Capacity and Skill Development of Farmers Through Video Extension

Building capacity among farmers is crucial and requires a good comprehension of what extension does in the learning relationship with farmers. For farmers to increase their field productivity, they need support in bringing them up to speed with current technologies and occupational skillsets.

> Digital Green pursued opportunities to use ICT devices to build capacity and increase outreach to poor farmers. Digital Green has been operating since 2006 in India, using video as its tool to reach farmers. The components of the Digital Green system include (1) a participatory process for local video production, (2) a human-mediated instruction model for video dissemination and training, (3) a hardware and software technology platform for exchanging data in areas with limited Internet and electrical grid connectivity, and (4) an iterative model to progressively better address the needs and interests of the community with analytical tools and interactive phone-based feedback channels.

Enabling hands-on learning and evolving farmers into peer teachers to impart knowledge would resolve many local problems. One effective way of achieving farmer-to-farmer experience sharing could be through video extension.

Even WhatsApp is creating impacts in many parts of India. In Punjab, a WhatsApp group started by Dr. Amrik Singh, an officer in the local agricultural office, allows a greater number of farmers to receive expert input on agricultural practices. On the level of daily interaction, the WhatsApp groups are successful at providing a sounding board of assistance and in motivating farmers. Capacity development videos can be effectively shared across local farmers' groups without losing the context in which such experiences could be adopted.

Key Learnings From This Section

The futuristic REA system should provide new opportunities for extension professionals and farmers to be more efficient and updated on various aspects of rice production and postproduction. The capacity of members could be developed through courses that provide understanding and practical skills in the latest rice production technologies from seed to market. Each state will identify issues and gaps in knowledge to be able to set the capacity-building agenda over the project period. The capacity-building syllabus can be built from the demand coming from farmers who are engaged with extension programs. The outcome of this capacity building will be to produce competent, practical-oriented advisors with access to the latest production knowledge and extension tools and an ability to work in a multichannel extension service.

A MOOC strategy aimed at unlimited participation and open access via the web can be deployed exclusively in the Indian rice sector. These sector-specific MOOCs should provide interactive user forums to support community interactions alongside the courses. Enabling hands-on learning and evolving farmers into peer teachers to impart knowledge would resolve many local problems. One effective way of achieving farmer-to-farmer experience sharing could be through video extension and the WhatsApp group approach.

STRENGTHENING THE VALUE CHAIN PROPOSITION IN THE RICE SECTOR

The Indian REA system has never been actively involved in integrating markets and value chain proposition into its activities. The present supply chain structure of rice in India works in the traditional framework, which involves many intermediaries at supply and distribution fronts. The traditional supply chain structure faces problems of inventory management, in which either overstocking exists, which results in obsolescence and increased supply chain costs, or stockouts of demanded varieties exist, resulting in lost sales. The supply chain of rice in India also faces problems related to procurement, distribution, intermediary collaboration, and logistics, and needs to be redesigned (Sharma et al., 2013).

Links with marketing and business processes will play an important role in future extension. Regional millers play a major role in marketing rice and they depend on a network of local agents and traders for their supply. To develop and strengthen a market-oriented extension system in the country, efforts have to be made at various levels. Marketing extension can adopt the following strategies:

1. Providing advice on product planning. Even for small farmers, the concept of product planning (i.e., the careful selection of the varieties to be grown with market ability in mind) is an important starting point. Providing this basic advice to farmers is essential to enable them to withstand competition in the market. From a practical point of view, branded rice consumption is rising in urban markets, and therefore raising awareness of this among farmers is a key to success.

2. Providing marketing information. Farmers need information on two aspects of marketing: current price and market arrival information, and market trends forecasting. This information also has to be supplemented with other information about reaching a particular market to obtain a particular price,

arrangements available in the market related to storage, transactional methods, quality requirements, postharvest handling requirements, etc. Alongside information on the spot market, forward and future market prices are required to be disseminated to farmers.
3. Securing markets for farmers. For grains to be sold to government procurement agencies, extension workers can advise on how, when, and at what price to sell the designated food grains to the government agencies.

DIGITAL TECHNOLOGIES AND THE LOCAL SUPPLY CHAIN

Digital technologies could reduce the number of steps between producers and consumers (thereby reducing the price spread). This refers to either a shorter distance or a smaller number of intermediaries between the production and consumption of the food. A short supply chain covers a broad range of organizational strategies, but some examples include

1. Farmers' markets: many of which have started to use social and digital marketing to increase the "reach" of their message.
2. Food hubs: many of which are moving to "virtual" methods of operation, using Internet and web technologies to coordinate market exchange between producers and consumers rather than using farmers' markets as a means to distribute produce.
3. Microenterprises: with dynamically organized supply chains, the application of mobile technologies will help to coordinate supply and demand and manage business processes.
4. mKrishi: The Mobile Agro Advisory System of Tata Consultancy Services uses a combination of mobile, web, IVR, and USSD services. The strategies focus on engaging with farmer-producer organizations at large and subscription- and module-based services right down to individual farmers and starting from the preparation of a crop planner, agroadvisory, cultivation practices, and marketing.

E-Procurement of Paddy: The Case of Chhattisgarh

Chhattisgarh (CG) state adopted the strategy of using ICT to control diversion and leakage in the delivery mechanism of the food supply chain. Based on the success of this strategy, the government of Chhattisgarh computerized the entire food grain supply chain in 2007–08. This encompassed the procurement of paddy at 1532 purchase centers and the transportation of public distribution system (PDS) commodities to 10,416 FPS for further distribution to 3.7 million ration card holders, covering 6 different organizations. As an outcome of the project, 780,000 farmers have received computer-generated checks without any delay and citizen participation has increased in PDS monitoring. The outcome of the project and some challenges faced are discussed.

Paddy procurement and the PDS are old schemes with complaints of diversion and leakage. Monitoring of the scheme is difficult partly because of the insufficiency of staff and partly because of their complicity. Innovative methods including bar-coded food coupons, food stamps, biometrically coded ration cards, etc., have been used to reduce leakage and diversion but none of these have been entirely successful. The project that was developed and implemented in Chhattisgarh is an end-to-end solution based on information technology and is giving very encouraging results.

The National Agriculture Market

The National Agriculture Market (NAM) is a pan-India electronic trading portal (www.enam.gov.in) that networks existing regulated markets to create a unified national market for agricultural commodities. The NAM portal provides a single window service for all market-related information and services. This includes commodity arrivals and prices, buy and sell trade offers, and provision to respond to trade offers, among other services. Although material flow (agricultural produce) continues to happen through *mandis*, an online market reduces transaction costs and information asymmetry.

Agricultural marketing is administered by the states as per their agrimarketing regulations. The state is divided into several market areas, each of which is administered by a separate Agricultural Produce Marketing Committee (APMC), which imposes its own marketing regulations (including fees). This fragmentation of markets, even within the state, hinders the free flow of agricultural commodities from one market area to another. Multiple handling of agricultural produce and multiple levels of market charges also result in escalating prices for consumers without commensurate benefit to farmers.

The NAM addresses these challenges by creating a unified market through an online trading platform, at both the state and national level. It promotes uniformity and streamlining of procedures across the integrated markets; removes information asymmetry between buyers and sellers; promotes real-time price discovery based on actual demand and supply; promotes transparency in auction processes and access to a nationwide market for farmers, with prices commensurate with the quality of their produce; and provides online payment and availability of better quality produce at more reasonable prices to consumers.

Key Learnings From This Section

The national e-market platform provides transparent sales transactions and price discovery in regulated markets. States are willing to accordingly enact suitable provisions in their APMC Act for the promotion of e-trading by their State Agricultural Marketing Board/APMC. The provision of soil testing laboratories in or near selected markets helps visiting farmers to access these facilities in the market itself. The REA system will have to integrate its efforts with the likes of eNAM to provide maximum benefit to rice farmers. Farmers' markets, many of which have started to use social and digital marketing to increase the reach of their message, could be effectively harnessed along with food hubs, many of which are moving to virtual methods of operation.

Using digital tools to control diversion and leakage in the delivery mechanism and its successful application in the computerization of the food grain supply chain will have a bearing on rice farming in the near future. Effective strategies should be in place regarding to what extent this should be integrated with the REA system in India.

STRENGTHENING EXISTING EXTENSION DELIVERY: CONVERGENCE OF IDEAS

The Ricecheck Approach in Extension Delivery

Ricecheck is an innovative farmers' participatory extension methodology with which farmers decide what the best practices are for their fields based on "experience" rather than recommendations from experts. The core activity under this approach is facilitating group discussions among farmers. Ricecheck is a procedure of extension-assisted farmer-group self-learning. It can be defined as a dynamic rice crop management system that presents key technology and management best practice as key checks, checks farmer practices with best practice to compare results, and learns through group discussion to sustain improvements in productivity, profitability, and environmental safety. Simply put, Ricecheck is learning by checking and sharing best farming practices.

India's Ricecheck program (participatory extension methodology) was piloted in four provinces by the IIRR, Hyderabad. Although everyone would like to see impact of new technologies on a wider scale, very few appreciate the need for adapting new technologies to meet the varied requirements of different socioeconomic, biophysical, organizational, and institutional settings.

As a part of the extension strategy, a customized Ricecheck program can be developed that will enable farmer organizations, grassroots NGOs, state agricultural extension, and private agencies to continuously improve production technologies with farmers. It lends itself to the localized development of communication tools for extension, for the implementation of ICT tools for precision agriculture, and for analytical feedback to service providers locally, statewise, and nationally. Farmers will be organized so that they will keep their own farm records following a jointly agreed format. These records will be linked with analytics and assessed periodically to understand progress and impact.

Reviving Frontline Demonstrations

Frontline demonstrations (FLDs) are considered to be the most effective and useful extension activities for demonstrating the latest technologies to farmers in their own fields. The principle of "seeing is believing" is operational in these demonstrations, as the farmers become easily convinced when they see the performance of new technologies in the fields of their neighboring farmers. FLDs are formulated by ICAR and funded by the Government of India.

In the past 8 years, 7007 FLDs (1 hectare each) have been conducted, benefiting 18,318 rice farmers directly. Some 2637 tons of seed of high-yielding varieties and hybrids were distributed to farmers through this program. The spin-off impacts of FLDs are many. The FLD program helped many stakeholders to obtain first-hand information about newly released varieties or hybrids that went into the value chains of the Indian rice sector. Field demonstrations are a long-term educational activity conducted in a systematic manner to show the value of a new practice or technology. The impressive yield advantage obtained in the various ecosystems proves that the FLDs conducted have been able to fulfill this objective.

The emerging extension strategy could include the following recommendations:

- There is a need for a paradigm shift in the FLD program. The funding-centric approach should be replaced by a need-centric approach. The yield advantage perspective should be replaced by "problem-solving" capabilities of new technologies. There should be continuous demonstration forums (resulting in regular dialog between FLD farmers and neighboring farmers) instead of a single field day. This will also promote continuous learning opportunities among the farmers.
- To make the most impact, it should be assured that a complete package of practices is adopted in the demonstration plots. In the absence of a complete package, the potential yield of the demonstrated technology will not be perceived by the farmers. Whenever possible, farmers' innovations can be included along with the demonstrated package of practices.
- FLD programs are planned and implemented by FLD centers (research stations). Other stakeholders who are responsible for scaling up the demonstrated technologies (e.g., pheromone traps) are not involved in planning and implementation. The involvement of stakeholder organizations (e.g., the agency involved in manufacturing and marketing of pheromone traps) will help to maximize the impacts of FLDs.
- Integrating the digital tools with FLDs will help in maximizing the impact of technologies. The integration should be in the form of documenting the various activities undertaken in the demonstration plots, such as method demonstrations, field days, etc.

Technology Backstopping, Agroadvisory, and Market Intelligence

General guidelines should be prepared for enterprise development within rice farming systems of resource-poor farmers so as to ensure profit maximization while safeguarding livelihood options. In view of this, knowledge and technology should be provided in the form of "capsules" rather than in an isolated manner.

Market intelligence and agribusiness management advice should also be provided. Various stakeholders, such as public, private, NGO, farmers', and farm women's organizations, must work in tandem to form need-based, location-specific, commodity-specific, and potential consortia. Modalities to be worked out for consortia-based approaches involve taking into account issues such as institutional arrangements, needs, purpose for convergence, drawing of a convergence model, operational steps for convergence, coordinating the identified activities, etc.

Scientist–farmer interaction needs to be a functional activity on a relatively continuous basis at critical stages of agrienterprise decision making. This will lead to respectable net benefit for entrepreneurs and farmers. Such interaction has to be direct or through a credible and dependable alternative, such as well-rewarded local resource persons or farm innovators. In recent years, ICAR has been emphasizing more "scientist–farmer linkage" so as to achieve faster technology outreach.

Group-based and participatory approaches to providing advisory services are gaining ground. These methods have the potential to overcome barriers to participation, foster inclusiveness, and lead to more demand-driven services. For strengthening the control over farmers, especially smallholders in backward and forward linkages, facilitation should be provided to form producers' organizations to ensure economic and institutional sustainability. Producers' organizations have greater strength in integrating process elements across the value chain for a given commodity or enterprise, thereby substantially improving cost efficiency and reducing transaction costs while improving net profit to farmers. This also provides ample scope for direct producer–consumer linkages in production-marketing, thereby minimizing the intermediaries. The approach to be evolved should use these kinds of producers' organizations for farmer-led extension by creating a model farm and in turn function as knowledge and resource experts for fellow farmers in their social system.

A favorable policy environment, research prioritization, and technology backstopping are essential for mainstreaming gender in agriculture. Reorientation of extension management approaches and methodologies is essential for developing knowledge management, capacity building, reducing drudgery, and safeguarding the concerns of women. Furthermore, enabling women as facilitators of agricultural development is needed wherein any woman can learn and share her knowledge, wisdom, and experience with fellow women.

BUILDING INNOVATION SYSTEMS: INTEGRATING KNOWLEDGE, TECHNOLOGY, AND MARKETS, AND ACCELERATING IMPACT

In a vibrant innovation system, agricultural development results from efforts to combine technological improvements in production, processing, and distribution with organizational improvements in how information and knowledge are exchanged among various members in these systems, alongside policy changes that create favorable incentives and institutions to promote change (Davis, 2008).

Strategies for building innovation systems include building pluralistic extension systems, providing business development services within rice farming, encouraging new roles for extension and advisory services, and leveraging the strength of extension systems as innovation brokers in the rice sector.

Based on the extensive review presented in this chapter and the empirical research conducted across India in the extension advisory system, the authors would like to propose a framework for an innovation system. Such an innovation system will have strategies for accelerating impact by integrating knowledge, technology, and markets.

Components of the Innovation System Proposed

The core efforts to combine technological improvements in rice production, processing, and distribution will be organizational improvements in extension advisory. Such improvements must aim to integrate information, knowledge, technologies, and markets from various members in these systems. Although the authors propose a new framework in terms of functional adjustments, they do not suggest structural transformation or the establishment of a new set of extension organizations.

AEA systems for the rice sector can make a big difference if they can (1) improve the access of farmers to improved seed varieties to enhance productivity at the farmer level; (2) set up sustainable mechanisms for providing improved agronomic skills and enhanced farmers' organizations; (3) strengthen marketing and business linkages with private-sector members (i.e., agroprocessors) and the input industry (seeds and fertilizers); and (4) synergize and cooperate with other members performing a facilitating role in the rice sector.

All four of these critical dimensions need to flow through three important levels: the policy subsystem, the research-extension-farmer subsystem, and the value-chain subsystem. Although there is no clear-cut demarcation across these three subsystems, for ease of comprehension, a few core activities are suggested under each of these subsystems.

Integrating knowledge, technology, and markets to provide better, faster, and cheaper solutions to reach out to rice farmers requires effectively harnessing digital technologies. It is suggested to harness digital technologies at various hierarchical levels across organizations and also among rice farmers. The critical questions in adapting a 3R3G kind of campaign strategy are the transferability of the concept in the Indian context and the effectiveness of incorporating digital technologies into functional components of all the stakeholder organizations. Given the penetration of digital technologies among rice farmers, the 3R3G campaign can be refined to suit the Indian context. This can be a starting point for introducing and piloting the concepts discussed in the section "Impact Acceleration in Rice: Cases." In India, yield-increasing and cost-saving technologies can be identified and disseminated through a campaign mode, synergizing efforts with national and provincial programs, such as the National Food Security Mission. ATMA can form the core of an Indian AEA innovation framework at the district level.

In the innovation system for extension advisory services, the research-extension-farmers continuum can never be a linear relation. Even though formal research organizations will have a greater say in technology development, technology evolution through collaborative workflows across organizations (including farmers' organizations) cannot be ignored. Adding desirable traits from farmer-preferred varieties into a widely accepted variety to accelerate product delivery and varietal targeting by leveraging digital technology (e.g., geospatial mapping tools) will be a crucial paradigm shift in the near future.

Within the context of such a paradigm shift, ATMA may have to take on a new role by coordinating diverse AEA services and facilitating linkages among research, extension, the private sector, NGOs, and farmers, at the district (subprovincial) level. KVKs may have to take the lead in operationalizing emerging extension philosophies, such as Ricecheck and refined FLDs that will result in evolving participatory extension advisory approaches.

The STRASA model focuses on organized, large-scale, prerelease seed multiplication and dissemination to ensure seed availability and meeting of demands upon commercial release. Toward this end, partnerships in the public, private, and civil society sectors can be encouraged. This means that successfully engaging communities on a small scale (e.g., subprovincial level) can circumvent bureaucracy and get the right partners on board to help scale up. This approach is not a blanket recommended for all varieties and hybrids; instead, it is for low-hanging and high-value varieties (such as DRRH 42, Samba Mahsuri-Sub1, etc.).

Although high-end digital technologies (business intelligence, crop models, remote sensing, and sensor-based big data analytics) will form the core of extension advisory targeting, digital interventions will also be important at the farmers' level. Mobile phones could be used for credit, service provision, and market and technical know-how for farmers across the board. Opportunities for digital technologies in agriculture and rural development (hence the rice sector) are being viewed as an investment rather than an expenditure. In the rice sector, IRRI–ICAR collaboration could pilot a series of such interventions that would integrate knowledge, technology, and markets. These initiatives could be synergized with the mKisan project of the Government of India. There has been evidence of leveraging the potential investments in Digital India, for which the government is investing considerable resources in establishing the hardware and connectivity. From individual personalized advisory services, there is increasing recognition of the importance of linking digital initiatives to groups, such as farmers' groups, farmer–producer organizations, etc.

The futuristic REA system should provide new opportunities for these professionals to be more efficient and updated on various aspects of rice production and postproduction. The capacity of members can be developed through MOOCs and e-learning strategies. A skill development program for rice farmers can be taken up on a large scale with video extension approaches. Enabling hands-on learning and evolving farmers into peer teachers to impart knowledge would resolve many local problems. One effective way of achieving farmer-to-farmer experience sharing could be through video extension and the WhatsApp group approach.

Perhaps the most crucial challenge for any REA innovation system lies in linking farmers with inputs, markets, and value chain propositions. An REA system will have to integrate its efforts with the likes of eNAM to deliver maximum benefit to rice farmers. Farmers' markets, many of which have started to use social and digital marketing to increase the "reach" of their message, could be effectively harnessed along with food hubs, many of which are moving to "virtual" methods of operation.

Using ICT to control diversion and leakage in the delivery mechanism and computerize the food grain supply chain will have a bearing on rice farming in the near future. Effective strategies should be in place regarding how much should be integrated with the REA system in India (Fig. 16.1).

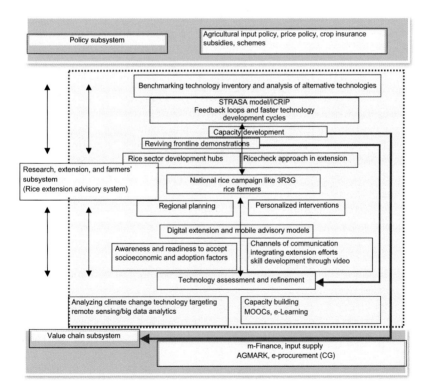

FIGURE 16.1   Framework for a rice extension advisory innovation system: integrating knowledge, technology, and markets, along with impact acceleration. *CG,* Chhattisgarh; *MOOC,* massive open online course; *3R3G,* Three Reductions, Three Gains; *STRASA,* Stress-Tolerant Rice for Africa and South Asia.

THE WAY FORWARD

The farmers and extension agents of tomorrow will need to operate in a far more complex and dynamic environment to meet the increasing targets for rice production. Multiple members will exist from public, private, and civil society organizations with stronger integration of the rice industry from seed to market. On the technology front, the Indian government has started several national programs to increase rice production: the National Food Security Mission, Rashtriya Krishi Vikas Yojana, Bringing Green Revolution to Eastern India, etc. However, strengthening of extension and advisory services is vital to the improvement of rice production and needs to occur alongside the technology interventions. Interest is increasing in leveraging the potential of investments in Digital India, for which the government is investing considerable resources in establishing the hardware and connectivity. Digital technologies in extension systems should go beyond providing advisories and technologies to farmers; they also need to transform financing and service coordination at the village level.

There is a need to develop a unique ICAR frontline mobile extension program to build sustainable models in which farm and crop management tools and financial services are "bundled" in affordable, unified platforms on mobile phone channels to promote mass uptake commercially. This approach can facilitate the development of a business model whereby the bundling process provides an increased value proposition for each partner, such as increased fee income, greater outreach, or reduced risks.

A series of e-voucher platforms across the country could enable extension agencies to provide specific noncash services. Such e-vouchers are much easier to track than cash vouchers and they also help to avoid fraud; a common problem with paper vouchers. Joining e-vouchers together with the online soil health cards, Nutrient Manager app, and optimum fertilizer recommendation would dramatically improve fertilizer demand and supply dynamics.

One of the areas in which extension systems can leverage the benefits from mobile phones is enabling easy, quick, and safe payments in emerging markets. A mobile and web application that integrates SMS and mobile money services will help savings and credit agencies. A mobile finance platform will help farmers to register with financiers, apply for loans and credit, and make repayments.

To enable all this work on mobile interfaces, we need to work on improving the existing rice semantic web portal, that is, the rice knowledge management portal (comprising dynamic location-specific, local language knowledge), local sharing mechanisms using social media (WhatsApp and Facebook groups), skill development platforms (video extension), ready-to-use scripts and experience sharing (community radio), big data platforms (GIS and remote sensing), market and input supply platforms (integration of private and public input supply inventory), etc.

New training courses for extension staff and subject matter specialists related to rice production technologies, seed production and certification, and enhancing technical and facilitation skills to effectively work with farmers need to be developed. These rice extension specialists will also link farmers to sources of technology, information, and support services. An Indian Rice Academy may help by serving as a central training facility to hone the skills of extension agronomists and subject matter specialists. Future Extension and Advisory Services (EAS) need to strategize, among many others, convergence of big data with disruptive technologies, such as mobile/cloud computing, Internet of Things, location-based social networks, etc., for a vibrant rice industry.

To meet the increasing targets for rice production, the farmers of tomorrow and extension agents will need to operate in a far more complex and dynamic environment than in the past. The demand-driven innovations and linking them to markets in a time-critical manner can be achieved only with innovations in rice extension advisory systems. The real challenge is to strengthen partnerships among multiple actors from public, private, and civil society organizations with stronger integration to farmers' needs.

REFERENCES

Africa Rice Center, 2011. Boosting Africa's Rice Sector: A Research for Development Strategy 2011-2020. Africa Rice Center, Cotonou, Benin.

Birner, R., Davis, K., Pender, J., Nkonya, E., Anandajayasekeram, P., Ekboir, J., Mbabu, A., Spielman, D., Horna, D., Benin, S., 2009. From best practice to best fit: a framework for analyzing agricultural advisory services worldwide. J. Int. Agr. Extens. Edu. 15 (4), 341–355.

Christoplos, I., 2010. Mobilizing the Potential of Rural and Agricultural Extension. Food and Agriculture Organization of the United Nations (FAO) and the Global Forum for Rural Advisory Services (GFRAS), Rome.

Davis, K., 2008. Extension in sub-Saharan Africa: overview and assessment of past and current models and future prospects. J. Int. Agr. Extens. Edu. 15 (3), 15–28.

Evenson, R., 1998. The economic contributions of agricultural extension to agricultural and rural development. In: Burton, E., Swanson, R., Bentz, P., Sofranko, A.J. (Eds.), Improving Agricultural Extension: A Reference Manual. FAO, Rome, M-67.

Ferroni, M., Zhou, Y., 2011. Review of Agricultural Extension in India. Syngenta Foundation for Sustainable Agriculture, Basel, Switzerland, pp. 1–46.

Ghimire, N., Koundinya, V., Holz-Clause, M., 2014. Government run vs. university managed agricultural extension: a review of Nepal, India and the United States. Asian J. Agr. Extens. Econ. Sociol. 3 (5), 461–472.

Huan, N.H., Thiet, L.V., Chien, H.V., Heong, K.L., 2005. Farmers' evaluation of reducing pesticides, fertilizers and seed rates in rice farming through participatory research in the Mekong Delta. Crop Prot. 24, 457–464.

Sharma, V., Giri, S., Rai, S.S., 2013. Supply chain management of rice in India: a rice processing company's perspective. Int. J Manag. Value Supply Chains (IJMVSC) 4 (1), 25–36.

Spielman, D.J., Ekboir, J., Davis, K., Ochieng, C.M.O., 2008. An innovation systems perspective on strengthening agricultural education and training in sub-Saharan Africa. Agric. Syst. 98, 1–9.

Srivastava, M.P., 2013. Plant clinic towards plant health and food security. ESci. J. Plant Pathol. 2 (3), 193–203.

van Mele, P., 2006. Zooming-in zooming-out: a novel method to scale up local innovations and sustainable technologies. Int. J. Agr. Sustain. 4 (2), 131–142.

World Bank, 2007. World Development Report 2008 Agriculture for Development. Washington, DC.

FURTHER READING

Bagchi, B., Emerick-Bool, M., 2012. Patterns of adoption of improved rice varieties and farm level impacts in stress-prone rainfed areas of West Bengal, India. In: Pandey, S. et al., (Eds.), Patterns of Adoption of Improved Rice Varieties and Farm-Level Impacts in Stress-Prone Rainfed Areas in South Asia. International Rice Research Institute, Los Baños, Philippines.

Devaux, A., Horton, D., Velasco, C., Thiele, G., Lopez, G., Bernet, T., Reinoso, I., Ordinola, M., 2009. Collective action for market chain innovation in the Andes. Food Policy 34 (1), 31–38.

FAO, 2014. A regional rice strategy for sustainable food security in Asia and the Pacific. Final edition RAP Publication 2014/05. Food and Agriculture Organization of the United Nations Regional Office for Asia and the Pacific, Bangkok.

Watts, J., Mackay, R., Horton, D., Hall, A., Douthwaite, B., Chambers, R., Acosta, A., 2003. Institutional learning and change: an introduction. ISNAR Discussion Paper No. 03–10. IFPRI and ISNAR, The Hague, Netherlands.

Chapter 17

The Synergic Role of the All India Coordinated Rice Improvement Project (AICRIP) in the Development of India's Rice Sector: Emerging Challenges and Policy Issues

Vemuri Ravindra Babu*, Aldas Janaiah**, Shaik N. Meera*

Indian Institute of Rice Research, ICAR, Hyderabad, Telangana, India; **International Rice Research Institute, New Delhi, India

Over the past 50 years, the rice research program in India has largely centered on improving the yield frontier, which substantially contributed to increased rice supplies and thus food security (CRRI, 1996). Several studies indicated high payoffs to rice research in India (Evenson and McKinsey, 1991 as quoted in Janaiah et al., 2005, 2006; Kumar and Rosegrant, 1994; Pingali and Hossain, 1999; Pingali et al., 1997). The Green Revolution began in the late 1960s in India with the first high-yielding rice cultivar (IR8) introduced in 1966. The release of Jaya and Padma in 1968, as well as many other high-yielding varieties (HYVs), enabled India to achieve food self-sufficiency and security during the mid-1980s. One of the strategies to achieve food security was the initiation of coordinated R&D networks in the national agricultural research and extension system (NARES). In view of the diverse production environments in the country, multiple research centers were established nationwide to permit multilocation and multidisciplinary research to be performed.

India's NARES has a fairly organized institutional framework in the public sector. The NARES has two tiers with around 32,000 scientists (excluding technical staff) in total. At the federal level, around 120 research centers, institutes, or directorates are under the control of the Indian Council of Agricultural Research (ICAR). These centers focus primarily on problems of regional and national importance. In addition, 70 state agricultural universities (SAUs) are

The Future Rice Strategy for India. http://dx.doi.org/10.1016/B978-0-12-805374-4.00017-8

under the control of the state governments and they concentrate on state-specific problems, as well as offer education in agricultural and allied sciences.

India has the world's largest area under rice cultivation (~45 million ha) and is the second-largest producer of rice, with nearly 154 million tons of paddy (unmilled rice) produced during 2014–15 (Directorate of Economics and Statistics, 2016). This chapter examines the synergic role of the All India Coordinated Rice Improvement Project (AICRIP) in the improvement of the rice sector in India and critically analyzes the impact of AICRIP as an institutional partnership for the promotion of rice improvement in the country.

The future rice strategy in India cannot afford to ignore the synergic role of the AICRIP. At the same time, some critical challenges need to be addressed in view of intellectual property rights and plant breeders' rights. Policy reforms that would improve the structural and functional setup of the AICRIP should also be explored. This chapter is divided into four sections: (1) "AICRIP: The Historical Perspective," (2) "AICRIP: An Institutional Innovation in NARES," (3) "The Synergic Role of AICRIP in Technological Breakthroughs," and (4) "Key Challenges and the Way Forward."

AICRIP: THE HISTORICAL PERSPECTIVE

The success of the coordinated maize improvement pilot project initiated by ICAR in 1957 led to the institutional innovation of India's NARES. This subsequently led to the initiation of AICRIP in 1965, and to many other crop and research areas of agriculture. To date approximately 90 All India Coordinated Research Projects (AICRPs) have been initiated by ICAR, linking both ICAR institutions and SAUs, and covering all major crops, livestock, dairy, fisheries, and natural resources.

The AICRIP was launched in 1965. The system was designed to bring all rice researchers and rice research centers from across Indian under one umbrella. The aims of the project were to collectively plan for rice breeding priorities, generate breeding material accordingly, coordinate the evaluation of the material generated, and decide rationally what and where to recommend it for commercial planting. Conceptually, the underlying strategy of rapid identification of test entries for a given situation is based on the scientific principle of multilocation evaluation for a few seasons. This helps to generate the same kind of data for understanding genotype × environment interactions as did the earlier practice of limited location testing for many seasons or years for precisely identifying varieties with general and specific adaptability.

The AICRIP system organizes the interdisciplinary and interinstitutional three-tier multilocation testing of elite breeding lines and improves management practices for all rice ecosystems. Over the years, AICRIP has become an institutionalized R&D network linking more than 100 research centers of both SAUs and ICAR institutions. There are now more than 500 scientists working for AICRIP in 45 funded centers and more than 100 voluntary centers that plan, conduct, and monitor all rice research activities. This network facilitates the free movement of information, breeding material, and modern rice cultivars across the states, and promotes multilocation testing of improved germplasm through multi-institutional and multidisciplinary approaches. It also collaborates with the International Rice Research Institute (IRRI) for the collection, evaluation, and use of improved germplasm. The Indian NARES has a close partnership with international agricultural research centers (IARCs) through the AICRPs. The AICRP was used as a model for several countries, such as Zimbabwe, Bangladesh, the Philippines, and Kenya, to initiate similar coordinated multilocation research networks.

AICRIP also became a model for IRRI to start the International Rice Testing Program (IRTP). This program evaluates international rice nurseries and facilitates access to and use of exotic germplasm of value for either direct introduction or use as a donor source in breeding in participating countries. Many varieties (IR8, IR20, IR36, IR64, IR70, IR72) with high yield, multiple pest resistance, and better grain quality have come to India through IRTP nurseries, as have hundreds of valuable germplasm accessions that continue to add strength to the country's breeding research. It was the extensive adoption of HYVs (developed within India and introduced from IRRI) that brought about stability and incremental growth in rice production and productivity even in years of erratic monsoons.

Another role for which AICRIP is responsible is the conceptualization, formulation, and coordination of research networks on problems of national importance that warrant immediate, focused, aggressive, and product-oriented research involving all relevant institutions and scientists. Two success stories worth mentioning are the development of high-yielding basmati rice and hybrid rice technology.

AICRIP: AN INSTITUTIONAL INNOVATION IN NARES

Over a period of 50 years, a few institutional mechanisms evolved in the rice research system (NARES), one of which was the partnership among public-sector research institutions.

The AICRIP led to institutionalization of partnership research in the public sector through which scientists from all parts of the country were allowed to freely share information and breeding material. Furthermore, the Indian NARES has a close partnership with IARCs through AICRPs.

A system for varietal improvement has evolved and been standardized across key stakeholders in the rice sector. The identified elite lines of improved cultivars (after the initial breeding process for 6–7 years) are pooled at AICRIP headquarters by all participating R&D institutions. These lines are then tested in several locations around the country, for three to four crop seasons at research station sites and for one season in the field (evaluation trials). A Central Variety Release Committee (CVRC) under the Ministry of Agriculture examines the performance of the tested cultivars based on the coordinated trials and approves them for release as a modern variety for the country as a whole if they are found suitable for more than one state.

Examples of how a state has significantly benefited from other states or institutions through partnership research under AICRIP are aplenty. For instance, the most popular modern variety in India at present is Swarna (MTU 7029, which was developed at a small rice research station of the Acharya N.G. Ranga Agricultural University in Maruteru Village in Andhra Pradesh and released in 1982). It is currently being grown in 11 states mostly in the eastern and northeastern regions, covering about 12% of India's rice area. Another example is in Assam, from where HYV Ranjit was released in 1991. It covers approximately one-third of the state's HYV area of rice. This variety was originally developed at Tamil Nadu Agricultural University for an irrigated ecosystem, but moved to Assam (a rainfed ecosystem) through the AICRIP network. Another HYV, Tella Hamsa, became a popular dry-season variety in Andhra Pradesh, but was actually developed at the Indian Agricultural Research Institute (IARI) in New Delhi (Janaiah and Hossain, 2004).

Interstate movement of HYVs is another innovation process that evolved because of the AICRIP system. If a cultivar (in AICRIP trials) is found suitable in only one state, the concerned SAU would approve its release as a HYV for that state. An elite or improved cultivar that is developed and identified in a state or institution may also be released by other states if it is found suitable through a coordinated variety testing mechanism. This institutionalized network, which has a built-in transparent mechanism for the free sharing of improved germplasm among scientists and institutions, is the most powerful means to facilitate the movement of HYVs from one state or institution to another. Another means by which HYVs can move from one state to another is through on-farm or frontline demonstration programs being conducted by the AICRIP network. All potential HYVs (mostly released ones) are identified from various states and their seeds are distributed to all targeted states for frontline demonstrations in farmers' fields (irrespective of where these HYVs were released or developed). This is done in consultation with the state's Department of Agriculture. Participating states are allowed to promote any HYVs if they were found to be suitable in farmers' fields through frontline demonstrations.

THE SYNERGIC ROLE OF AICRIP IN TECHNOLOGICAL BREAKTHROUGHS

The AICRIP network has four major activities: varietal improvement (plant breeding), plant protection, natural resource management, and transfer of technologies. Through nationwide on-farm and frontline demonstrations of new varieties, the AICRIP network has been instrumental in the testing and release of 1084 varieties (including 72 hybrids) from both Central and State Variety Release Committees in all rice ecosystems (Table 17.1). Of these varieties, 501

TABLE 17.1 High-Yielding Rice Varieties and Hybrids Released Through the Central and State Release System (1966–2014)

Total no. of varieties	Varieties	Hybrids	Total
CVRC	129	44	173
SVRC	883	28	911
Total	1012	72	1084

CVRC, Central Variety Release Committee; SVRC, State Variety Release Committee.

were for irrigated areas, 134 for rainfed uplands, 192 for rainfed lowlands, 44 for semideepwater conditions, 18 for deepwater conditions, 53 for high altitudes, 42 for saline and alkaline areas, 10 for aerobic areas, 18 for the boro season, and 72 are aromatic long- and short-grain varieties (Tables 17.2 and 17.3 and Fig. 17.1).

TABLE 17.2 Rice Varieties and Hybrids for Various Ecosystems Released Through the Central and State Release System (1965–2014)

	Number of varieties		
Ecosystems	CVRC	SVRC	Total
Irrigated			
Irrigated early	23	142	165
Irrigated midearly	27	123	150
Irrigated medium	31	155	186
Irrigated hills	6	39	45
Boro	5	13	18
Scented	16	56	72
Saline and alkaline soils	14	28	42
Rainfed			
Rainfed upland	17	117	134
Rainfed shallow lowland	17	175	192
Semideepwater	8	36	44
Deepwater	3	15	18
Upland hills	2	6	8
Aerobic	4	6	10
Total	173	911	1084

TABLE 17.3 Rice Varieties and Hybrids Released Through State Release Committees (1966–2014)

SVRC	Varieties	Hybrids	Total
Andhra Pradesh (including Telangana)	112	3	115
Assam	33	—	33
Bihar	38	—	38
Chhattisgarh	19	2	21
Delhi	2	—	2
Gujarat	32	—	32
Haryana	25	1	26
Himachal Pradesh	6	—	6
Jharkhand	19	—	19
Jammu and Kashmir	18	—	18
Karnataka	47	1	48
Kerala	62	—	62
Meghalaya	12	—	12
Manipur	14	—	14
Odisha	130	4	134
Punjab	23	—	23
Puducherry	8	—	8
Rajasthan	7	—	7
Tamil Nadu	69	5	74
Tripura	12	—	12
Uttar Pradesh	53	2	55
Uttarakhand	23	2	25
West Bengal	53	1	54
Madhya Pradesh	9	3	12
Maharashtra	57	4	61
Total	883	28	911

The AICRIP became a role model for IRRI and it used this model to initiate the IRTP (with India as a major partner) in 1975. The IRTP was later renamed as the International Network for Genetic Evaluation of Rice (INGER) and this program has facilitated easy and free exchange of genetic material among global rice researchers.

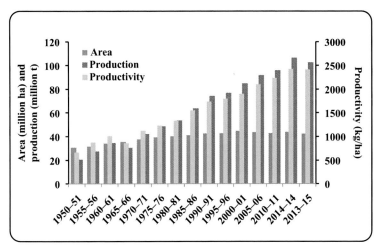

FIGURE 17.1 **Overview of impacts: All India Coordinated Rice Improvement Project (AICRIP) vis-à-vis the rice production scenario.** Five decades of area, production, and productivity of rice in India.

Through INGER, rice varieties that were bred and released in India after multilocation testing in the AICRIP have been directly released for commercial exploitation in many countries covering various ecosystems. For example, 46 Indian varieties have been released by other countries for farmers' cultivation: 25 in sub-Saharan Africa, 10 in South Asia, 5 in Latin America and the Caribbean, 3 in Southeast Asia, 2 in West and North Africa, and 1 in East Africa. The use of India-bred varieties has occurred in all rice ecosystems: 8 varietie in rainfed upland areas, 6 in rainfed lowlands, and 32 in irrigated areas. These have been released under different names in different countries. The global acceptance and adoption of 46 rice varieties of Indian origin vindicates the strength of the varietal testing program of AICRIP and INGER.

Several elite breeding lines that were generated in India under the AICRIP program have been widely used by many countries as donors for desirable traits, such as resistance to biotic and abiotic stresses, yield, and quality.

A pioneering study by Janaiah and Hossain (2004) reported that the AICRIP, through free interstate and interinstitutional movement of improved germplasm, has played a remarkable role in accelerating the overall impact of India's rice R&D. Nearly 28% of the total released HYVs in India were contributed by the interstate and interinstitutional flow of improved germplasm through AICRIP. Approximately 41% (14.3 million ha) of the total HYV area in 14 major rice-producing states was attributed to HYVs that were developed in other states or institutions. The MVs of rice that were developed at IRRI and released in India after multilocation evaluation through the AICRIP have covered about 12% of the total area in India. Similarly, rice varieties of other-country origin (exotic) moved to India through IRRI, and released for cultivation after multilocation evaluation through the AICRIP, have been grown on ~5% of India's rice area. The MVs released by ICAR institutions have been planted on ~13% of India's rice area (Janaiah and Hossain, 2004; Janaiah et al., 2006).

Andhra Pradesh is the biggest contributor in rice improvement in India through the AICRIP network, followed by IRRI (including exotic varieties) and ICAR institutions. This indicates the wider acceptance of some popular HYVs, such as Swarna (MTU 7029), Vijetha (MTU 1001), and Samba Mahsuri (BPT 5204), that originated in Andhra Pradesh. IR8, IR20, IR36, and IR64 were introduced from IRRI and Jaya from ICAR. The states in Eastern India have substantially benefited from HYVs developed in other states (particularly those from irrigated regions, such as Andhra Pradesh), and also from ICAR institutions and IRRI. This clearly implies that the benefits from research efforts and investments made in irrigated regions and in other states also percolated into the rainfed region and other states from domestic spillover through the AICRIP network (Janaiah and Hossain, 2004).

Basmati Rice

Basmati rice of the Indian subcontinent is the most highly priced and sought after specialty rice in both domestic and international markets. After 23 years of research efforts, the IARI in New Delhi finally succeeded in evolving the world's first high-yielding semidwarf basmati variety, Pusa Basmati 1. Along with IRRI-bred miracle variety IR8 and hybrid rice from China, Pusa Basmati 1 is widely considered as one of the 20th century's landmark rice achievements, as its yield is approximately twofold higher than traditional tall-statured basmati varieties (i.e., Basmati 370 and Taroari Basmati) with no compromise on basmati quality. Since its release in 1995, it continues to account for 40%–60% of all basmati rice production. This enabled the country to increase the volume of basmati exports from 67,000 tons in 1978–79 (INR 32 crores) to 1.18 million tons in 2007–08 (INR 4,334 crores). Aside from its own place of importance in production and exports, it provides the ideal (and only) plant type and quality donor-cum-agronomic base for the development of higher-yielding and early-maturing basmati quality varieties, such as Pusa sugandh 3, Pusa sugandh 4, Pusa sugandh 5, Pusa 1401, etc., and the first basmati hybrid, Pusa RH 10.

Hybrid Rice

The AICRIP evolved another network approach for hybrid rice during the 1990s. This strategy involved 10 major rice research centers and led to the development and release of five first-generation hybrids by 1994. Incidentally, the IRRI-sourced male sterile line IR58025A, which is widely used all over Asia, is based on the India-bred Pusa 167. As many as 70 hybrids have now been released and are commercially planted; 80% of these are from the private sector and 90% of the hybrid seed requirement is met by these only. The liberal national policy in this regard has promoted active involvement of the private sector in hybrid breeding and hybrid seed production and supply. Although both the public and private sector made considerable progress in the development and release of rice hybrids over the past 20 years, the large-scale adoption of hybrid rice is very limited (only 6% of India's rice area, particularly in eastern and northern states) because of the poor grain and cooking quality (Janaiah and Xie, 2010).

KEY CHALLENGES AND THE WAY FORWARD

One of the challenges faced by the rice sector is the declining trend in real profitability in rice farming (Janaiah et al., 2005). Rice farming can be economically sustained through further yield increases with new HYVs, cost reductions, or grain quality improvements (Janaiah and Hossain, 2003). AICRIP strategies should focus on developing rice varieties for different ecosystems with higher yield potential (i.e., exceeding the yield level of the highest yielder, Jaya) and improved tolerance of abiotic stresses, such as drought, submergence, stagnant and flash flooding, salinity, high temperature, and major biotic stresses in the context of climate change. To do this, it will need to integrate conventional and innovative approaches. The existing pace of harnessing marker-assisted breeding, heterosis breeding, and doubled-haploid breeding, functional genomics, and transgenics may not be enough.

The development of super rice, improvement of photosynthetic efficiency through the C_4 mechanism, development of nutrient-rich rice varieties, and improvement of grain quality parameters should receive high priority. The AICRIP system should develop a careful strategy geared toward effectively harnessing innovations from across the world. For trait-specific germplasm evaluation, the identification of promising donors for target traits; identification, fine mapping, and introgression of new genes/QTLs; and state-of-the-art high-throughput phenomics and genomics facilities are essential.

Research on crop production should focus on improving factor productivity, leading to efficient use of water, fertilizers, pesticides, and other agricultural inputs, all of which will decrease the cost of cultivation, maximize profit, and secure a cleaner environment. Research on rice-based cropping systems and crop diversification, boro rice, flood-prone rice, organic scented rice, aerobic rice, climate-resilient production technologies, soil resilience, integrated nutrient management, integrated crop management, etc., needs thorough analysis. Postharvest technology and value addition to maintain consumption, internal trade, and exports should form an integral part of AICRIP research; however, no clear strategy is in place yet. The quality of rice soil and its produce along with ecosystem carrying capacity are becoming more important in the context of growing concerns over food safety, international trade, and climate change (CRRI, 2013).

The need of the hour is prebreeding for widening the genetic base of rice breeding programs. The use of wild and weedy relatives of rice for the introgression of genes for yield, yield components, and biotic and abiotic stresses assumes greater significance and should be taken up as a priority. The segregating populations that are produced should be shared with cooperating centers for further selection of region-specific traits. To widen the variability of the maintainer and restorer gene pools, an intensive prebreeding program has been launched with other key AICRIP partners of hybrid rice. The variability available from both primary and secondary genepools of rice is being gainfully used through the deployment of biotechnological tools. This program needs to be strengthened further in a mission-mode approach.

It is very important that the breeding lines being identified for release as varieties undergo rigorous evaluation against biotic and abiotic stresses as insurance against unexpected stress conditions. For this purpose, a few selected AICRIP centers have to be strengthened with the development of artificial screening facilities.

The fine-tuning of efficient agronomic practices [direct-seeded rice, the system of rice intensification (SRI), aerobic rice, alternate wetting and drying (AWD), and microirrigation methods] will be critical for adaptation to climate change. Water-saving technologies, such as aerobic rice, SRI, direct seeding, AWD, etc., need to be intensively evaluated in various AICRIP centers and their adoption should be facilitated. Concomitant changes in rice production systems, including cropping and farming systems, need to be made. Any improvement at this level would provide farmers with a higher and more sustainable family income and economic security. Part of the focus of AICRIP may be on a holistic approach involving other areas of the farming system. Understanding soil health impacts in relation to climate change is possible through the use of indicators that relate soil physical, chemical, and biological properties to ecological functions and that can be monitored in the context of sustainable land management and climate change.

Selective mechanization for timely operations under labor shortage, to reduce human drudgery (for value addition), and for postharvest processing is essential. Thus, it is recognized that agricultural mechanization is crucial in the fight against hunger and poverty; however, at the same time, it raises questions over environmental and health concerns.

Climate, cultivation methods, genotypes, cropping systems, and the dynamics of insect pests and diseases are all changing. Taking this into account, short-term site-specific IPM packages need to be developed. Once complete, long-term studies on pest surveillance and forecasting methods encompassing GIS and remote-sensing technologies should be carried out. Developments in scientific research related to biotechnology, technological innovations, and new pesticide delivery systems, as well as changes in markets and the policy environment, offer new opportunities for reducing dependency on chemical pesticides and the evolution of holistic integrated pest management for sustainable rice production.

Proven and knowledge-intensive technologies need to be aggressively pushed for adoption of ICT tools. Fostering strong partnerships between the public and private sectors is going to be the basis for achieving the desired goals.

The AICRIP setup needs to be revamped to meet future demands and transformed from routine field-testing centers to technology-developing and highly proficient evaluating centers. Fewer, well-equipped centers could concentrate on technology development and larger voluntary centers could meet the requirement of validating technologies.

The institutional reforms within AICRIP are guided by structural adjustments to adapt to new challenges and the changing environment. Greater involvement of other stakeholders, especially the private sector, will help to further the cause of the AICRIP system. Keeping in mind the IPR regime with respect to Plant Breeders' Rights, a suitable monitoring mechanism has to be created using ICT tools to protect IP rights in the AICRIP system. This could be possible with a few small, yet effective, structural adjustments to the way multilocation testing is performed.

The AICRIP intranet should be strengthened to automate the whole process of AICRIP data. This should start from centers, cooperators, trials, technical programs, seed dispatch and confirmation, and crop conditions to final summary tables for the reaction of genotypes to abiotic and biotic stresses. Establishing centers of excellence with adequate infrastructure for high-end basic and strategic research, in addition to the many well-equipped centers for technology testing and validation, would help achieve the long-term goals of AICRIP. Location-specific thrust areas and research strategies, which are necessary for achieving the rice research targets, are highlighted in Table 17.4. Although this is not an exhaustive list, it will surely give future direction to location-specific rice research.

The AICRIP has been in effective service for more than 50 years. Established in 1965 with just 12 main centers, it has now grown to 45 funded centers and more than 100 voluntary participating centers. AICRIP became synonymous with rice research in India: it is committed to maintaining its leadership and is responsive, vibrant, and sensitive to the changing scenario and needs of its stakeholders.

The AICRIP setup needs to be revamped to meet future demands and transformed from routine field-testing centers to technology-developing and highly proficient evaluating centers. The strategies suggested in this chapter may help in redefining the NARES in general and the AICRIP system in particular.

TABLE 17.4 Thrust Areas for Rice Improvement by Location

Study number	Locations	Thrust areas
1	Aduthurai	Breeding for high-yielding quality rice varieties with pest resistance
2	Arundhutinagar	Development of varieties for aus, aman, and boro seasons
3	Bankura	Development of drought-tolerant and boro rice varieties
4	Brahmavar	Development of gall midge–resistant HYVs
5	Chatha	Development of quality rice varieties for hills
6	Chinsurah	Development of HYVs for irrigated, rainfed/flash flood, submergence-prone, drought-prone, and salinity ecosystems coupled with iron enrichment and arsenic tolerance
7	Chiplima	Development of drought-tolerant varieties for upland areas and suitable varieties for low-lying areas of Odisha
8	Coimbatore	Development of varieties and hybrids with biotic stress tolerance
9	Masodha	Development of varieties for shallow lowland and saline soils
10	Gangavathi	Development of quality rice for the Northern Karnataka area and varieties for inland salinity
11	Ghaghraghat	Development of rice varieties for deep water
12	Jagdalpur	Development of varieties for upland, midland, and lowlands
13	Jeypore	Development of varieties for rainfed shallow lowlands
14	Kanpur	Development of varieties and better management practices for sodic and saline soils
15	Karjat	Development of varieties and hybrids for the Konkan region
16	Kaul	Improvement of basmati rice, focus on nutrient depletion, and micronutrient deficiencies
17	Khudwani	Development of high-yielding cold-tolerant varieties for hills
18	Kohima	Development of varieties for Nagaland
19	Kota	Improvement of basmati rice for the region
20	Ludhiana	Breeding for high-yielding quality rice varieties with pest resistance
21	Malan	Development of varieties with cold and blast tolerance for hills

TABLE 17.4 Location-Specific Thrust Areas for Improvement in Rice Traits (*cont.*)

Study number	Locations	Thrust areas
22	Mandya	Development of varieties and hybrids with quality and resistance
23	Maruteru	Development of varieties and hybrids for quality and biotic stress tolerance; studies on nutrient depletion and micronutrient deficiencies
24	Moncompu	Development of varieties with biotic stress tolerance and for alkaline soils
25	Mugad	Development of upland rice varieties
26	Nagina	Development of basmati-quality rice varieties
27	Navsari	Breeding high-yielding rice varieties resistant to biotic stresses and tolerant of coastal salinity
28	Nawagam	Development of HYVs with stress tolerance
29	Pantnagar	Development of varieties and hybrids for plains of Uttarakhand, nutrient deficiency and management, and crop residue management
30	Patna	Development of HYVs for rainfed shallow lowland
31	Pattambi	Breeding varieties for biotic stress tolerance
32	Ponnampet	Development of blast-resistant rice varieties
33	Puducherry	Rice varieties for kuruvai and sornavari seasons
34	Pusa	Development of deepwater rice and aromatic varieties; focus on micronutrient deficiencies
35	Raipur	Development of varieties and hybrids for rainfed shallow lowlands with built-in resistance to pests
36	Ranchi	Development of varieties for rainfed uplands
37	Rajendranagar	Development of varieties with quality and cold tolerance
38	Rewa	Development of rainfed upland varieties
39	Sakoli	Gall midge–resistant rice varieties for Vidarbha
40	Titabar	Development of varieties for uplands, midlands, and lowlands. Focus on acidic and low-productivity soils
41	Tuljapur	Development of rainfed upland varieties
42	Upper Shillong	Development of varieties for NEH region
43	Varanasi	Development of medium-duration scented varieties
44	Wangbal	Development of gall midge (biotype 6)–resistant varieties
45	Warangal	Breeding of quality rice for biotic stress resistance

HYV, High-yielding variety.

REFERENCES

Central Rice Research Institute (CRRI), 1996. Fifty Years of Research at CRRI. CRRI, Cuttack, Odisha.

Central Rice Research Institute (CRRI), 2013. Vision 2050. CRRI, Cuttack, Odisha, (26 p.).

Directorate of Economics and Statistics, 2016. Agriculture Statistics at a Glance. Ministry of Agriculture and Farmers' Welfare, New Delhi.

Evenson, R.E., McKinsey, J.W., 1991. Research, extension, infrastructure, and productivity change in Indian agriculture. In: Evenson, R.E., Pray, C.E. (Eds.), Research and Productivity in Asian Agriculture. Cornell University Press, Ithaca, NY.

Janaiah, A., Hossain, M., 2003. Can hybrid rice technology help productivity growth in Asian tropics? Farmers' experiences. Econ. Polit. Wkly. 38 (25), 2492–2501.

Janaiah, A., Hossain, M., 2004. Partnership in the public sector agricultural R&D: evidence from India. Econ. Polit. Wkly. 39 (50), 5327–5334.

Janaiah, A., Xie, F., 2010. Hybrid Rice Adoption in India: Farm-Level Impacts and Challenges. Technical Bulletin 14. International Rice Research Institute, Los Baños, Philippines.

Janaiah, A., Hossain, M., Otsuka, K., 2006. Productivity impact of the modern varieties of rice in India. Dev. Econ. 44 (2), 190–207.

Janaiah, A., Otsuka, K., Hossain, M., 2005. Is the productivity impact of the Green Revolution in rice vanishing? Empirical evidence from TFP analysis for rice. Econ. Polit. Wkly. 40 (53), 5596–5600.

Kumar, P., Rosegrant, M.W., 1994. Productivity and sources of growth for rice in India. Econ. Polit. Wkly. 29 (53), A183–A188.

Pingali, P.L., Hossain, M. (Eds.), 1999. Impact of Rice Research. Thailand Development Research Institute; International Rice Research Institute, Bangkok, Thailand; Los Baños, Philippines.

Pingali, P.L., Hossain, M., Gerpacio, R., 1997. Asian Rice Bowls: The Returning Crisis. CAB International and International Rice Research Institute, Los Baños, Philippines.

Index

Printed in the United States
By Bookmasters